T0215792

Yuxiao Dong · Dunja Mladenić ·
Craig Saunders (Eds.)

Machine Learning and Knowledge Discovery in Databases

Applied Data Science Track

European Conference, ECML PKDD 2020
Ghent, Belgium, September 14–18, 2020
Proceedings, Part IV

Editors
Yuxiao Dong
Microsoft Research
Redmond, WA, USA

Dunja Mladenić
Jožef Stefan Institute
Ljubljana, Slovenia

Craig Saunders
Amazon Alexa Knowledge
Cambridge, UK

ISSN 0302-9743   ISSN 1611-3349 (electronic)
Lecture Notes in Artificial Intelligence
ISBN 978-3-030-67666-7   ISBN 978-3-030-67667-4   (eBook)
https://doi.org/10.1007/978-3-030-67667-4

LNCS Sublibrary: SL7 – Artificial Intelligence

This Springer imprint is published by the registered company Springer Nature Switzerland AG
The registered company address is: Gewerbestrasse 11, 6330 Cham, Switzerland

Preface

This edition of the European Conference on Machine Learning and Principles and Practice of Knowledge Discovery in Databases (ECML PKDD 2020) is one that we will not easily forget. Due to the emergence of a global pandemic, our lives changed, including many aspects of the conference. Because of this, we are perhaps more proud and happy than ever to present these proceedings to you.

ECML PKDD is an annual conference that provides an international forum for the latest research in all areas related to machine learning and knowledge discovery in databases, including innovative applications. It is the leading European machine learning and data mining conference and builds upon a very successful series of ECML PKDD conferences.

Scheduled to take place in Ghent, Belgium, due to the SARS-CoV-2 pandemic, ECML PKDD 2020 was the first edition to be held fully virtually, from the 14th to the 18th of September 2020. The conference attracted over 1000 participants from all over the world. New this year was a joint event with local industry on Thursday afternoon, the AI4Growth industry track. More generally, the conference received substantial attention from industry through sponsorship, participation, and the revived industry track at the conference.

The main conference programme consisted of presentations of 220 accepted papers and five keynote talks (in order of appearance): Max Welling (University of Amsterdam), Been Kim (Google Brain), Gemma Galdon-Clavell (Eticas Research & Consulting), Stephan Günnemann (Technical University of Munich), and Doina Precup (McGill University & DeepMind Montreal).

In addition, there were 23 workshops, nine tutorials, two combined workshop-tutorials, the PhD Forum, and a discovery challenge.

Papers presented during the three main conference days were organized in four different tracks:

- Research Track: research or methodology papers from all areas in machine learning, knowledge discovery, and data mining;
- Applied Data Science Track: papers on novel applications of machine learning, data mining, and knowledge discovery to solve real-world use cases, thereby bridging the gap between practice and current theory;
- Journal Track: papers that were published in special issues of the journals *Machine Learning* and *Data Mining and Knowledge Discovery*;
- Demo Track: short papers that introduce a new system that goes beyond the state of the art, accompanied with a video of the demo.

We received a record number of 687 and 235 submissions for the Research and Applied Data Science Tracks respectively. We accepted 130 (19%) and 65 (28%) of these. In addition, there were 25 papers from the Journal Track, and 10 demo papers (out of 25 submissions). All in all, the high-quality submissions allowed us to put together an exceptionally rich and exciting program.

The Awards Committee selected research papers that were considered to be of exceptional quality and worthy of special recognition:

- Data Mining best paper award: "Revisiting Wedge Sampling for Budgeted Maximum Inner Product Search", by Stephan S. Lorenzen and Ninh Pham.
- Data Mining best student paper award: "SpecGreedy: Unified Dense Subgraph Detection", by Wenjie Feng, Shenghua Liu, Danai Koutra, Huawei Shen, and Xueqi Cheng.
- Machine Learning best (student) paper award: "Robust Domain Adaptation: Representations, Weights and Inductive Bias", by Victor Bouvier, Philippe Very, Clément Chastagnol, Myriam Tami, and Céline Hudelot.
- Machine Learning best (student) paper runner-up award: "A Principle of Least Action for the Training of Neural Networks", by Skander Karkar, Ibrahim Ayed, Emmanuel de Bézenac, and Patrick Gallinari.
- Best Applied Data Science Track paper: "Learning to Simulate on Sparse Trajectory Data", by Hua Wei, Chacha Chen, Chang Liu, Guanjie Zheng, and Zhenhui Li.
- Best Applied Data Science Track paper runner-up: "Learning a Contextual and Topological Representation of Areas-of-Interest for On-Demand Delivery Application", by Mingxuan Yue, Tianshu Sun, Fan Wu, Lixia Wu, Yinghui Xu, and Cyrus Shahabi.
- Test of Time Award for highest-impact paper from ECML PKDD 2010: "Three Naive Bayes Approaches for Discrimination-Free Classification", by Toon Calders and Sicco Verwer.

We would like to wholeheartedly thank all participants, authors, PC members, area chairs, session chairs, volunteers, co-organizers, and organizers of workshops and tutorials for their contributions that helped make ECML PKDD 2020 a great success. Special thanks go to Vicky, Inge, and Eneko, and the volunteer and virtual conference platform chairs from the UGent AIDA group, who did an amazing job to make the online event feasible. We would also like to thank the ECML PKDD Steering Committee and all sponsors.

October 2020

Tijl De Bie
Craig Saunders
Dunja Mladenić
Yuxiao Dong
Frank Hutter
Isabel Valera
Jeffrey Lijffijt
Kristian Kersting
Georgiana Ifrim
Sofie Van Hoecke

Organization

General Chair

Tijl De Bie   Ghent University, Belgium

Research Track Program Chairs

Frank Hutter   University of Freiburg & Bosch Center for AI,
           Germany
Isabel Valera   Max Planck Institute for Intelligent Systems, Germany
Jefrey Lijffijt   Ghent University, Belgium
Kristian Kersting   TU Darmstadt, Germany

Applied Data Science Track Program Chairs

Craig Saunders   Amazon Alexa Knowledge, UK
Dunja Mladenić   Jožef Stefan Institute, Slovenia
Yuxiao Dong   Microsoft Research, USA

Journal Track Chairs

Aristides Gionis   KTH, Sweden
Carlotta Domeniconi   George Mason University, USA
Eyke Hüllermeier   Paderborn University, Germany
Ira Assent   Aarhus University, Denmark

Discovery Challenge Chair

Andreas Hotho   University of Würzburg, Germany

Workshop and Tutorial Chairs

Myra Spiliopoulou   Otto von Guericke University Magdeburg, Germany
Willem Waegeman   Ghent University, Belgium

Demonstration Chairs

Georgiana Ifrim   University College Dublin, Ireland
Sofie Van Hoecke   Ghent University, Belgium

Nectar Track Chairs

Jie Tang	Tsinghua University, China
Siegfried Nijssen	Université catholique de Louvain, Belgium
Yizhou Sun	University of California, Los Angeles, USA

Industry Track Chairs

Alexander Ypma	ASML, the Netherlands
Arindam Mallik	imec, Belgium
Luis Moreira-Matias	Kreditech, Germany

PhD Forum Chairs

Marinka Zitnik	Harvard University, USA
Robert West	EPFL, Switzerland

Publicity and Public Relations Chairs

Albrecht Zimmermann	Université de Caen Normandie, France
Samantha Monty	Universität Würzburg, Germany

Awards Chairs

Danai Koutra	University of Michigan, USA
José Hernández-Orallo	Universitat Politècnica de València, Spain

Inclusion and Diversity Chairs

Peter Steinbach	Helmholtz-Zentrum Dresden-Rossendorf, Germany
Heidi Seibold	Ludwig-Maximilians-Universität München, Germany
Oliver Guhr	Hochschule für Technik und Wirtschaft Dresden, Germany
Michele Berlingerio	Novartis, Ireland

Local Chairs

Eneko Illarramendi Lerchundi	Ghent University, Belgium
Inge Lason	Ghent University, Belgium
Vicky Wandels	Ghent University, Belgium

Proceedings Chair

Wouter Duivesteijn	Technische Universiteit Eindhoven, the Netherlands

Sponsorship Chairs

Luis Moreira-Matias   Kreditech, Germany
Vicky Wandels   Ghent University, Belgium

Volunteering Chairs

Junning Deng   Ghent University, Belgium
Len Vande Veire   Ghent University, Belgium
Maarten Buyl   Ghent University, Belgium
Raphaël Romero   Ghent University, Belgium
Robin Vandaele   Ghent University, Belgium
Xi Chen   Ghent University, Belgium

Virtual Conference Platform Chairs

Ahmad Mel   Ghent University, Belgium
Alexandru Cristian Mara   Ghent University, Belgium
Bo Kang   Ghent University, Belgium
Dieter De Witte   Ghent University, Belgium
Yoosof Mashayekhi   Ghent University, Belgium

Web Chair

Bo Kang   Ghent University, Belgium

ECML PKDD Steering Committee

Andrea Passerini   University of Trento, Italy
Francesco Bonchi   ISI Foundation, Italy
Albert Bifet   Télécom Paris, France
Sašo Džeroski   Jožef Stefan Institute, Slovenia
Katharina Morik   TU Dortmund, Germany
Arno Siebes   Utrecht University, the Netherlands
Siegfried Nijssen   Université catholique de Louvain, Belgium
Michelangelo Ceci   University of Bari Aldo Moro, Italy
Myra Spiliopoulou   Otto von Guericke University Magdeburg, Germany
Jaakko Hollmen   Aalto University, Finland
Georgiana Ifrim   University College Dublin, Ireland
Thomas Gärtner   University of Nottinghem, UK
Neil Hurley   University College Dublin, Ireland
Michele Berlingerio   IBM Research, Ireland
Elisa Fromont   Université de Rennes 1, France
Arno Knobbe   Universiteit Leiden, the Netherlands
Ulf Brefeld   Leuphana Universität Luneburg, Germany
Andreas Hotho   Julius-Maximilians-Universität Würzburg, Germany

Program Committees

Guest Editorial Board, Journal Track

Ana Paula Appel	IBM Research – Brazil
Annalisa Appice	University of Bari Aldo Moro
Martin Atzmüller	Tilburg University
Anthony Bagnall	University of East Anglia
James Bailey	The University of Melbourne
José Luis Balcázar	Universitat Politècnica de Catalunya
Mitra Baratchi	University of Twente
Srikanta Bedathur	IIT Delhi
Vaishak Belle	The University of Edinburgh
Viktor Bengs	Paderborn University
Batista Biggio	University of Cagliari
Hendrik Blockeel	KU Leuven
Francesco Bonchi	ISI Foundation
Ilaria Bordino	UniCredit R&D
Ulf Brefeld	Leuphana Universität Lüneburg
Klemens Böhm	Karlsruhe Institute of Technology
Remy Cazabet	Claude Bernard University Lyon 1
Michelangelo Ceci	University of Bari Aldo Moro
Loïc Cerf	Universidade Federal de Minas Gerais
Laetitia Chapel	IRISA
Marc Deisenroth	Aarhus University
Wouter Duivesteijn	Technische Universiteit Eindhoven
Tapio Elomaa	Tampere University
Stefano Ferilli	University of Bari Aldo Moro
Cesar Ferri	Universitat Politècnica de València
Maurizio Filippone	EURECOM
Germain Forestier	Université de Haute-Alsace
Marco Frasca	University of Milan
Ricardo José Gabrielli Barreto Campello	University of Newcastle
Esther Galbrun	University of Eastern Finland
Joao Gama	University of Porto
Josif Grabocka	University of Hildesheim
Derek Greene	University College Dublin
Francesco Gullo	UniCredit
Tias Guns	VUB Brussels
Stephan Günnemann	Technical University of Munich
Jose Hernandez-Orallo	Universitat Politècnica de València
Jaakko Hollmén	Aalto University
Georgiana Ifrim	University College Dublin
Mahdi Jalili	RMIT University
Szymon Jaroszewicz	Polish Academy of Sciences

Michael Kamp	Monash University
Mehdi Kaytoue	Infologic
Marius Kloft	TU Kaiserslautern
Dragi Kocev	Jožef Stefan Institute
Peer Kröger	Ludwig-Maximilians-Universität Munich
Meelis Kull	University of Tartu
Ondrej Kuzelka	KU Leuven
Mark Last	Ben-Gurion University of the Negev
Matthijs van Leeuwen	Leiden University
Marco Lippi	University of Modena and Reggio Emilia
Claudio Lucchese	Ca' Foscari University of Venice
Brian Mac Namee	University College Dublin
Gjorgji Madjarov	Ss. Cyril and Methodius University of Skopje
Fabrizio Maria Maggi	Free University of Bozen-Bolzano
Giuseppe Manco	ICAR-CNR
Ernestina Menasalvas	Universidad Politécnica de Madrid
Aditya Menon	Google Research
Katharina Morik	TU Dortmund
Davide Mottin	Aarhus University
Animesh Mukherjee	Indian Institute of Technology Kharagpur
Amedeo Napoli	LORIA
Siegfried Nijssen	Université catholique de Louvain
Eirini Ntoutsi	Leibniz University Hannover
Bruno Ordozgoiti	Aalto University
Pance Panov	Jožef Stefan Institute
Panagiotis Papapetrou	Stockholm University
Srinivasan Parthasarathy	Ohio State University
Andrea Passerini	University of Trento
Mykola Pechenizkiy	Technische Universiteit Eindhoven
Charlotte Pelletier	Univ. Bretagne Sud/IRISA
Ruggero Pensa	University of Turin
Francois Petitjean	Monash University
Nico Piatkowski	TU Dortmund
Evaggelia Pitoura	Univ. of Ioannina
Marc Plantevit	Claude Bernard University Lyon 1
Kai Puolamäki	University of Helsinki
Chedy Raïssi	Inria
Matteo Riondato	Amherst College
Joerg Sander	University of Alberta
Pierre Schaus	UCLouvain
Lars Schmidt-Thieme	University of Hildesheim
Matthias Schubert	LMU Munich
Thomas Seidl	LMU Munich
Gerasimos Spanakis	Maastricht University
Myra Spiliopoulou	Otto von Guericke University Magdeburg
Jerzy Stefanowski	Poznań University of Technology

Giovanni Stilo	Università degli Studi dell'Aquila
Mahito Sugiyama	National Institute of Informatics
Andrea Tagarelli	University of Calabria
Chang Wei Tan	Monash University
Nikolaj Tatti	University of Helsinki
Alexandre Termier	Univ. Rennes 1
Marc Tommasi	University of Lille
Ivor Tsang	University of Technology Sydney
Panayiotis Tsaparas	University of Ioannina
Steffen Udluft	Siemens
Celine Vens	KU Leuven
Antonio Vergari	University of California, Los Angeles
Michalis Vlachos	University of Lausanne
Christel Vrain	LIFO, Université d'Orléans
Jilles Vreeken	Helmholtz Center for Information Security
Willem Waegeman	Ghent University
Marcel Wever	Paderborn University
Stefan Wrobel	Univ. Bonn and Fraunhofer IAIS
Yinchong Yang	Sicmens AG
Guoxian Yu	Southwest University
Bianca Zadrozny	IBM
Ye Zhu	Monash University
Arthur Zimek	University of Southern Denmark
Albrecht Zimmermann	Université de Caen Normandie
Marinka Zitnik	Harvard University

Area Chairs, Research Track

Cuneyt Gurcan Akcora	The University of Texas at Dallas
Carlos M. Alaíz	Universidad Autónoma de Madrid
Fabrizio Angiulli	University of Calabria
Georgios Arvanitidis	Max Planck Institute for Intelligent Systems
Roberto Bayardo	Google
Michele Berlingerio	IBM
Michael Berthold	University of Konstanz
Albert Bifet	Télécom Paris
Hendrik Blockeel	Katholieke Universiteit Leuven
Mario Boley	MPI Informatics
Francesco Bonchi	Fondazione ISI
Ulf Brefeld	Leuphana Universität Lüneburg
Michelangelo Ceci	Università degli Studi di Bari Aldo Moro
Duen Horng Chau	Georgia Institute of Technology
Nicolas Courty	Université de Bretagne Sud/IRISA
Bruno Cremilleux	Université de Caen Normandie
Andre de Carvalho	University of São Paulo
Patrick De Causmaecker	Katholieke Universiteit Leuven

Nicola Di Mauro	Università degli Studi di Bari Aldo Moro
Tapio Elomaa	Tampere University
Amir-Massoud Farahmand	Vector Institute & University of Toronto
Ángela Fernández	Universidad Autónoma de Madrid
Germain Forestier	Université de Haute-Alsace
Elisa Fromont	Université de Rennes 1
Johannes Fürnkranz	Johannes Kepler University Linz
Patrick Gallinari	Sorbonne University
Joao Gama	University of Porto
Thomas Gärtner	TU Wien
Pierre Geurts	University of Liège
Manuel Gomez Rodriguez	MPI for Software Systems
Przemyslaw Grabowicz	University of Massachusetts Amherst
Stephan Günnemann	Technical University of Munich
Allan Hanbury	Vienna University of Technology
Daniel Hernández-Lobato	Universidad Autónoma de Madrid
Jose Hernandez-Orallo	Universitat Politècnica de València
Jaakko Hollmén	Aalto University
Andreas Hotho	University of Würzburg
Neil Hurley	University College Dublin
Georgiana Ifrim	University College Dublin
Alipio M. Jorge	University of Porto
Arno Knobbe	Universiteit Leiden
Dragi Kocev	Jožef Stefan Institute
Lars Kotthoff	University of Wyoming
Nick Koudas	University of Toronto
Stefan Kramer	Johannes Gutenberg University Mainz
Meelis Kull	University of Tartu
Niels Landwehr	University of Potsdam
Sébastien Lefèvre	Université de Bretagne Sud
Daniel Lemire	Université du Québec
Matthijs van Leeuwen	Leiden University
Marius Lindauer	Leibniz University Hannover
Jörg Lücke	University of Oldenburg
Donato Malerba	Università degli Studi di Bari "Aldo Moro"
Giuseppe Manco	ICAR-CNR
Pauli Miettinen	University of Eastern Finland
Anna Monreale	University of Pisa
Katharina Morik	TU Dortmund
Emmanuel Müller	University of Bonn
Sriraam Natarajan	Indiana University Bloomington
Alfredo Nazábal	The Alan Turing Institute
Siegfried Nijssen	Université catholique de Louvain
Barry O'Sullivan	University College Cork
Pablo Olmos	University Carlos III of Madrid
Panagiotis Papapetrou	Stockholm University

Andrea Passerini	University of Turin
Mykola Pechenizkiy	Technische Universiteit Eindhoven
Ruggero G. Pensa	University of Torino
Francois Petitjean	Monash University
Claudia Plant	University of Vienna
Marc Plantevit	Université Claude Bernard Lyon 1
Philippe Preux	Université de Lille
Rita Ribeiro	University of Porto
Celine Robardet	INSA Lyon
Elmar Rueckert	University of Lübeck
Marian Scuturici	LIRIS-INSA de Lyon
Michèle Sebag	Univ. Paris-Sud
Thomas Seidl	Ludwig-Maximilians-Universität Muenchen
Arno Siebes	Utrecht University
Alessandro Sperduti	University of Padua
Myra Spiliopoulou	Otto von Guericke University Magdeburg
Jerzy Stefanowski	Poznań University of Technology
Yizhou Sun	University of California, Los Angeles
Einoshin Suzuki	Kyushu University
Acar Tamersoy	Symantec Research Labs
Jie Tang	Tsinghua University
Grigorios Tsoumakas	Aristotle University of Thessaloniki
Celine Vens	KU Leuven
Antonio Vergari	University of California, Los Angeles
Herna Viktor	University of Ottawa
Christel Vrain	University of Orléans
Jilles Vreeken	Helmholtz Center for Information Security
Willem Waegeman	Ghent University
Wendy Hui Wang	Stevens Institute of Technology
Stefan Wrobel	Fraunhofer IAIS & Univ. of Bonn
Han-Jia Ye	Nanjing University
Guoxian Yu	Southwest University
Min-Ling Zhang	Southeast University
Albrecht Zimmermann	Université de Caen Normandie

Area Chairs, Applied Data Science Track

Michelangelo Ceci	Università degli Studi di Bari Aldo Moro
Tom Diethe	Amazon
Faisal Farooq	IBM
Johannes Fürnkranz	Johannes Kepler University Linz
Rayid Ghani	Carnegie Mellon University
Ahmed Hassan Awadallah	Microsoft
Xiangnan He	University of Science and Technology of China
Georgiana Ifrim	University College Dublin
Anne Kao	Boeing

Javier Latorre	Apple
Hao Ma	Facebook AI
Gabor Melli	Sony PlayStation
Luis Moreira-Matias	Kreditech
Alessandro Moschitti	Amazon
Kitsuchart Pasupa	King Mongkut's Institute of Technology Ladkrabang
Mykola Pechenizkiy	Technische Universiteit Eindhoven
Julien Perez	NAVER LABS Europe
Xing Xie	Microsoft
Chenyan Xiong	Microsoft Research
Yang Yang	Zhejiang University

Program Committee Members, Research Track

Moloud Abdar	Deakin University
Linara Adilova	Fraunhofer IAIS
Florian Adriaens	Ghent University
Zahra Ahmadi	Johannes Gutenberg University Mainz
M. Eren Akbiyik	IBM Germany Research and Development GmbH
Youhei Akimoto	University of Tsukuba
Ömer Deniz Akyildiz	University of Warwick and The Alan Turing Institute
Francesco Alesiani	NEC Laboratories Europe
Alexandre Alves	Universidade Federal de Uberlândia
Maryam Amir Haeri	Technische Universität Kaiserslautern
Alessandro Antonucci	IDSIA
Muhammad Umer Anwaar	Mercateo AG
Xiang Ao	Institute of Computing Technology, Chinese Academy of Sciences
Sunil Aryal	Deakin University
Thushari Atapattu	The University of Adelaide
Arthur Aubret	LIRIS
Julien Audiffren	Fribourg University
Murat Seckin Ayhan	Eberhard Karls Universität Tübingen
Dario Azzimonti	Istituto Dalle Molle di Studi sull'Intelligenza Artificiale
Behrouz Babaki	Polytechnique Montréal
Rohit Babbar	Aalto University
Housam Babiker	University of Alberta
Davide Bacciu	University of Pisa
Thomas Baeck	Leiden University
Abdelkader Baggag	Qatar Computing Research Institute
Zilong Bai	University of California, Davis
Jiyang Bai	Florida State University
Sambaran Bandyopadhyay	IBM
Mitra Baratchi	University of Twente
Christian Beecks	University of Münster
Anna Beer	Ludwig Maximilian University of Munich

Adnene Belfodil	Munic Car Data
Aimene Belfodil	INSA Lyon
Ines Ben Kraiem	UT2J-IRIT
Anes Bendimerad	LIRIS
Christoph Bergmeir	Monash University
Max Berrendorf	Ludwig Maximilian University of Munich
Louis Béthune	ENS de Lyon
Anton Björklund	University of Helsinki
Alexandre Blanché	Université de Lorraine
Laurens Bliek	Delft University of Technology
Isabelle Bloch	ENST - CNRS UMR 5141 LTCI
Gianluca Bontempi	Université Libre de Bruxelles
Felix Borutta	Ludwig-Maximilians-Universität München
Ahcène Boubekki	Leuphana Universität Lüneburg
Tanya Braun	University of Lübeck
Wieland Brendel	University of Tübingen
Klaus Brinker	Hamm-Lippstadt University of Applied Sciences
David Browne	Insight Centre for Data Analytics
Sebastian Bruckert	Otto Friedrich University Bamberg
Mirko Bunse	TU Dortmund University
Sophie Burkhardt	University of Mainz
Haipeng Cai	Washington State University
Lele Cao	Tsinghua University
Manliang Cao	Fudan University
Defu Cao	Peking University
Antonio Carta	University of Pisa
Remy Cazabet	Université Lyon 1
Abdulkadir Celikkanat	CentraleSupelec, Paris-Saclay University
Christophe Cerisara	LORIA
Carlos Cernuda	Mondragon University
Vitor Cerqueira	LIAAD-INESCTEC
Mattia Cerrato	Università di Torino
Ricardo Cerri	Federal University of São Carlos
Laetitia Chapel	IRISA
Vaggos Chatziafratis	Stanford University
El Vaigh Cheikh Brahim	Inria/IRISA Rennes
Yifei Chen	University of Groningen
Junyang Chen	University of Macau
Jiaoyan Chen	University of Oxford
Huiyuan Chen	Case Western Reserve University
Run-Qing Chen	Xiamen University
Tianyi Chen	Microsoft
Lingwei Chen	The Pennsylvania State University
Senpeng Chen	UESTC
Liheng Chen	Shanghai Jiao Tong University
Siming Chen	Frauenhofer IAIS

Liang Chen	Sun Yat-sen University
Dawei Cheng	Shanghai Jiao Tong University
Wei Cheng	NEC Labs America
Wen-Hao Chiang	Indiana University - Purdue University Indianapolis
Feng Chong	Beijing Institute of Technology
Pantelis Chronis	Athena Research Center
Victor W. Chu	The University of New South Wales
Xin Cong	Institute of Information Engineering, Chinese Academy of Sciences
Roberto Corizzo	UNIBA
Mustafa Coskun	Case Western Reserve University
Gustavo De Assis Costa	Instituto Federal de Educação, Ciência e Tecnologia de Goiás
Fabrizio Costa	University of Exeter
Miguel Couceiro	Inria
Shiyao Cui	Institute of Information Engineering, Chinese Academy of Sciences
Bertrand Cuissart	GREYC
Mohamad H. Danesh	Oregon State University
Thi-Bich-Hanh Dao	University of Orléans
Cedric De Boom	Ghent University
Marcos Luiz de Paula Bueno	Technische Universiteit Eindhoven
Matteo Dell'Amico	NortonLifeLock
Qi Deng	Shanghai University of Finance and Economics
Andreas Dengel	German Research Center for Artificial Intelligence
Sourya Dey	University of Southern California
Yao Di	Institute of Computing Technology, Chinese Academy of Sciences
Stefano Di Frischia	University of L'Aquila
Jilles Dibangoye	INSA Lyon
Felix Dietrich	Technical University of Munich
Jiahao Ding	University of Houston
Yao-Xiang Ding	Nanjing University
Tianyu Ding	Johns Hopkins University
Rui Ding	Microsoft
Thang Doan	McGill University
Carola Doerr	Sorbonne University, CNRS
Xiao Dong	The University of Queensland
Wei Du	University of Arkansas
Xin Du	Technische Universiteit Eindhoven
Yuntao Du	Nanjing University
Stefan Duffner	LIRIS
Sebastijan Dumancic	Katholieke Universiteit Leuven
Valentin Durand de Gevigney	IRISA

Saso Dzeroski	Jožef Stefan Institute
Mohamed Elati	Université d'Evry
Lukas Enderich	Robert Bosch GmbH
Dominik Endres	Philipps-Universität Marburg
Francisco Escolano	University of Alicante
Bjoern Eskofier	Friedrich-Alexander University Erlangen-Nürnberg
Roberto Esposito	Università di Torino
Georgios Exarchakis	Institut de la Vision
Melanie F. Pradier	Harvard University
Samuel G. Fadel	Universidade Estadual de Campinas
Evgeniy Faerman	Ludwig Maximilian University of Munich
Yujie Fan	Case Western Reserve University
Elaine Faria	Federal University of Uberlândia
Golnoosh Farnadi	Mila/University of Montreal
Fabio Fassetti	University of Calabria
Ad Feelders	Utrecht University
Yu Fei	Harbin Institute of Technology
Wenjie Feng	The Institute of Computing Technology, Chinese Academy of Sciences
Zunlei Feng	Zhejiang University
Cesar Ferri	Universitat Politècnica de València
Raul Fidalgo-Merino	European Commission Joint Research Centre
Murat Firat	Technische Universiteit Eindhoven
Francoise Fogelman-Soulié	Tianjin University
Vincent Fortuin	ETH Zurich
Iordanis Fostiropoulos	University of Southern California
Eibe Frank	University of Waikato
Benoît Frénay	Université de Namur
Nikolaos Freris	University of Science and Technology of China
Moshe Gabel	University of Toronto
Ricardo José Gabrielli Barreto Campello	University of Newcastle
Esther Galbrun	University of Eastern Finland
Claudio Gallicchio	University of Pisa
Yuanning Gao	Shanghai Jiao Tong University
Alberto Garcia-Duran	Ecole Polytechnique Fédérale de Lausanne
Eduardo Garrido	Universidad Autónoma de Madrid
Clément Gautrais	KU Leuven
Arne Gevaert	Ghent University
Giorgos Giannopoulos	IMSI, "Athena" Research Center
C. Lee Giles	The Pennsylvania State University
Ioana Giurgiu	IBM Research - Zurich
Thomas Goerttler	TU Berlin
Heitor Murilo Gomes	University of Waikato
Chen Gong	Shanghai Jiao Tong University
Zhiguo Gong	University of Macau

Hongyu Gong	University of Illinois at Urbana-Champaign
Pietro Gori	Télécom Paris
James Goulding	University of Nottingham
Kshitij Goyal	Katholieke Universiteit Leuven
Dmitry Grishchenko	Université Grenoble Alpes
Moritz Grosse-Wentrup	University of Vienna
Sebastian Gruber	Siemens AG
John Grundy	Monash University
Kang Gu	Dartmouth College
Jindong Gu	Siemens
Riccardo Guidotti	University of Pisa
Tias Guns	Vrije Universiteit Brussel
Ruocheng Guo	Arizona State University
Yiluan Guo	Singapore University of Technology and Design
Xiaobo Guo	University of Chinese Academy of Sciences
Thomas Guyet	IRISA
Jiawei Han	University of Illinois at Urbana-Champaign
Zhiwei Han	fortiss GmbH
Tom Hanika	University of Kassel
Shonosuke Harada	Kyoto University
Marwan Hassani	Technische Universiteit Eindhoven
Jianhao He	Sun Yat-sen University
Deniu He	Chongqing University of Posts and Telecommunications
Dongxiao He	Tianjin University
Stefan Heidekrueger	Technical University of Munich
Nandyala Hemachandra	Indian Institute of Technology Bombay
Till Hendrik Schulz	University of Bonn
Alexander Hepburn	University of Bristol
Sibylle Hess	Technische Universiteit Eindhoven
Javad Heydari	LG Electronics
Joyce Ho	Emory University
Shunsuke Horii	Waseda University
Tamas Horvath	University of Bonn and Fraunhofer IAIS
Mehran Hossein Zadeh Bazargani	University College Dublin
Robert Hu	University of Oxford
Weipeng Huang	Insight
Jun Huang	University of Tokyo
Haojie Huang	The University of New South Wales
Hong Huang	UGoe
Shenyang Huang	McGill University
Vân Anh Huynh-Thu	University of Liège
Dino Ienco	INRAE
Siohoi Ieng	Institut de la Vision
Angelo Impedovo	Università "Aldo Moro" degli studi di Bari

Muhammad Imran Razzak	Deakin University
Vasileios Iosifidis	Leibniz University Hannover
Joseph Isaac	Indian Institute of Technology Madras
Md Islam	Washington State University
Ziyu Jia	Beijing Jiaotong University
Lili Jiang	Umeå University
Yao Jiangchao	Alibaba
Tan Jianlong	Institute of Information Engineering, Chinese Academy of Sciences
Baihong Jin	University of California, Berkeley
Di Jin	Tianjin University
Wei Jing	Xi'an Jiaotong University
Jonathan Jouanne	ARIADNEXT
Ata Kaban	University of Birmingham
Tomasz Kajdanowicz	Wrocław University of Science and Technology
Sandesh Kamath	Chennai Mathematical Institute
Keegan Kang	Singapore University of Technology and Design
Bo Kang	Ghent University
Isak Karlsson	Stockholm University
Panagiotis Karras	Aarhus University
Nikos Katzouris	NCSR Demokritos
Uzay Kaymak	Technische Universiteit Eindhoven
Mehdi Kaytoue	Infologic
Pascal Kerschke	University of Münster
Jungtaek Kim	Pohang University of Science and Technology
Minyoung Kim	Samsung AI Center Cambridge
Masahiro Kimura	Ryukoku University
Uday Kiran	The University of Tokyo
Bogdan Kirillov	ITMO University
Péter Kiss	ELTE
Gerhard Klassen	Heinrich Heine University Düsseldorf
Dmitry Kobak	Eberhard Karls University of Tübingen
Masahiro Kohjima	NTT
Ziyi Kou	University of Rochester
Wouter Kouw	Technische Universiteit Eindhoven
Fumiya Kudo	Hitachi, Ltd.
Piotr Kulczycki	Systems Research Institute, Polish Academy of Sciences
Ilona Kulikovskikh	Samara State Aerospace University
Rajiv Kumar	IIT Bombay
Pawan Kumar	IIT Kanpur
Suhansanu Kumar	University of Illinois, Urbana-Champaign
Abhishek Kumar	University of Helsinki
Gautam Kunapuli	The University of Texas at Dallas
Takeshi Kurashima	NTT
Vladimir Kuzmanovski	Jožef Stefan Institute

Anisio Lacerda	Centro Federal de Educação Tecnológica de Minas Gerais
Patricia Ladret	GIPSA-lab
Fabrizio Lamberti	Politecnico di Torino
James Large	University of East Anglia
Duc-Trong Le	University of Engineering and Technology, VNU Hanoi
Trung Le	Monash University
Luce le Gorrec	University of Strathclyde
Antoine Ledent	TU Kaiserslautern
Kangwook Lee	University of Wisconsin-Madison
Felix Leibfried	PROWLER.io
Florian Lemmerich	RWTH Aachen University
Carson Leung	University of Manitoba
Edouard Leurent	Inria
Naiqi Li	Tsinghua-UC Berkeley Shenzhen Institute
Suyi Li	The Hong Kong University of Science and Technology
Jundong Li	University of Virginia
Yidong Li	Beijing Jiaotong University
Xiaoting Li	The Pennsylvania State University
Yaoman Li	CUHK
Rui Li	Inspur Group
Wenye Li	The Chinese University of Hong Kong (Shenzhen)
Mingming Li	Institute of Information Engineering, Chinese Academy of Sciences
Yexin Li	Hong Kong University of Science and Technology
Qinghua Li	Renmin University of China
Yaohang Li	Old Dominion University
Yuxuan Liang	National University of Singapore
Zhimin Liang	Institute of Computing Technology, Chinese Academy of Sciences
Hongwei Liang	Microsoft
Nengli Lim	Singapore University of Technology and Design
Suwen Lin	University of Notre Dame
Yangxin Lin	Peking University
Aldo Lipani	University College London
Marco Lippi	University of Modena and Reggio Emilia
Alexei Lisitsa	University of Liverpool
Lin Liu	Taiyuan University of Technology
Weiwen Liu	The Chinese University of Hong Kong
Yang Liu	JD
Huan Liu	Arizona State University
Tianbo Liu	Thomas Jefferson National Accelerator Facility
Tongliang Liu	The University of Sydney
Weidong Liu	Inner Mongolia University
Kai Liu	Colorado School of Mines

Shiwei Liu	Technische Universiteit Eindhoven
Shenghua Liu	Institute of Computing Technology, Chinese Academy of Sciences
Corrado Loglisci	University of Bari Aldo Moro
Andrey Lokhov	Los Alamos National Laboratory
Yijun Lu	Alibaba Cloud
Xuequan Lu	Deakin University
Szymon Lukasik	AGH University of Science and Technology
Phuc Luong	Deakin University
Jianming Lv	South China University of Technology
Gengyu Lyu	Beijing Jiaotong University
Vijaikumar M.	Indian Institute of Science
Jing Ma	Emory University
Nan Ma	Shanghai Jiao Tong University
Sebastian Mair	Leuphana University Lüneburg
Marjan Mansourvar	University of Southern Denmark
Vincent Margot	Advestis
Fernando Martínez-Plumed	Joint Research Centre - European Commission
Florent Masseglia	Inria
Romain Mathonat	Université de Lyon
Deepak Maurya	Indian Institute of Technology Madras
Christian Medeiros Adriano	Hasso-Plattner-Institut
Purvanshi Mehta	University of Rochester
Tobias Meisen	Bergische Universität Wuppertal
Luciano Melodia	Friedrich-Alexander Universität Erlangen-Nürnberg
Ernestina Menasalvas	Universidad Politécnica de Madrid
Vlado Menkovski	Technische Universiteit Eindhoven
Engelbert Mephu Nguifo	Université Clermont Auvergne
Alberto Maria Metelli	Politecnico di Milano
Donald Metzler	Google
Anke Meyer-Baese	Florida State University
Richard Meyes	University of Wuppertal
Haithem Mezni	University of Jendouba
Paolo Mignone	Università degli Studi di Bari Aldo Moro
Matej Mihelčić	University of Zagreb
Decebal Constantin Mocanu	University of Twente
Christoph Molnar	Ludwig Maximilian University of Munich
Lia Morra	Politecnico di Torino
Christopher Morris	TU Dortmund University
Tadeusz Morzy	Poznań University of Technology
Henry Moss	Lancaster University
Tetsuya Motokawa	University of Tsukuba
Mathilde Mougeot	Université Paris-Saclay
Tingting Mu	The University of Manchester
Andreas Mueller	NYU
Tanmoy Mukherjee	Queen Mary University of London

Ksenia Mukhina	ITMO University
Peter Müllner	Know-Center
Guido Muscioni	University of Illinois at Chicago
Waleed Mustafa	TU Kaiserslautern
Mohamed Nadif	University of Paris
Ankur Nahar	Indian Institute of Technology Jodhpur
Kei Nakagawa	Nomura Asset Management Co., Ltd.
Haïfa Nakouri	University of Tunis
Mirco Nanni	KDD-Lab ISTI-CNR Pisa
Nicolo' Navarin	University of Padova
Richi Nayak	Queensland University of Technology
Mojtaba Nayyeri	University of Bonn
Daniel Neider	MPI SWS
Nan Neng	Institute of Information Engineering, Chinese Academy of Sciences
Stefan Neumann	University of Vienna
Dang Nguyen	Deakin University
Kien Duy Nguyen	University of Southern California
Jingchao Ni	NEC Laboratories America
Vlad Niculae	Instituto de Telecomunicações
Sofia Maria Nikolakaki	Boston University
Kun Niu	Beijing University of Posts and Telecommunications
Ryo Nomura	Waseda University
Eirini Ntoutsi	Leibniz University Hannover
Andreas Nuernberger	Otto von Guericke University of Magdeburg
Tsuyoshi Okita	Kyushu Institute of Technology
Maria Oliver Parera	GIPSA-lab
Bruno Ordozgoiti	Aalto University
Sindhu Padakandla	Indian Institute of Science
Tapio Pahikkala	University of Turku
Joao Palotti	Qatar Computing Research Institute
Guansong Pang	The University of Adelaide
Pance Panov	Jožef Stefan Institute
Konstantinos Papangelou	The University of Manchester
Yulong Pei	Technische Universiteit Eindhoven
Nikos Pelekis	University of Piraeus
Thomas Pellegrini	Université Toulouse III - Paul Sabatier
Charlotte Pelletier	Univ. Bretagne Sud
Jaakko Peltonen	Aalto University and Tampere University
Shaowen Peng	Kyushu University
Siqi Peng	Kyoto University
Bo Peng	The Ohio State University
Lukas Pensel	Johannes Gutenberg University Mainz
Aritz Pérez Martínez	Basque Center for Applied Mathematics
Lorenzo Perini	KU Leuven
Matej Petković	Jožef Stefan Institute

Bernhard Pfahringer	University of Waikato
Weiguo Pian	Chongqing University
Francesco Piccialli	University of Naples Federico II
Sebastian Pineda Arango	University of Hildesheim
Gianvito Pio	University of Bari "Aldo Moro"
Giuseppe Pirrò	Sapienza University of Rome
Anastasia Podosinnikova	Massachusetts Institute of Technology
Sebastian Pölsterl	Ludwig Maximilian University of Munich
Vamsi Potluru	JP Morgan AI Research
Rafael Poyiadzi	University of Bristol
Surya Prakash	University of Canberra
Paul Prasse	University of Potsdam
Rameshwar Pratap	Indian Institute of Technology Mandi
Jonas Prellberg	University of Oldenburg
Hugo Proenca	Leiden Institute of Advanced Computer Science
Ricardo Prudencio	Federal University of Pernambuco
Petr Pulc	Institute of Computer Science of the Czech Academy of Sciences
Lei Qi	Iowa State University
Zhenyue Qin	The Australian National University
Rahul Ragesh	PES University
Tahrima Rahman	The University of Texas at Dallas
Zana Rashidi	York University
S. S. Ravi	University of Virginia and University at Albany – SUNY
Ambrish Rawat	IBM
Henry Reeve	University of Birmingham
Reza Refaei Afshar	Technische Universiteit Eindhoven
Navid Rekabsaz	Johannes Kepler University Linz
Yongjian Ren	Shandong University
Zhiyun Ren	The Ohio State University
Guohua Ren	LG Electronics
Yuxiang Ren	Florida State University
Xavier Renard	AXA
Martí Renedo Mirambell	Universitat Politècnica de Catalunya
Gavin Rens	Katholiek Universiteit Leuven
Matthias Renz	Christian-Albrechts-Universität zu Kiel
Guillaume Richard	EDF R&D
Matteo Riondato	Amherst College
Niklas Risse	Bielefeld University
Lars Rosenbaum	Robert Bosch GmbH
Celine Rouveirol	Université Sorbonne Paris Nord
Shoumik Roychoudhury	Temple University
Polina Rozenshtein	Aalto University
Peter Rubbens	Flanders Marine Institute (VLIZ)
David Ruegamer	LMU Munich

Matteo Ruffini	ToolsGroup
Ellen Rushe	Insight Centre for Data Analytics
Amal Saadallah	TU Dortmund
Yogish Sabharwal	IBM Research - India
Mandana Saebi	University of Notre Dame
Aadirupa Saha	IISc
Seyed Erfan Sajjadi	Brunel University
Durgesh Samariya	Federation University
Md Samiullah	Monash University
Mark Sandler	Google
Raul Santos-Rodriguez	University of Bristol
Yucel Saygin	Sabancı University
Pierre Schaus	UCLouvain
Fabian Scheipl	Ludwig Maximilian University of Munich
Katerina Schindlerova	University of Vienna
Ute Schmid	University of Bamberg
Daniel Schmidt	Monash University
Sebastian Schmoll	Ludwig Maximilian University of Munich
Johannes Schneider	University of Liechtenstein
Marc Schoenauer	Inria Saclay Île-de-France
Jonas Schouterden	Katholieke Universiteit Leuven
Leo Schwinn	Friedrich-Alexander-Universität Erlangen-Nürnberg
Florian Seiffarth	University of Bonn
Nan Serra	NEC Laboratories Europe GmbH
Rowland Seymour	Univeristy of Nottingham
Ammar Shaker	NEC Laboratories Europe
Ali Shakiba	Vali-e-Asr University of Rafsanjan
Junming Shao	University of Science and Technology of China
Zhou Shao	Tsinghua University
Manali Sharma	Samsung Semiconductor Inc.
Jiaming Shen	University of Illinois at Urbana-Champaign
Ying Shen	Sun Yat-sen University
Hao Shen	fortiss GmbH
Tao Shen	University of Technology Sydney
Ge Shi	Beijing Institute of Technology
Ziqiang Shi	Fujitsu Research & Development Center
Masumi Shirakawa	hapicom Inc./Osaka University
Kai Shu	Arizona State University
Amila Silva	The University of Melbourne
Edwin Simpson	University of Bristol
Dinesh Singh	RIKEN Center for Advanced Intelligence Project
Jasprect Singh	L3S Research Centre
Spiros Skiadopoulos	University of the Peloponnese
Gavin Smith	University of Nottingham
Miguel A. Solinas	CEA
Dongjin Song	NEC Labs America

Arnaud Soulet	Université de Tours
Marvin Ssemambo	Makerere University
Michiel Stock	Ghent University
Filipo Studzinski Perotto	Institut de Recherche en Informatique de Toulouse
Adisak Sukul	Iowa State University
Lijuan Sun	Beijing Jiaotong University
Tao Sun	National University of Defense Technology
Ke Sun	Peking University
Yue Sun	Beijing Jiaotong University
Hari Sundaram	University of Illinois at Urbana-Champaign
Gero Szepannek	Stralsund University of Applied Sciences
Jacek Tabor	Jagiellonian University
Jianwei Tai	IIE, CAS
Naoya Takeishi	RIKEN Center for Advanced Intelligence Project
Chang Wei Tan	Monash University
Jinghua Tan	Southwestern University of Finance and Economics
Zeeshan Tariq	Ulster University
Bouadi Tassadit	IRISA-Université de Rennes 1
Maryam Tavakol	TU Dortmund
Romain Tavenard	Univ. Rennes 2/LETG-COSTEL/IRISA-OBELIX
Alexandre Termier	Université de Rennes 1
Janek Thomas	Fraunhofer Institute for Integrated Circuits IIS
Manoj Thulasidas	Singapore Management University
Hao Tian	Syracuse University
Hiroyuki Toda	NTT
Jussi Tohka	University of Eastern Finland
Ricardo Torres	Norwegian University of Science and Technology
Isaac Triguero Velázquez	University of Nottingham
Sandhya Tripathi	Indian Institute of Technology Bombay
Holger Trittenbach	Karlsruhe Institute of Technology
Peter van der Putten	Leiden University & Pegasystems
Elia Van Wolputte	KU Leuven
Fabio Vandin	University of Padova
Titouan Vayer	IRISA
Ashish Verma	IBM Research - US
Bouvier Victor	Sidetrade MICS
Julia Vogt	University of Basel
Tim Vor der Brück	Lucerne University of Applied Sciences and Arts
Yb W.	Chongqing University
Krishna Wadhwani	Indian Institute of Technology Bombay
Huaiyu Wan	Beijing Jiaotong University
Qunbo Wang	Beihang University
Beilun Wang	Southeast University
Yiwei Wang	National University of Singapore
Bin Wang	Xiaomi AI Lab

Jiong Wang	Institute of Information Engineering, Chinese Academy of Sciences
Xiaobao Wang	Tianjin University
Shuheng Wang	Nanjing University of Science and Technology
Jihu Wang	Shandong University
Haobo Wang	Zhejiang University
Xianzhi Wang	University of Technology Sydney
Chao Wang	Shanghai Jiao Tong University
Jun Wang	Southwest University
Jing Wang	Beijing Jiaotong University
Di Wang	Nanyang Technological University
Yashen Wang	China Academy of Electronics and Information Technology of CETC
Qinglong Wang	McGill University
Sen Wang	University of Queensland
Di Wang	State University of New York at Buffalo
Qing Wang	Information Science Research Centre
Guoyin Wang	Chongqing University of Posts and Telecommunications
Thomas Weber	Ludwig-Maximilians-Universität München
Lingwei Wei	University of Chinese Academy of Sciences; Institute of Information Engineering, CAS
Tong Wei	Nanjing University
Pascal Welke	University of Bonn
Yang Wen	University of Science and Technology of China
Yanlong Wen	Nankai University
Paul Weng	UM-SJTU Joint Institute
Matthias Werner	ETAS GmbH, Bosch Group
Joerg Wicker	The University of Auckland
Uffe Wiil	University of Southern Denmark
Paul Wimmer	University of Lübeck; Robert Bosch GmbH
Martin Wistuba	University of Hildesheim
Feijie Wu	The Hong Kong Polytechnic University
Xian Wu	University of Notre Dame
Hang Wu	Georgia Institute of Technology
Yubao Wu	Georgia State University
Yichao Wu	SenseTime Group Limited
Xi-Zhu Wu	Nanjing University
Jia Wu	Macquarie University
Yang Xiaofei	Harbin Institute of Technology, Shenzhen
Yuan Xin	University of Science and Technology of China
Liu Xinshun	VIVO
Taufik Xu	Tsinghua University
Jinhui Xu	State University of New York at Buffalo
Depeng Xu	University of Arkansas
Peipei Xu	University of Liverpool

Yichen Xu	Beijing University of Posts and Telecommunications
Bo Xu	Donghua University
Hansheng Xue	Harbin Institute of Technology, Shenzhen
Naganand Yadati	Indian Institute of Science
Akihiro Yamaguchi	Toshiba Corporation
Haitian Yang	Institute of Information Engineering, Chinese Academy of Sciences
Hongxia Yang	Alibaba Group
Longqi Yang	HPCL
Xiaochen Yang	University College London
Yuhan Yang	Shanghai Jiao Tong University
Ya Zhou Yang	National University of Defense Technology
Feidiao Yang	Institute of Computing Technology, Chinese Academy of Sciences
Liu Yang	Tianjin University
Chaoqi Yang	University of Illinois at Urbana-Champaign
Carl Yang	University of Illinois at Urbana-Champaign
Guanyu Yang	Xi'an Jiaotong - Liverpool University
Yang Yang	Nanjing University
Weicheng Ye	Carnegie Mellon University
Wei Ye	Peking University
Yanfang Ye	Case Western Reserve University
Kejiang Ye	SIAT, Chinese Academy of Sciences
Florian Yger	Université Paris-Dauphine
Yunfei Yin	Chongqing University
Lu Yin	Technische Universiteit Eindhoven
Wang Yingkui	Tianjin University
Kristina Yordanova	University of Rostock
Tao You	Northwestern Polytechnical University
Hong Qing Yu	University of Bedfordshire
Bowen Yu	Institute of Information Engineering, Chinese Academy of Sciences
Donghan Yu	Carnegie Mellon University
Yipeng Yu	Tencent
Shujian Yu	NEC Laboratories Europe
Jiadi Yu	Shanghai Jiao Tong University
Wenchao Yu	University of California, Los Angeles
Feng Yuan	The University of New South Wales
Chunyuan Yuan	Institute of Information Engineering, Chinese Academy of Sciences
Sha Yuan	Tsinghua University
Farzad Zafarani	Purdue University
Marco Zaffalon	IDSIA
Nayyar Zaidi	Monash University
Tianzi Zang	Shanghai Jiao Tong University
Gerson Zaverucha	Federal University of Rio de Janeiro

Javier Zazo	Harvard University
Albin Zehe	University of Würzburg
Yuri Zelenkov	National Research University Higher School of Economics
Amber Zelvelder	Umeå University
Mingyu Zhai	NARI Group Corporation
Donglin Zhan	Sichuan University
Yu Zhang	Southeast University
Wenbin Zhang	University of Maryland
Qiuchen Zhang	Emory University
Tong Zhang	PKU
Jianfei Zhang	Case Western Reserve University
Nailong Zhang	MassMutual
Yi Zhang	Nanjing University
Xiangliang Zhang	King Abdullah University of Science and Technology
Ya Zhang	Shanghai Jiao Tong University
Zongzhang Zhang	Nanjing University
Lei Zhang	Institute of Information Engineering, Chinese Academy of Sciences
Jing Zhang	Renmin University of China
Xianchao Zhang	Dalian University of Technology
Jiangwei Zhang	National University of Singapore
Fengpan Zhao	Georgia State University
Lin Zhao	Institute of Information Engineering, Chinese Academy of Sciences
Long Zheng	Huazhong University of Science and Technology
Zuowu Zheng	Shanghai Jiao Tong University
Tongya Zheng	Zhejiang University
Runkai Zheng	Jinan University
Cheng Zheng	University of California, Los Angeles
Wenbo Zheng	Xi'an Jiaotong University
Zhiqiang Zhong	University of Luxembourg
Caiming Zhong	Ningbo University
Ding Zhou	Columbia University
Yilun Zhou	MIT
Ming Zhou	Shanghai Jiao Tong University
Yanqiao Zhu	Institute of Automation, Chinese Academy of Sciences
Wenfei Zhu	King
Wanzheng Zhu	University of Illinois at Urbana-Champaign
Fuqing Zhu	Institute of Information Engineering, Chinese Academy of Sciences
Markus Zopf	TU Darmstadt
Weidong Zou	Beijing Institute of Technology
Jingwei Zuo	UVSQ

Program Committee Members, Applied Data Science Track

Deepak Ajwani	Nokia Bell Labs
Nawaf Alharbi	Kansas State University
Rares Ambrus	Toyota Research Institute
Maryam Amir Haeri	Technische Universität Kaiserslautern
Jean-Marc Andreoli	Naverlabs Europe
Cecilio Angulo	Universitat Politècnica de Catalunya
Stefanos Antaris	KTH Royal Institute of Technology
Nino Antulov-Fantulin	ETH Zurich
Francisco Antunes	University of Coimbra
Muhammad Umer Anwaar	Technical University of Munich
Cristian Axenie	Audi Konfuzius-Institut Ingolstadt/Technical University of Ingolstadt
Mehmet Cem Aytekin	Sabancı University
Anthony Bagnall	University of East Anglia
Marco Baldan	Leibniz University Hannover
Maria Bampa	Stockholm University
Karin Becker	UFRGS
Swarup Ranjan Behera	Indian Institute of Technology Guwahati
Michael Berthold	University of Konstanz
Antonio Bevilacqua	Insight Centre for Data Analytics
Ananth Reddy Bhimireddy	Indiana University Purdue University - Indianapolis
Haixia Bi	University of Bristol
Wu Bin	Zhengzhou University
Thibault Blanc Beyne	INP Toulouse
Andrzej Bobyk	Maria Curie-Skłodowska University
Antonio Bonafonte	Amazon
Ludovico Boratto	Eurecat
Massimiliano Botticelli	Robert Bosch GmbH
Maria Brbic	Stanford University
Sebastian Buschjäger	TU Dortmund
Rui Camacho	University of Porto
Doina Caragea	Kansas State University
Nicolas Carrara	University of Toronto
Michele Catasta	Stanford University
Oded Cats	Delft University of Technology
Tania Cerquitelli	Politecnico di Torino
Fabricio Ceschin	Federal University of Paraná
Jeremy Charlier	University of Luxembourg
Anveshi Charuvaka	GE Global Research
Liang Chen	Sun Yat-sen University
Zhiyong Cheng	Shandong Artificial Intelligence Institute

Silvia Chiusano	Politecnico di Torino
Cristian Consonni	Eurecat - Centre Tecnòlogic de Catalunya
Laure Crochepierre	RTE
Henggang Cui	Uber ATG
Tiago Cunha	University of Porto
Elena Daraio	Politecnico di Torino
Hugo De Oliveira	HEVA/Mines Saint-Étienne
Tom Decroos	Katholieke Universiteit Leuven
Himel Dev	University of Illinois at Urbana-Champaign
Eustache Diemert	Criteo AI Lab
Nat Dilokthanakul	Vidyasirimedhi Institute of Science and Technology
Daizong Ding	Fudan University
Kaize Ding	ASU
Ming Ding	Tsinghua University
Xiaowen Dong	University of Oxford
Sourav Dutta	Huawei Research
Madeleine Ellis	University of Nottingham
Benjamin Evans	Brunel University London
Francesco Fabbri	Universitat Pompeu Fabra
Benjamin Fauber	Dell Technologies
Fuli Feng	National University of Singapore
Oluwaseyi Feyisetan	Amazon
Ferdinando Fioretto	Georgia Institute of Technology
Caio Flexa	Federal University of Pará
Germain Forestier	Université de Haute-Alsace
Blaz Fortuna	Qlector
Enrique Frias-Martinez	Telefónica Research and Development
Zuohui Fu	Rutgers University
Takahiro Fukushige	Nissan Motor Co., Ltd.
Chen Gao	Tsinghua University
Johan Garcia	Karlstad University
Marco Gärtler	ABB Corporate Research Center
Kanishka Ghosh Dastidar	Universität Passau
Biraja Ghoshal	Brunel University London
Lovedeep Gondara	Simon Fraser University
Severin Gsponer	Science Foundation Ireland
Xinyu Guan	Xi'an Jiaotong University
Karthik Gurumoorthy	Amazon
Marina Haliem	Purdue University
Massinissa Hamidi	Laboratoire LIPN-UMR CNRS 7030, Sorbonne Paris Cité
Junheng Hao	University of California, Los Angeles
Khadidja Henni	Université TÉLUQ

Manali Sharma	Samsung Semiconductor Inc.
Jiaming Shen	University of Illinois at Urbana-Champaign
Dash Shi	LinkedIn
Ashish Sinha	IIT Roorkee
Yorick Spenrath	Technische Universiteit Eindhoven
Simon Stieber	University of Augsburg
Hendra Suryanto	Rich Data Corporation
Raunak Swarnkar	IIT Gandhinagar
Imen Trabelsi	National Engineering School of Tunis
Alexander Treiss	Karlsruhe Institute of Technology
Rahul Tripathi	Amazon
Dries Van Daele	Katholieke Universiteit Leuven
Ranga Raju Vatsavai	North Carolina State University
Vishnu Venkataraman	Credit Karma
Sergio Viademonte	Vale Institute of Technology, Vale SA
Yue Wang	Microsoft Research
Changzhou Wang	The Boeing Company
Xiang Wang	National University of Singapore
Hongwei Wang	Shanghai Jiao Tong University
Wenjie Wang	Emory University
Zirui Wang	Carnegie Mellon University
Shen Wang	University of Illinois at Chicago
Dingxian Wang	East China Normal University
Yoshikazu Washizawa	The University of Electro-Communications
Chrys Watson Ross	University of New Mexico
Dilusha Weeraddana	CSIRO
Ying Wei	The Hong Kong University of Science and Technology
Laksri Wijerathna	Monash University
Le Wu	Hefei University of Technology
Yikun Xian	Rutgers University
Jian Xu	Citadel
Haiqin Yang	Ping An Life
Yang Yang	Northwestern University
Carl Yang	University of Illinois at Urbana-Champaign
Chin-Chia Michael Yeh	Visa Research
Shujian Yu	NEC Laboratories Europe
Chung-Hsien Yu	University of Massachusetts Boston
Jun Yuan	The Boeing Company
Stella Zevio	LIPN
Hanwen Zha	University of California, Santa Barbara
Chuxu Zhang	University of Notre Dame
Fanjin Zhang	Tsinghua University
Xiaohan Zhang	Sony Interactive Entertainment
Xinyang Zhang	University of Illinois at Urbana-Champaign
Mia Zhao	Airbnb
Qi Zhu	University of Illinois at Urbana-Champaign

Hengshu Zhu   Baidu Inc.
Tommaso Zoppi   University of Florence
Lan Zou   Carnegie Mellon University

Program Committee Members, Demo Track

Deepak Ajwani   Nokia Bell Labs
Rares Ambrus   Toyota Research Institute
Jean-Marc Andreoli   NAVER LABS Europe
Ludovico Boratto   Eurecat
Nicolas Carrara   University of Toronto
Michelangelo Ceci   Università degli Studi di Bari Aldo Moro
Tania Cerquitelli   Politecnico di Torino
Liang Chen   Sun Yat-sen University
Jiawei Chen   Zhejiang University
Zhiyong Cheng   Shandong Artificial Intelligence Institute
Silvia Chiusano   Politecnico di Torino
Henggang Cui   Uber ATG
Tiago Cunha   University of Porto
Chris Develder   Ghent University
Nat Dilokthanakul   Vidyasirimedhi Institute of Science and Technology
Daizong Ding   Fudan University
Kaize Ding   ASU
Xiaowen Dong   University of Oxford
Fuli Feng   National University of Singapore
Enrique Frias-Martinez Telefónica Research and Development
Zuohui Fu   Rutgers University
Chen Gao   Tsinghua University
Thomas Gärtner   TU Wien
Derek Greene   University College Dublin
Severin Gsponer   University College Dublin
Xinyu Guan   Xi'an Jiaotong University
Junheng Hao   University of California, Los Angeles
Ziniu Hu   University of California, Los Angeles
Chao Huang   University of Notre Dame
Hong Huang   UGoe
Neil Hurley   University College Dublin
Guillaume Jacquet   Joint Research Centre - European Commission
Di Jiang   WeBank
Song Jiang   University of California, Los Angeles
Jihed Khiari   Johannes Kepler Universität Linz
Mark Last   Ben-Gurion University of the Negev
Thach Le Nguyen   The Insight Centre for Data Analytics
Vincent Lemaire   Orange Labs
Camelia Lemnaru   Universitatea Tehnică din Cluj-Napoca
Bowen Liu   Stanford University

Sponsors

Contents – Part IV

Applied Data Science: Web Mining

Applied Data Science: Transportation

Applied Data Science: Activity Recognition

Applied Data Science: Hardware and Manufacturing

Applied Data Science: Spatiotemporal Data

Applied Data Science: Recommendation

Social Influence Attentive Neural Network for Friend-Enhanced Recommendation

Yuanfu Lu[1,2], Ruobing Xie[2], Chuan Shi[1(✉)], Yuan Fang[3], Wei Wang[2], Xu Zhang[2], and Leyu Lin[2]

[1] Beijing University of Posts and Telecommunications, Beijing, China
{luyuanfu,shichuan}@bupt.edu.cn
[2] WeChat Search Application Department, Tencent Inc., Shenzhen, China
xrbsnowing@163.com, {unoywang,xuonezhang,goshawklin}@tencent.com
[3] Singapore Management University, Singapore, Singapore
yfang@smu.edu.sg

Abstract. With the thriving of online social networks, there emerges a new recommendation scenario in many social apps, called Friend Enhanced Recommendation (FER) in this paper. In FER, a user is recommended with items liked/shared by his/her friends (called a friend referral circle). These friend referrals are explicitly shown to users. Different from conventional social recommendation, the unique friend referral circle in FER may significantly change the recommendation paradigm, making users to pay more attention to enhanced social factors. In this paper, we first formulate the FER problem, and propose a novel Social Influence Attentive Neural network (SIAN) solution. In order to fuse rich heterogeneous information, the attentive feature aggregator in SIAN is designed to learn user and item representations at both node- and type-levels. More importantly, a social influence coupler is put forward to capture the influence of the friend referral circle in an attentive manner. Experimental results demonstrate that SIAN outperforms several state-of-the-art baselines on three real-world datasets. (Code and dataset are available at https://github.com/rootlu/SIAN).

Keywords: Heterogeneous graph · Friend-enhanced recommendation · Social influence

1 Introduction

Nowadays, with the thriving of online social networks, people are more willing to actively express their opinions and share information with friends on social platforms. Friends become essential information sources and high-quality information filters. Items that friends have interacted with (shared, liked, etc.) have great impacts on users, which are likely to become users' future interests. There are lots of recommender systems that concentrate on social influences of friends

Y. Dong et al. (Eds.): ECML PKDD 2020, LNAI 12460, pp. 3–18, 2021.
https://doi.org/10.1007/978-3-030-67667-4_1

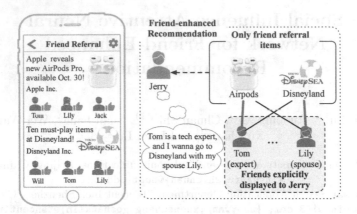

Fig. 1. A typical illustration of the friend-enhanced recommendation. The left shows the scenario that *Jerry* is recommended two articles, with friends (e.g., *Tom*) who have interacted with (shared, liked, etc.) them explicitly shown underneath. The right shows the formalization of the FER problem, where only friend referral items will be recommended and friends who interacted with the item are explicitly displayed to user.

(e.g., following feed in YouTube and Top Stories in WeChat). Some social recommendation algorithms also consider social factors for personalization [4,16].

Impressed by the great successes of social influence in recommendation, we propose a novel scenario named **Friend-Enhanced Recommendation (FER)**, which multiplies the influence of friends in social recommendation. FER has two major differences from the classical social recommendation: (1) FER only recommends to the user what his/her **friends have interacted with**, regarding friends as high-quality information filters to provide more high-quality items. (2) All friends who have interacted with the item are **explicitly displayed** to the user attached to the recommended item, which highlights the critical importance of explicit social factors and improves the interpretability for user behaviors.

In recent years, FER systems are blooming and have been widely-used by hundreds of millions of users. Figure 1 gives a typical illustration of a real-world FER. For each user-item pair, FER explicitly shows the friend set having interacted with the item, which is defined as the **Friend Referral Circle (FRC)** of the user to the item. For instance, the FRC of *Jerry* to the article about *Air-Pods* is { *Tom, Lily, Jack* }. Such a FRC drastically highlights the social influence of friends and their roles, which makes FER more complicated and relevant. It has even changed the recommendation paradigm compared to classical social recommendation. Taking Fig. 1 as an example, in classical social recommendation, *Jerry* would have no idea about the FRC (which is not displayed to him), hence he may read an article based on his own interest. However, in our FER, in addition to the attractiveness of the item itself, the influence of friends may be the main reason for the click. Here the FRC is explicitly displayed to *Jerry*, so the more likely reason why he clicks the article about *AirPods* is because *Tom* (a tech-expert friend) has read it. It is also entirely possible that *Jerry* reads the article about *Disneyland* because his spouse *Lily* has read it. Furthermore, when the article is related to technology, the coupling between the expert and technology may have a greater impact on Jerry than that between his spouse and technology, whereas the opposite scenario may happen w.r.t. entertainment. Hence, in FER, multiple factors contribute to user clicks. The reasons for a user clicking an article may come from (1) his interests in item contents (item), (2) the recommendation of an expert (item-friend combination), or even (3) the concerns on his friends themselves (friend). In FER, users have the tendency to see *what their friends have read*, rather than to merely see *what themselves are interested in*. It could even say that social recommendation focuses on bringing social information to better recommend items, while FER aims to recommend the combination of both items and friend referrals.

As the critical characteristic of FER, the explicit FRC brings in two challenges: (1) *How to extract key information from multifaceted heterogeneous factors?* FER involves multiple heterogeneous factors such as item contents, friend referrals and their interactions. The impacts of these factors vary in different scenarios with different combinations of users, items and friend referrals. FER is much more challenging since it is required not only to learn user preferences on items, but also to predict users' concerns towards different factors. (2) *How to exploit explicit friend referral information?* The explicit friend referrals greatly emphasize the importance of social information in recommendation, which are crucial in FER. However, there is few work that has explored the performances and characteristics of FRCs in real-world recommendation. A deliberate strategy is desired to make full use of the explicit friend referral information in FER.

To solve these issues, we propose a novel **S**ocial **I**nfluence **A**ttentive **N**eural network (**SIAN**). Specifically, we define the FER as a user-item interaction prediction task on a heterogeneous social graph, which flexibly integrates rich information in heterogeneous objects and their interactions. First, we design an attentive feature aggregator with both node- and type-level aggregations to learn user and item representations, without being restricted to pre-defined meta-paths in some previous efforts [3,19]. Next, we implement a social influence coupler to model the coupled influence diffusing through the explicit friend referral circles, which combines the influences of multiple factors (e.g., friends and items) with an attentive mechanism. Overall, SIAN captures valuable multifaceted factors in FER, which successfully distills the most essential preferences of users from a heterogeneous graph and friend referral circles. In experiments, SIAN significantly outperforms all competitive baselines in multiple metrics on three large, real-world datasets. Further quantitative analyses on attentive aggregation and social influence also reveal impressive sociological discoveries. We summarize the contributions as follows:

- We are the first to study the widely-adopted recommendation scenario named friend-enhanced recommendation (FER), where friend referrals are attached to items and explicitly exposed to users.

- We propose a novel Social Influence Attentive Neural network (SIAN) for FER. It uses a novel attentive feature aggregator to extract useful multi-faceted information, and leverages a social influence coupler to judge the significance of different friend referrals.
- Experiments on three real-world datasets verify the effectiveness and robustness of SIAN. Further quantitative analyses also reveal valuable sociological patterns, reflecting the changes and interpretability of user behaviors when social influence becomes more significant.

2   Preliminaries

Definition 1. Heterogeneous Social Graph (HSG). *A heterogeneous social graph is denoted as $\mathcal{G} = (\mathcal{V}, \mathcal{E})$, where $\mathcal{V} = \mathcal{V}_U \cup \mathcal{V}_I$ and $\mathcal{E} = \mathcal{E}_F \cup \mathcal{E}_R$ are the sets of nodes and edges. Here $\mathcal{V}_U$ and $\mathcal{V}_I$ are the sets of users and items. For $u, v \in \mathcal{V}_U$, $\langle u, v \rangle \in \mathcal{E}_F$ represents the friendship between users. For $u \in \mathcal{V}_U$ and $i \in \mathcal{V}_I$, $\langle u, i \rangle \in \mathcal{E}_R$ is the interaction relation between u and i.*

It not difficult to extend the HSG by adding attribute features or link relations as a Heterogeneous Information Network (HIN) [14]. Figure 1 shows an HSG containing three types of nodes, i.e., {*User, Article, Media*}, and multiple relations, e.g., {*User-User, User-Article, User-Media, Article-Media*}.

Definition 2. Friend Referral Circle (FRC). *Given an HSG $\mathcal{G} = (\mathcal{V}, \mathcal{E})$, we define the friend referral circle of a user u w.r.t. a non-interacting item i (i.e., $\langle u, i \rangle \notin \mathcal{E}_R$) as $\mathcal{C}_u(i) = \{v | \langle u, v \rangle \in \mathcal{E}_F \cap \langle v, i \rangle \in \mathcal{E}_R\}$. Here v is called an* **influential friend** *of user u.*

Taking Fig. 1 as an example, the friend referral circle of *Jerry* w.r.t. the non-interacting article about *AirPods* is {*Tom, Lily, Jack*}, while the FRC in terms of the article about *Disneyland* is $\mathcal{C}_{Jerry}(Disneyland) = \{Will, Tom, Lily\}$.

Definition 3. Friend-Enhanced Recommendation (FER). *Given an HSG $\mathcal{G} = (\mathcal{V}, \mathcal{E})$ and the FRC $\mathcal{C}_u(i)$ of a user u w.r.t. a non-interacting item i, the FER aims to predict whether user u has a potential preference to item i. That is, a prediction function $\hat{y}_{ui} = \mathcal{F}(\mathcal{G}, \mathcal{C}_u(i); \Theta)$ is to be learned, where $\hat{y}_{ui}$ is the probability that user u will interact with item i, and Θ is the model parameters.*

3   The Proposed Model

3.1   Model Overview

As illustrated in Fig. 2, SIAN models the FER with an HSG. In addition to the user and item representations (e.g., $\mathbf{h}_u$ for *Jerry* and $\mathbf{h}_i$ for the *Disneyland* article), SIAN learns a social influence representation (e.g., $\mathbf{h}_{ui}$) by coupling each influential friend (e.g., *Tom*) with the item. They are jointly responsible for predicting the probability $\hat{y}_{ui}$ of interaction between user u and item i.

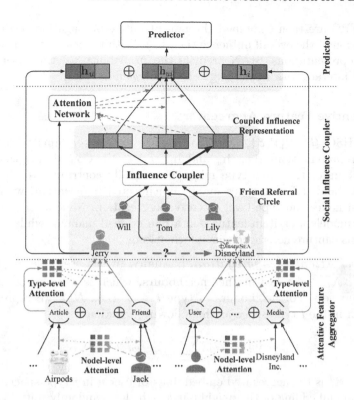

Fig. 2. The overall architecture of SIAN. The attentive feature aggregator hierarchically aggregates heterogeneous neighbour features with node- and type-level attention, and outputs the representations of users and items (i.e., $\mathbf{h}_u$ and $\mathbf{h}_i$). The social influence coupler couples the influence of each influential friends and the item, to encode the explicit social influence into the representation (i.e., $\mathbf{h}_{ui}$).

First, each user or item node is equipped with an attentive feature aggregator with node- and type-level aggregations, which is designed to exploit multifaceted information. At the node level, the features from the neighbours of the same type (e.g., articles that *Jerry* liked) will be aggregated in the current type space; at the type level, the representations from different type spaces will be further aggregated to encode multifaceted information. At each level, an attention mechanism is employed to differentiate and capture the latent relevance of the neighbors and types, respectively. Such a hierarchical attentive design enables SIAN to encode the fine-grained relevance of multifaceted information, and the dual attention mechanism allows it to delicately capture the effect of different factors. Unlike some previous works [3,19], SIAN does not require any manual selection of meta-paths, so that it is expected to yield a better performance.

Second, the influence from an influential friend (e.g., *Tom*) and an item (e.g., the *Disneyland* article) is jointly captured with a social influence coupler, which quantifies the degree of their coupled influence. Multiple coupled influences from the FRC are then combined through attentive propagation to derive the representation of the overall influence (i.e., $\mathbf{h}_c$). With the learned user, item and influence representations, SIAN predicts the probability $\hat{y}_{ui}$ that user u (e.g., *Jerry*) will interact with item i (e.g., the *Disneyland* article).

3.2   Attentive Feature Aggregator

Given an HSG $\mathcal{G} = \{\mathcal{V}, \mathcal{E}\}$, attentive feature aggregator aims to learn user and item representations (i.e., $\mathbf{h}_u$ and $\mathbf{h}_i$, $u, i \in \mathcal{V}$). Considering that different neighbours of the same type might not equally contribute to the feature aggregation, and different types entail multifaceted information, we design a hierarchical node- and type-level attentive aggregation. Node-level aggregation separately models user/item features in a fine-grained manner, while type-level aggregations capture heterogeneous information.

Node-Level Attentive Aggregation. Formally, given a user u, let $\mathcal{N}_u = \mathcal{N}_u^{t_1} \cup \mathcal{N}_u^{t_2} \cup \cdots \cup \mathcal{N}_u^{t_{|\mathcal{T}|}}$ denotes his/her neighbours, which is a union of $|\mathcal{T}|$ types of neighbour sets. For neighbours of type $t \in \mathcal{T}$ (i.e., $\mathcal{N}_u^t$), we represent the aggregation in the t type space as the following function:

$$\mathbf{p}_u^t = \mathrm{ReLU}(\mathbf{W}_p(\textstyle\sum_{k \in \mathcal{N}_u^t} \alpha_{ku}\mathbf{x}_k) + \mathbf{b}_p), \tag{1}$$

where $\mathbf{p}_u^t \in \mathbb{R}^d$ is the aggregated embeddings of user u in t type space. $\mathbf{x}_k \in \mathbb{R}^d$ is the initial embedding of the neighbour k, which is randomly initialized. Here $\mathbf{W}_p \in \mathbb{R}^{d \times d}$ and $\mathbf{b}_p \in \mathbb{R}^d$ are the weight and bias of a neural network. α_{ku} is the attentive contribution of neighbour k to the feature aggregation of u,

$$\alpha_{ku} = \frac{\exp(f([\mathbf{x}_k \oplus \mathbf{x}_u]))}{\sum_{k' \in \mathcal{N}_u^t} \exp(f([\mathbf{x}_{k'} \oplus \mathbf{x}_u]))}, \tag{2}$$

where $f(\cdot)$ is a two-layer neural network activated with ReLu function and $\oplus$ denotes the concatenation operation. Obviously, the larger α_{ku}, the greater contribution of neighbour k to the feature aggregation of user u.

Given multiple types of neighbours, we can get multiple embeddings for u in various type spaces, denoted as $\{\mathbf{p}_u^{t_1}, \cdots, \mathbf{p}_u^{t_{|\mathcal{T}|}}\}$.

Type-Level Attentive Aggregation. Intuitively, different types of neighbours indicate various aspects of information and a node is likely to have different preferences for multiple aspects. Given a user u and his/her node-level aggregated embeddings in different type spaces, we aggregate them as follows:

$$\mathbf{h}_u = \mathrm{ReLU}(\mathbf{W}_h \textstyle\sum_{t \in \mathcal{T}} \beta_{tu}\mathbf{p}_u^t + \mathbf{b}_h), \tag{3}$$

where $\mathbf{h}_u \in \mathbb{R}^d$ is the latent representation of user u. $\{\mathbf{W}_h \in \mathbb{R}^{d \times d}, \mathbf{b}_h \in \mathbb{R}^d\}$ are parameters of a neural network. β_{tu} is the attentive preferences of type t w.r.t. the feature aggregation of user u, as various types of neighbours contain multifaceted information and are expected to collaborate with each other. For user u, we concatenate the aggregated representations of all neighbour types, and define the following weight:

$$\beta_{tu} = \frac{\exp(\mathbf{a}_t^\top [\mathbf{p}_u^{t_1} \oplus \mathbf{p}_u^{t_2} \oplus \cdots \oplus \mathbf{p}_u^{t_{|\mathcal{T}|}}])}{\sum_{t' \in \mathcal{T}} \exp(\mathbf{a}_{t'}^\top [\mathbf{p}_u^{t_1} \oplus \mathbf{p}_u^{t_2} \oplus \cdots \oplus \mathbf{p}_u^{t_{|\mathcal{T}|}}])}, \tag{4}$$

where $\mathbf{a}_t \in \mathbb{R}^{|\mathcal{T}|d}$ is a type-aware attention vector shared by all users. With Eq. (4), the concatenation of various neighbour types captures multifaceted information for a user, and $\mathbf{a}_t$ encodes the global preference of each type.

Similarly, for each item i, the attentive feature aggregator takes the neighbours of i as input, and outputs the latent representation of i, denoted as $\mathbf{h}_i$.

3.3   Social Influence Coupler

To exploit the FRCs and capture the effects of influential friends, we propose a social influence coupler. The differential impact of the influential friends and the item on social behaviors is first coupled together, and then we attentively represent the overall influence in the FRC.

Coupled Influence Representation. Following [7], human behaviors are affected by various factors. In FER, whether u interacts with i is not simply driven by only the item itself or only the friends. More likely, the co-occurrence of friends and the item have a significant impact. As in the previous example (Fig. 1), when it is technology-related, the coupling between the expert (e.g. *Tom*) and the item (e.g. *AirPods*) has a greater impact than the coupling between the spouse and a tech-item, but the opposite scenario may happen for entertainment-related items. Hence, given user u, item i, and the FRC $\mathcal{C}_u(i)$, we couple the influence of each friend $v \in \mathcal{C}_u(i)$ and item i as following:

$$\mathbf{c}_{\langle v,i \rangle} = \sigma(\mathbf{W}_c \phi(\mathbf{h}_v, \mathbf{h}_i) + \mathbf{b}_c), \tag{5}$$

where $\mathbf{h}_v$ and $\mathbf{h}_i$ are aggregated representations of user v and item i. $\phi(\cdot, \cdot)$ serves as a fusion function, which can be element-wise product or concatenation (here we adopt concatenation). σ is the ReLU function. Obviously, Eq. (5) couples the features of item i and the influential friend v, capturing the influence of both.

Attentive Influence Degree. With the coupled influence representation $\mathbf{c}_{\langle v,i \rangle}$, our next goal is to obtain the influence degree of $\mathbf{c}_{\langle v,i \rangle}$ on the user u. Since the influence score depends on user u, we incorporate the representation of user u (i.e., $\mathbf{h}_u$) into the influence score calculation with a two-layer neural network parameterized by $\{\mathbf{W}_1, \mathbf{W}_2, \mathbf{b}_1, \mathbf{b}_2\}$:

$$d'_{u \leftarrow \langle v,i \rangle} = \sigma(\mathbf{W}_2(\sigma(\mathbf{W}_1 \phi(\mathbf{c}_{v,i}, \mathbf{h}_u) + \mathbf{b}_1)) + \mathbf{b}_2). \tag{6}$$

Then, the attentive influence degree is obtained by normalizing $d'_{u \leftarrow \langle v,i \rangle}$, which can be interpreted as the impact of the influential friend v on the user behavior:

$$d_{u \leftarrow \langle v,i \rangle} = \frac{\exp(d'_{u \leftarrow \langle v,i \rangle})}{\sum_{v' \in \mathcal{C}_u(i)} \exp(d'_{u \leftarrow \langle v',i \rangle})}. \tag{7}$$

Since the influences of friends propagate from the FRC, we attentively sum the coupled influences of the influential friends and item v on user u:

$$\mathbf{h}_{ui} = \sum\nolimits_{v \in \mathcal{C}_u(i)} d_{u \leftarrow \langle v,i \rangle} \mathbf{c}_{\langle v,i \rangle}. \tag{8}$$

As the coupled influence representation $\mathbf{c}_{\langle v,i \rangle}$ incorporates the latent factors of the influential friend and the item, Eq. (8) guarantees that the social influence propagating among them can be encoded into the latent representation $\mathbf{h}_{ui}$.

3.4   Behavior Prediction and Model Learning

With the representations of user, item and the coupled influence (i.e., $\mathbf{h}_u$, $\mathbf{h}_i$ and $\mathbf{h}_{ui}$), we concatenate them and then feed it into a two-layer neural network:

$$\mathbf{h}_o = \sigma(\mathbf{W}_{o_2}(\sigma(\mathbf{W}_{o_1}([\mathbf{h}_u \oplus \mathbf{h}_{ui} \oplus \mathbf{h}_i]) + \mathbf{b}_{o_1}) + \mathbf{b}_{o_2}). \tag{9}$$

Then, the predicted probability of a user-item pair is obtained via a regression layer with a weight vector $\mathbf{w}_y$ and bias b_y:

$$\hat{y}_{ui} = \text{sigmoid}(\mathbf{w}_y^\top \mathbf{h}_o + b_y). \tag{10}$$

Finally, to estimate model parameters Θ of SIAN, we optimize the following cross-entropy loss, where y_{ui} is the ground truth and λ is the L2-regularization parameter for reducing overfitting:

$$-\sum_{\langle u,i \rangle \in \mathcal{E}_R} (y_{ui} \log \hat{y}_{ui} + (1 - y_{ui}) \log (1 - \hat{y}_{ui})) + \lambda \|\Theta\|_2^2. \tag{11}$$

4   Experiments

We conduct comprehensive experiments on three real-world datasets, demonstrating superior performance and revealing interesting sociological patterns.

4.1   Datasets

Yelp and Douban are classical open datasets widely used in recommendation, for which we build FRCs for each user-item pair to simulate the FER scenarios. FWD is extracted from a deployed live FER system with real FRCs displayed to users. The detailed statistics of datasets are shown in Table 1.

- **Yelp**[1] is a business review dataset containing both interactions and social relations. We first sample a set of users. For each user u, we construct a set of FRCs based on the given user-user relations and user-item interactions. Interactions with an empty FRC are filtered from the data. To get the initial feature vector of a node, we learn the word embeddings with word2vec using the review texts, and average the learned vectors for each user or item.

---

[1] https://www.yelp.com/dataset/challenge.

Table 1. Statistics of datasets.

Datasets	Nodes	#Nodes	Relations	#Relations
Yelp	User (U)	8,163	User-User	92,248
	Item (I)	7,900	User-Item	36,571
Douban	User (U)	12,748	User-User	169,150
	Book (B)	13,342	User-Book	224,175
FWD	User (U)	72,371	User-User	8,639,884
	Article (A)	22,218	User-Article	2,465,675
	Media (M)	218,887	User-Media	1,368,868
			Article-Media	22,218

- **Douban**[2] is a social network related to sharing books, which including friendships between users and interaction records between users and items. As pre-processes done for Yelp, we construct a set of FRCs based on the given user-user relations and user-item interactions. We take book descriptions and user reviews as input of word2vec, and then output the feature vectors of books and users. We predict the interaction probability between users and books.
- **Friends Watching Data (FWD)** is extracted from a real-world live FER system named WeChat Top Stories after data masking, where FRCs are explicitly displayed. Based on FWD, we construct a HSG containing nearly 313 thousand nodes and 12 million edges. Each user or item is associated with some given features (e.g., age or content vectors). We predict the interaction probability between users and articles.

4.2 Experimental Settings

Baselines. We compare the proposed SIAN against four types of methods, including feature/structure-based methods (i.e., MLP, DeepWalk, node2vec and metapath2vec), fusion of feature/structure-based methods (i.e., DeepWalk+fea, node2vec+fea and metapath2vec+fea), graph neural network methods (i.e., GCN, GAT and HAN) and social recommendation methods (i.e., TrustMF and DiffNet).

- **MLP** [10] is the most simple baselines, which is implemented with the same architecture as the prediction layer in SIAN. It takes the concatenation of feature vectors of users and items as input, and output the prediction probability of the interaction. Here we vary the size of feature vector with $\{32, 64\}$.
- **DeepWalk, node2vec** and **metapath2vec**. DeepWalk [12] and node2vec [5] are two homogeneous network embedding methods. metapath2vec [3] is a heterogeneous network embedding method based on meta-paths [15]. Here we adopt meta-paths shorter than 4 and report the best performance. We

---
[2] https://book.douban.com.

feed the embeddings of users and items into a logistic regression classifier to predict the probability of interaction. The MLP as in SIAN is also be applied here, but the performance is worse. Thus, we use the logistic regression here.

- **DeepWalk+fea, node2vec+fea** and **metapath2vec+fea**. With the learned embeddings , we further respectively concatenate them with the features of users and items, and use the logistic regression to evaluate performances, which derives DeepWalk+fea, node2vec+fea and metapath2vec+fea.
- **GCN, GAT** and **HAN**. GCN [9] and GAT [17] are graph convolutional networks designed for homogeneous graphs, while HAN [19] is designed for heterogeneous graphs. These methods take node features as input and output the node embeddings. We learn embeddings for users and items and then predict the probability of interactions as the above method. We test the same meta-paths used in metapath2vec for HAN and report the best performance.
- **TrustMF** and **DiffNet**. TrustMF [22] factorizes social trust networks and maps users into two spaces. Here we use it to learn embeddings for users and items. Then, we employ the aforementioned method to predict the interaction probability. DiffNet [20] is a social recommendation method, which takes social relations as input to enhance user embeddings. We learn the probability of the user-item interaction by modifying the output layer with the sigmoid function.

Parameters Settings. For each dataset, the ratio of training, validation and test set is 7:1:2. We adopt Adam optimizer [8] with the PyTorch implementation. The learning rate, batch size, and regularization parameter are set to 0.001, 1, 024 and 0.0005 using grid search [1], determined by optimizing AUC on the validation set. For random walk based baselines, we set the walk number, walk length and window size as 10, 50, and 5, respectively. For graph neural network based methods, the number of layers is set to 2. For DiffNet, we set the regularization parameter as 0.001. The depth parameter is set to 2 as recommended in [20]. For other parameters of baselines, we optimize them empirically under the guidance of literature. Finally, for all methods except MLP, we set the size of feature vector as 64 and report performances under different embedding dimensions $\{32, 64\}$.

4.3   Experimental Results

We adopt three widely used metrics AUC, F1 and Accuracy to evaluate performance. The results w.r.t. the dimension of latent representation are reported in Tables 2, from which we have the following findings.

(1) SIAN outperforms all baselines in all metrics on three datasets with statistical significance ($p < 0.01$) under paired t-test. It indicates that SIAN can well capture user core concerns from multifaceted factors in FER. The improvements derive from both high-quality node representations generated from node- and type-level attentive aggregations, and the social influence coupler that digs out what users are socially inclined to. Besides, the consistent improvements on different dimensions verify that SIAN is robust to the dimension.

Table 2. Results on three datasets. The best method is bolded, and the second best is underlined. * indicate the significance level of 0.01.

Dataset	Model	AUC		F1		Accuracy	
		$d=32$	$d=64$	$d=32$	$d=64$	$d=32$	$d=64$
Yelp	MLP	0.6704	0.6876	0.6001	0.6209	0.6589	0.6795
	DeepWalk	0.7693	0.7964	0.6024	0.6393	0.7001	0.7264
	node2vec	0.7903	0.8026	0.6287	0.6531	0.7102	0.7342
	metapath2vec	0.8194	0.8346	0.6309	0.6539	0.7076	0.7399
	DeepWalk+fea	0.7899	0.8067	0.6096	0.6391	0.7493	0.7629
	node2vec+fea	0.8011	0.8116	0.6634	0.6871	0.7215	0.7442
	metapath2vec+fea	0.8301	0.8427	0.6621	0.6804	0.7611	0.7856
	GCN	0.8022	0.8251	0.6779	0.6922	0.7602	0.7882
	GAT	0.8076	0.8456	0.6735	0.6945	0.7783	0.7934
	HAN	0.8218	0.8476	0.7003	0.7312	0.7893	0.8102
	TrustMF	0.8183	0.8301	0.6823	0.7093	0.7931	0.8027
	DiffNet	<u>0.8793</u>	<u>0.8929</u>	<u>0.8724</u>	<u>0.8923</u>	<u>0.8698</u>	<u>0.8905</u>
	SIAN	**0.9486***	**0.9571***	**0.8976***	**0.9128***	**0.9096***	**0.9295***
Douban	MLP	0.7689	0.7945	0.7567	0.7732	0.7641	0.7894
	DeepWalk	0.8084	0.8301	0.7005	0.8054	0.8295	0.8464
	node2vec	0.8545	0.8623	0.8304	0.8416	0.8578	0.8594
	metapath2vec	0.8709	0.8901	0.8593	0.8648	0.8609	0.8783
	DeepWalk+fea	0.8535	0.8795	0.8347	0.8578	0.8548	0.8693
	node2vec+fea	0.8994	0.9045	0.8732	0.8958	0.8896	0.8935
	metapath2vec+fea	0.9248	0.9309	0.8998	0.9134	0.8975	0.9104
	GCN	0.9032	0.9098	0.8934	0.9123	0.9032	0.9112
	GAT	0.9214	0.9385	0.8987	0.9103	0.8998	0.9145
	HAN	0.9321	0.9523	<u>0.9096</u>	0.9221	<u>0.9098</u>	0.9205
	TrustMF	0.9034	0.9342	0.8798	0.9054	0.9002	0.9145
	DiffNet	<u>0.9509</u>	<u>0.9634</u>	0.9005	<u>0.9259</u>	0.9024	<u>0.9301</u>
	SIAN	**0.9742***	**0.9873***	**0.9139***	**0.9429***	**0.9171***	**0.9457***
FWD	MLP	0.5094	0.5182	0.1883	0.1932	0.2205	0.2302
	DeepWalk	0.5587	0.5636	0.2673	0.2781	0.1997	0.2056
	node2vec	0.5632	0.5712	0.2674	0.2715	0.2699	0.2767
	metapath2vec	0.5744	0.5834	0.2651	0.2724	0.4152	0.4244
	DeepWalk+fea	0.5301	0.5433	0.2689	0.2799	0.2377	0.2495
	node2vec+fea	0.5672	0.5715	0.2691	0.2744	0.3547	0.3603
	metapath2vec+fea	0.5685	0.5871	0.2511	0.2635	0.4698	0.4935
	GCN	0.5875	0.5986	0.2607	0.2789	0.4782	0.4853
	GAT	0.5944	0.6006	0.2867	0.2912	0.4812	0.4936
	HAN	0.5913	0.6025	0.2932	0.3011	0.4807	0.4937
	TrustMF	0.6001	0.6023	0.3013	0.3154	0.5298	0.5404
	DiffNet	<u>0.6418</u>	<u>0.6594</u>	<u>0.3228</u>	<u>0.3379</u>	<u>0.6493</u>	<u>0.6576</u>
	SIAN	**0.6845***	**0.6928***	**0.3517***	**0.3651***	**0.6933***	**0.7018***

(2) Compared with the graph neural network methods, the impressive improvements of SIAN proves the effectiveness of the node- and type-level attentive aggregations. Especially, SIAN achieves better performances than HAN

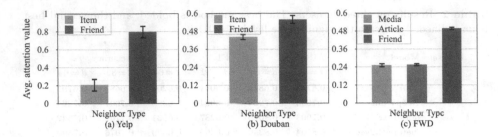

Fig. 3. Attentive aggregator analysis of *User*.

which is also designed for heterogeneous graphs with a two-level aggregation. It is because that the type-level attentive aggregation in SIAN captures heterogeneous information in multiple aspects, without being limited by the predefined meta-paths used in HAN. Moreover, the improvements also indicate the significance of our social influence coupler in FER.

(3) Social recommendation baselines also achieve promising performances, which further substantiates the importance of social influence in FER. Compared with baselines which only treat social relations as side information, the improvements imply that the friend referral factor may take the dominating position in FER, which should be carefully modeled. In particular, our SIAN achieves the best performance, reconfirming the capability of our social influence coupler in encoding diverse social factors for FER.

4.4   Impacts of Multifaceted Information

In attentive feature aggregator, each node embedding is aggregated from its neighbours of various types with different weights. We investigate the contribution of heterogeneous factors (e.g., friend, item, media), by finding the average type-level attention values (i.e., β in Eq. (4)) among all instances.

As shown in Fig. 3, the average attention value of the *Friend* type is significantly larger than that of other types. It is perhaps astonishing that the model pays more attention to users' social relationships, a notable departure from conventional recommendation where user-item interactions have thought to be more critical. This also justifies the proposed social influence coupler in SIAN, which plays an important role in extracting preferences from FRCs.

4.5   Analysis on Social Influence in FER

We have verified that FRC is the most essential factor in FER. However, a friend could impact user from different aspects (e.g., authority or similarity). Next, we show how different user attributes affect user behaviors in FER. Since we have detailed user attributes in FWD, here we conduct analysis on it.

Evaluation Protocol. The attention in social influence coupler reflects the importance of different friends. We assume that the friend v having the highest

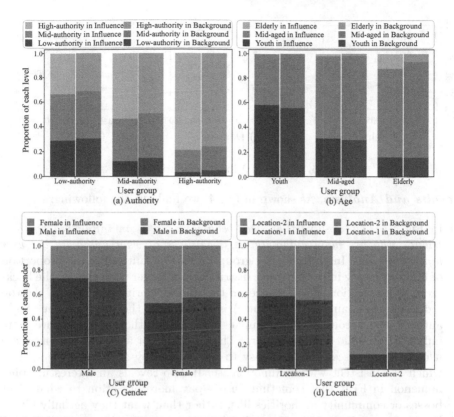

Fig. 4. Social influence analysis w.r.t user attributes. For each attribute and user group (e.g., the authority and the low-authority group in (a)), the left is the influence distribution while the right is the background distribution. In each bar, the height of each different-colored segment means the proportion of an attribute value in the influence or background distribution. Best read in color.

attention value (i.e., $d_{u\leftarrow\langle v,i\rangle}$ in Eq. (6)) is the most influential friend w.r.t. item i for user u, and all of v's attribute values are equally regarded as contributing to the influence. Given a user attribute and a user group, we define the *background distribution* by counting the attribute values of all friends in FRCs of users in this group, and also define the *influence distribution* by counting the attribute values of the most influential friends of users in the group. Thus, the background distribution represents the characteristics of general friends of this user group, while the influence distribution represents the characteristics of the most influential friends of this user group. If the two distributions perfectly agree with each other, this attribute is not a key social factor in influencing this user group. In contrast, the differences between the two distributions imply how much this attribute is a key social factor, and how its different values affect user behaviors.

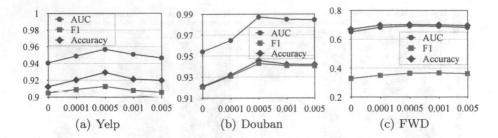

Fig. 5. Impact of λ in L2-regularization.

Results and Analysis. As shown in Fig. 4, we find out the following:

(1) In Fig. 4(a), we observe that *user behaviors are more influenced by their friends who are more authoritative, regardless of what authority the user him/herself has*. In all three user groups of varying authority, the proportion of high-authority in the influence distribution is larger than that in the background distribution. For instance, in the mid-authority user group, the top red block (high-authority influence) is larger than the top blue one (high-authority background), which implies that high-authority friends are more influential for mid-authority users. The result is not surprising as users are usually more susceptible and easy to be affected by authoritative persons, which is consistent with common sense. It also reveals an interesting phenomenon in FER that sometimes users pay more attention to what their bosses or community authorities like, rather than what they actually like.

(2) We also conduct several analyses on influences w.r.t. other user attributes. We find that *users are easy to be influenced by their friends which are similar to themselves*. Specifically, Fig. 4(b) shows that people like items recommended by their peers, especially for the youth and the elderly; meanwhile, Fig. 4(c) and (d) show that users tend to watch articles recommended by their friends with the same gender or location. Recommendation with user similarity, which has been widely assumed in collaborative filtering, is still classical even in FER.

In conclusion, while different social factors have various influences on the target user, none of them is dominating, which further establishes the complexity of FER. In this case, the promising improvements by SIAN demonstrate that it could well capture multifaceted social factors in FER, which could potentially contribute to the understanding of interpretable recommendation.

4.6   Parameters Analysis

Our SIAN involves two parameters, i.e., the embedding dimension $d \in \{32, 64\}$ and the L2-regularization parameter λ in Eq. (11). As we have reported model performance w.r.t. d in Sect. 4.3, here we vary λ in the set of $\{0, 0.0001, 0.0005, 0.001, 0.005\}$ to analyze its impact on model performance. As shown in Fig. 5, the optimal performance is obtained near $\lambda = 0.0005$, indicating that λ cannot be set too small or too large to prevent overfitting and underfitting.

5 Related Work

Social Recommendation. With the booming of social media, rich social information can be utilized for enhancing recommendation performance [2,6,11,13, 21], which motivates the advent of social recommendation. Specifically, SoRec [11] integrates collaborative filtering with social information by proposing a probabilistic matrix factorization model. [6] incorporates the trust influence on top of SVD++, which takes the social neighbours' preferences as the side information. TrustMF [22] factorizes social trust networks and maps users into two low-dimensional spaces: truster space and trustee space. Distinct from these methods merely treating social neighbours as side information, SIAN models the social information as first-class citizens based on the unique FRC formulation.

GNN-Based Social Recommendation. Recent advances in graph neural networks (GNN) have been crucial to modeling graph data [23]. Related to our work, HAN [19] embeds heterogeneous graphs with node- and semantic-level attentions, which heavily relies on the choice of predefined meta-paths. Besides, some works attempt to utilize GNNs to model user-item bipartite graphs or/and social networks. [18] integrates the knowledge graph into recommender systems, and [4] incorporates the social network into the learning of user and item latent factors. The recent DiffNet [20] models social influence with GCN. Although our SIAN also employs a GNN-based framework, it is tailored to capture multifaceted information diffusing from the FRCs through the novel node- and type-level attentive feature aggregator and social influence coupler.

6 Conclusion

In this paper, we first formulated a novel friend-enhanced recommendation problem, which is widely applicable to many social apps, and presented a social influence attentive neural network (SIAN). SIAN learns user and item representations with a two-level attentive aggregator and distills preferences from the unique friend referral circles with a social influence coupler. Experimental results demonstrate that SIAN significantly outperforms state-of-the-art baselines on three real-world datasets, and reveal interesting sociological patterns.

Acknowledgements. This work is supported in part by the National Natural Science Foundation of China (No. 61772082, 61806020, 61702296), the National Key Research and Development Program of China (2018YFB1402600), and the Tencent WeChat Rhino-Bird Focused Research Program. This work is also supported by the National Research Foundation, Singapore under its AI Singapore Programme (AISG Award No: AISG-RP-2018-001). Any opinions, findings and conclusions or recommendations expressed in this material are those of the author(s) and do not reflect the views of National Research Foundation, Singapore. Yuanfu Lu is also supported by 2019 Tencent Rhino-Bird Elite Training Program.

References

1. Bergstra, J.S., Bardenet, R., Bengio, Y., Kégl, B.: Algorithms for hyper-parameter optimization. In: NeurIPS, pp. 2546–2554 (2011)
2. Chen, C., Zhang, M., Ma, W., Zhang, Y., Liu, Y., Ma, S.: Efficient heterogeneous collaborative filtering without negative sampling for recommendation. In: AAAI (2020)
3. Dong, Y., Chawla, N.V., Swami, A.: metapath2vec: scalable representation learning for heterogeneous networks. In: SIGKDD, pp. 135–144. ACM (2017)
4. Fan, W., et al.: Graph neural networks for social recommendation. In: WWW, pp. 417–426 (2019)
5. Grover, A., Leskovec, J.: node2vec: scalable feature learning for networks. In: SIGKDD, pp. 855–864. ACM (2016)
6. Guo, G., Zhang, J., Yorke-Smith, N.: TrustSVD: collaborative filtering with both the explicit and implicit influence of user trust and of item ratings. In: AAAI (2015)
7. Jolly, A.: Lemur social behavior and primate intelligence. Science **153**, 501–506 (1966)
8. Kingma, D.P., Ba, J.: Adam: a method for stochastic optimization. arXiv preprint arXiv:1412.6980 (2014)
9. Kipf, T.N., Welling, M.: Semi-supervised classification with graph convolutional networks. In: ICLR (2017)
10. Kubat, M.: Neural networks: a comprehensive foundation by Simon Haykin, Macmillan, 1994, ISBN 0-02-352781-7. Knowl. Eng. Rev. **13**, 409–412 (1999)
11. Ma, H., Yang, H., Lyu, M.R., King, I.: SoRec: social recommendation using probabilistic matrix factorization. In: CIKM, pp. 931–940. ACM (2008)
12. Perozzi, B., Al-Rfou, R., Skiena, S.: DeepWalk: online learning of social representations. In: SIGKDD, pp. 701–710. ACM (2014)
13. Shen, T., et al.: Efficient heterogeneous collaborative filtering without negative sampling for recommendation. In: AAAI (2020)
14. Shi, C., Hu, B., Zhao, W.X., Philip, S.Y.: Heterogeneous information network embedding for recommendation. IEEE TKDE **31**, 357–370 (2018)
15. Sun, Y., Han, J., Yan, X., Yu, P.S., Wu, T.: PathSim: meta path-based top-k similarity search in heterogeneous information networks. VLDB **4**, 992–1003 (2011)
16. Tang, J., Hu, X., Liu, H.: Social recommendation: a review. Soc. Netw. Anal. Min. **3**, 1113–1133 (2013)
17. Velickovic, P., Cucurull, G., Casanova, A., Romero, A., Liò, P., Bengio, Y.: Graph attention networks. In: ICLR (2018)
18. Wang, H., et al.: RippleNet: propagating user preferences on the knowledge graph for recommender systems. In: CIKM, pp. 417–426. ACM (2018)
19. Wang, X., et al.: Heterogeneous graph attention network. In: WWW, pp. 2022–2032 (2019)
20. Wu, L., Sun, P., Fu, Y., Hong, R., Wang, X., Wang, M.: A neural influence diffusion model for social recommendation. In: SIGIR, pp. 235–244 (2019)
21. Xiao, W., Zhao, H., Pan, H., Song, Y., Zheng, V.W., Yang, Q.: Beyond personalization: social content recommendation for creator equality and consumer satisfaction. In: SIGKDD, pp. 235–245 (2019)
22. Yang, B., Lei, Y., Liu, J., Li, W.: Social collaborative filtering by trust. IEEE Trans. Pattern Anal. Mach. Intell. **39**, 1633–1647 (2016)
23. Zhang, Z., Cui, P., Zhu, W.: Deep learning on graphs: a survey. IEEE TKDE (2020)

Feedback-Guided Attributed Graph Embedding for Relevant Video Recommendation

Taofeng Xue[1,2], Xinzhou Dong[1,2], Wei Zhuo[3], Beihong Jin[1,2(✉)], He Chen[3], Wenhai Pan[3], Beibei Li[1,2], and Xuejian Zhang[3]

[1] State Key Laboratory of Computer Sciences, Institute of Software, Chinese Academy of Sciences, Beijing, China
Beihong@iscas.ac.cn
[2] University of Chinese Academy of Sciences, Beijing, China
[3] MX Media Co., Ltd., Singapore, Singapore

Abstract. Representation learning on graphs, as alternatives to traditional feature engineering, has been exploited in many application domains, ranging from e-commerce to computational biology. However, generating satisfactory video embeddings and putting them into practical use to improve the performance of recommendation tasks remains a challenge. In this paper, we present a video embedding approach named Equuleus, which learns video embeddings from user interaction behaviors. In Equuleus, we carefully incorporate user behavior characteristics into the construction of the video graph and the generation of node sequences. To accurately quantify the contributions of different attributes to embeddings, we propose a particular attributed encoder network, which employs an attention mechanism to aggregate different attributes in a distinguishable way. Moreover, we also leverage the user feedback as a guide to correct the generation of embeddings. Video embeddings generated by Equuleus have been used for relevant recommendation of videos in MX Player. Based on real data from MX Player, extensive offline experiments and online A/B test are conducted. Both experimental results and online CTRs illustrate that Equuleus can generate high-quality video embeddings and it can work effectively in a real-world production environment.

Keywords: Recommender system · Representation learning · Graph embedding · User behavior mining

1 Introduction

In recent years, with the rapid advancements in representation learning on graphs [10,19,22], node embeddings have been applied to multiple downstream tasks such as node classification, link prediction, visualization and pattern discovery, and thus adopted by many applications. For recommender systems, the key

© Springer Nature Switzerland AG 2021
Y. Dong et al. (Eds.): ECML PKDD 2020, LNAI 12460, pp. 19–35, 2021.
https://doi.org/10.1007/978-3-030-67667-4_2

elements, i.e., users, items and their relationships, can be modeled as a graph, therefore, graph embeddings have recently come into use with the expectation of mining the complicated relations between numerous users and massive items.

MX Player is one of India's largest streaming platforms and reaches more than 100 million daily active users from around the world. In the MX Player App, there are four major types of videos: movie, music video, short video and show, and video distribution and impression are heavily dependent on its internal recommender system. For example, based on the video that a user is currently watching, the recommender system in MX Player will recommend the related videos to the user immediately. In order to enhance user experiences, we try to apply the representation learning on graphs to MX Player for recommending various videos.

Specifically, we have observed that there exists some inherent characteristics in user behaviors of watching videos on the streaming App. First of all, while watching videos continuously, users have some specific behaviors, e.g., they are inclined to view the videos of the same type (e.g., movies). Secondly, users are apt to choose the videos from one or several aspects of video characterics, in other words, while users select what they will watch, some attributes of videos play more important roles in user decisions and some attributes show no sense of existence. Finally, compared with the other recommendation scenarios, in our scenario, we can get the numbers of user clicking the videos and real duration of watching videos, which indicate the user preferences. The above user behavior features should be fused into the video embeddings. However unfortunately, existing approaches cannot meet these requirements.

Therefore, for the video recommendation, we present a graph embedding approach named Equuleus. Equuleus constructs the video graph from the user-video interaction data, incorporates with the behavior features of users, and learns the video embeddings. Further, backed by the video embeddings, we implement the relevant video recommendation.

The main contributions of our work are summarized as follows.

- We give a detailed description for constructing a video graph, in which user's long-term preference and short-term preference are implied; we design a behavior-driven random walk, in which user behavior patterns are embedded intentionally.
- We design a node attributed encoder network, which employs an attention mechanism to aggregate different attributes in a distinguishable way; we also optimize the objective function by adding the user feedback as supervision information.
- We conduct extensive offline experiments on real datasets. Experimental results show that Equuleus can generate high-quality video embeddings, and behaves better than several state-of-the-art methods for relevant recommendation in terms of recall, NDCG (Normalized Discounted Cumulative Gain) and MRR (Mean Reciprocal Rank). We also conduct online A/B test in MX Player, the improvement of CTRs (Click Through Rates) shows that Equuleus is effective in the live production environment.

The rest of the paper is organized as follows. Section 2 introduces the related work. Section 3 gives the formulation of graph embedding to be solved. Section 4 describes our Equuleus approach in detail. Section 5 gives the experimental evaluation, including the results of online A/B test. Finally, the paper is concluded in Sect. 6.

2 Related Work

Existing methods of representation learning on graphs can be roughly classified into two categories. The first class, including DeepWalk [13], LINE [15], node2vec [8], GraRep [2], struc2vec [14], etc., generates the node embeddings by capturing the structure similarity of nodes. The second class of methods obtains node embeddings by mining both the graph structure and node features. GCNs (Graph Convolution Networks) such as spectral convolution methods [5,11], GraphSAGE [9], and GATs [16] fall into this category.

DeepWalk [13] is a representative method of representation learning on graphs, which borrows the idea from word2vec [12] in natural language processing. By the depth-first random walk, DeepWalk can effectively capture the spatial location similarity of nodes. Differing from DeepWalk, LINE [15] and GraRep [2] employ the breadth-first-like random walks. They calculate the n-order proximity of the nodes and obtain the spatial location similarity of the nodes. Further, node2vec [8] designs a biased random walk strategy, combining the depth-first and breadth-first walks. We note that these methods mainly consider the neighbors of nodes when generating node embeddings, that is, the more similar the neighbors of two nodes are, the more similar the node embeddings will be. However, struc2vec [14] gives the solution from another point of view, that is, it pays more attention to the topological structure similarity of nodes instead of the distance between two nodes. Thus, as long as the topological structures of nodes are similar in the graph, the final node embeddings will be similar.

Obviously, only depending on the graph structure to generate node embeddings is not good enough. For effectively learning from both the graph structure and node features, some spectral convolution methods and spatial convolution methods are presented. In general, spectral convolution methods apply the Fourier transform to the original graph, and then carry out the convolution operation in the spectral space [5,11]. By contrast, spatial convolution methods directly aggregate the features of the neighborhood nodes by different aggregation functions which can be Mean, MLP, LSTM, or Pooling, etc. Taking Graph-SAGE [9] as an example, it samples a fixed number of neighborhood nodes for each node, and then aggregates the feature of neighborhood nodes through the aggregation functions to learn the embedding of nodes. In particular, Graph-SAGE does not directly learn the node embeddings but a set of aggregation functions, so it can adapt to the dynamic changes of graph structures, which makes it an inductive learning method. Based on GraphSAGE, GATs [16] introduce a self-attention mechanism, which can dynamically calculate the intensities between the node and its neighbors, and then use the intensities as the weights of the aggregation functions for further weighted aggregation.

More recently, some work has successfully applied the representation learning on graphs to recommender systems [3,7,17,18,21]. For example, Alibaba develops EGES [17] which integrates the auxiliary information of the items into the node embeddings. By EGES, Alibaba learns the embeddings of billions of products, and then use the learned embeddings for similar product recommendation. Moreover, Pinterest [21] proposes a graph convolutional neural network named PinSage to simultaneously integrate the structure information of graphs and feature information of nodes. The generated embeddings of nodes have been used to recommend items (pins) of billions of scales.

For video recommendation, existing work largely exploits the rich contents of the videos. CER [6] incorporates content features with user-video interactions to make effective video recommendation. The limitation is that they rely heavily on the high-quality but expensive manual tagging content, which is probably too coarse-grained to discover non-linear user-item relationships. Youtube [4] proposes the deep learning based video recommendation method to explore the non-linear user-item relationships, but it is unable to explicitly mine the multi-order connectivity of user-item relationships.

Although there are successful cases, we note there is a gap between the goal of representation learning on a graph and the goal of a real-world recommendation. Previous representation learning methods often lack user feedback to guide the learning process. Moreover, previous representation learning methods are often incapable of learning the user behavior patterns in recommendation scenarios.

Comparing with the previous work, in the paper, we give a video-oriented graph embedding approach Equuleus. Equuleus employs an attributed encoder network and fuses the content description features of videos, watching-behavior patterns and user feedback to learn the embeddings of videos.

3   Problem Formulation

Our task is to generate high-quality video embeddings and apply them to recommender systems, e.g., taking them as input of embedding-based similarity search to form recall results or serving as complementary features to empower the downstream ranking tasks. This paper will mainly describe how to generate high-quality video embeddings to improve the user satisfaction in the relevant video recommendation scenario, where the target video and its related videos are similar videos of the same type (e.g., movie).

Video Graph. The video graph is a graph $\mathcal{G} = (\mathcal{V}, E, \mathcal{E}, \mathcal{F})$, where $\mathcal{V} = \{v_i\}$ is the node set containing all types of the videos, $E = \{(v_i, v_j); v_i, v_j \in \mathcal{V}\}$ is the edge set, $\mathcal{E} = \{e_{ij}; v_i, v_j \in \mathcal{V}\}$ is the weight set of edges, and $\mathcal{F} = \{f_1, ..., f_F\}$ is the attribute (i.e., feature) set of nodes. Besides, for each node v_i, we define two node feature mapping functions. That is, $\phi_f(v_i), \forall f \in \mathcal{F}$, which maps the attribute f of the node v_i into the set of values (i.e., a node may have multiple values for each attribute); $\psi(v_i)$, which maps the node v_i into a behavior scalar, and indicates a typical user behavior pattern in a specific recommendation scenario.

Problem (Graph Embedding). Given the graph $\mathcal{G} = (\mathcal{V}, E, \mathcal{E}, \mathcal{F})$ and feature mapping functions $\phi_f(\cdot)$ and $\psi(\cdot)$, the problem to be solved is to get the node embeddings $\{z_i\}$ by $z_i = \Phi(v_i)$ where $v_i \in \mathcal{V}$, $\Phi(\cdot) : v_i \to \mathbb{R}^k, k \ll |\mathcal{V}|$. $\Phi(\cdot)$ maps the node v_i in the original graph to z_i in the low dimensional embedding space.

Note that a node attributed encoder network is acted as $\Phi(\cdot)$ in our approach.

4   Equuleus Approach

In this section, we present the Equuleus approach. For learning the video embeddings, we carefully construct a video graph from the log and design a novel node attributed encoder network. Further, we leverage user feedback as a guide to correct the generation of embeddings. Finally, we design the watching-behavior driven random walk to produce the node sequences which act as training data to obtain video embeddings.

4.1   Construction of Graph

We construct the graph $\mathcal{G}$ with videos as nodes based on the user-video click log. Note that we build the video graph for all types of videos, it is because besides same types of videos, different types of videos also exist strong collaborative information. For example, a movie is related to some music videos that have the themes of the movie. Thus, taking the movie as a bridge, these music videos can be related to each other.

For constructing the graph, we filter out invalid user-video click records whose watching duration is less than 3 s, and split the log by day. We denote the log of the current date as ℓ_0, the τ-day log before the current date as ℓ_τ, and the log within n days from the current date as $L_n = \{\ell_0, \ell_1, ..., \ell_{n-1}\}$.

Then we give the definition of basic similarity between two videos.

$$sim_{Log}(v_i, v_j) = \sum_{a \in U_i \cap U_j} \sum_{b \in U_i \cap U_j} \frac{\exp\left(-(d_{a,b,i} + d_{a,b,j})/\beta\right)}{\alpha + |I_a \cap I_b|} \tag{1}$$

In Eq. (1), U_i, U_j denote the set of users who have clicked v_i and v_j, respectively. I_a, I_b represent the set of videos clicked by users a and b, respectively. $d_{a,b,i}$ and $d_{a,b,j}$ represent the date span between user a and b clicking video v_i and v_j, respectively. Specifically, α is a smoothing factor of the click number, and β is a decay factor of the date span.

Intuitively, if the interest of user a quite differs from user b (i.e., $1/|I_a \cap I_b|$ is very large), then, in general, videos clicked by user a or b also quite differ from each other. Coincidentally, if a and b both clicked some common videos, e.g., v_i and v_j, there is obviously some potential relationship between v_i and v_j, where the relationship can be indirectly reflected by the degree of interest difference between a and b. This is the rationality for the $1/(\alpha + |I_a \cap I_b|)$ in Eq. (1). For example, a likes sports and b likes entertainment. Accidentally, both a and b

have watched some videos about the gossip of sports stars, which indicates that there is a strong potential correlation between these videos. This strong potential correlation may originate from the inherent content relevance and can be easily reflected by the interest difference of users, i.e., $1/(\alpha + |I_a \cap I_b|)$.

Besides, the shorter the interection date span of a and b on v_i and v_j is, the greater the impact of interest difference on the similarity of the videos v_i and v_j should be. Therefore, in Eq. (1), we apply the exponential decay to the date span by $\exp\left(-(d_{a,b,i} + d_{a,b,j})/\beta\right)$.

Based on Eq. (1), we compute the following similarity of each pair (v_i, v_j), where the long-term similarity and short-term similarity of each video pair (v_i, v_j) capture the long-standing and short-lived impacts of the interactions between users and videos, respectively, and the compound similarity is the combination of long-term similarity and short-term similarity. Then, we treat the value of compound similarity of the video pair (v_i, v_j) as the weight e_{ij} of edge (v_i, v_j):

- Long-term similarity between the video pair (v_i, v_j) , i.e., $lsim(v_i, v_j) := sim_{L_n}(v_i, v_j)$.
- Short-term similarity between the video pair (v_i, v_j), i.e., $ssim(v_i, v_j) := \sum_{\tau=0}^{r-1} \exp(-\tau/\gamma) sim_{\ell_\tau}(v_i, v_j)$, where r is the number of the recent days.
- Compound similarity, i.e., $sim(v_i, v_j) := ssim(v_i, v_j) + lsim(v_i, v_j)$.

From the above process, we obtain the video graph $\mathcal{G} = (\mathcal{V}, E, \mathcal{E}, \mathcal{F})$. The building process of the video graph coarsely exploits the low-order connectivity of videos within the interaction data. We will discover the multi-order connectivity of videos by the following work.

4.2   Node Attributed Encoder Network

The goal of the encoder network is to encode the nodes from the high dimensional space $(O(|\mathcal{V}|))$ to the low dimensional space $(O(k), k \ll |\mathcal{V}|)$, and maximally preserve the original graph structure. We propose a novel node attributed encoder network that employs an attention mechanism to aggregate attributes of nodes into the node embeddings.

Feature Embedding. Let $P \in \mathbb{R}^{|\mathcal{V}| \times k}$ denote the base (i.e., id) embedding matrix of nodes. Thus, the base embedding of node v_i is p_i. Besides, for each sparse feature $f \in \mathcal{F}$, we set the embedding matrix for them, respectively, i.e., $W^1 \in \mathbb{R}^{S_1 \times k}, ..., W^f \in \mathbb{R}^{S_f \times k}, ..., W^F \in \mathbb{R}^{S_F \times k}$, where W^f denotes the embedding matrix of feature f and S_f is the number of the distinct values of f. P and $\{W^f, \forall f \in \mathcal{F}\}$ are the randomly initialized model parameters that need to be learned.

A node may have multiple values for each sparse feature. For the node v_i and the sparse feature f, we obtain the value set by $X_i^f = \phi_f(v_i)$. We conduct the mean pooling operation to map the feature f of node v_i into the k-dimension embedding. i.e.,

$$q_i^f = \text{Sparse-Embedding}(\phi_f(v_i), \boldsymbol{W}^f) = \frac{1}{|X_i^f|} \sum_{x \in X_i^f} \boldsymbol{w}_x^f \qquad (2)$$

where $\boldsymbol{w}_x^f$ is the row vector corresponding to the value x of feature f in $\boldsymbol{W}^f$.

Attributed Encoder Network. Intuitively, the base embedding of each node inherently reflects rich information of the graph structure. Besides, the attributes of nodes contain rich semantic information, which need to be incorporated into the node embeddings to capture the potential content connection between nodes.

In video recommendation scenarios, the information rooted in each sparse feature of a node is completely different. For example, the values of the "genre" attribute are fine-grained enough to better characterize the richness of information, while the "release year" attribute is coarse-grained one which lacks useful information.

It requires us to identify which attributes are more important and contribute more to the semantic representation of nodes. To meet the above requirements, we design an attributed encoder network $\Phi(\cdot)$ with an attention mechanism that is capable of quantifying the contributions of different attributes. As a result, $\Phi(\cdot)$ aggregates the base embedding and sparse feature embeddings of the nodes into the unified node embedding. The detailed architecture is shown in Fig. 1.

As shown in Fig. 1, this encoder network consists of an input layer, an embedding layer, a pooling layer, an attention based attribute aggregator and a layer normalization. Specifically, each sparse feature f of node v_i is fed into the mapping function in Eq. (2) to obtain $\{q_i^f, \forall f \in \mathcal{F}\}$. Besides, each node v_i is fed into the embedding layer to obtain the base embedding $\boldsymbol{p}_i$. After that, $\boldsymbol{p}_i, \{q_i^f, \forall f \in \mathcal{F}\}$ will be regarded as the input to the attention based attribute aggregator to obtain the embedding $\boldsymbol{g}_i$, which is the addition of the base embedding and the aggregated feature embedding. Finally, the layer normalization is applied to $\boldsymbol{g}_i$ to output the unified node embedding $\boldsymbol{z}_i = \Phi(v_i)$. Formally,

$$\boldsymbol{z}_i = \Phi(v_i) = \text{Encoder}(v_i) = \text{LayerNorm}(\boldsymbol{g}_i)$$

$$= \text{LayerNorm}(\text{Aggregator}(\boldsymbol{p}_i, \{q_i^f, \forall f \in \mathcal{F}\}))$$

$$= \text{LayerNorm}(\boldsymbol{p}_i + \sum_{f=1}^{F} \boldsymbol{a}_A(\boldsymbol{p}_i, q_i^f) \cdot q_i^f) \qquad (3)$$

For each node, the base embedding $\boldsymbol{p}_i$ serves as the connection across different attributes. Specifically, the aggregator aggregates the feature embeddings into the unified node embeddings based on the quantified contribution of different attributes to node base embeddings, where the contribution is measured by the delicately designed attention network $\boldsymbol{a}_A(\boldsymbol{p}_i, q_i^f)$ with $\boldsymbol{A} \in \mathbb{R}^{F \times k \times k}$ as network parameters. Formally,

$$a_A(\boldsymbol{p}_i, q_i^f) = \text{softmax}(\frac{{q_i^f}^T \boldsymbol{A}^f \boldsymbol{p}_i}{\sqrt{k}}) = \frac{\exp({q_i^f}^T \boldsymbol{A}^f \boldsymbol{p}_i / \sqrt{k})}{\sum_{f'} \exp({q_i^{f'}}^T \boldsymbol{A}^{f'} \boldsymbol{p}_i / \sqrt{k})}, \forall f = 1, ..., F$$

$$(4)$$

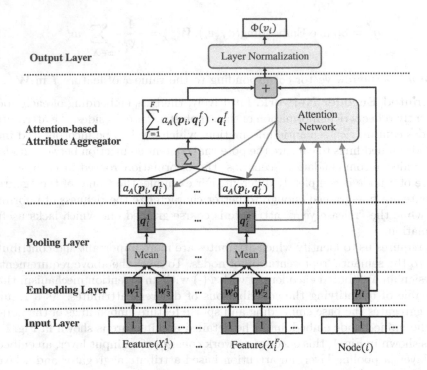

Fig. 1. The architecture of the attributed node encoder network

where $1/\sqrt{k}$ is the scaling factor, and $\boldsymbol{A}^f \in \mathbb{R}^{k \times k}$ is f-specific network parameters originated from the slice of $\boldsymbol{A}$. As such, the base embedding $\boldsymbol{p}_i$ essentially bridges the gap between feature-specific representations, and propagates information across features during the gradient back-propagation process.

The layer normalization, referring to [1,20], is used here to stabilize training.

4.3 Feedback-Guided Learning

In the actual recommendation scenario, we can collect some implicit or explicit user feedback data from the user-video click log, which could be regarded as the special scenario-related attributes of nodes and are probably able to improve the quality of embeddings if being incorporated into the learning objective. In our relevant recommendation scenario, for each video pair, i.e., a target video and the recommended related video, we collect the aggregated co-click feedback data to construct the feedback-guided learning objective.

Specifically, for each target node, the learning objective consists of three parts: a) the positively correlated objective of the implicit positive samples based on the behavior-driven random walk introduced in Sect. 4.4. b) the positively correlated objective of the explicit positive samples which are the top-$\mathcal{M}$ co-click feedback samples related to the target node. c) the negatively correlated objective of the negative samples.

Formally, the final objective for each pair (v_i, u_j) is as follows.

$$J(\Phi) = -\lambda_1 \log\left(\sigma(\Phi'(u_j)^T \Phi(v_i))\right) - \lambda_2 \sum_{m=1}^{M} \mathbb{E}_{y_m \sim p_i(y)} \log\left(\sigma(\Phi'(y_m)^T \Phi(v_i))\right)$$

$$- \sum_{t=1}^{T} \mathbb{E}_{s_t \sim p(s)} \log\left(\sigma(-\Phi'(s_t)^T \Phi(v_i))\right)$$

$$(5)$$

In Eq. (5), v_i is the target node and u_j is the context node of node v_i, which are constructed from the node sequences of the behavior-driven random walk with a h-size window in Sect. 4.4. $\Phi(\cdot)$ is the target node encoder introduced in Sect. 4.2 and $\Phi'(\cdot)$ is the context node encoder with the same architecture but different parameters to $\Phi(\cdot)$. λ_1 measures the importance of implicit positive samples and λ_2 measures the importance of explicit positive samples from feedback data. $p_i(y)$ is the distribution of the explicit positive samples of v_i. Given v_i and top-$\mathcal{H}$ samples $\{y_1, y_2, ..., y_m, ..., y_{\mathcal{H}}\}$ ranked by the co-click number, $p_i(y_m|v_i) = e^{\mathcal{H}-m}/\sum_{m'=1}^{\mathcal{H}} e^{\mathcal{H}-m'}$. $M(< \mathcal{H})$ is the sampled number of the explicit positive data. $p(s)$ is the distribution of negative samples. We randomly sample T nodes on the entire graph based on the uniform distribution as negative samples of v_i.

4.4 Behavior-Driven Random Walk

In real-world scenarios, user behaviors often show some regularities. For example, in the relevant recommendation scenario, there is usually a strong correlation between the next recommended related video and the last clicked video from the video attribute perspective (e.g., of the same types or languages). Therefore, we design a behavior-driven random walk to generate node sequences from the video graph. In detail, for the target node v_i, we design two meta-path schemes to obtain the node sequences starting from v_i.

- $\mathcal{P}_1 : v_i \rightarrow u_1 \rightarrow u_2 \rightarrow ... \rightarrow u_j... \rightarrow u_l$, where l is the walk length. Each node u_j in the path is independent of the previous node, i.e., $\psi(u_j) \in \mathbb{R}$. In this situation, it is equivalent to performing the random walk on the full graph according to the probability distribution calculated by the edge weights. The transition probability of each node is as follows,

$$p(u_{j+1}|u_j, \mathcal{P}_1) = \begin{cases} \dfrac{e_{j,j+1}}{\sum_{o \in \mathcal{N}(u_j)} e_{j,o}} & (u_j, u_{j+1}) \in E \\ 0 & (u_j, u_{j+1}) \notin E \end{cases} \tag{6}$$

where $u_0 = v_i$, $\mathcal{N}(u_j)$ is the neighborhoods of u_j.

- $\mathcal{P}_2 : v_i \xrightarrow{\psi(v_i)} u_1 \xrightarrow{\psi(u_1)} u_2 \xrightarrow{\psi(u_2)} ... \xrightarrow{\psi(u_{j-1})} u_j \xrightarrow{\psi(u_j)} ... \xrightarrow{\psi(u_{l-1})} u_l$, where we let $\psi(u_i) = u_i.c$. That is, the video types of the nodes $(u_i.c)$ will be used to characterize the behavior that a user continuously watches videos of the same type at the related recommendation pages. In this situation, we conduct the random walk along the nodes of same type to v_i, i.e., $u_j.c = u_{j-1}.c = v_i.c$. The detailed transition probability of each node is as follows,

$$p(u_{j+1}|u_j, \mathcal{P}_2) = \begin{cases} \dfrac{e_{j,j+1}}{\sum_{o \in \mathcal{N}(u_j) \cap \mathcal{V}_{u_j.c}} e_{j,o}} & (u_j, u_{j+1}) \in E, \ u_{j+1}.c = u_j.c \\ 0 & (u_j, u_{j+1}) \in E, u_{j+1}.c \neq u_j.c \\ 0 & (u_j, u_{j+1}) \notin E \end{cases} \quad (7)$$

where $\mathcal{V}_{u_j.c}$ is the set of nodes with video type $u_j.c$, and $\mathcal{N}(u_j) \cap \mathcal{V}_{u_j.c}$ is the set of nodes with video type $u_j.c$ among the neighbors of node u_j.

We construct the target-context node pairs with a h-size window over the node sequences of the behavior-driven random walk and feed them into the feedback-guided learning process in Sect. 4.3 to learn embeddings.

If needed to mimic the other behave patterns, we can customize ψ to be other functions.

5 Evaluation

To evaluate the effectiveness of Equuleus, we conduct extensive experiments. First, we conduct an ablation study to observe the contributions of different core components in Equuleus, including the behavior-driven random walk, the attributed encoder network and the feedback-guided learning. Then, in order to evaluate the quality of node embeddings obtained by Equuleus, we perform the dimension reduction and visualization of embeddings to observe whether videos of similar attributes are close to each other in the embedding space. Then, by the embedding-based similarity search for top-K recommendation, we compare Equuleus with some state-of-the-art methods. Finally, we conduct the A/B test to observe the performance of the embedding-boosting ranking for relevant video recommendations in a live production environment.

5.1 Experimental Setting

Dataset. We collect nearly 1 billion valid records occurred between Oct. 14, 2019 and Nov. 14, 2019 from MX Player. For experiments, we randomly choose the starting date g between Oct. 14, 2019 and Nov. 14, 2019 and use the data between $[g, g + 14)$ to form the training dataset and the data on the $g + 14$ day to form the testing dataset. In total, we form five groups of data, each with a training dataset and a testing dataset.

Next, we set the hyperparameters used in the process of constructing the graph. By default, $\alpha = 10$, $\beta = 7$, $\gamma = 7$, $n = 14$ and $r = 3$. The constructed graph serves as the same input to all experimental methods.

We construct the video graphs from training datasets, and the resulting five video graphs reach 181,628 nodes and 31,416,170 edges on average. Note that the following experimental results are the average over the results of five groups of data.

Embedding-Based Similarity Search for Top-K Recommendation. With the embeddings in hand, we calculate the cosine similarity between two embeddings and search the top-K related videos as the recommendation results. In experiments, we focus on four relevant video recommendation scenarios, i.e., recommending relevant movies to videos of movie type, recommending relevant shows to videos of show type, recommending relevant music videos to videos of music video type and recommending relevant short videos to videos of short video type. They are denoted as Movie, Show, Music Video and Short Video, respectively.

Evaluation Metric. We adopt three typical ranking metrics in top-K recommendation, i.e., Recall@K, NDCG@K (Normalized Discounted Cumulative Gain) and MRR@K (Mean Reciprocal Rank) where K is set to 50 by default. For each video, we use the embedding-based similarity search to generate the predicted top K recommendation results based on the learned embeddings from the training dataset. Then, we regard the real user feedback in the testing dataset as the ground truth. Specifically, for each video in the testing dataset, we rank the relevant videos by the total numbers of user clicks on the videos to form the ground truth ranking list. Similarly, the ground truth score for each recommended video in NDCG@K is measured by the number of clicks.

5.2   Ablation Study

In this section, we conduct an ablation study to observe the effectiveness and contribution of each core components in Equuleus, including the behavior-driven random walk (denoted as C1), the feedback-guided learning (denoted as C2) and the attributed encoder network (denoted as C3). We take the original DeepWalk method as the Baseline (denoted as BL), and successively replace the part in the Baseline with the proposed components to form BL+C1,BL+C2, BL+C3 and BL+C1+C2. BL+C1+C2+C3 forms Equuleus. Table 1 lists their performance in terms of Recall@50, NDCG@50 and MRR@50. The percentage in parentheses after each result in Table 1 is the ratio of the improvement relative to the Baseline.

From Table 1, individually replacing the part in the Baseline with each component proposed in this paper would contribute to the recommendation performance in terms of Recall@50, NDCG@50 and MRR@50. Specifically, the feedback-guided learning leads to the significant improvement in movie recommendation scenarios, i.e., increasing 44.95% in Recall@50, 45.67% in NDCG@50, and 15.81% in MRR@50. Moreover, the attributed encoder network improves significantly on Music Video and Short Video, which shows that the fusion of attributes can greatly deepen the exploration of related videos. Taking the music video recommendation as an example, the attributed encoder network improves the recall@50 by 58.54% and NDCG@50 by 54.53%. Above all, Equuleus shows

Table 1. Ablation study over core components in Equuleus

Videos	Metrics	BL	BL+C1	BL+C2	BL+C3	BL+C1+C2	Equuleus
Movie	Recall@50	0.1942	0.2455 (26.42%)	0.2815 (44.95%)	0.2422 (24.72%)	0.2900(49.33%)	**0.3058** (57.47%)
	NDCG@50	0.3063	0.3565 (16.39%)	0.4462 (45.67%)	0.3606 (17.73%)	**0.4524**(47.70%)	0.4211 (37.48%)
	MRR@50	0.1733	0.1894 (9.29%)	**0.2007** (15.81%)	0.1799 (3.81%)	0.1994(15.06%)	0.1964 (13.33%)
Show	Recall@50	0.5047	0.5559 (10.14%)	0.5458 (8.14%)	0.5481 (8.60%)	0.5768 (14.29%)	**0.6058** (20.03%)
	NDCG@50	0.4624	0.4912 (6.23%)	0.5309 (14.81%)	0.5117 (10.66%)	0.5458 (18.04%)	**0.5465** (18.19%)
	MRR@50	0.2786	0.2843 (2.05%)	0.2866 (2.87%)	0.2792 (0.22%)	**0.2889** (3.70%)	0.2880 (3.37%)
Music Video	Recall@50	0.2243	0.2984 (33.04%)	0.3108 (38.56%)	0.3556 (58.54%)	0.3811(69.91%)	**0.4347** (93.80%)
	NDCG@50	0.2208	0.2820 (27.72%)	0.3417 (54.76%)	0.3412 (54.53%)	0.3584(62.32%)	**0.3828** (73.37%)
	MRR@50	0.2074	0.2264 (9.16%)	0.2037 (-1.78%)	0.2242 (8.10%)	0.2122(2.31%)	**0.2306** (11.19%)
Short Video	Recall@50	0.1812	0.2443 (34.82%)	0.3298 (82.01%)	0.3342 (84.44%)	0.3666 (102.32%)	**0.3924** (116.56%)
	NDCG@50	0.1669	0.2099 (25.76%)	0.3034 (81.79%)	0.2708 (62.25%)	0.3168 (89.81%)	**0.3182** (90.65%)
	MRR@50	0.2032	0.2189 (7.73%)	0.1962 (-3.44%)	0.2224 (9.45%)	0.2022 (-0.49 %)	**0.2352** (15.75%)

the excellent results, which illustrates the effectiveness of the combination of three components.

5.3 Visualization

With the help of the t-SNE (t-distributed Stochastic Neighbor Embedding) method, we conduct the dimension reduction and visualization of embeddings obtained by Equuleus. Through the visualization, we are able to observe whether the content-related videos (e.g., the videos with the same genre or language) are clustered closely.

Figure 2 shows the language distribution of movies in a two-dimension form. From Fig. 2, videos of the different languages, including English, Tamil, Malayalam, Kannada, Telugu and Hindi videos, are separately clustered.

Further, as shown in Fig. 3, we give a visualization instance of shows from the three-dimension perspective. From Fig. 3, we can see the shows of similar style posters are clustered closely. Then we zoom the two regions and find that the cartoon shows are clustered closely and the drama and romance shows are clustered closely.

These visualization results vividly demonstrate that Equuleus can generate high-quality video embeddings.

5.4 Performance Comparison

We select following seven methods as competitors to conduct comparative experiments:

- DeepWalk [13]. DeepWalk is one of the earliest random walk based graph embedding approach.
- LINE [15]. LINE optimizes the objective function that preserves both the local and global network structures by first- and second-order proximities.
- node2vec [8]. node2vec designs a biased random walk to efficiently explore diverse neighborhoods.
- GraphSAGE [9]. GraphSAGE utilizes the node attributes and aggregates the local neighborhoods into the node embeddings.

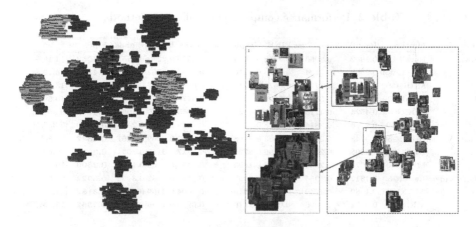

Fig. 2. The visualization of language distribution

Fig. 3. The visualization of posters of shows

- GATs [16]. GATs introduces the multi-head self-attention mechanism to dynamically aggregate the neighborhoods based on the attention scores between target node and neighborhood nodes.
- EGES [17]. EGES proposed by Alibaba differentially fuses the attributes of nodes by adaptive coefficients to obtain the node embeddings.
- PinSage [21]. PinSage is a random walk based GCN in Pinterest, which is trained in a supervised fashion using a max-margin ranking loss. We regard the co-click video pairs in the feedback data (described in Sect. 4.3) as the set of labeled pairs in our scenarios.

The above methods use the graph $\mathcal{G}$ constructed in Sect. 4.1 as the input. For each method, we set the values of dimensions k for unique id embedding and attribute embeddings to 64, the length of the random walk l to 20, the window size h to 5 and the number of negative samples T to 50. We utilize the grid search method to find the best values for the remaining parameters. That is, for node2vec, we set its own parameters $q = 2, p = 4$. For GraphSAGE, we use a two-layer mean aggregators with 25 and 10 sampled neighbors for each layer, respectively. For GATs, we use two attentional layers, each with 2 attention heads and 10 sampled neighbors. For PinSage, we use a two-layer GCN with neighborhood size 20. For Equuleus, $\lambda_1 = 1.0, \lambda_2 = 2.0, M = 5$. For fair comparison, GraphSAGE, GATs, EGES, PinSage and Equuleus all use the ids and attributes as features. The parameters for training are kept the same for these methods. i.e., $lr = 3e{-}4$, batch size $= 1024$, epoch $= 10$. We employ AdamOptimizer to control the progress of learning procedures.

Table 2 shows the performance results of different methods for the four types of video recommendation scenarios. The last column of Table 2 gives the improvement proportions of Equuleus relative to the best method of the other seven methods.

Table 2. Performance comparison of different methods

Videos	Metrics	DeepWalk	node2vec	LINE	GraphSAGE	GATs	EGES	PinSage	Equuleus	% Improv.
Movie	Recall@50	0.1942	0.2060	0.2212	0.2363	0.2679	0.2232	0.2452	**0.3058**	14.15%
	NDCG@50	0.3063	0.3134	0.3093	0.3210	0.3560	0.3502	0.3554	**0.4211**	18.29%
	MRR@50	0.1733	0.1724	0.0621	0.1017	0.0906	0.1557	0.1315	**0.1064**	13.33%
Show	Recall@50	0.5047	0.5162	0.5299	0.5259	0.5478	0.5359	0.5259	**0.6058**	10.59%
	NDCG@50	0.4624	0.4715	0.4318	0.4824	0.4865	0.4970	0.4930	**0.5465**	9.96%
	MRR@50	0.2786	0.2814	0.1228	0.2113	0.1939	0.2416	0.2427	**0.2880**	2.35%
Music Video	Recall@50	0.2243	0.2391	0.2383	0.3108	0.3752	0.3209	0.3267	**0.4347**	15.86%
	NDCG@50	0.2208	0.2328	0.2096	0.2725	0.3227	0.3145	0.3041	**0.3828**	18.62%
	MRR@50	0.2074	0.2096	0.0672	0.1493	0.1451	0.2065	0.1774	**0.2306**	10.02%
Short Video	Recall@50	0.1812	0.1956	0.2393	0.3095	0.3588	0.2687	0.3192	**0.3924**	9.36%
	NDCG@50	0.1669	0.1749	0.1905	0.2315	0.2604	0.2409	0.2556	**0.3182**	22.20%
	MRR@50	0.2032	0.1996	0.0598	0.1176	0.0900	0.1986	0.1593	**0.2352**	15.75%

From Table 2, we find that for all scenarios, the top-K recommendation performance of Equuleus is better than that of other methods, which fully demonstrates the effectiveness of Equuleus. Further, we can obtain the following observations and inferences:

- LINE and node2vec generally outperform DeepWalk in terms of Recall@50. This is probably due to the fact that the first- and second-order proximities of LINE and the BFS in node2vec can better utilize the local neighborhoods of nodes to better explore the structural equivalence over the full graph.
- Compared with DeepWalk, node2vec and Equuleus perform better, which may be due to their biased random walks. Further, Equuleus outperforms node2vec. We believe this is because Equuleus better captures the scenario-specific characteristics by the designed behavior-driven random walk.
- GraphSAGE, EGES, GATs, PinSage, and Equuleus are generally superior to DeepWalk, node2vec, and LINE in terms of the Recall@50 and NDCG@50. These show that the attributes play significant roles in discovering the content-related videos and effectively promote the recommendation opportunities of low-exposed videos.
- Among these methods, the disparity in MRR@50 is observed. In terms of MRR@50, Equuleus behaves best, followed by EGES. GCNs (e.g., Graph-SAGE, GATs and PinSAGE) behave worse than some methods without utilizing the node attributes, such as DeepWalk and node2vec. It is probably due to the different ways of utilizing attribute information. GCNs treat all sparse attributes equally, i.e., concatenate the corresponding embeddings together and feed into a fully connected layer to get the node representation. As a result, they may find many content-related but unsatisfactory videos. Both EGES and Equuleus fuse the attributes of nodes in a distinguishable way, which makes it easier to distinguish the satisfactory videos from unsatisfactory ones. Further, the win of Equuleus demonstrates the attention mechanism in Equuleus can better guide which attributes contain more semantics and form the informative unified embeddings.

- Equuleus outperforms GCNs. It shows that the careful exploitation of the internal information of target nodes may be more effective than that of the information of neighboring nodes. Besides, the feedback-guided learning and behavior-driven random walk also contribute to the performance of Equuleus. It is worth noting that our proposed components may be able to give an extra power to GCNs. Further studies will be made in our future work.

5.5   Online A/B Test

We have applied the learned embeddings to relevant video recommendation scenarios in MX Player.

We apply Equuleus to this downstream recommendation task from two aspects. Firstly, we regard the Equuleus as a new recall method. That is, we conduct the embedding-based similarity search to recommend top 100 relevant candidates to each video. The top 100 videos are further regarded as a recall source of the ranking model. Secondly, we regard the embeddings of videos as the fixed features of the input to the ranking model. To this end, we set up two ranking models in the A/B test.

(a) The ranking model without the embeddings of Equuleus (neither as a recall source nor as input features), called Vanilla Ranking.
(b) The ranking model with the embeddings of Equuleus (both as a recall source and as input features), called Ranking with Embeddings.

The experimental environments of the two ranking methods are the same. Figure 4 shows the two-week CTRs on four types of relevant video recommendation scenearios from Nov. 26th, 2019 to Dec. 9th, 2019.

From Fig. 4, we can see that the ranking model using embeddings performs better on all relevant video recommendation scenarios than the ranking model without embeddings. Specifically, during the 2-week A/B test, the average CTRs on Movie, Show, Music Video and Short Video improve by 0.55%, 1.28%, 1.90% and 3.39%, respectively. These results illustrate the effectiveness of Equuleus in a live production environment.

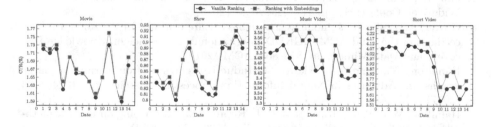

Fig. 4. A/B test on rankings (Nov. 26th, 2019 ~ Dec. 9th, 2019)

6   Conclusion

In this paper, we propose the Equuleus approach for recommending videos in MX Player. Equuleus employs an attributed encoder network, and fuses the attributes of videos, watching-behavior patterns and user feedback to learn the embeddings of videos. Results from both offline experiments and online A/B test demonstrate that Equuleus can generate the high-quality video embeddings. More practices on graph embedding for recommendation will be our future work. A possible attempt is to extend a GCN by fusing the components proposed in the paper, because in principle, these three components can strengthen the power of most existing graph-based methods including the GCNs.

Acknowledgments. This work was supported by the National Natural Science Foundation of China under Grant No. 61472408 and the 2019 joint project with MX Media.

References

1. Ba, J.L., Kiros, J.R., Hinton, G.E.: Layer normalization. arXiv preprint arXiv:1607.06450 (2016)
2. Cao, S., Lu, W., Xu, Q.: GraRep: learning graph representations with global structural information. In: Proceedings of the 24th ACM International on Conference on Information and Knowledge Management, pp. 891–900. ACM (2015)
3. Cen, Y., Zou, X., Zhang, J., Yang, H., Zhou, J., Tang, J.: Representation learning for attributed multiplex heterogeneous network. In: Proceedings of the 25th ACM SIGKDD International Conference on Knowledge Discovery and Data Mining (2019)
4. Covington, P., Adams, J., Sargin, E.: Deep neural networks for YouTube recommendations. In: Proceedings of the 10th ACM Conference on Recommender Systems, pp. 191–198 (2016)
5. Defferrard, M., Bresson, X., Vandergheynst, P.: Convolutional neural networks on graphs with fast localized spectral filtering. In: Advances in Neural Information Processing Systems, pp. 3844–3852 (2016)
6. Du, X., Yin, H., Chen, L., Wang, Y., Yang, Y., Zhou, X.: Personalized video recommendation using rich contents from videos. IEEE Trans. Knowl. Data Eng. **32**, 492–505 (2018)
7. Fan, W., et al.: Graph neural networks for social recommendation. In: The World Wide Web Conference, pp. 417–426. ACM (2019)
8. Grover, A., Leskovec, J.: node2vec: scalable feature learning for networks. In: Proceedings of the 22nd ACM SIGKDD International Conference on Knowledge Discovery and Data Mining, pp. 855–864. ACM (2016)
9. Hamilton, W., Ying, Z., Leskovec, J.: Inductive representation learning on large graphs. In: Advances in Neural Information Processing Systems, pp. 1024–1034 (2017)
10. Hamilton, W.L., Ying, R., Leskovec, J.: Representation learning on graphs: methods and applications. arXiv preprint arXiv:1709.05584 (2017)
11. Kipf, T.N., Welling, M.: Semi-supervised classification with graph convolutional networks. In: ICLR (2017)

12. Mikolov, T., Sutskever, I., Chen, K., Corrado, G.S., Dean, J.: Distributed representations of words and phrases and their compositionality. In: Advances in Neural Information Processing Systems, pp. 3111–3119 (2013)
13. Perozzi, B., Al-Rfou, R., Skiena, S.: DeepWalk: online learning of social representations. In: Proceedings of the 20th ACM SIGKDD International Conference on Knowledge Discovery and Data Mining, pp. 701–710. ACM (2014)
14. Ribeiro, L.F., Saverese, P.H., Figueiredo, D.R.: struc2vec: learning node representations from structural identity. In: Proceedings of the 23rd ACM SIGKDD International Conference on Knowledge Discovery and Data Mining, pp. 385–394. ACM (2017)
15. Tang, J., Qu, M., Wang, M., Zhang, M., Yan, J., Mei, Q.: Line: large-scale information network embedding. In: Proceedings of the 24th International Conference on World Wide Web, pp. 1067–1077. International World Wide Web Conferences Steering Committee (2015)
16. Veličković, P., Cucurull, G., Casanova, A., Romero, A., Liò, P., Bengio, Y.: Graph attention networks. In: International Conference on Learning Representations (2018)
17. Wang, J., Huang, P., Zhao, H., Zhang, Z., Zhao, B., Lee, D.L.: Billion-scale commodity embedding for e-commerce recommendation in Alibaba. In: Proceedings of the 24th ACM SIGKDD International Conference on Knowledge Discovery & Data Mining, pp. 839–848. ACM (2018)
18. Wang, X., He, X., Wang, M., Feng, F., Chua, T.S.: Neural graph collaborative filtering. arXiv preprint arXiv:1905.08108 (2019)
19. Wu, Z., Pan, S., Chen, F., Long, G., Zhang, C., Yu, P.S.: A comprehensive survey on graph neural networks. arXiv preprint arXiv:1901.00596 (2019)
20. Xu, J., Sun, X., Zhang, Z., Zhao, G., Lin, J.: Understanding and improving layer normalization. In: Advances in Neural Information Processing Systems, pp. 4383–4393 (2019)
21. Ying, R., He, R., Chen, K., Eksombatchai, P., Hamilton, W.L., Leskovec, J.: Graph convolutional neural networks for web-scale recommender systems. In: Proceedings of the 24th ACM SIGKDD International Conference on Knowledge Discovery & Data Mining, pp. 974–983. ACM (2018)
22. Zhou, J., Cui, G., Zhang, Z., Yang, C., Liu, Z., Sun, M.: Graph neural networks: a review of methods and applications. arXiv preprint arXiv:1812.08434 (2018)

Recommending Courses in MOOCs for Jobs: An Auto Weak Supervision Approach

Bowen Hao[1], Jing Zhang[1(✉)], Cuiping Li[1], Hong Chen[1], and Hongzhi Yin[2]

[1] Key Laboratory of Data Engineering and Knowledge Engineering of Ministry of Education, School of Information, Renmin University of China, Beijing, China
{jeremyhao,zhang-jing,licuiping,chong}@ruc.edu.cn
[2] School of Information Technology and Electrical Engineering,
The University of Queensland, Brisbane, China
h.yin1@uq.edu.au

Abstract. The proliferation of massive open online courses (MOOCs) demands an effective way of course recommendation for jobs posted in recruitment websites, especially for the people who take MOOCs to find new jobs. Despite the advances of supervised ranking models, the lack of enough supervised signals prevents us from directly learning a supervised ranking model. This paper proposes a general automated weak supervision framework (*AutoWeakS*) via reinforcement learning to solve the problem. On the one hand, the framework enables training multiple supervised ranking models upon the pseudo labels produced by multiple unsupervised ranking models. On the other hand, the framework enables automatically searching the optimal combination of these supervised and unsupervised models. Systematically, we evaluate the proposed model on several datasets of jobs from different recruitment websites and courses from a MOOCs platform. Experiments show that our model significantly outperforms the classical unsupervised, supervised and weak supervision baselines.

1 Introduction

Massive open online courses, or MOOCs, are attracting widespread interest as an alternative education model. Lots of MOOCs platforms such as Coursera, edX and Udacity have been built and provide low cost opportunities for anyone to access a massive number of courses from the worldwide top universities. As reported by Harvard business review[1], a primary goal of 52% of the people surveyed who takes MOOCs is to improve their current jobs or find new jobs. We call this group of MOOCs' users as career builders. Meanwhile, people usually resort to the online recruitment platforms such as LinkedIn.com and Job.com to seek jobs. However, there always exists a "Skill Gap" [1] between the career builders and the employers. The career builders who expect themselves to fit

---

[1] https://hbr.org/2015/09/whos-benefiting-from-moocs-and-why.

© Springer Nature Switzerland AG 2021
Y. Dong et al. (Eds.): ECML PKDD 2020, LNAI 12460, pp. 36–51, 2021.
https://doi.org/10.1007/978-3-030-67667-4_3

a job through taking MOOCs, need to deeply understand the demands of the job skills and then take the matchable courses. Clearly, to help career builders improve their skills for finding gainful jobs, it has been an essential task that is able to automatically match jobs with suitable courses.

Straightforwardly, to solve this problem, unsupervised methods such as BM25 [2], Word2vec [3] or the network embedding methods such as Deep-Walk [28] and LINE [6] can be used to calculate the relevance between a queried job and a candidate course. However, such unsupervised methods aim at modeling the implicit structures of the input data, i.e., the clustering of words in jobs and courses, while the ranking of different courses to a queried job cannot be obviously learned. In another word, the unsupervised models do not explicitly compare the relevance of the positive courses and the negative courses to a queried job. Although the supervised neural ranking models are demonstrated to have good performance in the information retrieval (IR) tasks [7,12], they cannot directly solve our problem, as the supervision signals about which courses can be recommended to a job are not easily available.

To alleviate the problem of lacking supervision signals, weak supervision models are proposed to train supervised IR models upon the pseudo labels provided by unsupervised models. For example, Dehghani et al. leverage the output of BM25 as the weak supervision signals [8] and Zamani and Croft extend a single pseudo signal to multiple signals to guide multiple supervised ranking models [9]. However, for different tasks, human efforts are demanded to determine the suitable weak signals and the supervised models. Even if each component is carefully selected by humans, their combination may not result in the best performance (which is also justified in our experiments). Thus, it is imperative to automatically identify an optimal combination of different components.

To address the above challenge, we propose a general automated weak supervision model *AutoWeakS*, which can automatically select the optimal combination of weak signals, supervised models and hyperparameters for a given ranking task and dataset. Specifically, the auto model trains a weak supervision model and a controller iteratively through reinforcement learning, where the weak supervision model aims to train a group of sampled supervised ranking models upon the pseudo labels (i.e., weak signals) provided by a group of sampled unsupervised ranking models, and the controller targets at automatically sampling an optimal configuration for the weak supervision model, i.e., it sequentially determines which unsupervised models should be sampled, how to set the hyperparameters for merging the unsupervised models, and which supervised models should be sampled. Our proposed model is a general framework to rank courses for jobs in this paper, but it is general enough to solve other ranking problems. Besides, we can incorporate any unsupervised and supervised models as candidate components to be selected by the controller.

Our contributions can be summarized as: (1) we are the first to explore the problem of recommending courses in MOOCs to jobs posted in online recruitment websites, which can help to eliminate the "Skill Gap" between the career builders who take MOOCs and the employers in the recruitment; (2) we pro-

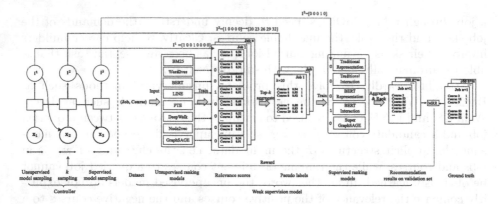

Fig. 1. Overview of the auto weak supervision model.

pose a general automated weak supervision model, *Auto WeakS*, to rank courses for jobs. With reinforcement joint training of the weak supervision model and the controller in *Auto WeakS*, we can automatically find the best configuration of the weak supervision model; (3) experiments on two real-world datasets of jobs and courses show that *Auto WeakS* significantly outperforms the classical unsupervised, supervised and weak supervision baselines.

2   The Auto Weak Supervision Model

We denote the jobs in a recruitment website as J and the courses in a MOOCs platform as C, where each job $j \in J$ contains maximal N words, i.e., $j = \{j_1, \cdots, j_N\}$, and each course $c \in C$ contains maximal M words, i.e., $c = \{c_1, \cdots, c_M\}$. The words are extracted from the descriptions of the jobs or courses. Given J and C, the goal is to learn a predictive function $\mathcal{F} : (J, C) \to Y$ to predict the label $y \in Y$ for each pair of a queried job j and a candidate course c, where y is a binary value with $y = 1$ if c is relevant to j and $y = 0$ otherwise.

2.1   Model Overview

In our problem, the set of the true labels Y about which courses should be recommended to a given job are not easily to be obtained, which motivates us to use a weak supervision method to solve this task, i.e., training supervised ranking models upon the pseudo labels provided by unsupervised models. However, selecting only one unsupervised model may suffer from the issue of ranking bias, while combing multiple unsupervised models may bring in additional noises. Besides, we also have many different choices for the supervised ranking models. Thus we explore an optimal combination of the pseudo labels from various unsupervised ranking models, together with various supervised ranking models.

Figure 1 illustrates the proposed auto weak supervision framework *Auto WeakS*, which consists of a weak supervision model and a controller, where the weak supervision model aggregates the results of multiple unsupervised ranking models as the pseudo labels and trains multiple supervised ranking models upon them, and the controller is responsible for automatically searching the optimal configuration of the unsupervised and the supervised models. Specifically, first, the controller is to sample the unsupervised models, sample the number k of the top ranked courses to a job after aggregating the results of the unsupervised models, and sample the supervised models to be trained upon the top-k pseudo labels, and since the above three sampling processes should be sequentially determined, we formalize the controller by a three-step LSTM model; second, the sampled supervised models are trained on the pseudo labels and evaluated on the validation data with a few human annotated labels; finally, the evaluation metric is returned as the reward signal to guide the training of the controller. When the whole training process converges, we can obtain an optimal combination of different components, which can be regarded as the final model to predict the courses for new jobs.

2.2   Weak Supervision Model

In the weak supervision model, we conduct N^u unsupervised ranking models to calculate N^u relevance scores for each pair of a queried job and a candidate course, aggregate the N^u relevance scores to generate the pseudo labels, and train N^s supervised ranking models on these pseudo labels. Although we select the following unsupervised and supervised models in our framework, the framework is general to incorporate any kinds of unsupervised and supervised models.

Unsupervised Ranking Model. We define two types of the unsupervised ranking models, namely unsupervised text-only matching models and unsupervised graph-based matching models. Unsupervised text-only matching models calculate a relevance score between a queried job and a candidate course based on their descriptions. For example, *BM25* [2] exactly matches the words between a job and a course. *Word2vec* [3] and *BERT* [4] first embed the descriptions of a course and a job into two vectors, and then calculate the cosine similarity between these two vectors.

Different from the unsupervised text-only matching models which represent the jobs and the courses independently, unsupervised graph-based matching models leverage the global correlations between the jobs and courses to represent them. Specifically, we first build a job-word-course heterogeneous graph $G = (V, E)$, which consists of three types of nodes, i.e., job, course and word, and two types of edges that connect courses and words, and connect jobs and words. Then we apply different unsupervised network embedding models to map each node in G into a low-dimensional vector to capture the structural properties. For example, *LINE* [6] and PTE [10] maximize the first-order and the second-order proximity between two nodes. *DeepWalk* [28] extends the first-order neighbors to distinct neighbors which can be reached by random walks. *Node2vec* [11] further proposes the biased random walks to balance the homophily by BFS search and the structural equivalence by DFS search. *GraphSAGE* [5] aggregates the

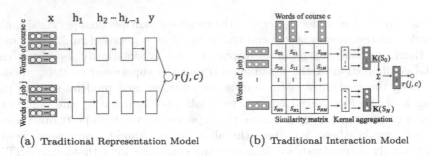

(a) Traditional Representation Model   (b) Traditional Interaction Model

Fig. 2. Traditional supervised text-only matching models.

neighbors' embeddings of the nodes to represent them. Finally, we can calculate the relevance based on the learned embeddings of job and course.

Pseudo Label Generator. To avoid the labeling bias from a single unsupervised model, we aggregate the results of the N^u unsupervised ranking models to generate the pseudo labels. Specifically, for each queried job, we average the N^u relevance scores for each candidate course, rank all the courses according to their average relevance scores, and then annotate the top-k courses as positive instances and the other courses as the negative instances for the queried job.

Supervised Ranking Model. We define two types of the supervised ranking models, including supervised text-only matching models and supervised graph-based matching models. For the supervised text-only matching models, we first explore two traditional models, namely traditional representation model and traditional interaction model, and inspired by the recently proposed pre-training model BERT [4], which has advanced the state-of-the-art in various NLP tasks, we further explore two BERT-based models, namely BERT representation model and BERT interaction model. Finally, we explore one supervised graph-based matching model, GraphSAGE.

Traditional Representation Model directly compares the embeddings of a queried job and a candidate course to capture their semantic relevance. Figure 2(a) illustrates the architecture of the model. We first transform the input word representations $\mathbf{x} \in \mathbb{R}^{N \times d_0}$ of a job into low-dimensional embeddings, and then apply multi-layer nonlinear projections on them to get the intermediate embeddings $\mathbf{h}_l \in \mathbb{R}^{d_l}$ and the final embedding $\mathbf{y} \in \mathbb{R}^{d_L}$ of a job, where N is the maximal number of words included in all the jobs, d_0, d_l and d_L represent the embedding dimensions. The paired inputs of courses are transformed in the same way. Finally, we estimate the relevance score $r(j, c)$ of c to j as the cosine similarity between the job embedding $\mathbf{y}_j$ and the course embedding $\mathbf{y}_c$.

Traditional Interaction Model compares each pair of the words in a queried job and a candidate course. Figure 2(b) illustrates the model. Inspired by [13], we first build a similarity matrix $\mathbf{S}$ between the word embeddings of a queried job and a candidate course, where each element S_{ik} in the similarity matrix $\mathbf{S}$ stands for the cosine similarity between the embedding of the i-th word

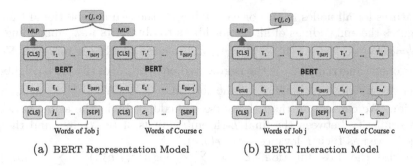

(a) BERT Representation Model   (b) BERT Interaction Model

Fig. 3. BERT-based supervised text-only matching models.

in job j and the embedding of the k-th word in course c. Then we transform the i-th row $\mathbf{S}_i = \{S_{i0}, \ldots, S_{iM}\}$ of the similarity matrix $\mathbf{S}$ into a feature vector $\mathbf{K}(\mathbf{S}_i) = \{K_1(\mathbf{S}_i), \ldots, K_H(\mathbf{S}_i)\}$, where $\mathbf{S}_i = \{S_{i0}, \ldots, S_{iM}\}$ represents the similarities between the i-th word of the queried job j and every word of the course c. Each of the h-th element is converted from $\mathbf{S}_i$ by the h-th RBF kernel with the mean value μ_h and the variance value σ_h, i.e., $K_h(\mathbf{S}_i) = \sum_{k=1}^{M} \exp[(S_{ik} - \mu_h)^2 / 2\sigma_h^2]$. Next, the similarity vectors of all the words in j are summed up into a similarity feature vector, i.e., $\sum_{i=1}^{N} \log \mathbf{K}(\mathbf{S}_i)$, which is then mapped into a one-dimension relevance score $r(j, c)$ to represent the relevance between the job j and the course c.

BERT Representation Model compares the embeddings of a queried job and a candidate course through independently encoding the descriptions of a queried job and a candidate course by the pre-training model BERT [4]. Figure 3(a) illustrates the proposed model. Specifically, for each job j, we take [CLS], $j_1, \cdots, j_N$, [SEP] as the input, where $j_1, \cdots, j_N$ represent job tokens, [CLS] and [SEP] are special tokens. Then we add a multi-layer perceptron (MLP) layer on top of the first output [CLS] embedding to get the representation of job j. A course c is represented in the same way. Finally, we calculate the cosine similarity between the embeddings of the job and the course to obtain the relevance score $r(j, c)$.

BERT Interaction Model compares each pair of the words in a queried job and a candidate course through a unified BERT model, where the multi-head attention in the BERT unit spans over the interactions between the job and the course so that the job-course interactions can be captured. Figure 3(b) illustrates the proposed model. Specifically, we take [CLS], $j_1, \cdots, j_N$, [SEP], $c_1, \cdots, c_M$ as the input, where $j_1, \cdots, j_N$ represent the job tokens and $c_1, \cdots, c_M$ represent the course tokens. Then we add a MLP layer on top of the first output [CLS] embedding to obtain the relevance score of the job-course pair.

SuperGraphSAGE Model aggregates the embeddings of the nodes' neighbors to represent them in a supervised end-to-end fashion. Specifically, given the job-word-course graph $G = (V, E)$, we invoke the bert-as-server API[2] to generate

---

[2] https://github.com/hanxiao/bert-as-service.

the features for all nodes in G. For each job j, at the l-th convolutional time, it aggregates the embeddings of all its neighbors to obtain its new embedding $\mathbf{h}_j^l$, i.e., $\mathbf{h}_j^l = \sigma(\mathbf{W}^l \cdot \text{CONCAT}(\mathbf{h}_j^{l-1}, \mathbf{h}_{\mathcal{N}(j)}^l))$, where σ is a nonlinear function, $\mathbf{W}^l$ is the parameter matrix, $\mathbf{h}_{\mathcal{N}(j)}^l$ is the aggregated embedding of the job's neighbors, $\mathbf{h}_j^{l-1}$ is the previous embedding of the job, and CONCAT is the concatenate operation. We can obtain the l-th course embedding $\mathbf{h}_c^l$ in the same way. The cosine similarity between the final L-th embeddings of the course and the job can be viewed as their relevance score $r(j, c)$.

We use the loss function, $\mathcal{L} = \sum_{j,c+} \log \sigma(r(j, c^+)) + \sum_{j,c-} \log(1 - \sigma(r(j, c^-)))$, to train the supervised models based on the pseudo labels provided by the unsupervised models. For a new (j, c) pair in the test set, we estimate their relevance $r(j, c)$ as the average of all the relevance scores predicted by the N^s supervised models, i.e., $r(j, c) = \frac{1}{N^s} \sum_{i=1}^{N^s} r_i(j, c)$.

2.3 Automated Model Search

To avoid human efforts to determine the suitable weak signals and the supervised models, in this paper, we propose to automatically search the optimal configuration of the weak supervision model. As the weak supervision model is a sequential process that first trains the unsupervised ranking models, then aggregates their outputs as the pseudo labels and finally trains the supervised ranking models based on the pseudo labels, we formalize the controller as a three-step LSTM model to sequentially determine which unsupervised models to select, which value of top-k to select and which supervised models to select. The controller maintains a representation for each choice of different components, i.e., each unsupervised model, each value of k and each supervised model. At step t, the representations of all the selections at $t-1$ are viewed as the input $\mathbf{x}_t \in \mathbb{R}^{d'}$, which is taken together with the previous hidden vector $\mathbf{h}_{t-1} \in \mathbb{R}^{d'}$ and the cell state $\mathbf{e}_{t-1} \in \mathbb{R}^{d'}$ to produce the hidden vector $\mathbf{h}_t$ and the cell state $\mathbf{e}_t$, i.e.,

$$\mathbf{h}_t, \mathbf{e}_t = \text{LSTM}(\mathbf{x}_t, \mathbf{h}_{t-1}, \mathbf{e}_{t-1}, \Phi), \tag{1}$$

where Φ are the parameters of LSTM and d' is the hidden vector dimension. Finally, the component at step t is determined according to the hidden vector $\mathbf{h}_t$ and the representations of all the choices of the component at time t. Now we present the details of the sampling process:

Step 1: Unsupervised model sampling is to sample unsupervised models. At the beginning, the controller selects none of the components and has no memory, thus we set the initial hidden state $\mathbf{h}_0$, the cell state $\mathbf{e}_0$ as empty embeddings, and randomly initialize the input $\mathbf{x}_1$. The controller takes $\mathbf{h}_0$, $\mathbf{e}_0$ and $\mathbf{x}_1$ as input, and output $\mathbf{h}_1$ and $\mathbf{e}_1$ by Eq. (1). Then given $\mathbf{h}_1$, the controller samples the unsupervised models. Intuitively, an unsupervised model is more likely to be sampled if it is more related to the hidden vector $\mathbf{h}_1$ at this step. We randomly initialize the representation $\mathbf{w}_i \in \mathbb{R}^{d'}$ for each unsupervised model, multiply $\mathbf{w}_i$ with $\mathbf{h}_1$ to represent their relevance, based on which we perform sampling:

$$I_i^1 \sim \text{softmax}(f(\mathbf{h}_1^T \times \mathbf{w}_i)), \tag{2}$$

where $\mathbf{h}_1^T \times \mathbf{w}_i$ is a d'-dimensional element-wise product between the two embeddings and f is a fully-connected layer that converts the product into a 2-dimensional vector. Softmax is used to convert the vector into a probability distribution, from which the indicator variable $I_i^1 \in \{0,1\}$ is sampled to represent whether the i-th unsupervised model should be selected or not. Essentially, we sample each model from a binomial distribution. After this step, we can get the indicator vector $\mathbf{I}^1 = [I_1^1, \cdots, I_{N^u}^1]$ to indicate the selected unsupervised models. For example in Fig. 1, $\mathbf{I}^1 = [10010000]$ indicates the controller selects BM25 and LINE as the unsupervised models.

Step2: k sampling is to sample the value of k for selecting the top-k ranked positive instances in the pseudo labels. After sampling the unsupervised models, we multiply the indicator vector $\mathbf{I}^1 = [I_1^1, \cdots, I_{N^u}^1]$ with the model representations $[\mathbf{w}_1, \cdots, \mathbf{w}_{N^u}]$ as the input $\mathbf{x}_2$ of the second step:

$$\mathbf{x}_2 = [I_1^1, \cdots, I_{N^u}^1] \cdot [\mathbf{w}_1, \cdots, \mathbf{w}_{N^u}]^T, \tag{3}$$

where $\mathbf{x}_2$ denotes the summation of the representations of all the sampled models. With $\mathbf{x}_2$, $\mathbf{h}_1$ and $\mathbf{e}_1$ as the input, we can obtain $\mathbf{h}_2$ and $\mathbf{e}_2$ by Eq. (1). Since the value space of k can be very large, to simplify the sampling process, we first categorize all the values of k into τ categories and sample one category for k. The sampling process is defined as:

$$\mathbf{I}^2 \sim \text{softmax}(g(\mathbf{h}_2)), \tag{4}$$

where g is a full-connected layer that converts the hidden vector $\mathbf{h}_2$ into a τ-dimensional vector. Softmax is used to convert the vector into a probability distribution, from which the indicator vector $\mathbf{I}^2 \in \{0,1\}^\tau$ is sampled to represent which category of k is selected. Note that $\mathbf{I}^2$ is a one-hot vector with only one dimension as one, whose index indicates the sampled category of k. Essentially, we sample the category of k from a multinomial distribution. For example in Fig. 1, $\mathbf{I}^2 = [10000]$ indicates the controller selects the first category for k and its corresponding value is 20.

Step 3: Supervised model sampling is to sample supervised models. After sampling k, we multiply the sampling indicator $\mathbf{I}^2$ with the concatenation of the k's category representations $[\mathbf{z}_1, \cdots, \mathbf{z}_\tau]$ as the input $\mathbf{x}_3$ of the third step:

$$\mathbf{x}_3 = [I_1^2, \cdots, I_\tau^2] \cdot [\mathbf{z}_1, \cdots, \mathbf{z}_\tau]^T, \tag{5}$$

where the category representations $[\mathbf{z}_1, \cdots, \mathbf{z}_\tau]$ for each category of k are randomly initialized. The input $\mathbf{x}_3$ denotes the representation of the selected category. With $\mathbf{x}_3$, $\mathbf{h}_2$ and $\mathbf{e}_2$ as the input, we can obtain $\mathbf{h}_3$ and $\mathbf{e}_3$ by Eq. (1). Given $\mathbf{h}_3$, we can sample the indicator $\mathbf{I}^3$ following the same sampling process of step 1 to determine which supervised models should be selected, i.e., $I_i^3 \sim \text{softmax}(q(\mathbf{h}_3^T \times \mathbf{u}_i))$, where q is a fully-connected layer that converts the

---

Input: A set of jobs J and a set of courses C.
Output: Parameters Θ of the weak supervision model and Φ of the controller.
Initialize $\Phi = \Phi^0$, $\Theta = \Theta^0$;
Pre-train the N^u unsupervised ranking models;
repeat

> Sample a weak supervision model $\mathbf{m}$ from $\pi(\mathbf{m}; \Phi)$;
> Train Θ of $\mathbf{m}$ by Eq. (6);
> Calculate $\mathcal{R}^s + \mathcal{R}^u$ of $\mathbf{m}$ on the validation set;
> Update Φ in the controller by REINFORCE;

until *Convergence*;

Algorithm 1: Reinforcement Jointly Training

product $\mathbf{h}_3^T \times \mathbf{u}_i$ into a 2-dimensional vector, $\mathbf{u}_i \in \mathbb{R}^{d'}$ is a randomly initialized embedding for the i-th supervised model. As shown in Fig. 1, $\mathbf{I}^3 = [00010]$ indicates the controller selects BERT interaction model.

2.4   Reinforcement Joint Training

Once the controller finishes searching the configurations of the weak supervision model, i.e., the unsupervised models, the top-k value and the supervised models, a combination with this architecture is built and trained. When the searching architecture achieves convergence, it will get an accuracy $\mathcal{R}$ on a small hold-out annotated dataset (validation set). The accuracy $\mathcal{R}$ is viewed as reward and the parameters of the controller LSTM are then optimized in order to search the best configurations that can achieve the maximal expect validation accuracy. In this paper, we propose a reinforcement joint training process to update the parameters of the controller LSTM, denoted by Φ, and the parameters of the weak supervision model, denoted by Θ. The reinforcement joint training process consists of two interleaving phrases (Algorithm 1), the first phrase trains Θ, while the second phrase trains Φ, the details are as follows:

Training Θ. When training Θ, we fix the controller's sampling policy $\pi(\mathbf{m}; \Phi)$, i.e., the three-step sampling strategy, and perform stochastic gradient descent on Θ to maximize the expected loss $\mathbb{E}_{\mathbf{m} \sim \pi}[\mathcal{L}(\mathbf{m}; \Theta)]$, where $\mathcal{L}(\mathbf{m}; \Theta)$ is the loss computed on a minibatch of training data, with a weak supervision model $\mathbf{m}$ sampled from $\pi(\mathbf{m}; \Phi)$. The gradient is computed using Monte Carlo estimate:

$$\nabla_\Theta \mathbb{E}_{\mathbf{m} \sim \pi}[\mathcal{L}(\mathbf{m}; \Theta)] \approx \frac{1}{N_m} \sum_{p=1}^{N_m} \sum_{q=1}^{N_u' + N_s'} \nabla_\Theta \mathcal{L}_q(\mathbf{m}_p, \Theta), \tag{6}$$

where N_m is the sampling times of the weak supervision model in one epoch. It is empirically proven that $N_m = 1$ works just fine [14]. Notations N_u' and N_s' represent the number of the sampled unsupervised and the supervised models respectively. The whole loss $\mathcal{L}(\mathbf{m}_p, \Theta)$ is the summed losses of all the sampled models. No matter which unsupervised ranking models are sampled, they are always trained on the same training data. So we can pre-train each unsupervised ranking model on the training data, and directly fetch the relevance scores between jobs and courses from all the sampled unsupervised models during each update of Eq. (6). Thus after each sampling of $\mathbf{m}$, we only need to re-optimize the loss $\mathcal{L}_q$ of each sampled supervised model. As a result, the loss $\mathcal{L}(\mathbf{m}_p, \Theta)$ is the summed losses of all the sampled supervised models.

Training Φ. When training Φ, we fix Θ of the weak supervision model and perform REINFORCE algorithm [15] on Φ to maximize the expected reward $\mathbb{E}_{\mathbf{m} \sim \pi(\mathbf{m}; \Phi)}[\mathcal{R}(\mathbf{m}, \Theta)]$. The actions of the controller are to sample the unsupervised models, k, and the supervised models sequentially. The reward is regarded as the evaluated mean reciprocal rank (MRR) of the sampled supervised models on the validation set, which is the set of a few job-course pairs annotated by human beings. Besides, the MRR achieved by the aggregation results of the sampled unsupervised models can also be regarded as the additional reward to accelerate the training process [16]. Thus, the final reward is defined as $\mathcal{R} = \mathcal{R}^s + \mathcal{R}^u$, where $\mathcal{R}^s$ and $\mathcal{R}^u$ are the rewards from the supervised and unsupervised models.

3 Experiment

In this section, we evaluate our proposed model *AutoWeakS* against several unsupervised, supervised, and weak supervision baselines. We also explore whether the selections for each component in *AutoWeakS* (i.e., the selections for the unsupervised models, the supervised models and the top-k values) are necessary.

3.1 Experimental Setup

Dataset. We collect all the courses from XuetangX[3], one of the largest MOOCs in China, and this results in 1951 courses. The collected courses involve seven areas: computer science, economics, engineering, foreign language, math, physics, and social science. Each course contains 131 words in its descriptions on average. We also collect 706 job postings from the recruiting website operated by JD.com[4] (JD) and 2,456 job postings from the website owned by Tencent corporation[5] (Tencent). The collected job postings involve six areas: technical post, financial post, product post, design post, market post, supply chain and engineering post. Each job contains 107 and 151 words in its posting on average in JD and Tencent respectively. To evaluate the model performance, for both JD and Tencent dataset, we randomly select 200 jobs, and ask ten volunteers to annotate the relevant courses to the jobs. Specifically, for a queried job, we first use each unsupervised model in Sect. 2.2 to calculate a relevance score for each course, average all the scores over all the models, select top 60 candidate courses,

---

[3] http://www.xuetangx.com.
[4] http://campus.jd.com/home.
[5] https://hr.tencent.com/.

Table 1. Overall performance of recommending courses for jobs. We try different k for WeakS and report its best performance.

Model	JD-XuetangX			Tencent-XuetangX		
	HR@5	NDCG@5	MRR	HR@5	NDCG@5	MRR
BM25	0.162	0.151	0.173	0.072	0.046	0.070
Word2vec	0.301	0.212	0.217	0.142	0.107	0.114
BERT	0.348	0.239	0.238	0.159	0.104	0.122
LINE	0.489	0.362	0.409	0.396	0.284	0.279
PTE	0.378	0.244	0.334	0.295	0.204	0.210
DeepWalk	0.390	0.249	0.258	0.370	0.262	0.261
Node2vec	0.374	0.279	0.284	0.386	0.282	0.277
GraphSAGE	0.312	0.252	0.232	0.186	0.121	0.139
Traditional Representation	0.407	0.261	0.262	0.201	0.125	0.148
Traditional Interaction	0.470	0.429	0.414	0.324	0.215	0.214
BERT Representation	0.350	0.232	0.231	0.294	0.195	0.204
BERT Interaction	0.564	0.537	0.497	0.405	0.254	0.222
SuperGraphSAGE	0.263	0.176	0.186	0.231	0.144	0.155
WeakS	0.704	0.548	0.592	0.370	0.255	0.227
LINE+AllS	0.736	0.550	0.624	0.408	0.289	0.236
AutoWeakS	**0.793**	**0.615**	**0.671**	**0.631**	**0.522**	**0.540**

annotate each candidate and obtain the ground truth by majority voting of all the volunteers' annotations. The Dataset and the code are online now[6].

Settings. For training the unsupervised ranking models, we use all the 706 job postings from JD and all the 2,456 job postings from Tencent to learn the embeddings of the jobs and courses. For the supervised ranking models, we hold out the human annotated jobs and only use the 506 unlabeled jobs from JD and 2,256 unlabeled jobs from Tencent for training. On each dataset, we averagely partition the annotated 200 jobs into a validation set and a test set and sample 99 negative instances for each positive instance (1 positive plus 99 negatives) [17]. We use Hit Ratio of top K items (HR@K), Normalized Discounted Cumulative Gain of top K items (NDCG@K) and Mean Reciprocal Rank (MRR) as the evaluation metrics for ranking, where K is set as 5.

3.2   Experimental Results

Comparison with Baselines. In this experiment, we evaluate our model *AutoWeakS* against the unsupervised and supervised models in Sect. 2.2, the weak supervision model WeakS, which includes all the unsupervised models

---

[6] https://github.com/jerryhao66/AutoWeakS.

and the supervised models without model search, and one competitive baseline, LINE+AllS, which includes LINE as the unsupervised model and all the
supervised models. Note that due to the lack of enough labeled data, we only use
a small annotation data (i.e., the validation set) to train the supervised models.

Table 1 shows the results on two datasets. We vary the value k for WeakS,
LINE+AllS and report its best performance in Table 1. From the results, we can
see that the proposed *AutoWeakS* performs clearly better than other baselines.
Compared with the unsupervised graph-based matching methods, unsupervised
text-only matching models perform worse, as only using the descriptive words
of the jobs and the courses can not capture high-order relationships between
the jobs and the courses. Some supervised methods such as BERT interaction
model and traditional interaction model perform better than the unsupervised
methods, as the unsupervised methods do not explicitly compare the relevance
of the positive courses and the negative courses to a queried job. However, due
to the lack of enough training labels, the performance of the supervised models
is worse than WeakS, LINE+AllS and our proposed method *AutoWeakS*.

Table 2. Performance of different choices of unsupervised models in AutoWeakS with
k and the supervised component fixed.

Unsuper. choices	JD-XuetangX			Tencent-XuetangX		
	HR@5	NDCG@5	MRR	HR@5	NDCG@5	MRR
BM25+	0.203	0.194	0.182	0.183	0.139	0.159
Word2vec+	0.435	0.392	0.336	0.333	0.312	0.321
BERT+	0.705	0.511	0.511	0.393	0.373	0.387
LINE+	0.722	0.559	0.516	0.589	0.478	0.499
PTE+	0.657	0.488	0.505	0.471	0.451	0.497
DeepWalk+	0.677	0.503	0.462	0.508	0.461	0.411
Node2vec+	0.684	0.507	0.451	0.534	0.449	0.463
GraphSAGE+	0.642	0.495	0.402	0.563	0.471	0.491
All unsupervised+	0.609	0.423	0.458	0.415	0.396	0.426
AutoWeakS	**0.793**	**0.615**	**0.671**	**0.631**	**0.522**	**0.540**

WeakS performs better than all the unsupervised models on JD-XuetangX,
as it explicitly learns the ranking of the candidate courses to queried jobs. However, on Tencent-XuetangX, WeakS underperforms several unsupervised models,
because BM25 performs particularly poorly on this dataset, which reduces the
effect of the aggregated pseudo labels from all the unsupervised models. This
also indicates that indiscriminately combing all the models may not result in
the best performance. Besides, *AutoWeakS* beats LINE+AllS, which implies
selecting only one unsupervised model may suffer from the issue of ranking bias.

We further remove BM25 from WeakS, name the model as WeakS-BM25 and
show the performance of WeakS, LINE+AllS and WeakS-BM25 in Fig. 4(a) and

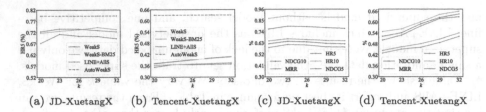

(a) JD-XuetangX   (b) Tencent-XuetangX   (c) JD-XuetangX   (d) Tencent-XuetangX

Fig. 4. (a), (b) show that the baselines with any choice of top-k underperform AutoWeakS with the automatically searched top-k value. (c), (d) further present the results of AutoWeakS under different top-k values, which indicates that the automatically searched top-k value performs the best against all the other top-k values.

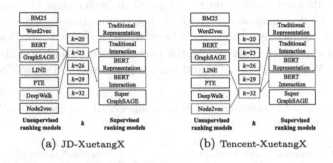

(a) JD-XuetangX   (b) Tencent-XuetangX

Fig. 5. The optimal selections of AutoWeakS.

Fig. 4(b). The k value of *AutoWeakS* is fixed as the automatically searched value. The results show that even if the worst performed BM25 is removed, given any value of k, WeakS-BM25 still underperforms *AutoWeakS*, which indicates the advantage of the automated model search in *AutoWeakS*.

Analysis of Unsupervised Component. We evaluate the performance of different choices of the unsupervised models, when fixing the sampled k and the supervised component in *AutoWeakS*. We name the model as BM25+ if only BM25 is chosen to produce pseudo labels. Other single model is named in the same way. All unsupervised+ means we combine the labels of all the unsupervised models. Figure 5(a) shows that on JD-XuetangX, *AutoWeakS* selects the combination of BERT, LINE, and DeepWalk, which performs better than any single unsupervised model and All unsupervised+ shown in Table 2. On Tencent-XuetangX, *AutoWeakS* also obtains the best performance, and it selects the combination of LINE and Node2vec as the unsupervised component. The results indicate the advantage of automatically searching the unsupervised models.

Analysis of k. We evaluate the performance of different choices of k to generate the pseudo labels, when fixing the sampled unsupervised and the supervised components in *AutoWeakS*. Figure 5 presents that the automatically searched k is 23 on JD-XuetangX and is 32 on Tencent-XuetangX. Figure 4(c) and Fig. 4(d)

Table 3. Performance of different choices of supervised models in AutoWeakS with the unsupervised component and k fixed.

Super. choices	JD-XuetangX			Tencent-XuetangX		
	HR@5	NDCG@5	MRR	HR@5	NDCG@5	MRR
Traditional Representation+	0.525	0.349	0.335	0.362	0.283	0.269
Traditional Interaction+	0.679	0.508	0.483	0.601	0.502	0.504
BERT Representation+	0.604	0.483	0.462	0.318	0.209	0.209
BERT Interaction+	0.729	0.537	0.609	0.576	0.464	0.481
SuperGraphSAGE+	0.348	0.212	0.368	0.265	0.181	0.193
All supervised+	0.652	0.482	0.507	0.402	0.317	0.301
AutoWeakS	**0.793**	**0.615**	**0.671**	**0.631**	**0.522**	**0.540**

show that *AutoWeakS* with other k values underperforms the searched k values. The results indicate the advantage of automatically searching k.

Analysis of Supervised Component. We evaluate the performance of different choices of the supervised models, when fixing the sampled unsupervised component and k in *AutoWeakS*. We name the model as BERT Interaction+ if only the BERT interaction model is trained. Other single supervised model is named in the same way. All supervised+ means we train all the supervised models. Figure 5(a) and Fig. 5(b) show that on both of the JD-XuetangX and the Tencent-XuetangX datasets, *AutoWeakS* selects the combination of the BERT interaction model and the traditional interaction model. The results show that *AutoWeakS* performs better than all the other choices shown in Table 3, which indicates the advantage of automatically searching the supervised models.

4   Related Work

Much effort has been made to provide better services for job seekers and recruiters through analyzing the flow of job seekers [29] or matching the job recruitment postings and the resumes of the job seekers [1]. The related works include:

Weak Supervision Model. Training neural ranking models on pseudo-labeled data has been attracted attentions. For example, Dehghani et al. [8] leverage the output of traditional IR models such as BM25 as the weak supervision signal to generate a large amount of pseudo labels to train effective neural ranking models. Zamani et al. [9] train a neural query performance predictor by multiple weak supervision signals, and they also provide a theoretical analysis of this weak supervision method [18]. The same idea is employed in [19,20]. However, for different tasks, human efforts are demanded to determine the suitable weak signals and the supervised models. Even if each signal is carefully selected by humans, their combination may not be optimal.

Automated Machine Learning (AutoML). The goal of AutoML is to automatically determine the optimal configurations such as selecting the optimal models [21,22], features [23,24], and neural architecture [25,26], which can help people use machine learning models easily. Different types of techniques are studied to search the optimal configuration. For example, Bayesian optimization methods such as Auto-sklearn [21] and Auto-Weka [22] model the relationship between a configuration and the corresponding performance in a probabilistic way. Reinforcement learning trains the optimal search policies according to the feedbacks of the searched configurations [27], where the search policy can be modeled by RNN [14,25]. Inspired by the above works, we propose a RL-based joint training framework to search an optimal combination of the unsupervised/supervised models and the hyperparameter k in the proposed weak supervision model for recommending courses for jobs.

5   Conclusion

We present the first attempt to solve the problem of recommending courses in MOOCs for jobs by a general automated weak supervision model. With reinforcement joint training of a weak supervision model for recommending courses and a controller for searching models, we can automatically find the best configuration of the weak supervision model. Experiments on two real-world datasets of jobs and courses show that the proposed AutoWeakS significantly outperforms the classical unsupervised, supervised and weak supervision baselines.

ACKNOWLEDGMENTS. This work is supported by National Key R&D Program of China (No. 2018YFB1004401) and NSFC (No. 61532021, 61772537, 61772536, 61702522).

References

1. Xu, T., Zhu, H., Zhu, C., Li, P., Xiong, H.: Measuring the popularity of job skills in recruitment market: a multi-criteria approach. In: AAAI 2017, pp. 2572–2579 (2017)
2. Robertson, S., Zaragoza, H.: The probabilistic relevance framework: BM25 and beyond. Found. Trends® Inf. Retr. **3**(4), 333–389 (2009)
3. Mikolov, T., Sutskever, I., Chen, K., Corrado, G.S., Dean, J.: Distributed representations of words and phrases and their compositionality. In: NeurIPS 2013 (2013)
4. Devlin, J., Chang, M., Lee, K., Toutanova, K.: BERT: pre-training of deep bidirectional transformers for language understanding. In: NAACL 2019, pp. 4171–4186 (2019)
5. Hamilton, W.L., Ying, Z., Leskovec, J.: Inductive representation learning on large graphs. In: NeurIPS 2017, pp. 1024–1034 (2017)
6. Tang, J., Qu, M., Wang, M., Zhang, M., Yan, J., Mei, Q.: Line: large-scale information network embedding. In: WWW 2015, pp. 1067–1077 (2015)
7. Qiu, X., Huang, X.: Convolutional neural tensor network architecture for community-based question answering. In: IJCAI 2015, pp. 1305–1311 (2015)

8. Dehghani, M., Zamani, H., Severyn, A., Kamps, J., Croft, W.B.: Neural ranking models with weak supervision. In: SIGIR 2017, pp. 65–74 (2017)
9. Zamani, H., Croft, W.B., Culpepper, J.S.: Neural query performance prediction using weak supervision from multiple signals. In: SIGIR 2018, pp. 105–114 (2018)
10. Tang, J., Qu, M., Mei, Q.: PTE: predictive text embedding through large-scale heterogeneous text networks. In: KDD 2015, pp. 1165–1174 (2015)
11. Grover, A., Leskovec, J.: node2vec: scalable feature learning for networks. In: SIGKDD 2016, pp. 855–864 (2016)
12. Pang, L., Lan, Y., Guo, J., Xu, J., Wan, S., Cheng, X.: Text matching as image recognition. In: AAAI 2016, pp. 2793–2799 (2016)
13. Xiong, C., Dai, Z., Callan, J., Liu, Z., Power, R.: End-to-end neural ad-hoc ranking with kernel pooling. In: SIGIR 2017, pp. 55–64 (2017)
14. Pham, H., Guan, M.Y., Zoph, B., Le, Q.V., Dean, J.: Efficient neural architecture search via parameter sharing. In: ICML 2018, pp. 4092–4101 (2018)
15. Williams, R.J.: Simple statistical gradient-following algorithms for connectionist reinforcement learning. Mach. Learn. **8**(3–4), 229–256 (1992)
16. Ghavamzadeh, M., Mahadevan, S.: Hierarchical policy gradient algorithms. Computer Science Department Faculty Publication Series, p. 173 (2003)
17. He, X., He, Z., Song, J., Liu, Z., Jiang, Y.G., Chua, T.S.: NAIS: neural attentive item similarity model for recommendation. IEEE TKDE **30**(12), 2354–2366 (2018)
18. Zamani, H., Croft, W.B.: On the theory of weak supervision for information retrieval. In: SIGIR 2018, pp. 147–154 (2018)
19. Dehghani, M., et al.: Avoiding your teacher's mistakes: training neural networks with controlled weak supervision. arXiv preprint arXiv:1711.00313 (2017)
20. Luo, C., Zheng, Y., Mao, J., Liu, Y., Zhang, M., Ma, S.: Training deep ranking model with weak relevance labels. In: Huang, Z., Xiao, X., Cao, X. (eds.) ADC 2017. LNCS, vol. 10538, pp. 205–216. Springer, Cham (2017). https://doi.org/10.1007/978-3-319-68155-9_16
21. Feurer, M., Klein, A., Eggensperger, K., Springenberg, J., Blum, M., Hutter, F.: Efficient and robust automated machine learning. In: NeurIPS 2015, pp. 2962–2970 (2015)
22. Kotthoff, L., Thornton, C., Hoos, H.H., Hutter, F., Leyton-Brown, K.: Auto-WEKA 2.0: automatic model selection and hyperparameter optimization in WEKA. J. Mach. Learn. Res. **18**(1), 826–830 (2017)
23. Katz, G., Shin, E.C.R., Song, D.: ExploreKit: automatic feature generation and selection. In: ICDM 2016, pp. 979–984 (2016)
24. Huang, S., Wang, C., Ding, B., Chaudhuri, S.: Efficient identification of approximate best configuration of training in large datasets. In: AAAI 2019, pp. 3862–3869 (2019)
25. Zoph, B., Le, Q.V.: Neural architecture search with reinforcement learning. In: ICLR 2017 (2017)
26. Liu, C., et al.: Progressive neural architecture search. In: ECCV 2018, pp. 19–34 (2018)
27. Sutton, R.S., Barto, A.G.: Reinforcement learning: an introduction. In: AI Magazine, pp. 15–34 (2011)
28. Perozzi, B., Al-Rfou, R., Skiena, S.: DeepWalk: online learning of social representations. In: SIGKDD 2014, pp. 701–710 (2014)
29. Oentaryo, R.J., Lim, E.-P., Ashok, X.J.S., Prasetyo, P.K., Ong, K.H., Lau, Z.Q.: Talent flow analytics in online professional network. Data Sci. Eng. **3**(3), 199–220 (2018). https://doi.org/10.1007/s41019-018-0070-8

Learning a Contextual and Topological Representation of Areas-of-Interest for On-Demand Delivery Application

Mingxuan Yue[1](✉), Tianshu Sun[1], Fan Wu[2], Lixia Wu[2], Yinghui Xu[2], and Cyrus Shahabi[1]

[1] University of Southern California, Los Angeles, USA
{mingxuay,shahabi}@usc.edu, tianshus@marshall.usc.edu
[2] Cainiao Network, Hangzhou, China
{wf118503,wallace.wulx}@cainiao.com

Abstract. A good representation of urban areas is of great importance in on-demand delivery services such as for ETA prediction. However, the existing representations learn either from sparse check-in histories or topological geometries, thus are either lacking coverage and violating the geographical law or ignoring contextual information from data. In this paper, we propose a novel representation learning framework for obtaining a unified representation of Area of Interest from both contextual data (trajectories) and topological data (graphs). The framework first encodes trajectories and graphs into homogeneous views, and then train a multi-view autoencoder to learn the representation of areas using a ranking-based loss. Experiments with real-world package delivery data on ETA prediction confirm the effectiveness of the model.

Keywords: Representation learning · Trajectories · Multi-view autoencoder

1 Introduction

In recent years, we witness the rapid growth of on-demand deliveries everywhere and every day (e.g., Amazon Prime Now). We deliver people, food, parcels by cars, bicycles, and foot from dawn to midnight and from city centers to suburbans. The explosion of E-commerce and recent advances in spatial crowdsourcing have prompted the surge of deliveries, and are still calling for better solutions.

A good representation of spatial units is of vital importance to all delivery-related services [13]. Various companies like Uber and DiDi utilize different spatial extents such as grids, hexagons, or polygons to partition the space into spatial units [10]. These spatial units, represented by their coordinates and other geometric features, are then used as sources and targets for delivery services. Such spatial units fully cover an entire space (e.g., a city) and have nice topological properties. Thus, the algorithms based on these units can accommodate any possible delivery request. However, such topological representation can only capture

Y. Dong et al. (Eds.): ECML PKDD 2020, LNAI 12460, pp. 52–68, 2021.
https://doi.org/10.1007/978-3-030-67667-4_4

spatial relationships between these units and ignore human's intuition and tacit knowledge on how to navigate between these regions. For example, when couriers deliver packages on foot and/or by bike, they mainly choose paths according to their knowledge and experience on real-world road conditions and connections such as shortcuts, bridges, crowded streets, and crossings with long traffic lights. In such cases, mere topological representation often fails to capture key information thus may not be sufficient for the real delivery tasks. Fortunately, human trajectories capture such tacit knowledge and experiences.

Towards this end, recent work [3, 15, 32] strives to add such contextual data to Point Of Interest(POI) representation from check-in histories by adopting NLP models like Word2vec [17]. However, these studies mainly focus on recommending POI to users. Hence, the representation of POIs usually does not cover the entire space, thus they cannot be directly applied to delivery systems that requires every points in space to be reachable. Besides, the learned representation may also lose the topological property and conflict with the Tobler's First Law of Geography [26] which says "Everything is related to everything else, but near things are more related than distant things", due to the discrete locations of POIs and sampling bias in the collection of check-in histories.

Therefore, the best representation should learn from both topological and contextual data to take advantage of the best of the two worlds. To achieve this, we propose a novel Deep Multi-view informAtion-encoding RanKing-based network (DeepMARK) to learn a representation of spatial regions. Rather than regular-shaped regions, we consider the spatial regions to be geographically partitioned by map segmentation, i.e., the Areas of Interests (AOI) used in this paper. AOIs are non-overlapping irregular polygons that fully partition (and hence cover) the space and each AOI captures its individual context. For example, while hexagons or grids may split a school into two units or may have a unit containing multiple land uses, each AOI represents a single context.

To learn both the topological and contextual features of these AOIs, our proposed framework DeepMARK consists of three components: one to learn the topological representation, the second one to learn the contextual representation and finally the third to unify the first two components.

Contextual Representation Component: In the field of NLP, contextual representations are usually learned based on the distributional hypothesis [21] from real-world language sequences, i.e., human utterance. Analogous to NLP, for "spatial" context, we consider location sequences, i.e., human trajectories, as the data source from which we learn the contextual representation of AOIs. The trajectory data is selected for its relevance and scalability in learning contextual representations for delivery problems: 1) trajectories preserve human's knowledge and preferences in traveling between AOIs. 2) with the ubiquity of mobile devices and the prevalence of spatial crowdsourcing apps, trajectories can be easily collected at scale. To learn contextual representation from trajectories, we model the spatial distributional hypothesis using Pointwise Mutual Information (PMI) between AOIs calculated from trajectories. Subsequently, we learn a distributed representation based on the PMI using an autoencoder framework.

Topological Representation Component: To model topological properties of irregular-shaped AOIs, we define Euclidean graph and Adjacency graph to capture the spatial relationships of the AOIs. Lately, to learn representations from graphs, researchers proposed various graph embedding approaches [2, 6, 7, 20, 27]. Popular methods like Deepwalk [20] and Node2vec [6] are based on random walks and train the network on randomly generated samples. However, in our problem, such a process cannot be easily trained with trajectories jointly. Therefore, we propose to estimate the node-wise mutual information in graphs and use the same autoencoder framework as used for trajectories to align the learning of the two heterogeneous views.

Unified Representation Component: Finally, to combine the two heterogeneous views, previous studies employ different strategies to model the correlation between views and control the learning across views [4, 14, 24]. However, none of these approaches could be directly applied to our problem because most of them are designed for text and image data. To the best of our knowledge, we are the first to study the joint learning of AOI representation using both trajectory and graph data. To join the learning of trajectories and graphs, we propose a novel multi-view autoencoder neural network that takes the PMI matrices generated by the previous two components and utilizes an innovative ranking-weighted loss to dynamically balance the learning between views.

We evaluated our representation with a large real-world package delivery data acquired from Cainiao Network. Our representation approach is shown to have up to 20% reduction of errors as compared to the adapted baseline approaches in predicting Estimated Time of Arrival (ETA) of real-world deliveries.

The remainder of the paper is organized as follows. Section 2 clarifies some basic definitions and important notations used in this paper. Section 3 presents the details of our proposed framework. In Sect. 4, we show the evaluation of our approach on real-world data. Finally, Sect. 5 introduces the related work followed by our conclusion in Sect. 6.

2 Preliminaries

In this section, we introduce some important concepts followed by the formal problem definition.

Definition 1 (Area Of Interest (AOI)). *An AOI is a minimum geographical unit in the form of a polygon. The raw AOIs are generated by partitioning a space with fine-grain road networks and geometric boundaries (e.g., roads, rivers, railways) using map segmentation techniques.*

By definition, the boundaries of AOIs, i.e., the irregular polygons can have different sizes and numbers of edges, which differentiate them from those of the conventional space partitioning techniques using regular shapes (e.g., hexagons). Moreover, our AOIs still do cover the entire space and each AOI captures a single context (e.g., a school). Later in Sect. 3.2 we show how we add latent features (learned from topological representations) to each AOI to enforce the Tobler's First Law of Geography.

Definition 2 (Trajectory). *A trajectory s is a sequence of spatio-temporal tuples* $s = [s(1), s(2), \ldots, s(k), \ldots]$, *where* $s(k)$ *is represented by a tuple consisting of the AOI v that contains the GPS point and a timestamp t, i.e.,* $s(k) = (v, t)$.

We derive the modeling of contextual representation from the analogy in language models. Most word representation models explicitly or implicitly follow the distributional hypothesis introduced by linguists [21]. The hypothesis is often stated as: *words which are similar in meaning occur in similar contexts.* In our problem, as sequences of AOIs (trajectories) are analogous to sequences of words (sentences), we make the following assumption:

Assumption 1 (Contextual representation of AOIs). *A contextual representation of AOIs follows the spatial distributional hypothesis, that AOIs have similar contextual representations are usually visited closely and in a trip.*

Given the above definitions, we define our problem of learning a contextual and topological representation of AOIs as below.

Definition 3 (Learning a Contextual and Topological Representation of AOIs (CTRA) Problem). *Given a set of raw AOIs (i.e., without latent features) V, and a set of trajectories S, s.t.* $\forall s \in S, \forall(v, t) \in s, v \in V$, *the objective is to learn a mapping* $V \to Z$, *s.t., it generates a latent representation* $z \in Z$ *for each AOI* $v \in V$, *that follows the spatial distributional hypothesis, and Tobler's First Law of Geography.*

3 Methodology

We propose a Deep Multi-view informAtion-based RanKing network (Deep-MARK) to solve the CTRA problem. DeepMARK consists of three parts: learning contextual representation, learning topological representation and jointly learning of both representations, which are elaborated in the following sections.

3.1 Learning Contextual Representation from Trajectories

Modeling Spatial Distributional Hypothesis. The learning of contextual representation of AOIs in trajectories is analogous to the learning of word embeddings from sentences. To model the distributional hypothesis, word embedding techniques usually describe similarities between words using their contexts and then map words to hidden embeddings according to such similarities. For example, the word2vec model [17] maximizes the log probability as in Eq. 1. The modeling of the context similarity is implicitly computed by predicting the context words (w_{t+j}) of a target word (w_t), which usually requires a sampling-based training process, e.g., negative sampling.

$$\frac{1}{T}\sum_{t=1}^{T}\sum_{-c\leq j\leq c,j\neq 0} \log p(w_{t+j}|w_t) \tag{1}$$

In this paper, rather than use the sampling-based training and objective, we propose to use Pointwise Mutual Information (PMI) to describe the contextual similarity between AOIs and learn the representation by decomposing the similarities using neural networks. We believe such approach has better compliance with CTRA problem because of the following reasons.

1. *The similarity is symmetric.* In word2vec models, people choose a center word and its context word to describe the similarity. In this case, the similarity of "A to B" might be different that of "B to A", when choosing A or B as the center word. However, in delivery scenarios, we concern more about whether the 2 places are likely to be visited from each other. So we expect the similarity to be symmetric, i.e., $similarity(A, B) = similarity(B, A)$, which is guaranteed in PMI.
2. *The decomposition of PMI has comparable performance and is implicitly equivalent to SGNS.* As shown in recent studies, the SGNS model is implicitly factorizing the shifted PMI matrix [11] and a good decomposition of PMI(PPMI) matrix is comparable with word2vec models in various tasks [12].
3. *The training process is easy for alignment in a multi-view learning framework.* Sampling-based training is hard to be extended to multi-view problems like CTRA. Even applying iterative training one cannot align different views well to the same training target (a single AOI) and train them jointly. However, the decomposition of PMI is easy for aligning the same AOI from different views which allows joint training described in Sect. 3.3.

Formally, given AOI v_i and v_j, we define the contextual similarity from the trajectory data as follows:

$$\mathrm{PMI}_{traj}(v_i, v_j) = log(\frac{p(v_i, v_j)}{p(v_i)p(v_j)}) \tag{2}$$

Here, $p(v_i)$ and $p(v_j)$ denote the probability of randomly visiting v_i and v_j, and $p(v_i, v_j)$ denotes the probability of visiting v_i and v_j together. we can interpret $\frac{p(v_i, v_j)}{p(v_i)p(v_j)}$ as: the ratio of *how likely people visit v_i and v_j together in the real world* to *how likely v_i and v_j are visited together at random*. Therefore, a large ratio means the two AOIs v_i and v_j are, rather than randomly visited together, co-visited for some real reason, e.g., they are easily accessible in human knowledge.

Computation of PMI in Trajectories. Now, to compute the PMI between AOIs, the remaining task is to define the computation of $p(v_i, v_j), p(v_i)$ and

$p(v_j)$ for AOIs in trajectories. For calculating $p(v_i, v_j)$, it's important to properly define the co-occurrence of AOI v_i and AOI v_j in the trajectories. Different from the skip-gram model, we define that two AOIs co-occur in a close context if they fall in a fixed-length temporal window in a trajectory. To count such co-occurrences, we apply a time sliding window to each trajectory: $[t, t + \Delta]$, where Δ is the window size. As shown in Fig. 1b, each sliding window may contain various numbers of AOIs (window T_1 has 2 AOIs while T_3 includes 3 AOIs) but the temporal length of each window is the same. In addition, we slide the windows with an offset of $\Delta/2$ to make the best use of the trajectories while avoid generating too many samples, similar to [31].

We use such temporal windows for defining co-occurrence because of the nature of trajectories. In detail, as depicted in Fig. 1a, if we adopt the way skip-gram model building context windows (C_1 to C_4), for each AOI in the trajectory, we have to extract a fixed number of preceding and succeeding AOIs as its co-occurring neighbors. However, in trajectories, consecutively collected spatio-temporal points usually have variant time differences, e.g., from 2 min to 20 min, because of the unstable signals and different mobile application settings. Consequently, if we adopt skip-gram and consider two consecutive but distant AOIs as a co-occurrence, it will mislead the model to produce similar embeddings between the two distant AOIs (e.g., in C_3, two distant nodes are counted in the same context window), which is not expected.

After the sliding windows are generated, we count any two AOIs in the same window as a co-occurring pair. The probabilities $p(v_i, v_j), p(v_i), p(v_j)$, and the PMI matrix between AOIs can be estimated by counting the co-occurring pairs as below.

$$\text{PMI}_{traj}(v_i, v_j) = log(\frac{p(v_i, v_j)}{p(v_i)p(v_j)})$$

$$= log(\frac{\#(v_i, v_j)/|C|}{(\#(v_i)/|C|) \cdot (\#(v_j)/|C|)})$$

$$= log(\frac{\#(v_i, v_j) \cdot |C|}{\#(v_i) \cdot \#(v_j)})$$

$$\text{where } |C| = \sum_{i'} \sum_{j'} \#(v_{i'}, v_{j'})$$

In the equation above, $\#(v_i, v_j)$ denotes the count of co-occurring pairs (v_i, v_j) from all windows, $\#(v_i)$ and $\#(v_j)$ denotes the count of pairs containing v_i and v_j respectively. C denotes the set of all co-occurring pairs and $|C|$ is the number of all pairs.

Learning a Distributed Representation Using Autoencoder. After the computation of PMI from trajectories, we propose to use autoencoder to decompose the PMI for a dense and distributed representation. Although for each AOI v_i, we can use its PMI similarity to all AOIs as its representation, i.e.,

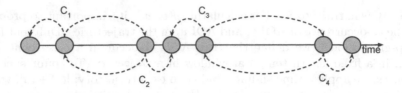

Count co-occurrences by skip-gram.

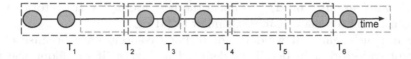

Count co-occurrences by sliding window.

Fig. 1. Count co-occurrences by skip-gram.

$[\mathrm{PMI}_{traj}(v_i, v_0), \mathrm{PMI}_{traj}(v_i, v_1), \ldots \mathrm{PMI}_{traj}(v_i, v_n)]$, we propose to apply low-rank decomposition by autoencoder on the sparse PMI matrix. Because a distributed representation [9](i.e., each element encodes multiple things) is always expressive and allows efficient activation in downstream training [1]. In addition, an autoencoder allows non-linear encoding, and thus could have more accurate reconstruction of the similarities. Specifically, the autoencoder consists of an encoder f and a decoder g. The encoder f takes the PMI vector of each AOI and learns a low-dimensional embedding. Then the decoder g takes the low-dimensional embedding and reconstructs the PMI vector with minimum error. The objective of the network is minimizing the reconstruction error $\mathbb{L}$ in Eq. 3.

$$\mathbb{L}_{traj} = \sum_i^n \|\mathrm{PMI}_{traj}(i), g(f(\mathrm{PMI}_{traj}(i)))\|^2 \tag{3}$$

In summary, as depicted in Fig. 2, DeepMARK first slides windows in trajectories, and then counts the co-occurring pairs in these windows. After that, the

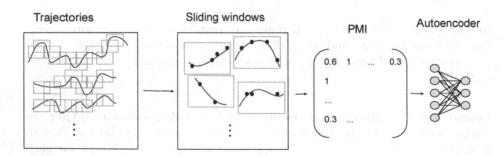

Fig. 2. Learn contextual representation from trajectories.

PMI matrix is computed based out of the counts, and is fed to an autoencoder for learning representations. Notice that here we actually employ the common practice of positive PMI [8, 11, 12] rather than PMI but we use PMI for simplicity.

3.2 Learning Topological Representation from Graphs

Given that our initial AOIs already cover the entire space (see Definition 1 in Sect. 2), here we would like to learn features (latent representation) per AOI to capture the spatial relationships among these AOIs that follow Tobler's first Law of Geography. Therefore, we use two graphs to capture the spatial relations between AOIs: Euclidean Graph G_{euc} and Adjacency Graph G_{adj}. Intuitively, the former graph captures Euclidean proximity to enforce Tobler's First Law between nearby AOIs and the latter uses adjacency relationships to enforce the law for adjacent AOIs.

Euclidean Graph G_{euc}. We define $G_{euc} = \{V, E_{euc}, W_{euc}\}$. V is the set of nodes i.e., AOIs. We define the weights $W = [w_{ij}] \in \mathbb{R}^{n \times n}$ representing the proximity between v_i and v_j as a function of their Euclidean distance $dist(v_i, v_j)$. In particular, we define the proximity function as a thresholded Gaussian kernel function [22] as in Eq. 4. Intuitively, the closer nodes, the larger weight is assigned to the edge between the nodes.

$$W_{ij} = \begin{cases} exp(-\frac{dist(v_i, v_j)^2}{\sigma^2}) & \text{if } dist(v_i, v_j) \leq \mathcal{K} \\ 0 & \text{otherwise} \end{cases} \quad (4)$$

Adjacency Graph G_{adj}. We model the adjacency between AOIs as a graph $G_{adj} = \{V, E_{adj}, W_{adj}\}$. V is the set of nodes, i.e., AOIs. The weights $W = [w_{ij}] \in \{0, 1\}$ represent the adjacency between AOIs, where w_{ij} is defined as below.

$$W_{ij} = \begin{cases} 1 & \text{if } v_i \text{ is adjacent to } v_j \\ 0 & \text{otherwise} \end{cases} \quad (5)$$

We expect the learning from the two graphs and from trajectories could have homogeneous processes for a flexible and alignable joint learning of topological and contextual representations. Therefore, we design PMI matrices for the graphs to be homogeneous with the trajectory view. For any two nodes v_i, v_j in a graph G given its weights W, to prepare the probabilities $p(v_i, v_j), p(v_i)$ and $p(v_j)$ in the graph, we define $p(v_i, v_j)$ as the proximity from v_i to v_j within K-step random walks. Specifically, we first define a transition matrix M^k, in which $M^k_{i,j}$ presents the probability of visiting v_j in a k step random walk from v_i with restart ratio η according to [25].

$$M^k = \eta \cdot \mathbb{I} + (1 - \eta) M^{(k-1)} \cdot (D^{-1}W),$$
$$\text{where } M^0 = \mathbb{I}$$

Here $\mathbb{I}$ is the identity matrix. D is a diagonal matrix, s.t., each element in the diagonal is the summation of the corresponding row in W, i.e., $D_{ii} = \sum_j W_{i,j}$.

And respectively, as depicted in Fig. 3, we can compute the proximity matrix P^K as the sum of random walks within K steps starting from any node: $P^K = \sum_{k=1}^{K} M^k$. Then $p(v_i, v_j)$ is defined as $P_{i,j}^K$, and accordingly, $p(v_i)$ is defined as $\sum_l P_{l,i}^K$.

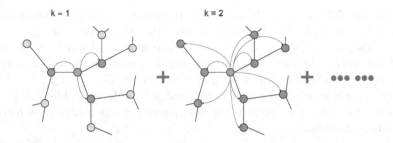

Fig. 3. Calculate proximity to other nodes by different-step random walks

Therefore we compute the PMI matrix for a graph $G = \{V, E, W\}$ with the maximum walking step k as below.

$$PMI_{graph}^K(v_i, v_j) = log(\frac{P_{i,j}^K}{\sum_l P_{l,i}^K \cdot \sum_l P_{l,j}^K})$$

$$\text{where } P^K = \sum_{k=1}^{K} M^k$$

After computing the PMI matrices for G_{euc} and G_{adj}, we can use the same autoencoder framework as for PMI_{traj} to learn the topological representations from the graphs. And the remaining task is to jointly learn a representation from all autoencoders.

3.3 Jointly Learning One Representation by a Multi-view Ranking Autoencoder

After the heterogeneous data are transformed into homogeneous views through different PMI computations, we propose to use a multi-view autoencoder for jointly learning the PMI in both trajectory and graphs as described in previous sections. In detail, all views (the PMI matrices) are fed into separate encoders and share the same middle layers, which generate embedding for the AOIs. Then separate decoders take the outputs of the shared layers and reconstruct the views and minimizing all errors. The network structure is depicted in Fig. 4. For a straightforward multi-view autoencoder [18], the loss of our network could be written as a summation of all reconstruction losses:

$$\ell = (1 - \alpha - \beta) \cdot \mathbb{L}_{traj} + \alpha \cdot \mathbb{L}_{euc} + \beta \cdot \mathbb{L}_{adj} \tag{6}$$

Dynamic Ranking-Weighted Loss. In the Eq. 6, α, β are the weights of the two topological views (graphs). Rather than use static weights which require much effort in finding the optimal values and do not change during the training, we propose to dynamically change the weights according to the alignment between different views. Specifically, we expect the weight on the contextual view be correlated with the order-sensitive discordance between contextual and topological views. Below we provide the definition of such discordance:

Definition 4 (Order-sensitive Discordance). *For a given AOI v_i, an order-sensitive discordance between view A and view B happens if, the sorting of other AOIs by their similarities to v_i in view A is largely different from that in view B. In other words, AOI v_j ranks high in v_i's similarity sorted by view A, but ranks low in the sorting by view B.*

In this strategy, we introduce the inductive bias from real-world observations and domain knowledge. In detail, we observe that trajectories have different sampling density at different AOIs. Some AOIs and their neighbors are frequently visited in the trajectories. These AOIs have sufficient contextual semantics and are also consistent with the geography law. Then we want to learn more from (put more weight on) the contextual view. In contrast, some AOIs are rarely or never visited, and the learning from trajectories cannot learn a meaningful embedding from these AOIs. From the trajectory view, these AOIs cannot correctly order their relationships to other AOIs and would conflict the geographical law. In the latter case, we require more effort from the learning of graphs (higher weights on topological views) to ensure the law of geography.

Therefore, rather than use static values for α and β, we propose to use dynamic weights based on ListMLE [30], a list-wise ranking loss, computed between the graph PMI vectors and the trajectory PMI vector. If we denote x_{traj} as the reconstructed vector from trajectory view, y_{euc}, y_{adj} as the reconstructed vectors from G_{euc} and G_{adj}, $h_i(y_{euc})$ as the AOI index at the i_{th} largest value of y_{euc}, the ListMLE-based weights can be written in Eq. 7. Intuitively, α and β are large if y_{euc}, y_{adj} have a different ranking of elements from x_{traj}. In other words, the largest element in y_{euc} might be the smallest in x_{traj}. A possible example could be when both v_1 and v_2 are not visited in trajectories, the ranking of their similarity is based on a default value which could conflict with the ranking of the topological similarity learned from y_{euc} and y_{adj}. In this case, DeepMARK puts more weights on G_{euc} and G_{adj}.

$$\alpha = -\frac{1}{2} \sum_{j=1}^{n} log \frac{e^{x_{traj}(h_j(y_{euc}))}}{\sum_{k=j}^{n} e^{x_{traj}(h_k(y_{euc}))}} \tag{7}$$

$$\beta = -\frac{1}{2} \sum_{i=1}^{n} log \frac{e^{x_{traj}(h_j(y_{adj}))}}{\sum_{k=j}^{n} e^{x_{traj}(h_k(y_{adj}))}} \tag{8}$$

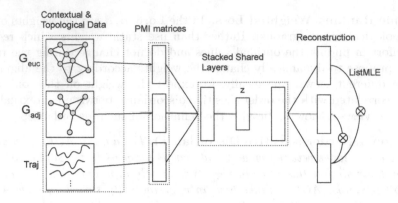

Fig. 4. DeepMARK Network Structure

4   Experiments

In this section, we evaluate the proposed model with real-world delivery datasets and task. We also include visualizations of interpretive results to help understand the model and the effect of different modules.

4.1   Dataset

We conduct the experiments on the package delivery data collected by Cainiao Network, handling more than a hundred million packages per day. In the experiment, the trajectory data is from a dispatching region from July 1, 2019 to Aug 31, 2019. The trajectories are pre-processed by removal of outliers, proper aggregation and mapped to AOIs. The original form of trajectories are GPS coordinates and timestamps, and after pre-processing, the input of this paper is sequences of AOIs and timestamps.

4.2   Experimental Settings

Adapted Baseline Algorithms Since there is no existing work on learning a contextual and topological AOI representation from trajectories and graphs, we adapted various approaches to our problem and compare them with DeepMARK. Here we list these adapted baseline approaches:

– Topological-only baselines:
  • **GeoHash** [19] is a general encoding of spatial objects. It maps the coordinates to fixed-length vectors in which common prefix usually infers close locations.
  • **Deepwalk** [20] and **Node2vec** [6] are state-of-the-art graph representation models which learn node embedding by skip-gram model from generated random walks.

- Contextual-only baseline:
  - **Word2vec** [17] is a word embedding approach that learns word representation from sentences. We adapt the model to our problem by treating trajectories as sentences and AOIs as words.
- Homogeneously integrated baselines:
  - We create two straightforward baselines **word2vec + deepwalk** and **word2vec + node2vec**, which are concatenations of word2vec embedding and graph embeddings. Thus these approaches also have the same input information as PTE (described below) and DeepMARK for a fair comparison.
- Heterogeneously integrated baseline:
  - **PTE** [24] is a heterogeneous embedding model that learns word embedding from both sentences and graphs based on a Heterogeneous Information Network Embedding (HINE) approach. We adapt this model to our problem by treating the trajectories of AOIs as the sentences of words and replacing their graphs with our graphs.

Parameter Settings For all random walk generations from graphs in Deepwalk, Node2vec and PTE, the walking length is set to 30, and walks per node is set to 30. Specifically, for Node2vec, p and q are set to 4 and 1. In DeepMARK, the revisiting ratio η for G_{euc} and G_{adj} is set to 0.1. The sliding window size is set to 20 min and the sliding offset is 10 min.

Training, Validation and Testing. Following the principle of time-related prediction, we use the latter data for testing and the earlier for training and validation using 80-20 splits.

4.3   Evaluation with ETA Prediction

We evaluate our embedding framework in the prediction of the Estimated Time of Arrival(ETA) in the last-mile package delivery task. Predicting ETA in the last-mile deliveries is challenging because the couriers usually travel by non-motor vehicles and the environments are very complex. In this task, we use deepETA [29] as the prediction model and replace the spatial representation of AOIs (by default Geohash in deepETA) with embeddings from the listed approaches to evaluate their performances.

Evaluation Metrics. We utilize Rooted Mean Squared Error (RMSE) and Mean Absolute Error (MAE) to evaluate the prediction performance on different embeddings. The smaller value indicates better performance in the prediction of ETA.

Comparison Results. In Table 1 we compare the errors of ETA prediction using deepETA with different representations. We can observe that DeepMARK has a significant advance over all other baselines, inducing up to 20% reduction of errors. Its variant DeepMARK$_{static}$ which uses fixed weights on multiple views is worse than DeepMARK but slightly better than others. We also draw the curves

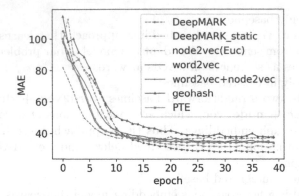

Fig. 5. Learning curves of ETA prediction by different representations

Table 1. ETA prediction performance

Model	RMSE (min)	MAE (min)
Geohash	55.22	40.98
word2vec	53.43	36.91
Deepwalk$_{euc}$	56.52	37.35
Deepwalk$_{adj}$	55.36	37.44
Node2vec$_{euc}$	54.76	37.12
Node2vec$_{adj}$	55.23	38.56
word2vec + deepwalk	54.15	37.41
word2vec + node2vec	53.86	36.53
PTE	53.17	36.52
DeepMARK$_{static}$	51.78	34.87
DeepMARK	**48.68**	**32.61**

of MAE of validation set versus the training epochs in deepETA using different representations. We can observe that the embedding by DeepMARK enables the model converge to the lowest validation error. And we can observe that PTE has a similar performance with word2vec+node2vec. Both have little control of coordinating different views, thus induce larger errors than DeepMARK (Fig. 5).

4.4 Model Interpretation

Visualization of the Effect of Joint Learning. We utilize t-SNE [16] to visualize the embeddings of AOIs by Word2vec on trajectories, Node2vec on G_{euc} and DeepMARK on both views in Fig. 6. The colors of the points are based on Geohash values. That means points in similar colors are close in the real world. We can observe that in the Word2vec result, many distant AOIs are embedded closely (light yellow points and dark blue points). The Node2vec result has a smooth color transition from yellow to blue which indicates a nicely consistency with the law of geography, but the colors are almost evenly distributed which means it does not reveal any human knowledge on the AOIs. On the contrary, in DeepMARK result, the points have some variances in colors and shapes (holes and clusters in the figure) while overall the color transition is also smooth. This reflects that DeepMARK can learn human knowledge and meanwhile maintain the law of geography.

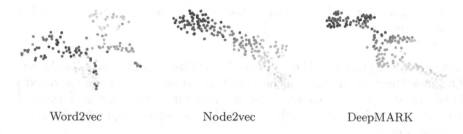

Word2vec   Node2vec   DeepMARK

Fig. 6. Embedding visualization by t-SNE

Visualization of the Changes of Ranking-Based Weights. To understand how the listMLE losses direct the training process of DeepMARK, we visualize the change of β (the ranking loss between the trajectory view and G_{adj} view) in Fig. 7. In each plot, the x-axis and y-axis are the geometric coordinates, i.e., longitude and latitude. Each dot in the figures representing an AOI and its color denotes the value of β calculated for this AOI. dark blue indicates large β and shallow green indicates small β. In Fig. 7 we show the calculated β of all AOIs at different training stages, i.e., epoch = 1, 30, 100. We can observe that: (a) The β for all AOIs have little difference at epoch 1 because of the randomness caused by initial parameters of the neural network; (b) The β of some AOIs get larger and others get smaller as the training proceeds to epoch 30; (c) A few AOIs have relatively large enough *beta* while the majority of AOIs gain low *beta* when the network is well-trained at epoch 100. Such change from epoch 0 to epoch 100 indicates the intuition behind our ranking-based weight strategy. Specifically, the ranking losses don't affect much in the reconstruction of all views at the early stages. Therefore it allows the model to have a warm start on roughly learning the representation of all views. However, when each view gets well trained and the topological views and contextual view become inconsistent in the ordering perspective, the ranking losses (α and β) start to regularize these disordering according to our inductive bias until the reconstructions and the ranking losses across different views are balanced.

5   Related Work

Point of Interest Recommendation. Lately, a few researchers spot their light on employing trajectory or (mostly) user check-in data in recommendation

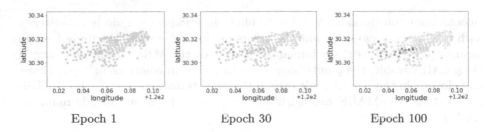

Epoch 1   Epoch 30   Epoch 100

Fig. 7. Visualize the change of ranking loss between the topological view and the contextual view

services, e.g., in [3,15,32]. These approaches utilize word2vec ideas for learning POI embeddings in the recommendation task. However, they are restricted to study embeddings for POIs which are discrete places thus not fully cover the space. And they didn't sufficiently consider the law of geography for the learned representation.

Graph Representation Learning. As diverse real-world data could be formulated as graph structures, recent researchers study how to learn a graph representation to support different prediction and recommendation tasks. Inspired by word2vec models from studies in NLP, many researchers learn the node embedding in a graph by conducting random walks in the graph and treat the walks as corpus [6,20]. In addition, some recent researches utilize deep neural networks to learn more neighboring information through an autoencoder framework [2,27].

Multi-view Representation Learning. Since real-world problems always involve different views of data, such as audio/video, image/text, multi-view representation learning attracts more attention in recent studies [14,28]. In this area, researchers usually align different views by similarities or correlations or adopt different parameter sharing strategies to learn a representation [5,18,23]. Researchers also discover representation learning from heterogeneous graphs such as in [4]. Based on the heterogeneous graph embedding approaches, a recent study [24] proposed an approach that learns from sentences as well as graphs for text representation. However, the adaption of these approaches may not perform as good in delivery tasks due to the domain-specific alignment requirements in the spatial scenario.

6   Conclusion

In this paper, we introduced DeepMARK, an innovative deep multi-view autoencoder framework which learns a representation of AOI from trajectories and graphs data. The framework learns embedding of AOI that takes the best of both contextual and topological representations, i.e., incorporates data-driven contextual information and follows the Tobler's First Law of Geography. Deep-MARK is evaluated in real-world package delivery ETA prediction and achieved a better performance than various adapted baselines.

Acknowledgments. This research has been funded in part by the USC Integrated Media Systems Center (IMSC) and Alibaba Group through Alibaba Research Fellowship Program. Any opinions, findings, and conclusions or recommendations expressed in this material are those of the authors and do not necessarily reflect the views of the sponsor.

References

1. Bengio, Y., Courville, A., Vincent, P.: Representation learning: a review and new perspectives. IEEE Trans. Pattern Anal. Mach. Intell. **35**(8), 1798–1828 (2013)
2. Cao, S., Lu, W., Xu, Q.: Deep neural networks for learning graph representations. In: Thirtieth AAAI Conference on Artificial Intelligence (2016)
3. Chang, B., Park, Y., Park, D., Kim, S., Kang, J.: Content-aware hierarchical point-of-interest embedding model for successive poi recommendation. In: IJCAI, pp. 3301–3307 (2018)
4. Chang, S., Han, W., Tang, J., Qi, G.J., Aggarwal, C.C., Huang, T.S.: Heterogeneous network embedding via deep architectures. In: Proceedings of the 21th ACM SIGKDD International Conference on Knowledge Discovery and Data Mining, pp. 119–128 (2015)
5. Dhillon, P., Foster, D.P., Ungar, L.H.: Multi-view learning of word embeddings via CCA. In: Advances in Neural Information Processing Systems, pp. 199–207 (2011)
6. Grover, A., Leskovec, J.: node2vec: scalable feature learning for networks. In: Proceedings of the 22nd ACM SIGKDD International Conference on Knowledge Discovery and Data Mining, pp. 855–864 (2016)
7. Hamilton, W., Ying, Z., Leskovec, J.: Inductive representation learning on large graphs. In: Advances in Neural Information Processing Systems, pp. 1024–1034 (2017)
8. Hamilton, W.L., Leskovec, J., Jurafsky, D.: Diachronic word embeddings reveal statistical laws of semantic change. arXiv preprint arXiv:1605.09096 (2016)
9. Hinton, G.E., McClelland, J.L., Rumelhart, D.E., et al.: Distributed Representations. Carnegie-Mellon University, Pittsburgh (1984)
10. Ke, J., et al.: Hexagon-based convolutional neural network for supply-demand forecasting of ride-sourcing services. IEEE Trans. Intell. Transp. Syst. **20**, 4160–4173 (2018)
11. Levy, O., Goldberg, Y.: Neural word embedding as implicit matrix factorization. In: Advances in Neural Information Processing Systems, pp. 2177–2185 (2014)
12. Levy, O., Goldberg, Y., Dagan, I.: Improving distributional similarity with lessons learned from word embeddings. Trans. Assoc. Comput. Linguist. **3**, 211–225 (2015)
13. Li, Y., Fu, K., Wang, Z., Shahabi, C., Ye, J., Liu, Y.: Multi-task representation learning for travel time estimation. In: Proceedings of the 24th ACM SIGKDD International Conference on Knowledge Discovery & Data Mining, pp. 1695–1704 (2018)
14. Li, Y., Yang, M., Zhang, Z.: A survey of multi-view representation learning. IEEE Trans. Knowl. Data Eng. **31**(10), 1863–1883 (2018)
15. Liu, X., Liu, Y., Li, X.: Exploring the context of locations for personalized location recommendations. In: IJCAI, pp. 1188–1194 (2016)
16. van der Maaten, L., Hinton, G.: Visualizing data using t-SNE. J. Mach. Learn. Res. **9**(Nov), 2579–2605 (2008)

17. Mikolov, T., Sutskever, I., Chen, K., Corrado, G.S., Dean, J.: Distributed representations of words and phrases and their compositionality. In: Advances in Neural Information Processing Systems, pp. 3111–3119 (2013)
18. Ngiam, J., Khosla, A., Kim, M., Nam, J., Lee, H., Ng, A.Y.: Multimodal deep learning. In: ICML, pp. 689–696 (2011)
19. Niemeyer, G.: Geohash (2008)
20. Perozzi, B., Al-Rfou, R., Skiena, S.: DeepWalk: online learning of social representations. In: Proceedings of the 20th ACM SIGKDD International Conference on Knowledge Discovery and Data Mining, pp. 701–710 (2014)
21. Sahlgren, M.: The distributional hypothesis. Ital. J. Disabil. Stud. **20**, 33–53 (2008)
22. Shuman, D.I., Narang, S.K., Frossard, P., Ortega, A., Vandergheynst, P.: The emerging field of signal processing on graphs: Extending high-dimensional data analysis to networks and other irregular domains. IEEE Signal Process. Mag. **30**(3), 83–98 (2013)
23. Su, H., Maji, S., Kalogerakis, E., Learned-Miller, E.: Multi-view convolutional neural networks for 3D shape recognition. In: Proceedings of the IEEE International Conference on Computer Vision, pp. 945–953 (2015)
24. Tang, J., Qu, M., Mei, Q.: PTE: predictive text embedding through large-scale heterogeneous text networks. In: Proceedings of the 21th ACM SIGKDD International Conference on Knowledge Discovery and Data Mining, pp. 1165–1174 (2015)
25. Teng, S.H.: Scalable algorithms for data and network analysis. Found. Trends Theor. Comput. Sci. **12**(1–2), 1–274 (2016)
26. Tobler, W.R.: A computer movie simulating urban growth in the detroit region. Econ. Geogr. **46**, 234–240 (1970)
27. Wang, D., Cui, P., Zhu, W.: Structural deep network embedding. In: Proceedings of the 22nd ACM SIGKDD International Conference on Knowledge Discovery and Data Mining, pp. 1225–1234 (2016)
28. Wang, W., Arora, R., Livescu, K., Bilmes, J.: On deep multi-view representation learning. In: International Conference on Machine Learning, pp. 1083–1092 (2015)
29. Wu, F., Wu, L.: DeepETA: a spatial-temporal sequential neural network model for estimating time of arrival in package delivery system. In: Proceedings of the AAAI Conference on Artificial Intelligence, vol. 33, pp. 774–781 (2019)
30. Xia, F., Liu, T.Y., Wang, J., Zhang, W., Li, H.: Listwise approach to learning to rank: theory and algorithm. In: Proceedings of the 25th International Conference on Machine Learning, pp. 1192–1199. ACM (2008)
31. Yao, D., Zhang, C., Zhu, Z., Huang, J., Bi, J.: Trajectory clustering via deep representation learning. In: 2017 International Joint Conference on Neural Networks (IJCNN), pp. 3880–3887. IEEE (2017)
32. Zhang, J.D., Chow, C.Y., Li, Y.: iGeoRec: a personalized and efficient geographical location recommendation framework. IEEE Trans. Serv. Comput. **8**(5), 701–714 (2014)

Strategic and Crowd-Aware Itinerary Recommendation

Junhua Liu[1,2]($\boxtimes$), Kristin L. Wood[1,3], and Kwan Hui Lim[1]

[1] Singapore University of Technology and Design, Singapore, Singapore
{kristinwood,kwanhui_lim}@sutd.edu.sg
[2] Forth AI, Singapore, Singapore
j@forth.ai
[3] University of Colorado Denver, Denver, USA

Abstract. There is a rapidly growing demand for itinerary planning in tourism but this task remains complex and difficult, especially when considering the need to optimize for queuing time and crowd levels for multiple users. This difficulty is further complicated by the large number of parameters involved, i.e., attraction popularity, queuing time, walking time, operating hours, etc. Many recent works propose solutions based on the single-person perspective, but otherwise do not address real-world problems resulting from natural crowd behavior, such as the Selfish Routing problem, which describes the consequence of ineffective network and sub-optimal social outcome by leaving agents to decide freely. In this work, we propose the Strategic and Crowd-Aware Itinerary Recommendation (SCAIR) algorithm which optimizes social welfare in real-world situations. We formulate the strategy of route recommendation as Markov chains which enables our simulations to be carried out in poly-time. We then evaluate our proposed algorithm against various competitive and realistic baselines using a theme park dataset. Our simulation results highlight the existence of the Selfish Routing problem and show that SCAIR outperforms the baselines in handling this issue.

Keywords: Tour recommendations · Trip planning · Recommendation systems · Sequence modelling

1 Introduction

Itinerary recommendation has seen a rapid growth in recent years due to its importance in various domains and applications, such as in planning tour itineraries for tourism purposes. Itinerary recommendation and planning is especially complex and challenging where it involves multiple points of interest (POIs), which have varying levels of popularity and crowdedness. For instance, while visiting a theme park, the visitor's route can include POIs such as roller coasters, water rides, and other attractions or events. The itinerary recommendation problem can be modelled as an utility optimization problem that maximizes

© Springer Nature Switzerland AG 2021
Y. Dong et al. (Eds.): ECML PKDD 2020, LNAI 12460, pp. 69–85, 2021.
https://doi.org/10.1007/978-3-030-67667-4_5

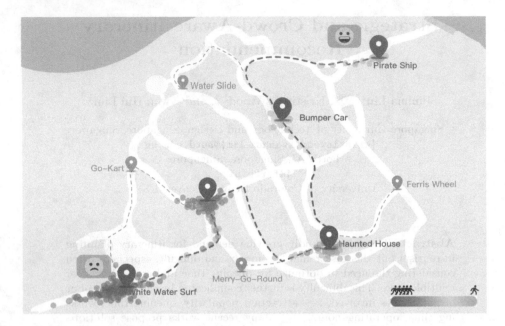

Fig. 1. Existing itinerary recommendation problems leverage data-driven approaches with a single-person perspective. In real life, this will result in the Selfish Routing problem, where leaving all agents free to act according to their own interests results in a sub-optimal social welfare. As illustrated, the recommended path performed sub-optimally, where the closer the POIs are to the start of the route, the more crowded they would be, while leaving all other POIs (in grey) not utilized.

the number of facilities visited and the popularity of these facilities[1], while minimizing the queuing time and travel time from one facility to the other. Facilities in a theme park come with different properties such as popularity, duration, location and a dynamic queuing time. Visitors are often constrained by a time budget that limits the number of facilities one could visit in a single trip. While many algorithms have been developed [4,7,18,26,40], they mostly aim to recommend itineraries for individual travellers, whereas a real-life itinerary is also affected by the actions of other travellers, such as lengthening the queuing time at a facility.

Many works focus on constructing a single optimal path for the individual traveller, solely based on historical data. While this approach works for the individual traveller, it leads to a sub-optimal itinerary when all travellers are given the same recommendation. Consider a recommender system that recommends an itinerary comprising the most popular POIs with the least queuing time based on such historical data. In a real-life scenario with multiple travellers, all travellers will follow the same recommended itinerary with the shortest historical queuing time, resulting in an expected queuing time that would grow with each

---

[1] The terms "POIs", "attractions" and "facilities" are used interchangeably.

new arrival, as illustrated in Fig. 1. In other words, the later an agent[2] arrives to the system, the longer her expected queuing time will be. As a result, the social welfare or the collective utility of all agents has failed to be optimized. As an individual traveller, it is extremely difficult for an agent to gain knowledge of the system state, i.e., the people who are visiting the park and their respective paths. As a result, letting the agent find an optimal strategy that maximizes her expected utility is unrealistic without considering the actions of other agents.

To address this problem, we propose the Strategic and Crowd-Aware Itinerary Recommendation (SCAIR) algorithm, which is a recommender system that maintains an internal information of all recommended routes and leverages on this internal information to make routing recommendations to its arriving agents. In other words, we take a game-theoretic approach to address the problem and formulate a crowd-aware itinerary recommendation algorithm having in mind the Selfish Routing problem [32], i.e., allowing agents act freely results in a sub-optimal social welfare. Concretely, we model the itinerary recommendation problem into a strategic game [25], where the system, i.e., a theme park, defines a set of allocation rules to allocate route to each player in the system, instead of leaving the agents a high degree of freedom to choose their own path. Experiments show that our approach is effective in optimizing utility of all agents.

2 Main Contributions

Our main contributions are as follows:

- We introduce and formulate the crowd-aware itinerary recommendation problem as a social welfare optimization problem that considers the actions of multiple travellers, in contrast to existing works that only consider the perspective of the single traveller (Sect. 4).
- To address this crowd-aware itinerary recommendation problem, we propose the SCAIR algorithm which utilizes a game-theoretic approach to recommend itineraries for multiple travellers (Sect. 5).
- Using a theme park dataset, we compare our SCAIR algorithm against various competitive and realistic baselines and show how SCAIR outperforms these baselines with a large reduction in queuing times (Sects. 6 and 7).

For the rest of the paper, Sect. 3 discusses related works and how our research differs from these earlier works. Section 8 summarizes this paper and introduces some future research directions. Next, we introduce the problem formulation of this crowd-aware itinerary recommendation problem.

3 Related Work

There have been numerous works that aim to solve the itinerary recommendation problem and other related tourism recommendation problems. In this literature review, we cover various related work from the Operations Research and Information Retrieval communities.

---

[2] We use the terms "travellers", "visitors" and "agents" interchangeably.

3.1 Itinerary and Tourism-Related Recommendation

Many works have modelled the itinerary recommendation problem as a variant of the Orienteering problem [4,7,17]. In the Orienteering problem [33], the recommendation aims to optimize social welfare with a global reward such as popularity, with respect to budget constraints such as travel time or distance among attractions in an itinerary. This approach typically does not take into consideration the trade-off between the duration in a facility and its popularity, which may contribute substantially to the global profit. For more details on the Orienteering problem and its variants, we refer the interested reader to the various survey papers on this topic [8,34].

There have been various approaches proposed to solve this itinerary recommendation problem based on variants of the Orienteering problem. For example, Zhang et al. proposed the use of heuristic approximation to solve a variant of this problem that involves POI opening hours and incorporating uncertainty in different travel modes [40,41]. Others have used variations of the Ant Colony System to solve the itinerary recommendation problem [19], and variants that incorporate the additional consideration of crowd levels [36]. Another approach is to solve this itinerary recommendation problem using integer programming, to optimize for user interests based on the amount of time tourists spend at POIs [18]. Similarly, there are also various works that aim to recommend routes that are deemed more attractive to tourists [10,35,42]. In the Information Retrieval community, a popular research topic is on item recommendations and this problem can be easily extended for recommending POIs. For example, many works have used matrix factorization or collaborative filtering approaches for finding a ranked list of top locations, which is known as top-k POI recommendation [14,15,37–39].

3.2 Discussion

These earlier works face a major limitation where the recommendation algorithms are constructed based on a single person's perspective. Despite some recent works exploring the effects of group or crowd behavior [1,6,9,11,36], the algorithms treat the system as a static environment where properties such as queuing time only depend on historical data. Simulating an optimal path in such a static environment has a natural disadvantage where self-interested agents prioritize personal objective functions which may result in ineffective social welfare. For instance, when everyone visiting the theme park follow the same recommended path, the queuing time will increase dramatically, and the optimality of such recommendation algorithms will then collapse. Roughgarden's work [32] discusses this problem extensively, defined as Selfish Routing problem, where giving agents the freedom to act according to their own interests results in a sub-optimal social welfare.

The Selfish Routing problem was studied in the area of Game Theory and Mechanism Design [12,28,32]. The inefficiency of achieving the optimize natural objective is quantitatively measured by Price of Anarchy, which was first defined as the ratio between the worst-case Nash equilibrium and the optimum sum of payoffs in game-theoretic environments [12,28]. Braess's Paradox for traffic flow [3] describe the phenomenon where adding a new link to a transportation network might not improve the operation of the system, in the sense of reducing the total vehicle-minutes of travel in the system [30]. To break out from this phenomenon, a system operator can manually interfere with or change agents' actions to provide policies or economic incentives with well designed strategies. Our proposed game-theoretic, dynamic itinerary recommendation algorithm in this paper is an instance of such strategy.

To address these limitations, we propose the Strategic and Crowd-Aware Itinerary Recommendation (SCAIR) algorithm to address the ineffectiveness of welfare optimization due to the lack of centralized control [31]. The proposed recommendation algorithm takes into consideration all visits in an itinerary planning scenario (e.g., a theme park), and makes recommendations to the next visitor with the knowledge of other visitors' paths in the park. Furthermore, the queuing time at all facilities at a certain hour is dynamically modelled according to the expected number of visitors in the same place at the same hour.

4   Crowd-Aware Itinerary Recommendation Problem

In this section, we first give an overview of our general approach, followed by formulating our crowd-aware itinerary recommendation problem, before showing the NP-hardness of this proposed problem.

4.1   General Approach

In this work, we view the itinerary recommendation problem from a global perspective and formulate it as a strategic game where the system designs and distributes the optimal path to every agent on arrival, based on the existing agents in the system and their respective paths. In the context of a theme park, one can think of this entity as the theme park operator that gives out the recommendation of various itineraries to visit the attractions to different visitors. We propose the SCAIR algorithm that dynamically recommends routes taking into consideration all existing agents in the system.

The crowd-aware itinerary recommendation problem aims to maximize the sum of all agents' utility in the system. This turns out to be a social welfare optimization problem that is NP-hard [24]. Furthermore, simulating or solving the problem is also empirically challenging. One has to take into consideration the entire history of existing visitors results in exponential space-complexity with respect to the number of agents, and exponential time-complexity with respect to the number of facilities in a path.

To overcome these challenges, we propose a simplified version which models the recommendation problem as a finite markov chains and is known to be in NC [29] and decidable in poly-logarithmic time [2]. The simplified model makes an assumption that each decision embeds information of the immediate last decision and the model as a result is able to provide a snapshot of the entire history. Next we will discuss the formulation of the problem.

4.2 Problem Formulation

We formulate the crowd-aware itinerary recommendation problem to be a finite markov chain and impose constraints such as (1) fixing the starting point, (2) setting a time budget for the path, and (3) limiting the distance between two stations. These constraints reflect real-life considerations closely, such as fixed starting point near the entrance; visitors having limited time to tour; and dissatisfaction arising with long walking distance among facilities.

Concretely, we model the theme park comprising numerous tourist attractions as a fully connected graph $G(F, C)$, where $F = \{f_1, ..., f_n\}$ is the collection of n facilities in the system, and $C = [c_{ij}]$ is the set of connections from f_i to f_j. Each connection c_x is associated with the properties of distance $Dist(c_{ij})$ and travel time $Trav(c_{ij})$ in minutes. Each facility f_x is associated with a set of properties including coordinates $(lat_x, long_x)$, duration of visit $Dur(f_x)$ in minutes, capacity $Cap(f_x)$ and popularity $Pop(f_x)$.

We formulate the agents' visits as m states $S = \{s_1, ..., s_m\}$, where each state s_x is associated with a feasible path $p_x = [f_1^{(x)}, ..., f_{n_x}^{(x)}]$ with n facilities $[f_1^{(x)}, ..., f^{(x_n)}]$. The total time TT_x of path p_x is defined as:

$$TT_x = \sum_{i=1}^{n_x} Dur(f_i^{(x)}) + \sum_{i=1}^{n_x-1} Trav(c_{i,i+1}) \tag{1}$$

We model the utility of the agents with respect to the popularity of each facility visit normalized by the expected waiting time at each facility. Our assumption is that higher popularity of a facility indicates a greater attractiveness to visitors, subjected to how long they have to wait for that facility. Concretely, we define the utility function U_x for path x with n nodes as follows:

$$U_x = \frac{\sum_{f \in p_j} Pop(f)}{Q(p_x | p_{x-1})} \tag{2}$$

where $Q(p_x | p_{x-1})$ is the expected queuing time at path p_x given p_{x-1}, and $Pop(p_x)$ is the sum of popularity of all facilities in the path. The path's expected queuing time $Q(p_x | p_{x-1})$ is calculated by summing up the queuing time at all facilities:

$$Q(f_i) = \frac{1}{Cap(f_y)} Dur(f_y) \delta(f_{y,h}^{(x)} = f_{y,h}^{(x-1)}) \tag{3}$$

where $\delta(f_{y,h}^{(x)} = f_{y,h}^{(x-1)}) = 1$ if the facility appears to overlap between paths p_x and p_{x-1} within the same hour h. Capacity $Cap(f_x)$ is set to be a constant for simplicity. Finally, the transition matrix T is defined as:

$$T_{ij} = \frac{\sum_{f \in p_j} Pop(f)}{Q(p_j | p_{j-1=i})} \tag{4}$$

The transition matrix is then normalized by:

$$T_{ij} := \frac{T_{ij}}{\sum_j T_{ij}} \qquad (5)$$

The set of feasible paths, i.e., total search space, is determined by solving an optimization problem, as follows:

$$\text{maximize} \quad TT_x = \sum_{i=1}^{n_x} Dur(f_i) + \sum_{j=1}^{n_x-1} Trav(c_{j,j+1}) \qquad (6)$$

$$\text{subject to} \quad Dist(c_{j,j+1}) \leq s, \ TT_x \leq t$$

for n facilities in the path, with a constant time budget t.

Finally, we model the strategic itinerary recommendation problem as a social welfare optimization problem as follows:

$$\text{maximize} \quad W = \sum_x U_x p_x \qquad (7)$$

$$\text{subject to} \quad \sum_x TT_x \leq t, \ x \in \{1, ..., n\}$$

for n agents and time budget t.

4.3 Proof of NP-Hardness

We further investigate the NP-hardness of various sub-problems and show the respective proofs in this section.

Theorem 1. *The path finding problem defined in Eq. 6 is NP-hard.*

Proof. We prove the NP-hardness of the path finding problem by reduction from the 0-1 Knapsack problem which is known to be NP-hard [23]. Recall that the 0-1 Knapsack problem is a decision problem as follows:

$$\text{maximize} \quad z = \sum_i p_i x_i$$

$$\text{subject to} \quad \sum_i w_i x_i \leq c \qquad (8)$$

$$x_i \in \{0,1\}, \ i \in \{1, ..., n\}$$

for n available items where x_i represents the decision of packing item i, p_i is the profit of packing item i, w_i is the weight of item i, c is the capacity of the knapsack.

Intuitively, the path finding problem is a decision problem of allocating a set of facilities into a path with a capacity of time budget, where each facility comes with properties of profit and duration time.

Formally, we transform the minimization problem in Eq. 6 to an equivalent maximization problem. Concretely, the binary variable $f_i \in \{0, 1\}$ is included, where $f_i = 1$ if f_i is in path p_x, and 0 if otherwise. Furthermore, we define the profit of facility f_i as $p_i = -Dur(f_i)$ and set the travel time $Trav(c_{ij})$ to be a constant. Finally, the distance constant cap s is set to be infinity. The new problem formulation is represented as follows:

$$\text{maximize} \quad T'_{path} = \sum_i p_i f_i$$

$$\text{subject to} \quad \sum_i Dur(f_i) f_i \leq t \tag{9}$$

$$f_i \in \{0, 1\}, \quad i \in \{1, ..., n\}$$

In this formulation, a path is equivalent to the knapsack in the 0-1 Knapsack problem, where each facility has its profit of $Pop(p_i)$, and its cost of $Dur(f_i)$ that is equivalent to the profit and weight of an item respectively. The maximization problem is subjected to a constant time budget t which is equivalent to the capacity c in a 0-1 Knapsack problem.

As a result, for any instance of the 0-1 Knapsack problem (i.e. item allocation decisions), we are able to find an equivalent instance of the path finding problem (i.e. a facility allocation decisions). Therefore, a solution in the path finding problem yields an equivalent solution to the 0-1 Knapsack decision problem. As such, we have completed the proof of NP-hardness for our path finding problem to be NP-hard. □

Theorem 2. *The social welfare optimization problem defined in Eq. 7 is NP-hard.*

Proof. Once again, we prove the NP-hardness of our welfare optimization problem by reduction from the 0-1 Knapsack problem.

In Eq. 7, the set of paths assigned to agents in the system is equivalent to the set of items in 0-1 Knapsack problem; each path has its utility and total time, which are equivalent to the profit and weight of an item respectively; the maximization problem is subjected to a constant time budget t which is equivalent to the capacity c in a 0-1 Knapsack problem.

As a result, for any instance of the 0-1 Knapsack problem decisions, we are able to find an equivalent instance of a path assignment decision that yields a solution to the original Knapsack decision problem. As such, we conclude the proof of NP-hardness and have shown that our welfare recommendation problem is NP-hard. □

Next, we describe our proposed SCAIR algorithm for solving this crowd-aware itinerary recommendation problem.

5 Strategic and Crowd-Aware Itinerary Recommendation (SCAIR) Algorithm

In this section, we describe our proposed SCAIR algorithm, which comprises the main steps of finding feasible paths, generating a transition matrix and simulating traveller visits.

5.1 Finding Feasible Paths

Algorithm 1 shows the pseudocode of our path finding algorithm based on a breadth-first strategy. The input is a graph $G(F, C)$ that represent a theme park with the set of facilities F and connections C, time budget TT_{max}, and distance limit between two facilities $Dist_{max}$. This algorithm then generates and returns a collection of feasible paths, $Paths$, with respect to the provided input graph $G(F, C)$.

We iterate the collection of intermediate $Paths$, and call the $FindViableFacilities$ function to find viable facilities, where $f_{-1}^{(i)}$ is the last facility of the path, and $Dist_{max}$ is the maximum distance an agent wants to travel from one facility to another. We set the parameters of total time budget $T_{max} < 8\,$h and maximum allowed distance between two facilities $Dist_{max}(f_{current}, f_{next}) < 200\,$m. If there are no available facility that meets the distance constraint and the path has available time budget remaining, the agent proceeds to the next nearest facility. We also do not allow an agent to revisit a facility in the same trip.

Line 2. The algorithm starts with constructing a 2-dimensional array, where each row represent a path as a sequence of facilities visited. We then conduct a breadth-first search (line 3 to 25), starting with the first row with an element of the initial facility, i.e., the entrance of a theme park.

Line 6 to 11. If the algorithm is unable to find a facility within the feasible range, it will instead find the nearest facility that is not yet visited, and assign the new path into the $Paths$ collection if two conditions are met, namely (1) the new path's total time is within the visitor's time budget TT_{max}, and (2) no identical path exists in the $Paths$ collection. Eventually we remove the path the iteration started off.

Line 13 to 20. If the algorithm manages to find a set of viable facilities, it will then iterate through the set and execute a similar selection process.

Line 22 to 24. The algorithm breaks out from the infinite loop when any one of two conditions is met, namely (1) all paths in the $Paths$ collection have maximized its time budget i.e. any additional facility will make the total time of a path to be larger than the visitor's time budget; or (2) every path has included all available facilities.

Algorithm 1. SCAIR - FindFeasiblePaths()

Data: $f_i \in F, c_{ij} \in C, TT_{max}, Dist_{max}, f_0$
Result: *Paths*: the set of feasible paths

1 **begin**
2      $Paths = [[f_0]]$;
3   **while** *True* **do**
4   **for** $path_i \in Paths$ **do**
5              $VF = FindViableFacilities(f_{-1}^{(i)}, Dist_{max})$;
6   **if** $len(VF) == 0$ **then**
7                  $path_x = path_i + [FindNextNearest(f_{(-1)}^{(i)})]$;
8   **if** $TT_x < TT_{max}$ and $path_x \notin Paths$ **then**
9                      $Paths + = [path_x]$;
10                     $Paths.pop(path_i)$
11   **end**
12   **end**
13   **foreach** $vf \in VF$ **do**
14                 $path_x = path_i + [vf]$;
15   **if** $TT_x < TT_{max}$ and $path_x \notin Paths$ **then**
16                     $Paths + = [path_x]$;
17   **end**
18   **end**
19             $Paths.pop(path_i)$;
20   **end**
21   **if** $AllPathsMaxTimeBudget(Paths)$ **or**
            $AllPathsReachFullLength(Paths)$ **then**
22   break;
23   **end**
24   **end**
25 **end**

5.2 Transition Matrix

Using the set of feasible paths found (Sect. 5.1), we now construct a Transition Matrix T by calculating T_{ij} as the costs of taking path j given path $j - 1 = i$. The output of $FindCost()$ function varies based on the arrival interval λ because it affects the expected time of arrival for each facilities at $path_j$, which leads to different occurrence of overlapping facilities between $path_i$ and $path_j$.

5.3 Simulation

Algorithm 2 shows an overview of the simulation procedure, which involves iterating through the visit data of theme parks *Parks*, a list of time budgets *TimeBudgets*, and an array of arrival intervals *ArrivalIntervals*.

Line 6 to line 13. For each step, the $FindFeasiblePaths()$ function finds the set of feasible paths which enables the $ConstructTM()$ function to construct the

---

Algorithm 2. SCAIR - Simulate()

Data: $Parks, TimeBudgets, ArrivalIntervals$
Result: Export simulation data to a CSV file

1 **begin**
2    $Results = \{\}$;
3   **for** $Park \in Parks$ **do**
4   **for** $SimTime \in TimeBudgets$ **do**
5   **for** $\lambda \in ArrivalIntervals$ **do**
6             $Paths = FindFeasiblePaths(Park, SimTime)$;
7             $T = ConstructTM(Park, Paths)$;
8             $Qt, Pop, Utility = RunSimulation(Paths, \lambda, SimTime)$;
9             $Update(Results, [Qt, Pop, Utility])$;
10   **end**
11   **end**
12   **end**
13    $ExportCsvFromDict(Results)$;
14 **end**

---

transition matrix, with input parameters namely park data $Park$ and simulation time $SimTime$. The $RunSimulation()$ function then runs the simulation to find the total queuing time Qt, average sum of popularity among all facilities visited Pop, and the expected utility $Utility$ which is calculated as a function of Qt and Pop. Finally, we update the $Results$ dictionary (Line 9) and export the experimental data into CSV files (line 13) after completing the simulations.

6 Experimental Setup

In this section, we describe our dataset, evaluation process and baselines.

6.1 Dataset

We conduct our experiments using a publicly available theme park dataset from [16]. This dataset is based on more than 655k geo-tagged photos from Flickr and is the first that includes the queuing time distribution of attractions in various Disney theme parks in the United States. In our work, we perform our experiments and evaluation using the dataset of user visits in Epcot Theme Park and Disney Hollywood Studio.

6.2 Experimental Parameters

As previously described in Sect. 5.2, we denote the arrival interval of agents as λ which indicates the time between the arrival of two agents, measured in minute. In this work, λ is set to be a constant for simplicity. For a robust evaluation, we perform our evaluation using multiple values of the evaluation parameters, namely arrival interval $\lambda \in \{0.01, ...0.09, 0.1, ..., 1.0\}$, and simulation time T between 60 and 360 min in 30 min intervals (i.e. $T \in \{60, 90, ..., 360\}$).

6.3   Evaluation and Baselines

We compare our proposed SCAIR algorithm against three competitive and realistic baselines. The first two algorithms are based on intuitive strategies commonly used by visitors in real-life [18], while the third is a greedy algorithm used in [41]. In summary, the three baseline algorithms are:

1. Distance Optimization (denoted as $DisOp$) [18]. An iterative algorithm where agents always choose the facility with the shortest distance to the currently chosen one.
2. Popularity Optimization (denoted as $PopOp$) [18]. An iterative algorithm where agents always choose the next most popular facility that satisfies the specified distance constraint from the currently chosen one.
3. Popularity over Distance Optimization (denoted as $PodOp$) [41]. An iterative greedy approach that models utility as the popularity of the POI normalized by the distance from the current one, and iteratively chooses the POI with the highest utility.

Similar to many itinerary recommendation works [16,17], we adopt the following evaluation metrics:

1. Average Popularity of Itinerary (denoted as $AvgPop$). Defined as the average popularity of all attractions recommended in the itineraries.
2. Expected Queuing Time per Visitor (denoted as $AvgQt$). Defined as the average queuing time that each visitor spends waiting for attractions in the recommended itinerary.
3. Expected Utility (denoted as Uty). Defined as the average utility score for all users based on the recommended itineraries.

7   Results and Discussion

Figure 2 shows the experimental results of our proposed SCAIR algorithm compared to the three baseline algorithms. The x-axis indicates the time budget of visits and the y-axis indicates the queuing time, popularity and utility. To examine the effects of different user arrival frequency, multiple experiments are conducted based on different arrival intervals λ, i.e., from 0.01 to 0.1 with a step size of 0.01, and from 0.1 to 1.0 with a step size of 0.1. The values in the graph are averaged across all λ.

7.1   Queuing Time

In relative terms, we observe that SCAIR outperforms the baselines for both the queuing time and utility in both theme parks. SCAIR is able to maintain a low queuing time with different time budgets, while the baseline's queuing time increases with the growth of time budget. The observation is consistent for both theme parks. Table 1 shows the ratio of queuing time and time budget of visitors. SCAIR produces a queuing time ratio that is 78.9% to 93.4% shorter than that of the baselines, across both DisHolly and Epcot theme parks.

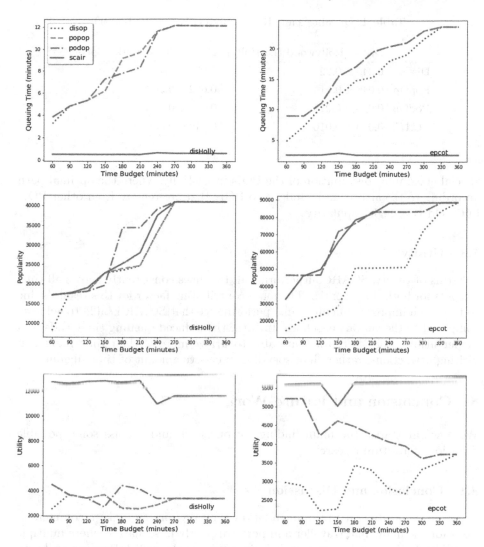

Fig. 2. The plots show how the queuing times, popularity and utility change with respect to simulation time T, over two theme parks data (Disney Hollywood and Epcot Theme Park). We observe that: (1) SCAIR's queuing time is consistently and significantly lower than the baselines. (2) Popularity of of all 4 algorithms perform similarly for DisHolly, while DisOp performs significantly poorer than the others for Epcot. (3) SCAIR's utility consistently outperforms the baselines.

7.2   Popularity

All four algorithms perform similarly for the DisHolly dataset, while PopOp, PodOp and SCAIR remain similar but outperform DisOp for the Epcot dataset. We observe that PodOp achieves a relatively high Popularity when time budget is equal to 180 min and 210 min. We observe that this phenomena is due to the

Table 1. Queuing Time Ratio (Smaller values are better)

	Disney Hollywood (DisHolly)	Epcot Theme Park (Epcot)
DisOp	0.045 ± 0.221	0.076 ± 0.414
PopOp	0.046 ± 0.215	0.092 ± 0.368
PodOp	0.045 ± 0.211	0.092 ± 0.368
SCAIR	0.003 ± 0.010	0.016 ± 0.006

special geographic distribution of the POIs in DisHolly, where the optimal path according to the algorithm includes two POIs that are remote from other POIs but yield very high popularity.

7.3   Utility

In terms of Utility, SCAIR outperforms all baselines consistently across all time budgets for both theme parks. The main contributing factor for this result is due to the much improved queuing time performance that SCAIR is able to achieve, compared to the various baselines. In turn, the reduced queuing time leads to a higher utility score as tourists are able to utilize more of their time budget in visiting attractions, rather than spending excessive amount of time queuing.

8   Conclusion and Future Work

We now summarize the main findings of our work and discuss some possible directions for future research.

8.1   Conclusion and Discussion

Prior works on itinerary recommendation typically aim to make recommendations for the individual traveller and perform poorly in scenarios where multiple travellers use the same recommended itinerary, i.e., the Selfish Routing problem. In this paper, we introduced the crowd-aware itinerary recommendation problem and highlighted this Selfish Routing problem where all self-interested agents aim to maximize their own utility which result in sub-optimal social welfare. For example, when all travellers are recommended the same POIs with a short queuing time based on historical data, those POIs then become congested and suffer from a long queuing time.

To address this problem, we proposed the SCAIR algorithm that takes into consideration crowd behavior and addresses the NP-hard Social Welfare Optimization problem with a finite markov chains, which is in NC and can be solved in poly-logarithmic time. We performed a series of experiments using a theme park dataset. Experimental results show that SCAIR outperforms various competitive baselines in terms of a reduced queuing time and improved utility, while offering similar levels of popularity scores. An investigation into the effects of user arrival rates show that the performance of SCAIR remains competitive, compared to the various baselines, regardless of the arrival rates.

8.2 Future Work

We intend to further investigate models that further simulate real-life situations. For instance, we can also locate the entrances and exits of the theme parks to initialize and end paths; we could also use soft-max instead of one-hot to simulate the choices of paths which simulates the probabilistic decisions visitors make in real-life. We will also attempt to improve the formulation of the multi-objective optimization problem, such as by assessing the Pareto efficiency of the two objectives. It is also worthwhile to look into modifying our strategic recommendation algorithm and applying them to other game-theoretic environments, such as knowledge acquisition [20], crisis management [22] and career path planning [5,21]. Finally, we intend to look further into prior works, such as [13,27] to explore Machine Learning approaches in solving time-variant path planning problems and attempt to enhance the solution and simulation performance.

Acknowledgements. This research is funded in part by the Singapore University of Technology and Design under grants SRG-ISTD-2018-140 and PIE-SGP-AI-2020-01.

References

1. Anagnostopoulos, A., Atassi, R., Becchetti, L., Fazzone, A., Silvestri, F.: Tour recommendation for groups. Data Min. Knowl. Disc. **31**(5), 1157–1188 (2017)
2. Arora, S., Barak, B.: Computational Complexity: A Modern Approach. Cambridge University Press, Cambridge (2009)
3. Braess, D.: Über ein paradoxon aus der verkehrsplanung, pp. 0042–0573. Physica-Verlag (1968). https://doi.org/10.1007/BF01918335
4. Choudhury, M.D., Feldman, M., Amer-Yahia, S., Golbandi, N., Lempel, R., Yu, C.: Automatic construction of travel itineraries using social breadcrumbs. In: Proceedings of HT 2010, pp. 35–44 (2010)
5. Dave, V.S., Zhang, B., Al Hasan, M., AlJadda, K., Korayem, M.: A combined representation learning approach for better job and skill recommendation. In: Proceedings of CIKM 2018, pp. 1997–2005. ACM (2018)
6. Garcia, I., Sebastia, L., Onaindia, E.: On the design of individual and group recommender systems for tourism. Expert Syst. Appl. **38**(6), 7683–7692 (2011)
7. Gionis, A., Lappas, T., Pelechrinis, K., Terzi, E.: Customized tour recommendations in urban areas. In: Proceedings of WSDM 2014, pp. 313–322 (2014)
8. Gunawan, A., Lau, H.C., Vansteenwegen, P.: Orienteering problem: a survey of recent variants, solution approaches and applications. Eur. J. Oper. Res. **255**(2), 315–332 (2016)
9. Gunawan, A., Yuan, Z., Lau, H.C.: A mathematical model and metaheuristics for time dependent orienteering problem. In: Proceedings of PATAT 2014, pp. 202–217 (2014)

10. Herzog, D., Sikander, S., Worndl, W.: Integrating route attractiveness attributes into tourist trip recommendations. In: Companion Proceedings of WWW 2019, pp. 96–101 (2019)

11. Hu, F., Huang, X., Gao, X., Chen, G.: AGREE: attention-based tour group recommendation with multi-modal data. In: Li, G., Yang, J., Gama, J., Natwichai, J., Tong, Y. (eds.) DASFAA 2019. LNCS, vol. 11448, pp. 314–318. Springer, Cham (2019). https://doi.org/10.1007/978-3-030-18590-9_36

12. Koutsoupias, E., Papadimitriou, C.: Worst-case equilibria. In: Meinel, C., Tison, S. (eds.) STACS 1999. LNCS, vol. 1563, pp. 404–413. Springer, Heidelberg (1999). https://doi.org/10.1007/3-540-49116-3_38

13. Lees-Miller, J., Anderson, F., Hoehn, B., Greiner, R.: Does Wikipedia information help Netflix predictions? In: Proceedings of ICMLA 2008, pp. 337–343. IEEE (2008)

14. Leung, K.W.T., Lee, D.L., Lee, W.C.: CLR: a collaborative location recommendation framework based on co-clustering. In: Proceedings of SIGIR 2011, pp. 305–314 (2011)

15. Li, X., Cong, G., Li, X.L., Pham, T.A.N., Krishnaswamy, S.: Rank-GeoFM: a ranking based geographical factorization method for point of interest recommendation. In: Proceedings of SIGIR 2015, pp. 433–442 (2015)

16. Lim, K.H., Chan, J., Karunasekera, S., Leckie, C.: Personalized itinerary recommendation with queuing time awareness. In: Proceedings of SIGIR 2017, pp. 325–334. ACM (2017)

17. Lim, K.H., Chan, J., Karunasekera, S., Leckie, C.: Tour recommendation and trip planning using location-based social media: a survey. Knowl. Inf. Syst. 60(3), 1247–1275 (2019)

18. Lim, K.H., Chan, J., Leckie, C., Karunasekera, S.: Personalized trip recommendation for tourists based on user interests, points of interest visit durations and visit recency. Knowl. Inf. Syst. 54(2), 375–406 (2018)

19. Lim, K.H., et al.: PersTour: a personalized tour recommendation and planning system. In: Extended Proceedings of HT 2016 (2016)

20. Liu, J., Loh, L., Ng, E., Chen, Y., Wood, K.L., Lim, K.H.: Self-evolving adaptive learning for personalized education. In: Conference Companion Publication of the 2020 on Computer Supported Cooperative Work and Social Computing, pp. 317–321 October 2020

21. Liu, J., Ng, Y.C., Wood, K.L., Lim, K.H.: IPOD: a large-scale industrial and professional occupation dataset. In: Conference Companion Publication of the 2020 on Computer Supported Cooperative Work and Social Computing, pp. 323–328 October 2020

22. Liu, J., Singhal, T., Blessing, L.T., Wood, K.L., Lim, K.H.: CrisisBERT: a robust transformer for crisis classification and contextual crisis embedding. arXiv preprint arXiv:2005.06627 (2020)

23. Martello, S., Pisinger, D., Toth, P.: Dynamic programming and strong bounds for the 0–1 knapsack problem. Manag. Sci. 45(3), 414–424 (1999)

24. Nguyen, T.T., Roos, M., Rothe, J.: A survey of approximability and inapproximability results for social welfare optimization in multiagent resource allocation. Ann. Math. Artif. Intell. 68(1–3), 65–90 (2013)

25. Osborne, M.J., Rubinstein, A.: A Course in Game Theory. MIT Press, Cambridge (1994)

26. Padia, P., Lim, K.H., Chan, J., Harwood, A.: Sentiment-aware and personalized tour recommendation. In: Proceedings of BigData (2019)

27. Palumbo, E., Rizzo, G., Troncy, R., Baralis, E.: Predicting your next stop-over from location-based social network data with recurrent neural networks. In: RecTour@ RecSys (2017)
28. Papadimitriou, C.H.: Algorithms, games, and the internet. In: ICALP 2001, pp. 1–3 (2001)
29. Papadimitriou, C.H., Tsitsiklis, J.N.: The complexity of Markov decision processes. Math. Oper. Res. **12**(3), 441–450 (1987)
30. Pas, E.I., Principio, S.L.: Braess' paradox: some new insights. 265–276 (1997)
31. Piliouras, G., Nikolova, E., Shamma, J.S.: Risk sensitivity of price of anarchy under uncertainty. ACM Trans. Econ. Comput. **5**(1), 1–27 (2016)
32. Roughgarden, T.: Selfish Routing and the Price of Anarchy, vol. 174. MIT Press, Cambridge (2005)
33. Tsiligirides, T.: Heuristic methods applied to orienteering. J. Oper. Res. Soc. **35**(9), 797–809 (1984)
34. Vansteenwegen, P., Souffriau, W., Oudheusden, D.V.: The orienteering problem: a survey. Eur. J. Oper. Res. **209**(1), 1–10 (2011)
35. Wakamiya, S., Siriaraya, P., Zhang, Y., Kawai, Y., Aramaki, E., Jatowt, A.: Pleasant route suggestion based on color and object rates. In: Proceedings of WSDM 2019, pp. 786–789 (2019)
36. Wang, X., Leckie, C., Chan, J., Lim, K.H., Vaithianathan, T.: Improving personalized trip recommendation to avoid crowds using pedestrian sensor data. In: Proceedings of CIKM 2016, pp. 25–34 (2016)
37. Yao, L., Sheng, Q.Z., Qin, Y., Wang, X., Shemshadi, A., He, Q.: Context-aware point-of-interest recommendation using tensor factorization with social regularization. In: Proceedings of SIGIR 2015, pp. 1007–1010 (2015)
38. Ye, M., Yin, P., Lee, W.C., Lee, D.L.: Exploiting geographical influence for collaborative point-of-interest recommendation. In: Proceedings of SIGIR 2011, pp. 325–334 (2011)
39. Yuan, Q., Cong, G., Ma, Z., Sun, A., Thalmann, N.M.: Time-aware point-of-interest recommendation. In: Proceedings of SIGIR 2013, pp. 363–372 (2013)
40. Zhang, C., Liang, H., Wang, K.: Trip recommendation meets real-world constraints: POI availability, diversity, and traveling time uncertainty. ACM TOIS **35**(1), 5 (2016)
41. Zhang, C., Liang, H., Wang, K., Sun, J.: Personalized trip recommendation with POI availability and uncertain traveling time. In: Proceedings of CIKM 2015, pp. 911–920 (2015)
42. Zhang, Y., Siriaraya, P., Wang, Y., Wakamiya, S., Kawai, Y., Jatowt, A.: Walking down a different path: route recommendation based on visual and facility based diversity. In: Companion Proceedings of WWW 2018, pp. 171–174 (2018)

Applied Data Science: Anomaly Detection

A Context-Aware Approach to Detect Abnormal Human Behaviors

Roghayeh Mojarad$^{(\boxtimes)}$ ⓘ, Ferhat Attal ⓘ, Abdelghani Chibani ⓘ,
and Yacine Amirat ⓘ

Univ Paris Est Creteil, LISSI, 94400 Vitry, France
{Contact-lissi,roghayeh.mojarad,ferhat.attal,
abdelghani.chibani,amirat}@u-pec.fr
http://www.lissi.fr/

Abstract. Abnormal human behaviors can be signs of a health issue
or the occurrence of a hazardous incident. Detecting such behaviors is
essential in Ambient Intelligent (AmI) systems to enhance the safety
of people. While detecting abnormalities has been extensively explored
in different domains, there are still some challenges for developing efficient approaches dealing with the limitations of data-driven approaches
to detect abnormal human behaviors in AmI systems. In this paper, a
novel approach is proposed to detect such behaviors exploiting the contextual information of human behaviors. Machine-learning models are
firstly used to recognize human activities, locations, and objects. Different contexts of human behaviors are then extracted in terms of the duration, frequency, time of the day, locations, used objects, and sequences
of the frequent recognized activities. An ontology, called Human ACtivity ONtology (HACON), is proposed to conceptualize the contexts of
human behaviors. Finally, a probabilistic version of ASP, a high-level
expressive logic-based formalism, is proposed to detect abnormal behaviors through a set of rules based on the HACON ontology. The proposed
approach is evaluated in terms of precision, recall, F-measure, and accuracy using two datasets, namely *Orange4Home dataset* and *HAR dataset
using smartphones*. The evaluation results demonstrate the ability of the
proposed approach to detect abnormal human behaviors.

Keywords: Context-aware approach · Human behavior analysis ·
Abnormal human behavior detection · Answer set programming

1 Introduction

An abnormal human behavior can be seen as any behavior that deviates from
typical or usual behaviors. This type of behavior can be signs of a health issue or
a hazardous incident [15]. Detecting such behaviors is essential in Ambient Intelligent (AmI) systems to provide smart and supportive services to enhance the
safety of people [5,13,22,23]. While detecting abnormalities has been extensively

© Springer Nature Switzerland AG 2021
Y. Dong et al. (Eds.): ECML PKDD 2020, LNAI 12460, pp. 89–104, 2021.
https://doi.org/10.1007/978-3-030-67667-4_6

explored in different domains [5,31], there are still some challenges for developing efficient approaches dealing with the limitations of data-driven approaches intended for detecting abnormal human behaviors in AmI systems. One of the main limitations of these approaches is their inability to consider the context of human behaviors [22]. Dey [8] defined context as "any information that can be used to characterize the situation of an entity. An entity is a person, place, or object that is considered relevant to the interaction between a user and an application, including the user and applications themselves". Moreover, developing efficient approaches to detect abnormal human behaviors requires a comprehensive and machine-understandable human behavior definition which considers different contexts of human behaviors.

Although there are some differences between human behaviors and human activities, these two terms are usually used interchangeably in the literature [27]. However, in a few studies [14,24], these two terms are defined differently; i.e., human behaviors are usually defined as repetitive human activities [4]. However, this definition is not comprehensive and does not consider the different contexts of human behaviors, such as location, duration, object, etc. Hence, in this paper, human behavior is defined as repetitive human activities in particular contexts. Human behavior is defined as a structure with six concepts of context: (i) frequent activities in particular locations, such as *eating in kitchen*, (ii) frequent activities with particular objects, such as *eating with fork*, (iii) frequent activities in particular times of the day, such as *eating at noon*, (iv) frequent activities within particular ranges of duration, such as *eating takes between d_{min} to d_{max} minutes, where d_{min} and d_{max} represent the minimum and maximum duration of eating activity, respectively*, (v) recurrent activities with particular frequencies per day, such as *frequency of eating activity per day is between f_{min} and f_{max}, where f_{min} and f_{max} represent the minimum and maximum frequency of eating activity, respectively*, and (vi) frequent sequences of activities, such as *the activity sequence eating- cleaning*. An abnormal human behavior is defined as unexpected or unusual behavior [9]. According to the proposed definition of human behavior in this study, abnormal human behavior can be classified into six abnormality types: (i) recurrent unexpected activities in particular locations, (ii) recurrent unexpected activities with particular objects, (iii) recurrent unexpected activities in particular times of the day, (iv) recurrent unexpected activities within particular ranges of duration, (v) recurrent unexpected activities with particular frequencies per day, and (vi) recurrent unexpected sequences of activities.

In this paper, a hybrid approach including knowledge-driven and data-driven methods is proposed to detect abnormal human daily living behaviors by exploiting the contextual information of human behaviors. This approach seeks to address the drawbacks of knowledge-driven and data-driven methods while leveraging their benefits, such as considering different human behavior contexts, handling a huge amount of data, and managing uncertain information. The proposed approach is composed of four main modules: (i) human activity, location, and object recognition, (ii) capturing human behavior contexts, (iii) mapping to an ontology, and (iv) abnormal human behavior detection. In the first module, machine-learning models are firstly used to recognize human activities, locations, and objects. The recognized ones are then used to analyze human behaviors in terms of the duration, frequency, time of the day, location, used objects, and sequences of the frequent activities to capture six concepts of human behavior contexts, which include the defined six human behavior concepts. An ontology, called Human ACtivity ONtology (HACON), is used to conceptualize the recognized human activities, behaviors, and their contexts. The captured human behavior contexts are mapped to the HACON ontology to conceptualize human behavior contexts. The predictions of human activities, locations, and objects are generally uncertain; Mapping uncertain information over ontology may lead to weak human behavior recognition [11]. Hence, in the fourth module, a probabilistic version of ASP, called Probabilistic Answer Set Programming (PASP), is proposed to detect abnormal human behaviors while handling uncertain information. The latter are considered in PASP by assigning probabilities to rules and their literals. The proposed approach is evaluated in terms of precision, recall, F-measure, and accuracy using two datasets, namely *Orange4Home dataset* [6] and *HAR dataset using smartphones* [1].

The rest of this paper is organized as follows: Sect. 2 is dedicated to a review of related works in the field of abnormal human behavior detection. The necessary background of this study is given in Sect. 3. The proposed context-aware approach is presented in Sect. 4. The evaluation results are provided and discussed in Sect. 5. The conclusion and research perspectives are presented in Sect. 6.

2   Related Works

One of the main challenges in the domain of human behavior analysis is abnormal human behavior detection, which has gained remarkable attention from researchers in different application domains, such as healthcare [16] and ambient assisted living systems [14]. Abnormal human behavior detection aims to detect unexpected human behaviors as they vary from the common behaviors [16]. The most existing approaches in this domain are vision-based [20,33]; these approaches present several disadvantages, such as visual occlusions and privacy issue. The existing abnormal human behavior detection approaches can be classified into three main categories: (i) data-driven approaches (ii) knowledge-driven approaches, and (iii) hybrid approaches.

In [2], a probabilistic spatio-temporal model is used to recognize daily behavior. A cross-entropy measure is then used to detect abnormalities, which are defined as significant changes from the learned behavioral model. In [37], a Dynamic Bayesian Network (DBN) is proposed to model each behavior pattern. An accumulative abnormality measure is then proposed to detect abnormal behaviors using a Likelihood Ratio Test (LRT) method. The most common data-driven approaches proposed for abnormal human behavior detection use machine-learning models, such as Support Vector Data Description (SVDD) [30], Support Vector Machine (SVM) [16], and Recurrent Neural Networks (RNN) [3].

The data-driven approaches strongly rely on data and do not consider the contexts of human behaviors, which may lead to consider some usual behaviors in specific situations as abnormal behaviors.

Unlike data-driven approaches, knowledge-driven approaches depend on knowledge of experts. These approaches commonly use logical axioms and rules. In [36], a rule-based abnormality detection algorithm is proposed to characterizes abnormal patterns. Normal behavior is defined based on captured events occurring on the five, six, seven, and eight weeks prior to the consideration day. This approach depends on the events that fit a certain rule for the current day and the number of cases matching the same rule from five to eight weeks ago. In [38], a rule-based approach is proposed to detect abnormality in *sleeping* behavior. The adopted rules are based on the location, time of the day, and the duration of human activities. The proposed rule-based approach is compared to an SVM model; the results show the superiority of the latter to the former one. In [19], an approach based on Intertransaction Association Rule (IAR) mining is proposed to detect abnormal behaviors. The major limitation of knowledge-driven approaches is their inability to handle uncertain data.

To overcome the disadvantages of data-driven and knowledge-driven approaches while exploiting their advantages, hybrid approaches have been proposed in the literature. In [26], Fine-grained Abnormal BEhavior Recognition (FABER) hybrid approach is proposed. In this approach, a Markov Logic Network (MLN) is used to detect the starting and finishing points of human activities. Abnormal human behaviors are detected by analyzing these points using a knowledge-based inference engine. In [10], an HMM-based approach is used to detect abnormalities in daily activities. A process is used to identify abnormality in human routines from statistical histories. A fuzzy rule-based model is then used to fuse the outputs of these models to detect abnormal human behaviors. In [31], a K-means model is used to recognize human activities. A sequential pattern mining is then used to analyze the recognized activities based on the time. From the recognized patterns, some properties and data types for the ontology are defined. The ontology is then used for semantic analysis of human activities. Abnormal patterns are detected using pattern analysis algorithms, such as the longest common subsequence algorithm.

The aforementioned studies deal mainly with the problem of abnormal human behavior detection either without considering human behavior contexts, such as location, objects, and sequences of human activities or without considering uncertain information. In this paper, a hybrid approach is proposed to detect abnormal human behavior while considering different contexts of human behaviors as well as handling uncertain information.

3   Background

ASP is a high-level expressive non-monotonic logic-based formalism which allows different reasoning, such as defeasible reasoning, causal reasoning, and diagnostic reasoning [17,18]. ASP represents knowledge through logical phrases and then derives new knowledge using reasoning. ASP programs are derived from the syntax of Prolog language [18]. An ASP program is consists of a set of rules in the following form:

$$h \leftarrow b_1, ..., b_k, not\, b_{k+1}, ..., not\, b_{k+n} \quad (n, k \geq 0). \tag{1}$$

where the left-hand and right-hand sides are known as head and body, respectively; and b_i are atoms of propositional language forming the body. The negation symbol, not, is used to depict epistemic negation [29]. The rules with empty heads are known as constraints; while the ones with empty bodies are called facts.

Let S be the set of ground atoms in the ASP program I in the form of (1). Then S satisfies the body of a rule when $\{b_1, ..., b_k\} \subseteq S$ and $S \cap \{b_{k+1}, ..., b_{k+n}\} = \varnothing$. Hence, S satisfies a rule with a non-empty head when S does not satisfy body or $h \in S$; moreover, S satisfies a constraint when it does not satisfy the body. Answer set is formalized as follows:

Suppose program I consists of ASP rules. S is a set of ground atoms obtained using grounding; the latter replaces variables used in the program I with ground atoms. A reduct I^S, which does not contain any negated atoms, is obtained from the program I using two steps: (i) for each atom $l \in S$, drop rules with $not\, l$ in their body, (ii) drop literals $not\, l$ from all other rules. The minimal model of (I^S) is the answer set S.

Let considering the following illustrative example, program I is composed of two facts and one rule, where a fact is a rule without body and with a single disjunct in the head. In this program, the predicate $act(activity, T)$ represents the fact that the user performs specific activity $activity$ at the timestamp T. The predicate $loc(location, T)$ describes the fact that the user is in a specific location $location$ at the timestamp T. The predicate $abnormalActLoc(activity, location, T)$ represents the fact that there is an abnormality at the timestamp T when $activity$ is performed in specific $location$. It is worth mentioning that terms starting with lowercase letter represent constants, e.g., $activity$; while terms starting with uppercase letter represent variables, e.g., T.

Program I:
  Facts :
      $act(eating, t)$.
      $loc(bedroom, t)$.
  Rule :
      $abnormalActLoc(eating, bedroom, T) : -act(eating, T), loc(bedroom, T)$.

The answer set for this program is as follows:

$act(eating, t)$, $loc(bedroom, t)$, $abnormalActLoc(eating, bedroom, t)$

The inferred information, $abnormalActLoc(eating, bedroom, t)$, is obtained using reasoning performed by an answer set solver [18]. ASP is suitable for representing commonsense knowledge and also modeling commonsense reasoning. The rich knowledge representation and efficient solvers are the main characteristics of ASP, which make it superior in comparison with other logic programming

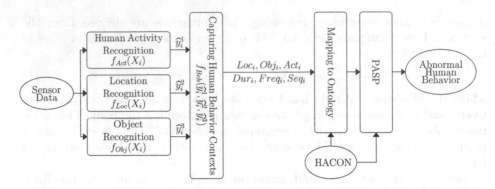

Fig. 1. Architecture of the proposed approach.

languages. However, the main limitation of ASP is its inability to handle uncertainty; i.e., ASP does not make any difference between answer sets that are more likely to be true with the others that are less likely to be true. To overcome this limitation, a probabilistic ASP, which allows integrating probabilistic reasoning with non-monotonic logic programs, is proposed.

4 Proposed Approach

Figure 1 shows the overall architecture of the proposed approach. Firstly, an LSTM model is used to recognize human activities, locations, and objects. This model allows sequential information modeling in the short and also long term, which is required to analyze human behaviors as the latter are characterized by time-series data. The different contexts of human behaviors are then extracted in terms of location, object, time of the day, duration, frequency, and sequences of activities. The captured contexts are mapped to the proposed HACON ontology to provide shared concepts about human activities and behaviors. Finally, PASP is used to detect abnormal human behaviors. PASP enables probabilistic inferences through a set of probabilistic rules about abnormal human behaviors. The latter is defined by experts according to the HACON ontology, e.g., shared concepts defined in HACON ontology are used to define predicates exploited in PASP rules. Consequently, each PASP rule weighted with a value corresponding to the true degree of rule, which is learned from data by optimizing a pseudo-likelihood measure [34].

4.1 Human Activity, Location, and Object Recognition

To classify sensor data into three labels, namely human activity, location, and object, describing the ongoing activities, a machine-learning model is used. Learning three models, namely activity recognition, location recognition, and object recognition, independently allows the proposed approach to detect abnormal human behavior even if the location and/or object models are not available.

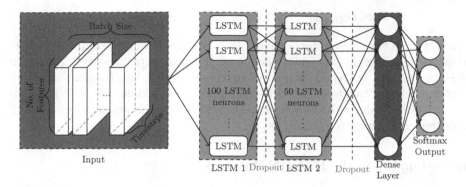

Fig. 2. Architecture of the LSTM model.

The human activity, location, and object recognition models are formalized as follows:

$$\widehat{y}_1 = f_{Act}(X_i)$$
$$\widehat{y}_2 = f_{Loc}(X_i) \qquad (2)$$
$$\widehat{y}_3 = f_{Obj}(X_i)$$

where $X_i = \{x_i^1, x_i^2, ..., x_i^d\}$ is the ith data sample. Each data sample includes d attributes. $\widehat{y}_1$, $\widehat{y}_2$, and $\widehat{y}_3$ represent the predicted labels in the case of human activity, location, and object, respectively. f_{Act}, f_{Loc}, and f_{Obj} represent the prediction function of human activity, location, and object recognition models, respectively. In this study, an LSTM model, a type of Recurrent Neural Networks (RNN) that consists of special units beside standard units, is used as activity, location, and object recognition models. The choice of LSTM can be explained by the fact that it allows modeling sequential data in the short and also long term, which is required in modeling human behaviors.

Figure 2 show the overall architecture of the LSTM model used in this study. This model consists of five layers: (i) LSTM layer with 100 neurons, (ii) dropout layer with a fraction rate 0.5, (iii) LSTM layer with 50 neurons, (iv) dropout layer with a fraction rate 0.5, and (v) dense layer with neuron number equals to the number of classes. The optimization and loss functions are respectively set to *Adaptive Moment Estimation (Adam)* and *categorical-crossentropy*.

4.2   Capturing Human Behavior Contexts

The human behaviors are analyzed to obtain the defined six concepts of human behavior context. A statistical algorithm is developed to extract these concepts. In this algorithm, for each activity, eight lists of hash maps are generated; these eight lists are associated with locations, objects, time of the day, minimum duration, maximum duration, minimum frequency, maximum frequency, and previous activity. This algorithm is formalized as a function f_{Beh} as follows:

$$Loc_i, Obj_i, Act_i, Dur_i, Freq_i, Seq_i = f_{Beh}(\hat{y}_1, \hat{y}_2, \hat{y}_3) \tag{3}$$

where Loc_i, Obj_i, Act_i, Dur_i, $Freq_i$, and Seq_i represent respectively lists of frequent activities in particular locations, frequent activities with particular objects, frequent activities at particular times of the day, frequent activities within particular ranges of duration (minimum and maximum duration), recurrent activities with particular frequencies (minimum and maximum frequencies), and frequent activity sequences.

4.3   Mapping to the HACON Ontology

The HACON ontology, inspired from the *ConceptNet* semantic network [32], is proposed to conceptualize human activities and behaviors. The *ConceptNet* is a knowledge graph with words and phrases connected using relationships, also called labeled edge; e.g., the words *oven* and *cooking* are linked using the *is used for* relationship. The HACON ontology consists of eight main human activity concepts, namely: *activity, date time, time of the day, duration, frequency, object, actor,* and *location.* These latter are connected using eight relationships, namely *has start time, has end time, has time, has duration, has frequency, has object, has actor,* and *has place.* Figure 3 shows the overview of the HACON ontology, which is modeled by the semantic Web Ontology Language (OWL) [21].

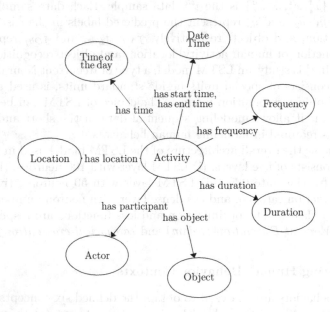

Fig. 3. Main human activity concepts and relationships used in the HACON ontology.

4.4 PASP

PASP exploits the output of mapping to the HACON ontology module to detect abnormal human behaviors while handling uncertainty of human activity, location, and object recognition models. The uncertainties are handled by assigning a probability to each literal of PASP rule. The PASP is formalized as follows:

$$(h, p(h)) \leftarrow (b_1, p(b_1)), (b_2, p(b_2)), ..., (b_n, p(b_n)) \ (n \geq 0). \tag{4}$$

where $(b_1, p(b_1)), (b_2, p(b_2)), ..., (b_n, p(b_n))$ represents the body of the rule and $(h, p(h))$ represents its head. $p(b_i)$ represents the probability of the literal b_i. Moreover, we assume that if the probability of the given literal b_i is $p(b_i)$, the probability of the negative literal *not* b_i is $1 - p(b_i)$. The probability of literals are assumed to be independent from each other. Therefore, the probability for the head of rules are calculated as follows:

$$p(h) = w_{rule} \times p(b_1) \times p(b_2) \times ... \times p(b_n) \tag{5}$$

where w_{rule} represents the normalized weight of the rule that can be defined by experts or weight-learning process [35]. In this study, the weights of rules are learned from data by optimizing a pseudo-likelihood measure [34]. Consider the following example with the PASP program II including two facts and one rule. The latter is defined based on the Knowledge Base (KB) provided by an expert. The predicate $act(activity, T, P)$ represents the fact that the user performs specific activity $activity$ at timestamp T with probability P. $loc(location, T, P)$ describes the fact that the user is in specific location $location$ at timestamp T with probability P. $abnormalActLoc(activity, location, T, P)$ represents the fact that there is an abnormality with probability P at timestamp T where $activity$ is performed in $location$.

Program II:
    $Facts$:
        $act(eating, t, p_{act})$.
        $loc(bedroom, t, p_{loc})$.
    $Rule$:
        $abnormalActLoc(eating, bedroom, T, (w \times P_{act} \times P_{loc}))$
        $: - act(eating, T, P_{act}), loc(bedroom, T, P_{loc})$.
The answer set for this program is as follows:
        $act(eating, t, p_{act}), loc(bedroom, t, p_{loc})$,
        $abnormalActLoc(eating, bedroom, t, w \times p_{act} \times p_{loc})$

The probability of predicates $abnormalActLoc$, which is the head of rule, is calculated using the multiplication of probabilities of literals in the body of the rule, P_{act} and P_{loc}, and the weight of the rule, w. Abnormal human behaviors are then detected when the probability of abnormal behavior, $(w \times P_{act} \times P_{loc})$, is greater than its complement, $1 - (w \times P_{act} \times P_{loc})$. The PASP rules allow detecting the six different types of abnormal human behaviors, see Table 1.

Table 1. List of abnormal behaviors with their predicates used in PASP.

Unexpected behaviors in particular locations	AbnormalActLoc(act, loc, time)
Unexpected behaviors with particular objects	AbnormalActObj(act, obj, time)
Unexpected behaviors in particular times of the day	AbnormalActTime(act, timeday, time)
Unexpected behaviors with particular duration	AbnormalActDur(act, dur, time)
Unexpected behaviors with particular frequencies	AbnormalActFreq(act, freq, time)
Unexpected behaviors with particular activity sequences	AbnormalSeqAct(act1, act2, time)

Unlike to ASP, PASP can handle uncertainty through the weight-learning process and probabilistic reasoning. In addition, PASP can also overcome the erroneous prediction of machine-learning models by assigning a weight to each rule and assigning a probability to each literal used in the rules.

5 Experiments, Results, and Discussion

In this section, the abnormal human behavior detection performances obtained using the proposed approach are presented and discussed. Accuracy, F-measure, recall, and precision are used as performance evaluation metrics. Moreover, the proposed approach is evaluated in comparison with two baseline approaches, namely MLN and SVM using two public datasets: *Orange4Home dataset* [6] and *HAR dataset using smartphones data* [1]. We firstly present the performance results of the human activity, location, and object recognition module and then we evaluate the performance of the abnormal human behavior detection module.

5.1 Human Activity, Location, and Object Recognition

The LSTM model used in the proposed approach is evaluated in terms of precision, recall, F-measure, and accuracy while the epoch number, batch size, and timestep number are set to 300 iterations, 50 instances, and 128 sequences, respectively. The hyper-parameters of the LSTM model are estimated using a grid search method.

Case 1: *Orange4Home* dataset [6]
In this dataset, the data are collected from 236 environmental sensors placed in different locations of an instrumented home to capture information, such as the state of electrical equipment and water consumption. This dataset is composed of routine daily living activities classified in 17 categories, such as *eating* and

Table 2. Performances obtained using the LSTM models in Case 1.

Evaluation metrics	Activity recognition	Location recognition
Precision	96.00	97.98
Recall	95.71	97.89
F-measure	95.63	97.83
Accuracy	95.71	97.90

Table 3. Comparison with the baseline approaches for activity recognition in Case 1.

Models	F-measure
Baseline Approaches [7]	
Home MLP	77.85
Home SVM	89.61
Place-based SVM	92.08
Place-based MLP	93.05
Proposed Approach	
Proposed LSTM model	95.63

cleaning, performed by one participant during working hours for four weeks. The interval of timestamps between records of this dataset is two seconds. This dataset includes three contexts, namely the time of the day, location, and human activity; therefore, two LSTM models are learned independently to recognize human activity and location. Table 2 presents the obtained results in terms of precision, recall, F-measure, and accuracy in the case of human activity and location recognition, respectively. It can be observed that the obtained rates in terms of all performance metrics are greater than 95% in case of activity recognition. However, rates around 97% are achieved for all performance metrics in the case of location recognition. It is worth noting that the performances in the case of location recognition are greater than those obtained in the case of activity recognition. This can be explained by the fact that the sensors used in this dataset are environmental which help distinguishing different classes of location (Table 3).

In addition, the LSTM model is compared with two baseline approaches considering the *Orange4Home* dataset, see Table 5. The baseline approaches are MultiLayer Perceptron (MLP) and the SVM model in two different conditions, namely *Home* and *Place-based*. The *Home* configuration is based on the decision of a single classifier using all sensors, while *Place-based* configuration is based on the fusion of eight different classifiers (one for each location). The results show the superiority of the proposed LSTM model in comparison with the baseline approaches. This is explained by the fact that the LSTM model is a well-suited model for time-series data, which is not the case of MLP and SVM models.

Table 4. Performances obtained using the LSTM models in Case 2.

Evaluation metrics	Activity recognition
Precision	94.08
Recall	94.03
F-measure	94.05
Accuracy	97.98

Table 5. Comparison with the baseline approaches in Case 2.

Models	F-measure
Baseline Approaches [28]	
KNN	90.16
SVM	93.79
Proposed Approach	
Proposed LSTM model	94.05

Case 2: *HAR dataset using smartphones* dataset [1]

In this dataset, a waist-mounted smartphone with embedded inertial sensors, such as accelerometers and gyroscopes, is used to collect the data. Six activities, namely: (i)Walking, (ii) Walking-upstairs, (iii) Walking-downstairs, (iv) Sitting, (v) Standing, and (vi) Laying, are performed by thirty participants wearing a smartphone (Samsung Galaxy S II) on the waist. Each participant performed each activity twice. Triaxial linear acceleration and angular velocity signals are collected at a sampling rate of 50 Hz. This dataset includes only human activity label; therefore, an LSTM model is used for human activity recognition. Table 4 presents the obtained results using the LSTM model in terms of precision, recall, F-measure, and accuracy. It is noticeable that the model obtains more than 94% on average in terms of all performance metrics.

In addition, the LSTM model is compared with two baseline models, namely K-Nearest Neighbors (KNN) and SVM. Table 5 shows that the LSTM achieves better performance in terms of average F-measure with 94.05% while KNN and SVM achieve 90.16% and 93.79%, respectively.

5.2 Abnormal Human Behavior Detection

PASP is implemented in Clingo [12], an ASP tool for grounding and solving logic programs, to detect abnormal human behaviors. PASP is evaluated considering different types of abnormal human behaviors in terms of precision, recall, F-measure, and accuracy. Since the *Orange4Home* dataset does not include the object label, abnormality with particular objects is not considered in the evaluation. The other five types of abnormal human behaviors, namely recurrent unexpected activities in specific loca-

Table 6. Performances obtained using SVM, MLN, and PASP in Case 1.

Abnormality types	SVM				MLN				PASP			
	Precision	Recall	F-measure	Accuracy	Precision	Recall	F-measure	Accuracy	Precision	Recall	F-measure	Accuracy
AbnormalActLoc	95.22	95.11	95.14	95.11	89.40	94.63	91.94	95.03	100	94.21	97.02	98.26
AbnormalActTime	98.66	98.66	98.66	98.66	94.98	99.53	97.20	98.28	100	98.10	99.04	99.82
AbnormalActDur	81.47	79.59	74.53	79.59	72.00	97.29	82.75	94.86	100	91.89	95.77	98.97
AbnormalActFreq	66.45	81.52	73.22	81.52	81.25	76.47	78.78	93.06	100	94.11	96.96	99.00
AbnormalSeqAct	76.61	72.85	72.71	72.85	87.76	70.37	78.11	86.12	100	88.45	93.87	95.93
Average	83.68	85.54	82.85	85.54	85.08	87.66	85.76	93.47	100	93.35	96.53	98.39

Table 7. Performances obtained using SVM, MLN, and PASP in Case 2.

Abnormality types	SVM				MLN				PASP			
	Precision	Recall	F-measure	Accuracy	Precision	Recall	F-measure	Accuracy	Precision	Recall	F-measure	Accuracy
AbnormalActDur	80.98	84.89	84.17	84.17	96.15	96.15	96.15	99.15	100	96.15	98.04	99.87
AbnormalSeqAct	62.04	77.31	70.24	70.24	90.44	94.53	92.44	94.69	100	100	100	100
Average	71.51	81.10	77.20	77.20	93.29	95.34	94.29	96.92	100	98.07	99.02	99.93

tions, *AbnormalActLoc*, recurrent unexpected activities in particular times of the day, *AbnormalActTime*, recurrent unexpected activities within particular ranges of duration, *AbnormalActDur*, recurrent unexpected activities with particular frequencies per day, *AbnormalActFreq*, and recurrent unexpected sequences of activities, *AbnormalSeqAct*, are used for the evaluation on this dataset, see Table 1. The *HAR dataset using smartphones* does not include object and location labels; therefore only two types of abnormal human behaviors, namely recurrent unexpected activities within particular ranges of duration, *AbnormalActDur* and recurrent unexpected sequences of activities, *AbnormalSeqAct*, are considered for the evaluation of the PASP, see Table 7. Abnormalities are randomly injected into these datasets to simulate the presence of abnormal human behaviors since the used datasets do not include abnormalities.

PASP is evaluated and compared with two baseline models, namely MLN and SVM, in terms of abnormal human behavior detection. The evaluation performance results obtained using PASP, MLN, and SVM on *Orange4Home* dataset are shown in Table 6. It can be noticed that PASP detects more than 93% on average in terms of precision, recall, F-measure, and accuracy. PASP obtains 100% in terms of precision while MLN and SVM achieve 85.08% and 83.68%, respectively. Table 4 shows the evaluation performance results achieved using PASP, MLN, and SVM on the *HAR dataset using smartphones*. As it can be observed, PASP outperforms MLN and SVM in terms of precision, recall, F-measure, and accuracy. PASP also obtains 100% in terms of average precision on this dataset while MLN and SVM achieve 93.29% and 71.51%, respectively. It is worth noticing that the SVM model is not able to handle uncertainty and does not consider the contexts of human behaviors, which explains its low performances. However, the proposed PASP handles uncertainty through probabilistic rules, such as uncertainties of human activity predictions. PASP also filters abnormalities with low probability in order to decrease false positive rate. MLN handles the uncertainty of rules; however, it is not able to consider uncertainty (probability) of human activity and location predictions. In MLN, a weight value is assigned to each rule while in PASP, beside of assigning a weight value to each rule, one probability is assigned to each literal used in the rules to enhance the detection performance. In some cases, such as *leaving* activity and *entrance* location, LSTM models achieve low accuracy. Hence detecting abnormal behaviors based on the outputs of LSTM models using MLN results in a high false-positive rate if the weight value of rules according to these activity and location is high. Otherwise, if this weight value is low, it may prevent detecting associated abnormal behaviors.

6   Conclusion and Future Works

In this paper, a comprehensive definition of human behavior considering different human behavior contexts is provided. A context-aware approach is proposed to detect abnormal human behaviors in AmI systems. Machine-learning models are firstly used to predict human activity, location, and object. The predicted labels are then analyzed to extract different concepts of human behavior contexts; six main concepts are extracted according to the duration, frequency, sequence, time of the day, object, and location of repetitive activities. The extracted contexts are then conceptualized using the HACON ontology. Finally, PASP is used to detect abnormal behaviors through a set of probabilistic rules, which are defined by experts considering the HACON ontology. The proposed approach is evaluated using the *Orange4Home* dataset and the *HAR dataset using smartphones*. In addition, it is compared with MLN and SVM models. The evaluation results show the superiority of the proposed approach in comparison with these models in terms of abnormal human behavior detection. As a research perspective, an interesting topic is to use the proposed approach to develop a recommendation system based on wellbeing guidelines. Another future research work is to extend the PASP rules with the dynamic probability of predicates based on real-time observations.

References

1. Anguita, D., Ghio, A., Oneto, L., Parra, X., Reyes-Ortiz, J.L.: A public domain dataset for human activity recognition using smartphones. In: Computational Intelligence, p. 6 (2013)
2. Aran, O., Sanchez-Cortes, D., Do, M.-T., Gatica-Perez, D.: Anomaly detection in elderly daily behavior in ambient sensing environments. In: Chetouani, M., Cohn, J., Salah, A.A. (eds.) HBU 2016. LNCS, vol. 9997, pp. 51–67. Springer, Cham (2016). https://doi.org/10.1007/978-3-319-46843-3_4
3. Arifoglu, D., Bouchachia, A.: Activity recognition and abnormal behaviour detection with recurrent neural networks. Procedia Comput. Sci. **110**, 86–93 (2017). 14th International Conference on Mobile Systems and Pervasive Computing (MobiSPC 2017)

4. Banovic, N., Buzali, T., Chevalier, F., Mankoff, J., Dey, A.K.: Modeling and understanding human routine behavior. In: Proceedings of the 2016 CHI Conference on Human Factors in Computing Systems, CHI 2016, pp. 248–260. ACM, New York (2016)
5. Bruno, B., Mastrogiovanni, F., Sgorbissa, A., Vernazza, T., Zaccaria, R.: Analysis of human behavior recognition algorithms based on acceleration data. In: IEEE International Conference on Robotics and Automation, pp. 1602–1607, May 2013
6. Cumin, J., Lefebvre, G., Ramparany, F., Crowley, J.L.: A dataset of routine daily activities in an instrumented home. In: Ochoa, S.F., Singh, P., Bravo, J. (eds.) UCAmI 2017. LNCS, vol. 10586, pp. 413–425. Springer, Cham (2017). https://doi.org/10.1007/978-3-319-67585-5_43
7. Cumin, J., Ramparany, F., Crowley, J.L., et al.: Inferring availability for communication in smart homes using context. In: IEEE International Conference on Pervasive Computing and Communications Workshops (PerCom Workshops), pp. 1–6. IEEE (2018)
8. Dey, A.K.: Understanding and using context. Pers. Ubiquit. Comput. 5(1), 4–7 (2001)
9. Durand, V.M., Barlow, D.H.: Essentials of abnormal psychology. Cengage Learning Boston (2012)
10. Forkan, A.R.M., Khalil, I., Tari, Z., Foufou, S., Bouras, A.: A context-aware approach for long-term behavioural change detection and abnormality prediction in ambient assisted living. Pattern Recogn. 48(3), 628–641 (2015)
11. Gayathri, K., Easwarakumar, K., Elias, S.: Probabilistic ontology based activity recognition in smart homes using Markov logic network. Knowl.-Based Syst. 121, 173–184 (2017)
12. Gebser, M., Kaminski, R., Kaufmann, B., Schaub, T.: Clingo = ASP + Control: Preliminary Report. arXiv:1405.3694 [cs], May 2014
13. Kordestani, H., et al.: Hapicare: a healthcare monitoring system with self-adaptive coaching using probabilistic reasoning. In: 2019 IEEE/ACS 16th International Conference on Computer Systems and Applications (AICCSA), pp. 1–8. IEEE (2019)
14. Lago, P., Jiménez-Guarín, C., Roncancio, C.: Contextualized behavior patterns for ambient assisted living. In: Salah, A.A., Kröse, B.J.A., Cook, D.J. (eds.) HBU 2015. LNCS, vol. 9277, pp. 132–145. Springer, Cham (2015). https://doi.org/10.1007/978-3-319-24195-1_10
15. Lawton, M.P., Brody, E.M.: Assessment of older people: self-maintaining and instrumental activities of daily living. Gerontologist 9(3_Part_1), 179–186 (1969)
16. Lentzas, A., Vrakas, D.: Non-intrusive human activity recognition and abnormal behavior detection on elderly people: a review. Artif. Intell. Rev. 53(3), 1975–2021 (2019). https://doi.org/10.1007/s10462-019-09724-5
17. Lifschitz, V.: Answer set planning. In: Gelfond, M., Leone, N., Pfeifer, G. (eds.) LPNMR 1999. LNCS (LNAI), vol. 1730, pp. 373–374. Springer, Heidelberg (1999). https://doi.org/10.1007/3-540-46767-X_28
18. Lifschitz, V.: Answer Set Programming. Springer, Heidelberg (2019). https://doi.org/10.1007/978-3-030-24658-7
19. Lühr, S., West, G., Venkatesh, S.: Recognition of emergent human behaviour in a smart home: a data mining approach. Pervasive Mob. Comput. 3(2), 95–116 (2007)
20. Mabrouk, A.B., Zagrouba, E.: Abnormal behavior recognition for intelligent video surveillance systems: a review. Expert Syst. Appl. 91, 480–491 (2018)
21. McGuinness, D.L., Van Harmelen, F., et al.: Owl web ontology language overview. W3C Recomm. 10(10) (2004)

22. Mojarad, R., Attal, F., Chibani, A., Fiorini, S.R., Amirat, Y.: Hybrid approach for human activity recognition by ubiquitous robots. In: IEEE/RSJ International Conference on Intelligent Robots and Systems (IROS), pp. 5660–5665, October 2018
23. Mojarad, R., Attal, F., Chibani, A., Amirat, Y.: Automatic classification error detection and correction for robust human activity recognition. IEEE Robot. Autom. Lett. 5(2), 2208–2215 (2020)
24. Monekosso, D.N., Remagnino, P.: Behavior analysis for assisted living. IEEE Trans. Autom. Sci. Eng. 7(4), 879–886 (2010)
25. Reiss, A., Hendeby, G., Stricker, D.: A competitive approach for human activity recognition on smartphones. In: European Symposium on Artificial Neural Networks, Computational Intelligence and Machine Learning, Bruges, Belgium, 24–26 April, pp. 455–460 (2013)
26. Riboni, D., Bettini, C., Civitarese, G., Janjua, Z.H., Helaoui, R.: Fine-grained recognition of abnormal behaviors for early detection of mild cognitive impairment. In: IEEE International Conference on Pervasive Computing and Communications (PerCom), pp. 149–154 (2015)
27. Rodríguez, N.D., Cuéllar, M.P., Lilius, J., Calvo-Flores, M.D.: A survey on ontologies for human behavior recognition. ACM Comput. Surv. 46(4), 43:1–43:33 (2014)
28. Romera-Paredes, B., Aung, M.S., Bianchi-Berthouze, N.: A one-vs-one classifier ensemble with majority voting for activity recognition. In: ESANN (2013)
29. Shen, Y.D., Eiter, T.: Evaluating epistemic negation in answer set programming. Artif. Intell. 237, 115–135 (2016)
30. Shin, J.H., Lee, B., Park, K.S.: Detection of abnormal living patterns for elderly living alone using support vector data description. Trans. Inf. Tech. Biomed. 15(3), 438–448 (2011)
31. Soto-Mendoza, V., García-Macías, J.A., Chávez, E., Gomez-Montalvo, J.R., Quintana, E.: Detecting abnormal behaviours of institutionalized older adults through a hybrid-inference approach. Pervasive Mob. Comput. 40, 708–723 (2017)
32. Speer, R., Chin, J., Havasi, C.: ConceptNet 5.5: an open multilingual graph of general knowledge. In: Thirty-First AAAI Conference on Artificial Intelligence (2017)
33. Sun, J., Shao, J., He, C.: Abnormal event detection for video surveillance using deep one-class learning. Multimed. Tools Appl. 78(3), 3633–3647 (2017). https://doi.org/10.1007/s11042-017-5244-2
34. Van Haaren, J., Van den Broeck, G., Meert, W., Davis, J.: Lifted generative learning of Markov logic networks. Mach. Learn. 103(1), 27–55 (2016)
35. Wang, Y., Zhu, R., Wang, Z., Zhang, X., Zhang, B.: Logic-based online complex event rule learning with weight optimization. In: 2019 IEEE Fourth International Conference on Data Science in Cyberspace (DSC), pp. 150–155 (2019)
36. Wong, W.K., Moore, A., Cooper, G., Wagner, M.: Rule-based anomaly pattern detection for detecting disease outbreaks. In: AAAI/IAAI, pp. 217–223 (2002)
37. Xiang, T., Gong, S.: Video behavior profiling for anomaly detection. IEEE Trans. Pattern Anal. Mach. Intell. 30(5), 893–908 (2008)
38. Zambrana, C., Palou, X.R., Vargiu, E.: Sleeping recognition to assist elderly people at home. Artif. Intell. Res. 5(2), 64–69 (2016)

RADAR: Recurrent Autoencoder Based Detector for Adversarial Examples on Temporal EHR

Wenjie Wang[1]([✉]), Pengfei Tang[1], Li Xiong[1], and Xiaoqian Jiang[2]

[1] Emory University, Atlanta, GA, USA
{wang.wenjie,pengfei.tang,lxiong}@emory.edu
[2] UTHealth, Houston, TX, USA
xiaoqian.jiang@uth.tmc.edu

Abstract. Leveraging the information-rich and large volume of Electronic Health Records (EHR), deep learning systems have shown great promise in assisting medical diagnosis and regulatory decisions. Although deep learning models have advantages over the traditional machine learning approaches in the medical domain, the discovery of adversarial examples has exposed great threats to the state-of-art deep learning medical systems. While most of the existing studies are focused on the impact of adversarial perturbation on medical images, few works have studied adversarial examples and potential defenses on temporal EHR data. In this work, we propose RADAR, a **R**ecurrent **A**utoencoder based **D**etector for **A**dversarial examples on temporal EHR data, which is the first effort to defend adversarial examples on temporal EHR data. We evaluate RADAR on a mortality classifier using the MIMIC-III dataset. Experiments show that RADAR can filter out more than 90% of adversarial examples and improve the target model accuracy by more than 90% and F1 score by 60%. Besides, we also propose an enhanced attack by introducing the distribution divergence into the loss function such that the adversarial examples are more realistic and difficult to detect.

Keywords: Adversarial example detection · Recurrent autoencoder · Temporal Electronic Health Records (EHR)

1 Introduction

Electronic Health Record (EHR) is the digital version of a patient's medical history including diagnoses, medications, physician summary and medical image. The automated and routine collection of EHR data not only improves the health care quality but also places great potential in clinical informatics research [26]. Leveraging the information-rich and large volume EHR data, deep learning systems have been applied for assisting medical diagnosis, predicting health trajectories and readmission rates, as well as supporting disease phenotyping [33]. Deep learning models have crucial advantages over the traditional machine learning

© Springer Nature Switzerland AG 2021
Y. Dong et al. (Eds.): ECML PKDD 2020, LNAI 12460, pp. 105–121, 2021.
https://doi.org/10.1007/978-3-030-67667-4_7

approaches including the capability of modeling complicated high-dimensional inter-feature relationship within data and capturing the time-series pattern and long-term dependency [30]. Taking advantage of a sufficient amount of training dataset, in some cases, complex neural networks can even exceed capabilities of experienced physicians in head-to-head comparisons [6].

However, recent studies show that the statistical boundary of deep learning model is vulnerable, allowing the creation of adversarial examples by adding imperceptible perturbations on input to mislead the classifier [10]. These adversarial threats are more severe in the medical domain. First, the sparse, noisy and high-dimensional nature of EHR data exposes more vulnerability to potential attackers. Second, some modalities of EHR data such as genetic panels and clinical summary may be generated by a third-party company that has a higher risk being attacked. Finally, medical machine learning systems may be uniquely susceptible to adversarial examples [8] due to high financial interests such as insurance claims.

Most research on adversarial examples in medical domain has been focused on medical images, such as X-ray and MRI image [20,32] which can be easily adapted from traditional image domain. The attack algorithms in the image domain aim to minimize the perturbation scale while mislead model predictions. This optimization problem can be either directly solved such as in C&W attack [4] or approximated with gradient method such as Fast Gradient Sign Method [10]. A few recent works have studied adversarial examples on temporal EHR data. Sun et al. [30] proposed a Recurrent Neural Network (RNN)-based time-preferential minimum attack strategy to identify susceptible locations on EHR data. An et al. [1] proposed LAVA, a saliency score based adversarial example generation approach that aims to minimize the number of perturbations. However, it only works for binary-coded features and is not applicable for general temporal EHR with continuous or categorical features.

Despite these two attempts on the attack algorithms for temporal EHR data, there is no study on potential defense techniques. The existing defense mechanisms in image domain can be categorized into adversarial training [27], image denoising [7] and detection mechanisms [21,22]. One of the most promising and state-of-the-art detection methods is MagNet [21], which is based on autoencoder and rejects examples with large autoencoder reconstruction errors. As MagNet can work with any pre-trained classifier, only requires clean data for training, and does not depend on specific image features, it has the potential to be adapted for temporal EHR data. However, there are several critical challenges due to the characteristics of temporal EHR data:

- *Multivariate temporal dependency.* The intuition of autoencoder based defense is to learn the representation from clean data. However learning the representation and capturing the pattern of time-series EHR data is more challenging than images due to the temporal dependency between time points in addition to the correlations between attributes. Besides, the significance of each timestamp on the prediction outcomes differ as more recent features may have a stronger influence.

– *Sparsity and high-dimensionality.* Sequential EHR data is extremely sparse, discrete and high-dimensional compared to image data. Therefore, the traditional distance metrics may not be effective for measuring the autoencoder reconstruction error which cannot capture the real similarity or validity of temporal EHR data.

In this work, we propose RADAR, a **R**ecurrent **A**utoencoder based **D**etector for **A**dversarial examples on temporal EHR data, which is the first effort to defend adversarial examples on temporal EHR data. Similar to MagNet, the intuition is that an autoencoder can learn the manifold of the clean examples. At the test phase, given an input, the autoencoder will reconstruct the input and push the reconstructed output closer to the manifold. As a result, clean examples will have lower reconstruction error since they are closer to the manifold while adversarial examples may have larger error because they have been strategically perturbed. Thus the reconstruction error and additional criteria can be used to detect adversarial examples.

Different from existing methods, RADAR has two main technical contributions addressing the challenges that are specific to temporal EHR data. First, in order to more effectively model the multivariate time series data, we build an autoencoder by integrating attention mechanism [2] with bi-directional LSTM cell to capture both past and future of the current time frame and their interdependence. By increasing the amount of input information available to the network, RADAR has a higher reconstruction ability which guarantees a higher detectability. Second, to address the sparsity and high dimensionality, besides l_p-norm reconstruction error and prediction divergence of the target classifier between the input and reconstructed output which are used in MagNet, our method introduces prediction uncertainty of the constructed output as an additional detection criteria. Our hypothesis is that autoencoder reconstructed output of adversarial examples can result in more uncertainty on the prediction due to its goal of flipping the original class label. This metric focuses on the downstream prediction rather than the data itself thus can overcome the sparsity challenge of EHR data, and provide a critical and complementary criteria for detecting adversarial examples.

Besides RADAR, we also propose an enhanced attack by introducing distribution divergence into the loss function, making the adversarial examples more realistic and difficult to detect. To our knowledge, RADAR is the first effort to propose defense techniques on temporal EHR data. We evaluate RADAR on a mortality classifier using the MIMIC-III [14] dataset against both existing and our enhanced attacks. Experiments show that RADAR can effectively filter out adversarial examples and significantly improve the target model performance.

2   Preliminaries and Related Work

Neural Networks for Sequential Data. Deep neural networks (DNN) have been increasingly applied to solve difficult real-world tasks. For time-sequence data, Recurrent Neural Network (RNN) is designed for capturing the temporal information among features. A variant of RNN, Long Short-Term Memory (LSTM) network [12] is proposed to capture not only the short-term dependency but also the long term dependency among temporal features. In order to model both forward (past to current) and backward (current to past) temporal correlation, Schuster et al. [25] proposed a bi-directional structure by feeding the reversed input into RNN model as well.

Autoencoder is a type of neural network architecture that learns the data representation in an unsupervised manner through dimension reduction [11]. Recurrent autoencoder refers to a type of autoencoder whose layers are RNN cells [29], which has been widely applied to sequence to sequence (seq2seq) tasks such as machine translation [5,31]. To solve the long-term dependency problem of recurrent autoencoder, Bahdanau et al. [2] proposed an attention mechanism that calculates the weights of states among all the time steps as the attention scores and computes an element-wise weighted sum of all the states as the context vector. Recurrent autoencoder without attention mechanisms has been applied for EHR data imputation and synthesization [35]. In this paper, we adopt a recurrent autoencoder with attention mechanism for the temporal EHR data and use it for adversarial example detection for the first time.

The applications of RNN on sequential EHR data range from mortality prediction, readmission prediction, to trajectory prediction [24,34,36]. Most works use different datasets with different pre-processing methods, and cannot be directly applied to our data. In this work, since our focus is not on the classification model, we adopt a single layer LSTM model as our target classifier to demonstrate the effectiveness of the proposed adversarial example detection method.

Adversarial Examples. Generating adversarial examples can be formulated as a constrained optimization problem. Given a clean input x, its label y and a classifier F, if $L_p(x, x_{adv}) < C$, such that $F(x_{adv}) \neq y$, x_{adv} is an adversarial example, where L_p represents the L_p-norm of the perturbation and C represents the perturbation constraint. This optimization problem can be either directly solved such as in C&W attack [4] or approximated with gradient method such as Fast Gradient Sign Method (FGSM) [10] and iterative FGSM [15].

Very recently, it has been pointed out that medical machine learning systems may be uniquely susceptible to adversarial examples [8]. Several works studied adversarial examples in medical image models [9,17,20,32]. A few works explored the adversarial examples on temporal sequential EHR data. Sun et al. [30] proposed an RNN-based time-preferential minimum attack strategy. Their attack algorithm is similar to the C&W attack in image domain. An et.al [1] proposed a saliency score based adversarial attack on longitudinal EHR data that requires a minimal number of perturbations and minimizes the likelihood of detection. The limitation of this work is that their medical features are binary coded so it is not applicable to continuous features. We propose an enhanced attack in this paper and compare it with the attack algorithm in Sun et al. [30]

Defenses Against Adversarial Examples. The existing defense methods against adversarial examples (mainly focused on the image domain) can be characterized into three categories:

- Image preprocessing and denoising such as image compression [7,13] which are image specific and autoencoder based denoiser (HGD) [19]. The drawback of HGD is that it requires a large number of adversarial samples to train the denoiser.
- Detection based defense mechanism. The traditional detection method is usually a binary classifier which is trained on both adversarial samples and clean samples [22]. However, these detectors failed to generalize across various attack schemes. More recently, Mend et al. [21] proposed an autoencoder based detector called MagNet, which rejects samples (as adversarial examples) with large reconstruction errors. One major advantage of MagNet is that it only requires clean examples for training the autoencoder, which significantly increases its generalization ability.
- Adversarial training. Adversarial training [27] utilizes adversarial examples and integrate them in model training. It can be also used in combination with gradient masking [3,23] which makes gradient-based attacks infeasible or difficult. The drawback of adversarial training is that it lacks the generalization ability to unseen adversarial examples and may compromise the model performance on clean examples. In addition, it requires a larger number of adversarial examples in the training stage.

Until now, there is no defense algorithms proposed for adversarial examples on sequential EHR data. The existing defense strategies for image data are either specific to the image domain, or require large volume of clean and adversarial training data, which is not suitable. MagNet has a strong generalization ability and does not depend on image characteristics. Besides, it does not require adversarial examples in training phase and is independent of the target classifier. In this work, we adapt this autoencoder based detection method and propose the first defense mechanism against adversarial examples on temporal EHR data.

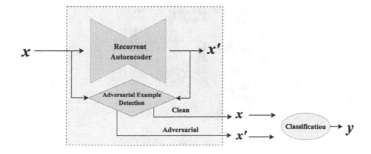

Fig. 1. RADAR pipeline

3   Methodology

In this section, we first give an overview of the RADAR framework. We then present the details of the recurrent autoencoder architecture, followed by the adversarial example detection criteria. Finally, we present our enhanced attack algorithm.

RADAR is an autoencoder based detector as shown in Fig. 1. A recurrent autoencoder consisting of encoder and decoder is trained on natural temporal examples and learns the manifold of the natural examples. At the test phase, given an input x, the autoencoder will push the reconstructed output x' closer to the manifold. Adversarially designed examples can be interpreted as out-of-manifold examples that are far away from natural example manifold. Therefore, when an adversarial example x is fed into a well trained autoencoder, the reconstruction distance between x and x' would be high. The stronger the adversarial perturbation, the larger the reconstruction distance. By contrast, as clean example itself is close to the manifold, the reconstruction distance would be small. Based on a set of carefully designed detection criteria including the reconstruction error, RADAR can detect adversarial examples. As autoencoder can push the reconstructed output closer to the manifold, it can play the role of a reformer. In other words, if an adversarial example is detected, its reconstructed output x' will be treated as reformed output and fed into the classifier.

3.1   Recurrent Autoencoder Architecture

Temporal EHR data is multivariate time series data. As our goal is to benefit from the autoencoder's reconstruction ability to distinguish adversarial examples and clean examples, it is crucial to build a recurrent autoencoder structure that is capable of learning both temporal correlations and feature correlations. In this work, we adopt the bidirectional-RNN with attention mechanism for temporal EHR. While the architecture is commonly used, the attention mechanism is first used for EHR data.

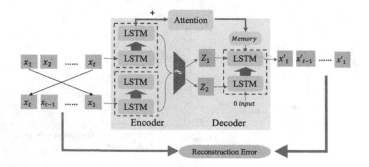

Fig. 2. BRNN-AE Architecture.

Our model is a bidirectional-RNN autoencoder which is shown in Fig. 2. For the RNN cell, we adopt a stacked LSTM cell designed to capture the long-term dependency and remember information for long periods of time. We feed into the bidirectional-RNN autoencoder with input $x_1, x_2, ..., x_t$ and reversed input $x_t, x_t - 1, ..., x_1$. The forward stacked LSTM of the encoder steps through forward input and encodes the input into hidden states h_{1f} for the first stack and h_{2f} for the second stack. Similarly, the backward stacked LSTM works on the reversed input and generates hidden states h_{1b} and h_{2b}. These hidden states are concatenated and a fully-connected layer is applied to form two fixed-length vectors z_1 and z_2. These two vectors are treated as the initial states of stacked LSTM cells in the decoder, feeding z_1 to the first stacked LSTM cell and z_2 to the second stacked LSTM cell, which enables the decoder to generate reconstructed output.

One limitation of this encoder and decoder structure is that when the input sequence is long, the fixed-length vector may fail to compress all the information. This issue is significant in temporal EHR data, as the duration of a patient's stay may vary and can be extremely long. To address this, we add the attention mechanism between the encoder and the decoder. Rather than encoding the input sequence into a fixed-length vector, attention forms a weighted sum of each hidden state, referred to as context vectors, allowing the decoder to focus on certain parts of the input when generating its output. In this work, we adopt Bahdanau attention [2] which uses weighted sum of attention weights and encoder hidden states to calculate context vectors and compute the final output of decoder.

We train the autoencoder on clean temporal EHR examples. The loss function is the reconstruction error between the input sequence and the generated output sequence, which is defined as:

$$L(x, x') = \|x, x'\|_2 + L_{reg}(\theta) \tag{1}$$

where L_{reg} denotes the L_1 regularization on parameters.

3.2   RADAR Detection Criteria

Given an input sequence and the reconstructed sequence, RADAR uses a set of detection criteria to distinguish between a clean example and an adversarial example. Considering the sparsity and high-dimensionality nature of EHR data, our detection criteria includes not only the reconstruction error and prediction divergence that are employed in MagNet, but also the prediction uncertainty of the target classifier.

Reconstruction Error. The reconstruction error between the original and reconstructed sequence is measured by the L_p-norm $L_p(x, x')$. Most commonly used L_p-norm is L_1 norm and L_∞ norm.

Prediction Divergence. In addition to the distance between x and x' in the data space, the prediction divergence between x and x' in their prediction output on the target classifier is also considered. The intuition is that clean examples should have a low divergence. Jensen Shannon Divergence (JSD), a symmetric measurement of the distribution similarity is applied to the target classifier's prediction logits, which is defined as:

$$JSD(l_x\|l_{x'}) = \frac{1}{2}KL(l_x\|\frac{1}{2}(l_x + l_{x'})) + \frac{1}{2}KL(l_{x'}\|\frac{1}{2}(l_x + l_{x'})) \qquad (2)$$

where l_x and $l_{x'}$ are the classifier's prediction logits of input x and reconstructed output x'. KL denotes the Kullback-Leibler divergence which is a non-symmetric measurement of the difference between two probability distributions. The lower value of JSD, the more similar two distributions are.

Prediction Uncertainty. In addition to the above two measures, we introduce a new criteria based on the prediction uncertainty of the reconstructed output on the target classifier. Our hypothesis is that the reconstructed output of an adversarial examples can result in more uncertainty on the prediction due to its goal of flipping the original class label. Prediction uncertainty focuses on the downstream prediction rather than the data itself thus can overcome the sparsity challenge of EHR data, and provide a critical and complementary criteria for detecting adversarial examples. Some existing works have proposed methods to measure neural network prediction uncertainty, such as entropy of predictive distribution [18], mutual information and differential entropy [28]. In this work, we use entropy of predictive distribution to reflect uncertainty, which is defined as:

$$Entropy(l_{x'}) = -\sum_{i=1}^{n} s_i log(s_i), \quad \text{where} \quad s_i = \frac{e^{l_{x'}^i}}{\sum_{j=1}^{n} e^{l_{x'}^j}} \qquad (3)$$

Here, n is the number of prediction classes, s_i is the softmax value of the ith class and $l_{x'}^i$ is the logits value of the ith class of x'.

Given an input x, RADAR detects it as an adversarial example if any one of the above three measurements is greater than a threshold: $M(x, x') > \delta^M$ where M represents reconstruction error, prediction divergence, and prediction uncertainty; and δ^M is the corresponding threshold. In practice, we can choose δ^M to allow a certain percentage of clean examples (e.g. 95%) to pass each criteria. We will study its tradeoff in the experiments section.

3.3 Enhanced Attack

In this paper, we also propose an enhanced attack algorithm that addresses the sparsity and high-dimensionality of sequential EHR data to generate more powerful adversarial examples.

Adversarial examples are designed by adding small perturbations to clean examples. For temporal EHR data, a clean example can be represented as $x \in \mathbb{R}^{t \times f} = \{x_1, x_2, ..., x_t\}$, where $x_i \in \mathbb{R}^f$ denotes the f-dimension feature space at the time step i. Given a classifier F, if x_{adv} satisfies that $F(x_{adv}) \neq F(x)$ and $L_p(x, x_{adv}) < C$, we say x_{adv} is the corresponding adversarial example of x. The attack algorithm that we applied to evaluate our proposed defense mechanism is similar to the method proposed in Sun et al. [30]. The purpose of the attack is to maximize the prediction logits on the position of targeted label (which equals to minimizing the logits on the position of true label) while minimizing the perturbation magnitude, which is formulated as:

$$\arg\min_{x_{adv}} L_y + \alpha L_x, \quad with \tag{4}$$

$$L_y = \max\{l(x_{adv})_{y_{true}} - l(x_{adv})_{y_{false}}, -k\} \quad and \quad L_x = ||x_{adv} - x||_p \tag{5}$$

where $l(\cdot)_{y_{true}}$ and $l(\cdot)_{y_{false}}$ denotes the logits on the position of true label and false label, as mortality prediction is a binary prediction. A positive value of k ensures a gap between true and adversarial label, which is commonly set to 0. α is a coefficient for the perturbation magnitude.

The L_p-norm is aimed to minimize the EHR location-wise similarity, which does not take into consideration the sparsity and high-dimensionality of sequential EHR data. Therefore, the adversarial examples generated by the attack algorithm can be easily detected by an autoencoder based detection. To craft more powerful adversarial examples, we introduce Gaussian observation [16] into the loss function to force the generated adversarial example to follow the same distribution as clean examples and less detectable by an autoencoder based detection. Gaussian observation is defined as the probability of clean example following the Gaussian distribution with mean as the corresponding adversarial examples and covariance as an identity matrix. Adding the objective of maximizing the Gaussian observation $N(x|x_{adv}, I)$, the attack algorithm can be formulated as a minimization problem:

$$\arg\min_{x_{adv}} L_y + \alpha L_x - \beta N(x|x_{adv}, I) \tag{6}$$

where α and β are the coefficients of the two parts of perturbation constraint. For the perturbation magnitude L_x, the L_1 norm induces sparsity on the perturbation and encourages the attack to be more focused on some specific location. By contrast, L_∞ norm encourages the perturbation to be more uniformly distributed with smaller magnitude on each location. In the experiments, we will compare the attack performance of L_1 norm and L_∞ norm with and without Gaussian observation.

4   Experimental Evaluation

In this section, we will first compare adversarial examples generated by our enhanced attack compared to existing works. Then, we will evaluate the detection performance of RADAR.

Dataset and Model Architecture. MIMIC-III (The Multiparameter Intelligent Monitoring in Intensive Care) dataset [14] is a publicly available clinic dataset containing thousands of de-identified intensive care unit patients' health care records. For mortality prediction, we directly adopt the processed MIMIC-III data from Sun et al. [30] The data contains 3177 positive samples and 30344 negative samples. Each sample consists of 48 timestamps and 19 features at each time step. These 19 variables include vital signs measurements such as heart rate, systolic blood pressure, temperature, and respiratory rate, as well as lab events such as carbon dioxide, calcium, and glucose. Missing features are imputed using average value across all timestamps and outliers are removed and imputed according to interquartile range (IQR) criteria. Then, each sequence is truncated or padded to the same length (48 h). After imputation and padding, each feature is normalized using min-max normalization.

The BRNN-AE architecture consists of an encoder with bi-directional two-stacked LSTM cells of units 32 and 64 respectively for both forward and backward LSTM, followed by two fully-connected layers of size 16 and 32 to form two fixed-length vectors as the input to decoder. The decoder consists of an attention layer of size 64 and two-stacked LSTM cells of size 16 and 32.

Pretrained Model Performance. Our target model is a mortality classifier. The network architecture is a simple LSTM of 128 units followed by a fully-connected layer of 32 units and a softmax layer. The 5-fold mean and standard deviation of the model performance is shown in Table 1.

Table 1. 5-fold cross validation performance of target classifier

Metric	Accuracy	AUC	F1	Precision	Recall
$Avg \pm STD$	0.894 ± 0.0124	0.812 ± 0.0187	0.603 ± 0.0279	0.536 ± 0.0548	0.702 ± 0.0564

4.1   Attack Performance

We use different distance metric to measure the similarity between adversarial examples and clean examples, including L_p-norm and KL divergence. L_p-norm aims to measure EHR location-wise similarity and KL divergence measures the distribution similarity over the whole set of adversarial examples and clean examples. A lower distance means a less detectable attack. In this experiment, the stop criteria for generating each adversarial example is when the prediction label is flipped. Only the successfully attacked examples will be used to calculate the L_p-norm and KL divergence.

Table 2 shows the distance metrics of the successfully flipped examples by different attacks. For the baseline attack with no distance optimization, the α and β in Eq. 6 are set to 0. For the L_1-norm attack (Sun et al.[30]) and L_∞-norm attack, α is set to 1 and β is set to 0. The last two columns correspond to our enhanced attacks with Gaussian observation. We observe that the no dist attack (that only aims to flip the label) has the highest distance as expected. Our enhanced attacks based on L_1 and L_∞ have the lowest L_1 and L_∞ distances respectively, and significantly outperform the existing L_1 and L_∞ based attacks.

Table 2. Attack performance comparison

Metric	Loss Func				
	No dist	L_1-norm	L_∞-norm	L_1-norm enhanced	L_∞-norm enhanced
L_1	3.672	0.815	0.920	**0.524**	0.792
L_∞	0.427	0.138	0.131	0.129	**0.119**
KL	6.521	0.736	0.817	0.811	**0.735**

This verifies the benefit of Gaussian observation in our enhanced attacks. By forcing the generated adversarial example to follow the same distribution as clean examples, it not only helps to decrease the KL divergence (in the case of L_∞ based attacks) but more importantly significantly decrease the L_p-norm. The comparison between L_1-norm and L_∞-norm enhanced attacks demonstrates that the L_∞-norm enhanced attack achieves smaller KL divergence, as it encourages the perturbation to be more uniformly distributed with smaller magnitude on each location.

The above results show the comparison of different attack methods for successfully flipped examples. To give a more comprehensive comparison, we also use varying perturbation magnitude as stopping criteria and compare the attack success rate and detection rate (by our detection approach) of different attack methods, which is shown in Fig. 3. In all cases, our enhanced attacks achieve a higher attack success rate and lower detection rate than the baseline attacks, which confirms the effectiveness of adding Gaussian observation as part of the minimization in the attack.

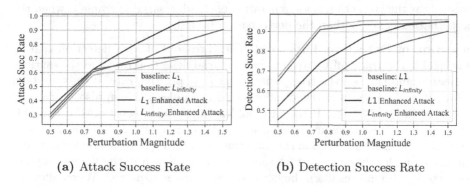

(a) Attack Success Rate   (b) Detection Success Rate

Fig. 3. Comparison between baseline attack and enhanced attack

To illustrate the perturbation introduced by the adversarial examples, we also show the mean perturbation for each of the feature-time points by our enhanced L_∞ attack added to the positive and negative clean examples respectively in Fig. 4. We observe that most of the perturbation is imposed on the recent time stamps. In addition, interestingly, it requires more perturbation to flip a positive example to negative than vice versa. The reason is that, for an imbalanced dataset, the confidence level is high when classifier predicts an example as positive, which means it requires more perturbation to flip its label.

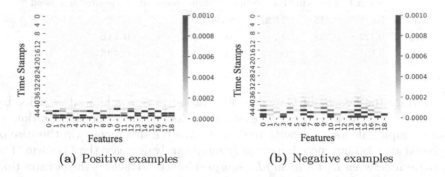

(a) Positive examples   (b) Negative examples

Fig. 4. Mean perturbation distribution

4.2   Detection Performance

In this section, we will first show the impact of varying detection threshold on the clean example pass rate and adversarial example detection rate, and then evaluate the detectability of RADAR in terms of detection rate and the accuracy of the classification model with the detection. We use L_∞-norm enhanced attack and apply varying perturbation bounds of 0.5, 0.75, 1.0, 1.25 and 1.5, which means that the stop criteria for generating each adversarial example is when the perturbation is larger than the perturbation bound.

Selection of Detection Threshold. The threshold of each detection criteria is crucial in the trade-off between the adversarial detection rate and the sacrifice of clean examples, i.e., the true positive and false positive rate. If the threshold is low, it can successfully detect adversarial examples but can also mistakenly filter out clean examples. If the threshold is high, the effectiveness of RADAR will be compromised. Figure 5 demonstrates this trade-off by showing the corresponding adversarial detection rate and the clean example pass rate for different thresholds under different perturbation bound. As shown in the figure, a higher perturbation bound results in higher detection rate as expected. When allowing more clean examples to pass, fewer adversarial examples can be detected. The optimal threshold would allow a majority of clean examples to pass while still remaining effective in detecting adversarial examples. In the following experiments, we select the threshold that allows 95% clean example pass rate.

Detection Success Rate. Figure 6 shows how much contribution each detection criterion makes to filter adversarial examples. It also compares RADAR (with all three criteria) and the existing MagNet approach (which uses the L-norm and JS Divergence only). With the increase of attack magnitude, the attack

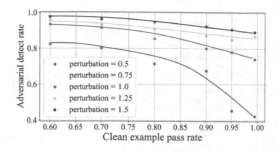

Fig. 5. The trade-off between adversarial detection rate and clean pass rate

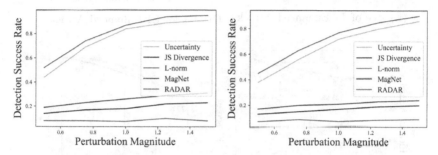

(a) RADAR performance under L_1 enhanced attack

(b) RADAR performance under L_∞ enhanced attack

Fig. 6. Contribution of each criterion and comparison of RADAR with MagNet

detection rate for all criteria/approaches increase as expected. Among the three criteria, our newly introduced prediction uncertainty makes the most and dominating contribution in detecting adversarial examples. As a result, RADAR dramatically outperforms MagNet.

Model Performance. We also evaluate the performance of RADAR in terms of the improvement of the target model's prediction accuracy and F1 score. Since any detection mechanism should not sacrifice the accuracy of clean examples, we report the accuracy of clean examples without RADAR (clean) and with RADAR (clean + RADAR). For the purpose of abalation study, we report the accuracy of adversarial examples under different scenarios: 1) when there is no defense (adv), 2) with detector only (adv + detector), 3) with reformer only (adv + reformer), and 4) with both detector and reformer (adv + RADAR). When the RADAR detector is used, if an example is detected as adversarial, we will flip its classification label and softmax output as the final prediction because our task is a binary classification. When only reformer is used, the autoencoder reconstructed output will be used for classification.

Figure 7 shows the target model accuracy and F1 score vs. varying perturbation magnitude for different methods under different attacks. For clean examples, employment of RADAR as a defense mechanism does not affect the prediction

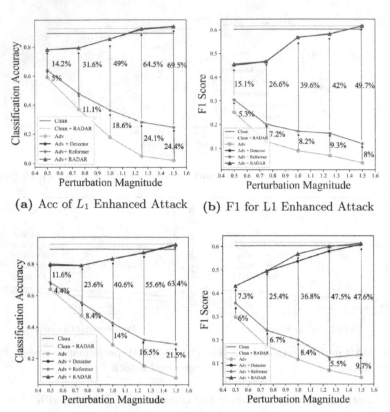

(a) Acc of L_1 Enhanced Attack   (b) F1 for L1 Enhanced Attack

(c) Acc of L_∞ Enhanced Attack   (d) F1 for L_∞ Enhanced Attack

Fig. 7. Performance improvement

performance and can even improve the accuracy. We speculate the reason is that the clean examples that are originally misclassified are usually close to the classification boundary or are outliers, hence may have a high prediction uncertainty or reconstruction error and be detected as adversarial examples. Once they are detected, their prediction will be automatically flipped, which will be correctly classified. Comparing the adversarial examples, only applying RADAR as a reformer can effectively reform the adversarial examples and improve the accuracy and F1 score by more than 10%. When RADAR works as both detector and reformer, it can additionally improve prediction accuracy by more than 60% and even exceeds the accuracy of clean examples. The F1 scores can also be improved by 40% when the perturbation magnitudes are larger than 1.0. The benefit of reformer on top of detector can be noticed in Fig. 7d. With increasing perturbation magnitude, the model accuracy and F1 score of adversarial examples with no defense and reformer drop dramatically due to the increasing attack power. However, interestingly, the model performance with the detection mechanism increases thanks to the increased detection rate as we have observed earlier. These experiments verify the significant improvement of the model performance and the effectiveness of the RADAR mechanism.

5   Conclusion

This paper is the first attempt to study potential defense methods for adversarial examples on temporal EHR data. We proposed a recurrent autoencoder based detection method called RADAR to detect adversarial examples according to autoencoder reconstruction error, prediction divergence, and prediction uncertainty. According to the evaluation on a mortality classifier, RADAR can effectively detect more than 90% of adversarial examples and improve the target model accuracy and F1 score by almost 90% and 60% respectively. Besides, we also introduced an enhanced adversarial attack by incorporating the distribution divergence into the loss function of the attack algorithm.

In the future, we plan to evaluate the performance of RADAR on other clinical deep learning systems such as readmission prediction models. In addition, the architecture of RADAR also has great potential to be improved by incorporating other deep learning models that are more powerful to model structural EHR data such as Graph Convolutional Networks (GCN).

Acknowledgement. This work is partially supported by the National Science Foundation (NSF) BigData award IIS 1838200, the Georgia Clinical & Translational Science Alliance under National Institutes of Health (NIH) CTSA Award UL1TR002378, and Air Force Office of Scientific Research (AFOSR) DDDAS award FA9550-12-1-0240. XJ is CPRIT Scholar in Cancer Research (RR180012), and he was supported in part by Christopher Sarofim Family Professorship, UT Stars award, UTHealth startup, the National Institute of Health (NIH) under award number R01AG066749, R01GM114612 and U01TR002062.

References

1. An, S., Xiao, C., Stewart, W.F., Sun, J.: Longitudinal adversarial attack on electronic health records data. In: The World Wide Web Conference (2019)
2. Bahdanau, D., Cho, K., Bengio, Y.: Neural machine translation by jointly learning to align and translate. arXiv preprint arXiv:1409.0473 (2014)
3. Buckman, J., Roy, A., Raffel, C., Goodfellow, I.: Thermometer encoding: one hot way to resist adversarial examples (2018)
4. Carlini, N., Wagner, D.: Towards evaluating the robustness of neural networks. In: 2017 IEEE Symposium on Security and Privacy (SP), pp. 39–57. IEEE (2017)
5. Cho, K., et al.: Learning phrase representations using RNN encoder-decoder for statistical machine translation. arXiv preprint arXiv:1406.1078 (2014)
6. Choi, E., Bahadori, M.T., Schuetz, A., Stewart, W.F., Sun, J.: Doctor AI: predicting clinical events via recurrent neural networks. In: Machine Learning for Healthcare Conference, pp. 301–318 (2016)
7. Das, N., et al.: Keeping the bad guys out: protecting and vaccinating deep learning with JPEG compression (2017)

8. Finlayson, S.G., Bowers, J.D., Ito, J., Zittrain, J.L., Beam, A.L., Kohane, I.S.: Adversarial attacks on medical machine learning. Science **363**, 1287–1289 (2019)

9. Finlayson, S.G., Chung, H.W., Kohane, I.S., Beam, A.L.: Adversarial attacks against medical deep learning systems. arXiv preprint arXiv:1804.05296 (2018)

10. Goodfellow, I.J., Shlens, J., Szegedy, C.: Explaining and harnessing adversarial examples. arXiv preprint arXiv:1412.6572 (2014)

11. Hinton, G.E., Salakhutdinov, R.R.: Reducing the dimensionality of data with neural networks. Science **313**(5786), 504–507 (2006)

12. Hochreiter, S., Schmidhuber, J.: Long short-term memory. Neural Comput. **9**(8), 1735–1780 (1997)

13. Jia, X., Wei, X., Cao, X., Foroosh, H.: ComDefend: an efficient image compression model to defend adversarial examples. In: Proceedings of the IEEE Conference on Computer Vision and Pattern Recognition, pp. 6084–6092 (2019)

14. Johnson, A.E., et al.: MIMIC-III, a freely accessible critical care database. Sci. Data **3**, 160035 (2016)

15. Kurakin, A., Goodfellow, I., Bengio, S.: Adversarial machine learning at scale. arXiv preprint arXiv:1611.01236 (2016)

16. Larsen, A.B.L., Sønderby, S.K., Larochelle, H., Winther, O.: Autoencoding beyond pixels using a learned similarity metric. arXiv preprint arXiv:1512.09300 (2015)

17. Li, Y., Zhang, H., Bermudez, C., Chen, Y., Landman, B.A., Vorobeychik, Y.: Anatomical context protects deep learning from adversarial perturbations in medical imaging. Neurocomputing **379**, 370–378 (2020)

18. Li, Y., Gal, Y.: Dropout inference in Bayesian neural networks with alpha-divergences. In: Proceedings of the 34th International Conference on Machine Learning, vol. 70, pp. 2052–2061. JMLR.org (2017)

19. Liao, F., Liang, M., Dong, Y., Pang, T., Hu, X., Zhu, J.: Defense against adversarial attacks using high-level representation guided denoiser. In: Proceedings of the IEEE Conference on Computer Vision and Pattern Recognition (2018)

20. Ma, X., et al.: Understanding adversarial attacks on deep learning based medical image analysis systems. arXiv preprint arXiv:1907.10456 (2019)

21. Meng, D., Chen, H.: MagNet: a two-pronged defense against adversarial examples. In: Proceedings of the 2017 ACM SIGSAC Conference on Computer and Communications Security, pp. 135–147. ACM (2017)

22. Metzen, J.H., Genewein, T., Fischer, V., Bischoff, B.: On detecting adversarial perturbations. arXiv preprint arXiv:1702.04267 (2017)

23. Papernot, N., McDaniel, P., Goodfellow, I., Jha, S., Celik, Z.B., Swami, A.: Practical black-box attacks against machine learning. In: Proceedings of the 2017 ACM on Asia Conference on Computer and Communications Security, pp. 506–519 (2017)

24. Pham, T., Tran, T., Phung, D., Venkatesh, S.: Predicting healthcare trajectories from medical records: a deep learning approach. J. Biomed. Inform. **69**, 218–229 (2017)

25. Schuster, M., Paliwal, K.K.: Bidirectional recurrent neural networks. IEEE Trans. Signal Process. **45**(11), 2673–2681 (1997)

26. Shickel, B., Tighe, P.J., Bihorac, A., Rashidi, P.: Deep EHR: a survey of recent advances in deep learning techniques for electronic health record (EHR) analysis. IEEE J. Biomed. Health Inform. **22**(5), 1589–1604 (2017)

27. Shrivastava, A., Pfister, T., Tuzel, O., Susskind, J., Wang, W., Webb, R.: Learning from simulated and unsupervised images through adversarial training. In: Proceedings of the IEEE Conference on Computer Vision and Pattern Recognition (2017)

28. Smith, L., Gal, Y.: Understanding measures of uncertainty for adversarial example detection. arXiv preprint arXiv:1803.08533 (2018)

29. Srivastava, N., Mansimov, E., Salakhudinov, R.: Unsupervised learning of video representations using LSTMs. In: International Conference on Machine Learning, pp. 843–852 (2015)
30. Sun, M., Tang, F., Yi, J., Wang, F., Zhou, J.: Identify susceptible locations in medical records via adversarial attacks on deep predictive models, pp. 793–801, July 2018. https://doi.org/10.1145/3219819.3219909
31. Sutskever, I., Vinyals, O., Le, Q.V.: Sequence to sequence learning with neural networks. In: Advances in Neural Information Processing Systems (2014)
32. Vatian, A., et al.: Impact of adversarial examples on the efficiency of interpretation and use of information from high-tech medical images. In: FRUCT (2019)
33. Wickramasinghe, N.: Deepr: a convolutional net for medical records (2017)
34. Zebin, T., Chaussalet, T.J.: Design and implementation of a deep recurrent model for prediction of readmission in urgent care using electronic health records. In: IEEE CIBCB (2019)
35. Zhang, J., Yin, P.: Multivariate time series missing data imputation using recurrent denoising autoencoder. In: 2019 IEEE BIBM, pp. 760–764. IEEE (2019)
36. Zheng, H., Shi, D.: Using a LSTM-RNN based deep learning framework for ICU mortality prediction. In: Meng, X., Li, R., Wang, K., Niu, B., Wang, X., Zhao, G. (eds.) WISA 2018. LNCS, vol. 11242, pp. 60–67. Springer, Cham (2018). https://doi.org/10.1007/978-3-030-02934-0_6

Self-supervised Log Parsing

Sasho Nedelkoski[1]($\boxtimes$), Jasmin Bogatinovski[1]($\boxtimes$), Alexander Acker[1],
Jorge Cardoso[2], and Odej Kao[1]

[1] Distributed Systems, TU Berlin, Berlin, Germany
{nedelkoski,jasmin.bogatinovski,alexander.acker,odej.kao}@tu-berlin.de
[2] Department of Informatics Engineering/CISUC, University of Coimbra,
Coimbra, Portugal
jcardoso@dei.uc.pt

Abstract. Logs are extensively used during the development and maintenance of software systems. They collect runtime events and allow tracking of code execution, which enables a variety of critical tasks such as
troubleshooting and fault detection. However, large-scale software systems generate massive volumes of semi-structured log records, posing a
major challenge for automated analysis. Parsing semi-structured records
with free-form text log messages into structured templates is the first
and crucial step that enables further analysis. Existing approaches rely
on log-specific heuristics or manual rule extraction. These are often specialized in parsing certain log types, and thus, limit performance scores
and generalization. We propose a novel parsing technique called NuLog
that utilizes a self-supervised learning model and formulates the parsing
task as masked language modeling (MLM). In the process of parsing,
the model extracts summarizations from the logs in the form of a vector embedding. This allows the coupling of the MLM as pre-training
with a downstream anomaly detection task. We evaluate the parsing
performance of NuLog on 10 real-world log datasets and compare the
results with 12 parsing techniques. The results show that NuLog outperforms existing methods in parsing accuracy with an average of 99% and
achieves the lowest edit distance to the ground truth templates. Additionally, two case studies are conducted to demonstrate the ability of
the approach for log-based anomaly detection in both supervised and
unsupervised scenario. The results show that NuLog can be successfully
used to support troubleshooting tasks. The implementation is available
at https://github.com/nulog/nulog.

Keywords: Representation learning · Log parsing · Transformers ·
Anomaly detection · IT systems

1 Introduction

Current IT systems are a combination of complex multi-layered software and
hardware. They enable applications of ever-increasing complexity and system

S. Nedelkoski and J. Bogatinovski—Equal contribution.

© Springer Nature Switzerland AG 2021
Y. Dong et al. (Eds.): ECML PKDD 2020, LNAI 12460, pp. 122–138, 2021.
https://doi.org/10.1007/978-3-030-67667-4_8

diversity, e.g., cloud platforms, where many technologies such as the Internet of Things (IoT), distributed processing frameworks, databases, and operating systems are used. The complexity and diversity of the systems relate to high managing and maintenance overhead for the operators to a point where they are no longer able to holistically operate and manage these systems. Therefore, service providers are introducing AI solutions for anomaly detection, error analysis, and recovery to the IT ecosystems [15,16]. The foundation for these data-driven troubleshooting solutions is the availability of data that describe the state of the systems. The large variety of used technologies leads to diverse data requiring the developed methods to generalize well over different applications, operating systems, or cloud infrastructure management tools.

One specific data source – the logs, are commonly used to inspect the behavior of an IT system. They represent interactions between data, files, services, or applications, which are typically utilized by developers, DevOps teams, and AI methods to understand system behaviors and to detect, localize, and resolve problems that may arise. The first step for understanding log information and their utilization for further automated analysis is to parse them. The content of a log record is semi-structured data which contains markers to separate semantic elements (e.g., timestamp, pid, and service name) and free-text written by software developers. The tagged data is relatively simple to process, while analyzing the text has always been a challenge[7,24]. The free text is a composition of constant string templates and variable values. The template is the logging instruction (e.g. *print()*, *log.info()*) from which the log message is produced. It records a specific system event. The general objective of a log parser is the transformation of the unstructured free-text into a structured log template and an associated list of variables. For example, the template *"Attempting claim: memory $\langle * \rangle$ MB, disk $\langle * \rangle$ GB, vcpus $\langle * \rangle$ CPU"* is associated with the variable list *["2048", "20", "1"]*. Here, $\langle * \rangle$ denotes the position of each variable and is connected with the positions of the values within the list. The variable list can be empty if a template does not contain variable parts.

Traditional log parsing techniques rely on regular expressions designed and maintained by human experts. Large systems consisting of diverse software and hardware components render it intricate to maintain this manual effort. Additionally, frequent software updates necessitate constant checking and adjusting of these statements, which is a tedious and error-prone task. Related log parsing methods [2,5,7,25] depend on parse trees, heuristics, and domain knowledge. They are either specialized to perform well on logs from specific systems or can reliably parse data with a low variety of unique templates. Analyzing the performance of existing log parsing methods on a variety of diverse systems reveals their lack of robustness to produce consistently good parsing results [24]. This implies the necessity to choose a parsing method for the application or system at hand and incorporating domain-specific knowledge. Operators of large IT infrastructures would end up with the overhead of managing different parsing methods for their components whereof each need to be accordingly understood. We state that log parsing methods have to be accurate on log data from various systems ranging from single applications over mobile operating systems to cloud infrastructure management platforms with minimal human intervention as possible.

Accurate parsing of log messages into templates is crucial for many log processing methods including anomaly detection and root cause analysis [3,9,17]. However, recent log anomaly detection approaches, are applying methods from natural language processing (NLP) to achieve improved results and provide robustness against changes of the underlying system (e.g. software updates) [10, 23]. Thereby, semantic embeddings for parsed log templates together with a defined distance measure enable a notion of similarity between those. The described procedures to generate word and log template embedding vectors introduce additional sources of uncertainty. It requires two additional methods (word and template embedding generation) that rely on good parsing results and are crucial for the subsequent anomaly detection task. Furthermore, the proposed semantic embedding models were trained on text datasets (e.g., Wikipedia) or news articles instead of log data, which renders their embedding results for log templates as not persuasive. Since a log parsing method needs to analyze the log dataset anyway, it is reasonable to incorporate the task of generating semantic log template embeddings into the parsing procedure. Such a log parsing approach would meet the requirements of recent log analysis methods and avoid the additional external embedding generators. To the best of our knowledge, no related log parsing method exists that can provide a representational embedding of the parsed log messages making it impossible to directly connect them to models for fine-tuning in downstream tasks that require those.

Contribution. We propose a self-supervised method for log parsing NuLog, which utilizes the transformer architecture [1,21]. Self-supervised learning is a form of unsupervised learning where parts of the data provide supervision. To build the model, the learning task is formulated such that the presence of a word on a particular position in a log message is conditioned on its context. This is done with masking the word which is predicted. The model is forced to learn the appearance of the word within its context. The key idea for parsing is that the correct prediction of the masked word means that the word isjkk]=] part of the log template. Otherwise it is a parameter of the log. The advantages of this approach are that it can produce both a log template and a numerical vector sumarization, while domain knowledge is not needed. Through exhaustive experimentation, we show that NuLog outperforms the previous state of the art log parsing methods and achieves the best scores overall. The model is robust and generalizes well across different datasets. Further, we illustrate two use cases, supervised and unsupervised, on how the model can be coupled with and fine-tuned for a downstream task like anomaly detection. The results suggest that the knowledge obtained during the masked language modeling is useful as a good prior knowledge for the downstream tasks.

2   Related Work

Automated log parsing is important due to its practical relevance for the maintenance and troubleshooting of software systems. A significant amount of research and development for automated log parsing methods has been published in both industry and academia [6, 24]. Parsing techniques can be distinguished in various aspects, including technological, operation mode, and preprocessing. In Fig. 1, we give an overview of the existing methods.

Clustering. The main assumption in these methods is that the message types coincide in similar groups. Various clustering methods with proper string matching distances have been used. Methods in this groups are SHISO, LenMa, Log-Mine, LKE, and LogSig [4, 5, 12, 18, 19].

Frequent pattern mining assumes that a message type is a frequent set of tokens that appear throughout the logs. The procedures involve creating frequent sets, grouping the log messages, and extraction of message types. Representative parsers for this group are SLCT, LFA, and LogCluster [13, 14, 22].

Evolutionary. Its member MoLFI [11] uses an evolutionary approach to find the Pareto optimal set of message templates.

Log-structure heuristics methods produce the best results among the different adopted techniques [6, 24]. They usually exploit different properties that emerge from the structure of the log. The state of the art algorithm Drain [7] assumes that at the beginning of the logs the words do not vary too much. It uses this assumption to create a tree of fixed depth which can be easily modified for new groups. Other parsing methods in this group are IPLoM and AEL [8, 22].

Longest-common sub-sequence uses the longest common subsequence algorithm to dynamically extract log patterns from incoming logs. Here the most representative parser is Spell [2].

Our method belongs to a new category called **Neural** in the taxonomy of log parsing methods. Different from the current state-of-the-art heuristic-based methods, our method does not require any domain knowledge. Through empirical results, we show that the model is robust and applicable to a range of log types in different systems. We believe that in future this category will have the most influence considering the advances of deep learning.

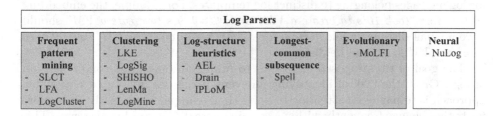

Log Parsers					
Frequent pattern mining	**Clustering** - LKE - LogSig - SHISHO - LenMa - LogMine	**Log-structure heuristics** - AEL - Drain - IPLoM	**Longest-common subsequence** - Spell	**Evolutionary** - MoLFI	**Neural** - NuLog
- SLCT - LFA - LogCluster					

Fig. 1. Taxonomy of log parses according to the underlying technology they adopt.

3   Neural Log Parsing

3.1   Preliminaries

We define a log as a sequence of temporally ordered unstructured text messages $L = (l_i : i = 1, 2, ...)$, where each message l_i is generated by a logging instruction (e.g. *printf()*, *log.info()*) within the software source code, and i is its positional index within the sequence. The log messages consist of a constant and an optional varying part, respectively referred to as log template and variables. We define log templates and variables as tuples $EV = ((e_i, v_i) : e \in \mathbb{E}, i = 1, 2, ...)$, where $\mathbb{E}$ is the finite set of all log event templates, $K = |\mathbb{E}|$ is the number of all unique templates and v_i is a list of variables for the respectively associated template. They are associated with its original log message by the positional index i.

The smallest inseparable singleton object within a log message is a token. Each log message consists of a bounded sequence of tokens, $\mathbf{t_i} = (t_j : t \in \mathbb{T}, j = 1, 2, ..., |\mathbf{t_i}|)$, where $\mathbb{T}$ is a set of all tokens, j is the positional index of a token within the log message l_i, and $|\mathbf{t_i}|$ is the total number of tokens in l_i. For different l_i, $|\mathbf{t_i}|$ can vary. Depending on the concrete tokenization method, t can be a word, word piece, or character. Therefore, tokenization is defined as a transformation function $\mathcal{T} : l_i \rightarrow \mathbf{t_i}, \forall i$.

With respect to our proposed log parsing method, the notions of context and embedding vector are additionally introduced. Given a token t_j, its context is defined by a preceding and subsequent sequence of tokens, i.e. a tuple of sequences: $C(t_j) = ((t_a, t_{a+1}, ..., t_{j-1}), (t_{j+1}, t_{j+2}, ..., t_b))$, where $a < j < b$. An embedding vector is a d-dimensional real valued vector representation $\mathbf{s} \in \mathbb{R}^d$ of either a token or a log message.

We establish a requirement and a property for the proposed log parsing method:

Requirement. Given a temporally ordered sequence of log messages L, generated from an unknown set $\mathbb{E}$ of distinct log templates, the log parsing method should provide a mapping function $f_1 : L \rightarrow EV$.

Property. The log parsing approach enables for vector representation of a log (log2vec), which leads to a possibility for addressing various downstream tasks like anomaly detection.

The generated vector representations should be closer embedding vectors for log messaged belonging to the same log template and distant embedding vectors for log messages belonging to distinct log templates. For example, the embedding vectors for *"Took 10 s to create a VM"* and *"Took 9 s to create a VM"* should have a small distance while vectors for *"Took 9 s to create a VM"* and *"Failed to create VM 3"* should be distant.

The goal of the proposed method is to mimic an operator's comprehension of logs. Given the task of identifying all event templates in a log, a reasonable approach is to pay close attention to parts that re-appear constantly and ignore parts that change frequently within a certain context (e.g. per log message). This can be modelled as a probability distribution for each token conditioned on its

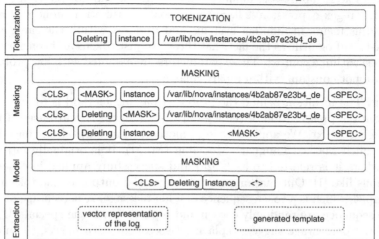

Fig. 2. Instance of parsing of a single log message with NuLog.

context, i.e. $P(t_j|C(t_j))$. Such probability distribution would allow the distinction of constant and varying tokens, referring to solving **Requirement**. The generation of log embedding vectors would naturally enable utilization of such representation for fine-tuning in downstream tasks. Moreover, the representation is obtained by focusing on constant parts of the log message, as they are more predictable, providing the necessary generalization for **Property**.

3.2 NuLog: Self-attentive Neural Parsing with Transformers

The proposed method is composed of preprocessing, model, and template extraction. The overall architecture based on an example log message input is depicted in Fig. 2.

The log preprocessor transforms the log messages into a suitable format for the model. It is composed of two main parts: tokenization and masking. Before the tokenization task, the meta-information from the logging frameworks is stripped, and the payload, i.e., the print statement, is used as input to the tokenization step.

Tokenization. Tokenization transforms each log message into a sequence of tokens. For NuLog, we utilize a simple filter based splitting criterion to perform a string split operation. We keep these filters short and simple, i.e. easy to construct. All concrete criteria are described in Sect. 4.1. In Fig. 2 we illustrate the tokenization of the log message "*Deleting instance /var/lib/nova/instances/4b2ab87e23b4de*". If a splitting criterion matches white spaces, then the log message is tokenized as a list of three tokens [*"Deleting"*, *"instance"*, *"/var/lib/nova/instances/4b2ab87e23b4de"*]. In contrast to several related approaches that use additional hand-crafted regular expressions to parses parameters like IP addresses, numbers, and URLs, we do not parse any parameters with a regex expression [24]. Such approaches are error-prone and require manual adjustments in different systems and updates within the same system. In contrast, NuLog utilizes the fact that these are just tokens and considers their appearance within a context. These parameters are assigned with low probability as they are not constant within a particular context.

Masking. The intuition behind the proposed parsing method is to learn a general semantic representation of the log data by analyzing occurrences of tokens within their context. We apply a general method from natural language (NLP) research called Masked Language Modeling (MLM). It is originally introduced in [20] (where it is referred to as *Cloze*) and successfully applied in other NLP publications like [1]. Our masking module takes the output of the tokenization step as input, which is a token sequence of a log message. A percentage of tokens from the sequence are randomly chosen and replaced with the special $\langle MASK \rangle$ token. If the percentage suggest replacing two tokens with masks, the masking module will create two samples where each of the words will be masked once as depicted in Fig. 2. The masked token sequence is used as input for the model, while the masked token acts as the prediction target. To denote the start and end of a log message, we prepend a special $\langle CLS \rangle$ and apply padding with $\langle SPEC \rangle$ tokens. The number of padding tokens for each log message is given by $M - |\mathbf{t_i}|$, where $M = \max(|\mathbf{t_i}|) + 1$, $\forall i$ is the maximal number of tokens across all log messages within the log dataset added by one, and $|\mathbf{t_i}|$ is the number of tokens in the *i-th* log message. Note, that the added one ensures that each log message is padded by at least one $\langle SPEC \rangle$ token.

Model. The method has two operation modes - offline and online. During the offline phase, log messages are used to tune all model parameters via backpropagation and optimal hyper-parameters are selected. During the online phase, every log message is passed forward through the model. This generates the respective log template and an embedding vector for each log message.

Figure 3 depicts the complete architecture. The model applies two operations on the input token vectors: token vectorization and positional encoding. The subsequent encoder structure takes the result of these operations as input. It is composed of two elements: self-attention layer and feedforward layer. The last model component is a single linear layer with a softmax activation overall tokens appearing in the logs. In the following, we provide a detailed explanation of each model element.

Since all subsequent elements of the model expect numerical inputs, we initially transform the tokens into randomly initialized numerical vectors $\mathbf{x} \in \mathbb{R}^d$. These vectors are referred to as token embeddings and are part of the training process, which means they are adjusted during training to represent the semantic meaning of tokens depending on their context. These numerical token embeddings are passed to the positional encoding block. In contrast to e.g., recurrent architectures, attention-based models do not contain any notion of input order. Therefore, this information needs to be explicitly encoded and merged with the input vectors to take their position within the log message into account. This

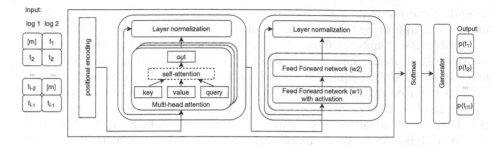

Fig. 3. Model architecture of NuLog for parsing of the logs.

block calculates a vector $\mathbf{p} \in \mathbb{R}^d$ representing the relative position of a token based on a sine and cosine function.

$$p_{2k} = sin\left(\frac{j}{10000^{\frac{2k}{v}}}\right), \quad p_{2k+1} = cos\left(\frac{j}{10000^{\frac{2k+1}{v}}}\right). \tag{1}$$

Here, $k = 0, 1, \ldots, d-1$ is the index of each element in $\mathbf{p}$ and $j - 1, 2, \ldots, M$ is the positional index of each token. Within the equations, the parameter k describes an exponential relationship between each value of vector $\mathbf{p}$. Additionally, a sine and cosine function are interchangeably applied. Both allow better discrimination of the respective values within a specific vector of $\mathbf{p}$. Furthermore, both functions have an approximately linear dependence on the position parameter j, which is hypothesized to make it easy for the model to attend to the respective positions. Finally, both vectors can be combined as $\mathbf{x}' = \mathbf{x} + \mathbf{p}$.

The encoder block of our model starts with a multi-head attention element, where a softmax distribution over the token embeddings is calculated. Intuitively, it describes the significance of each embedding vector for the prediction of the target masked token. We summarize all token embedding vectors as rows of a matrix X' and apply the following formula

$$X_l'' = softmax\left(\frac{Q_l \times K_l^T}{\sqrt{w}}\right) \times V_l, \text{ for } l = 1, 2, \ldots, L, \tag{2}$$

where L denotes the number of attention heads, $w = \frac{d}{L}$ and $d \bmod L = 0$. The parameters Q, K and V are matrices, that correspond to the query, key, and value elements in Fig. 3. They are obtained by applying matrix multiplications between the input X' and respective learnable weight matrices W_l^Q, W_l^K, W_l^V:

$$Q_l = X' \times W_l^Q, \ K_l = X' \times W_l^K, \ V_l = X' \times W_l^V, \tag{3}$$

where W_l^Q, W_l^K, $W_l^V \in \mathbb{R}^{M \times w}$. The division by $\sqrt{w}$ stabilizes the gradients during training. After that, the softmax function is applied and the result is used to scale each token embedding vector V_l. The scaled matrices X_l'' are concatenated to a single matrix X'' of size $M \times d$. As depicted in Fig. 3 there is a residual connection between the input token matrix X' and its respective attention transformation X'', followed by a normalization layer *norm*. These are used for improving the performance of the model by tackling different potential problems encountered during the learning such as small gradients and the covariate shift phenomena. Based on this, the original input is updated by the attention-transformed equivalent as $X' = norm(X' + X'')$.

The last element of the encoder consists of two feed-forward linear layers with a ReLU activation in between. It is applied individually on each row of X'. Thereby, identical weights for every row are used, which can be described as a convolution over each attention-transformed matrix row with kernel size one. This step serves as additional information enrichment for the embeddings. Again, a residual connection followed by a normalization layer between the input matrix and the output of both layers is employed. This model element preserves the dimensionality X'.

The final element of the model consists of a single linear layer. It receives the encoder result X' and extracts the token embedding vector of the $\langle CLS \rangle$ token. Since every log message token sequence is pre-padded by this special token, it is the first row of the matrix, i.e. $x_0' \in X'$. The linear layer maps this vector of size d to a vector whose size corresponds to the total number of tokens $|\mathbb{T}|$ in the dataset. The subsequent softmax is utilized to calculate a probability distribution over each element of $\mathbb{T}$. During training, the masked token is used as the target to be predicted. Since the last vector embedding of the $\langle CLS \rangle$ token is used for prediction, it is forced to summarize the log message. Otherwise, it would not be able to solve the masked token prediction task well enough across all tokens. We hypothesize that the constant part of log templates will constraint the model to learn similar $\langle CLS \rangle$ token embeddings when log messages are of the same template. This leads to a mapping of the log messages to their vector representation, which can after be used for diverse downstream tasks like anomaly detection. This log message embedding vector satisfies the proposed **Property** (see Sect. 3.1).

3.3   Log Template Extraction

The extraction of all log templates within a log dataset is executed online, after the model training. Therefore, we pass each log message as input and configure the masking module in a way that every token is masked consecutively, one at a time. We measure the model's ability to predict each token, and thus, decide whether the token is a constant part of the template or a variable. High confidence in the prediction of a specific token indicates a constant part of the template, while small confidence is interpreted as a variable. More specifically, for the results in this paper we employ the following procedure. If the prediction of a particular token is in the top ϵ predictions and doesn't contain numbers, we consider it as a constant part of the template, otherwise, it is considered to be a variable. For all variables, an indicator $\langle * \rangle$ is placed on its position within the log message. However, This addresses the **Requirement** proposed in Sect. 3.1.

4   Evaluation

To quantify the performance of the proposed method, we perform an exhaustive evaluation of the log parsing task on a set of 10 benchmark datasets and compare the results with 12 other log template parsing methods. The datasets together with the implementation of the other parsers were obtained from the log benchmark [24]. Furthermore, the model of NuLog provides log message vector embeddings. We show that these, along with the model, can be used for anomaly detection as downstream tasks.

4.1   Datasets

The log datasets employed in our experiments are summarized in Table 1. These real-world log data range from supercomputer logs (BGL and HPC), distributed system logs (HDFS, OpenStack, Spark), to standalone software logs (Apache, Windows, Mac, Android). To enable reproducibility, we follow the guidelines from [24] and utilize a random sample of 2000 log messages from each dataset, where the ground truth templates are available. The number of templates (#T) contained within each dataset is shown in Table 1.

The BGL dataset is collected by Lawrence Livermore National Labs (LLNL) from BlueGene/L supercomputer system. HPC logs are collected from a high-performance cluster, consisting of 49 nodes with 6,152 cores. HDFS is a log data set collected from the Hadoop distributed file system deployed on a cluster of 203 nodes within the Amazon EC2 platform. OpenStack is a result of a conducted anomaly experiment within CloudLab with one control node, one network node and eight compute nodes. Spark is an aggregation of logs from the Spark system deployed within the Chinese University of Hongkong, which comprises 32 machines. The Apache HTTP server dataset consists of access and error logs from the apache web server. Windows, Mac, and Android datasets consist of logs generated from single machines using the respectively named operating system. HealthApp contains logs from an Android health application, recorded over ten days on a single android smartphone.

As described in Sect. 3.2, the tokenization process of our method is implemented by splitting based on a filter. We list the applied splitting expressions for each dataset in Table 1. Besides, we also list the additional training parameters. The number of epochs is determined by an early stopping criterion, which terminated the learning when the loss converges. The hyperparameter ϵ is determined via cross-validation.

4.2   Evaluation Methods

To quantify the effectiveness of NuLog for log template generation from the presented 10 datasets, we compare it with 12 existing log parsing methods on parsing accuracy, edit distance, and robustness. We reproduced the results from Zhu et al. [24] for all known log parsers. Furthermore, we enriched the extensive benchmark reported by an additional metric, i.e., edit distance. Note, that all

Table 1. Datasets and *NuLog* hyperparameter setting.

System	#T	Tokenization filter	#epochs	ϵ
BGL	120	([|:|\(|\)|=|,])|(core.)|(\.{2,})	3	50
Android	166	([|:|\(|\)|=|,|"|\{|\}|@|\$|\[|\]|||;])	5	25
OpenStack	43	([|:|\(|\)|"|\{|\}|@|\$|\[|\]|||;])	6	5
HDFS	14	(\s+blk_)|(:)|(\s)	5	15
Apache	6	([ ])	5	12
HPC	46	([|=])	3	10
Windows	50	([ ])	5	95
HealthApp	75	([ ])	5	100
Mac	341	([ ])|([\w-]+\.){2,}[\w-]+	10	300
Spark	36	([ ])|(\d+\sB)|(\d+\sKB)|(\d+\.){3}\d+	3	50

methods we comparing with are described in detail in Sect. 2. To evaluate the log message embeddings for the anomaly detection downstream tasks, we use the common metrics accuracy, recall, precision, and F1 score. In the following, we describe each evaluation metric.

Parsing Accuracy. To enable comparability between our method to the previous work [24], we adopt their standard proposed parsing accuracy (PA) metric. It is defined as the ratio of correctly parsed log messages over the total number of log messages. After parsing, each log message is assigned to a log template. A log message is considered correctly parsed if its log template corresponds to the same group of log messages as the ground truth does. For example, if a log sequence $[e_1, e_2, e_2]$ is parsed to $[e_1, e_4, e_5]$, we get $PA = \frac{1}{3}$ since the second and third messages are not grouped together.

Edit Distance. The PA metric is considered as the standard for evaluation of log parsing methods, but it has limitations when it comes to evaluating the template extraction in terms of string comparison. Consider a particular group of logs produced from single *print("VM created successfully")* statement that is parsed with the word *VM*. As long as this is consistent over every occurrence of the templates from this group throughout the dataset, PA would still yield a perfect score for this template parsing result, regardless of the obvious error. Therefore, we introduce an additional evaluation metric: Levenshtein edit distance. This is a way of quantifying how dissimilar two log messages are to one another by counting the minimum number of operations required to transform one message into the other.

4.3 Parsing Results

Parsing Accuracy. This section presents and discusses the log parsing PA results of NuLog on the benchmark datasets and compares them with twelve

Table 2. Comparisons of log parsers and our method NuLog in parsing accuracy (PA).

Dataset	SLCT	AEL	LKE	LFA	LogSig	SHISHO	LogCluster	LenMa	LogMine	Spell	Drain	MoLFI	BoA	NuLog
HDFS	0.545	0.998	1.000	0.885	0.850	0.998	0.546	0.998	0.851	1.000	0.998	0.998	**1.000**	0.998
Spark	0.685	0.905	0.634	**0.994**	0.544	0.906	0.799	0.884	0.576	0.905	0.920	0.418	0.994	**1.000**
OpenStack	**0.867**	0.758	0.787	0.200	0.200	0.722	0.696	0.743	0.743	0.764	0.733	0.213	0.867	**0.990**
BGL	0.573	0.758	0.128	0.854	0.227	0.711	0.835	0.690	0.723	0.787	**0.963**	0.960	0.963	**0.980**
HPC	0.839	**0.903**	0.574	0.817	0.354	0.325	0.788	0.830	0.784	0.654	0.887	0.824	0.903	**0.945**
Windows	0.697	0.690	0.990	0.588	0.689	0.701	0.713	0.566	0.993	0.989	**0.997**	0.406	0.997	**0.998**
Mac	0.558	0.764	0.369	0.599	0.478	0.595	0.604	0.698	**0.872**	0.757	0.787	0.636	**0.872**	0.821
Android	0.882	0.682	0.909	0.616	0.548	0.585	0.798	0.880	0.504	**0.919**	0.911	0.788	**0.919**	0.827
HealthApp	0.331	0.568	0.592	0.549	0.235	0.397	0.531	0.174	0.684	0.639	**0.780**	0.440	0.780	**0.875**
Apache	0.731	1.000	1.000	1.000	1.000	1.000	0.709	1.000	1.000	1.000	1.000	1.000	**1.000**	1.000

other related methods. These are presented in Table 2. Specifically, each row contains the datasets while the compared methods are represented in the table columns. Additionally, the penultimate column contains the highest value of the first twelve columns - referred to as best of all - and the last column contains the results for NuLog. In the bold text, we highlight the best of the methods per dataset. HDFS and Apache datasets are most frequently parsed with 100% PA. This is because HDFS and Apache error logs have relatively unambiguous event templates that are simple to identify. On those, NuLog achieves comparable results. For the Spark, BGL and Windows dataset, the existing methods already achieve high PA values above 96% (BGL) or above 99% (Spark and Windows). Our proposed method can slightly outperform those. For the rather complex log data from OpenStack, HPC and HealthApp the baseline methods achieve a PA between 78% and 90%, which NuLog significantly outperforms by 4–13%.

PA Robustness. Employing a general parsing method in production requires a robust performance throughout different log datasets. With the proposed method, we explicitly aim at supporting a broad range of diverse log data types. Therefore, the robustness of NuLog is analyzed and compared to the related methods. Figure 4 shows the accuracy distribution of each log parser across the log datasets within a boxplot. From left to right in the figure, the log parsers are arranged in ascending order of the median PA. That is, LogSig has the lowest and NuLog obtains the highest parsing accuracy on the median. We postulate the criterion of achieving consistently high PA values across many different log types as crucial for their general use. However, it can be observed that, although most log parsing methods achieve high PA values of 90% for specific log datasets, they have a large variance when applied across all given log types. NuLog outperforms every other baseline method in terms of PA robustness yielding a median of 0.99, which even lies above the best of all median of 0.94.

Edit Distance. As an evaluation metric, PA measures how well the parsing method can match log templates with the respective log messages throughout the dataset. Additionally, we want to verify the correctness of the templates, e.g., whether all variables are correctly identified. To achieve this, the edit distance

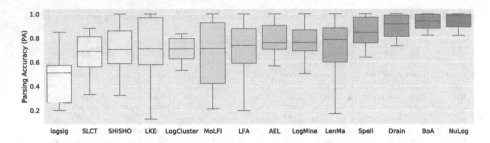

Fig. 4. Robustness evaluation on the parsing accuracy of the log parsers.

score is employed to measure the dissimilarity between the parsed and the ground truth log templates. Note that this indicates that the objective is to achieve low edit distance values. All edit distance scores are listed in Table 3. The table structure is the same as for PA results. In bold we highlight the best edit distance value across all tested methods per dataset. It can be seen that in terms of edit distance NuLog outperforms existing methods on the HDFS, Windows, Android, HealthApp and Mac datasets. It performs comparable on the BGL, HPC, Apache and OpenStack datasets and achieves a higher edit distance on the Spark log data.

Edit Distance Robustness. Similar to the PA robustness evaluation, we want to verify how consistent NuLog is performing in terms of edit distance across the different log datasets. Figure 5 shows a box-plot that indicates the edit distance distribution of each log parser for all log datasets. From left to right in the figure, the log parsing methods are arranged in descending order of the median edit distance. Again, it can be observed that although most log parsing methods achieve the minimal edit distance scores under 10, most of them have a large variance over different datasets and are therefore not generally applicable for diverse log data types. MoLFI has the highest median edit distance, while Spell and Drain perform constantly well - i.e. small median edit distance values - for multiple datasets. Again, our proposed parsing method outperforms the lowest edit distance values with a median of 5.00, which is smaller the best of all median of 7.22.

5   Case Study: Anomaly Detection as a Downstream Task

In previous sections, we demonstrate the effectiveness of NuLog in terms of accuracy, edit distance and robustness over different datasets. Although high accuracy is necessary for log parsing methods, it does not guarantee good performance in important subsequent log mining task like anomaly detection. Little parsing error may cause an order of magnitude performance degradation in log mining when the errors occur in log messages that are infrequent but essential for the log analysis task [6].

Table 3. Comparisons of log parsers and our method NuLog in edit distance.

Dataset	LogSig	LKE	MoLFI	SLCT	LFA	LogCluster	SHISHO	LogMine	LenMa	Spell	AEL	Drain	BoA	NuLog
HDFS	19.1595	17.9405	19.8430	13.6410	30.8190	28.3405	10.1145	16.2495	10.7620	9.2740	8.8200	8.8195	8.8195	**3.2040**
Spark	13.0615	41.9175	14.1880	6.0275	9.1785	17.0820	7.9100	16.0040	10.9450	6.1290	3.8610	**3.5325**	3.5325	12.0800
BGL	11.5420	12.5820	10.9250	9.8410	12.5240	12.9550	8.6305	19.2710	8.3730	7.9005	5.0140	**4.9295**	4.9295	5.5230
HPC	4.4475	7.6490	3.8710	2.6250	3.1825	3.5795	7.8535	3.2185	2.9055	5.1290	**1.4050**	2.0155	1.4050	2.9595
Windows	7.6645	11.8335	14.1630	7.0065	10.2385	6.9670	5.6245	6.9190	20.6615	4.4055	11.9750	6.1720	5.6245	**4.4860**
Android	16.9295	12.3505	39.2700	3.7580	9.9980	16.4175	10.1505	22.5325	3.2555	8.6680	6.6550	3.2210	3.2210	**1.1905**
HealthApp	17.1120	14.6675	21.6485	16.2365	20.2740	16.8455	24.4310	19.5045	16.5390	8.5345	19.0870	18.4965	14.6675	**6.2075**
Apache	14.4420	14.7115	18.4410	11.0260	10.3675	16.2765	12.4405	10.2655	13.5520	10.2335	10.2175	**10.2175**	10.2175	11.6915
OpenStack	21.8810	29.1730	67.8850	20.9855	28.1385	31.4860	18.5820	23.9795	18.5350	27.9840	**17.1425**	28.3855	17.1425	21.2605
Mac	27.9230	79.6790	28.7160	34.5600	41.8040	21.3275	19.8105	17.0620	19.9835	22.5930	19.5340	19.8815	17.062	**2.8920**

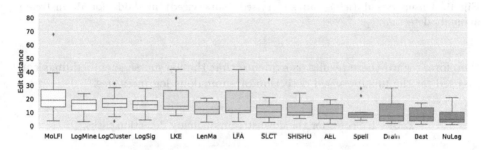

Fig. 5. Robustness evaluation on the edit distance of the log parsers.

Our model architecture allows for coupling of the parsing approach and a downstream anomaly detection task. The knowledge obtained during the log parsing phase is used as a good prior bias. The architecture allows to treat the problem of anomaly detection in both supervised and unsupervised way. To illustrate this, we designed two experimental case studies described in the following.

5.1   Unsupervised Anomaly Detection

We test the log message embeddings produced by NuLog for unsupervised log anomaly detection by employing a similar approach as during parsing. The model is trained for three epochs. Each token of a log message is masked and predicted based on the $\langle CLS \rangle$ token embedding. All respectively masked tokens that are not in the top-ϵ predictions are marked as anomalies. We compute the percentage of anomalous tokens within the log message to decide whether the whole log message is anomalous. If it is larger than a threshold δ, the log message is considered as an anomaly, otherwise as normal. This process is shown in the left part of Fig. 6.

To the best of our knowledge, only the BGL dataset contains anomaly labels for each individual log message, and is therefore suitable to evaluate the proposed anomaly detection approach. Due to its large volume, we use only the first 10% of it. For training 80% of that portion is utilized, while the rest is used for testing. In the first row of Table 4 we show the accuracy, recall, precision, and F1 score results. It can be seen that the method yields scores between 0.999 and 1.0. We,

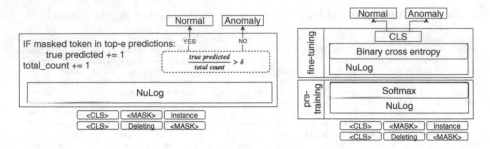

Fig. 6. Unsupervised (left) and supervised (supervised) methods for downstream anomaly detection.

therefore, regard these results as evidence that the log message embeddings can be used for the unsupervised detection of anomalous log messages.

Table 4. Scores for the downstream anomaly detection tasks.

	Accuracy	Recall	Precision	F1 Score
Unsupervised	0.999	0.999	1.000	0.999
Supervised	0.999	1.000	0.999	0.999

5.2 Supervised Anomaly Detection

For the second case study, we utilize log message embeddings as a feature for supervised anomaly detection. First, the model is trained on the self-supervised MLM task. Second, we replace the last softmax layer by a linear layer, that is adapted via supervised training of predicting a given $\langle CLS \rangle$ as either normal or anomaly, i.e. binary classification. For this downstream task, we applied a fine-tuning of two epochs.

Again, the first 10% of the BGL dataset were used for evaluation. Thereby, the model is trained on the first 80% and evaluated on the remaining 20%. The results are listed in the second row of Table 4 and show that two epochs of fine-tuning are sufficient to produce an F1 score of 0.99. It further adds evidence to the proposed hypothesize of enabling the application of the MLM based pre-training as a good initialization bias for different downstream tasks.

6 Conclusion

We addressed the log parsing problem with deep language modelling. The insight of having words appearing at a constant position of a log record implies that their correct prediction can be directly used to produce a log message type. An incorrect prediction indicates that a token is a parameter. The method also produces a numerical representation of the context of the log message, which primarily is utilized for parsing. This allows the model for utilization in downstream tasks such as anomaly detection.

To evaluate the effectiveness of NuLog, we conducted experiments on 10 real-world log datasets and evaluated it against 12 log parsers. The experimental results show that NuLog outperforms the existing log parsers in terms of accuracy, edit distance, and robustness. Furthermore, we conducted case studies on a real-world supervised and unsupervised anomaly detection task. The results show that the model and the learned log vector summarization can be utilized in both supervised and unsupervised scenario.

Our approach shows that log parsing can be performed with deep language modeling. This imply that future research in log parsing and anomaly detection should focus more into generalization across domains, transfer of knowledge, and learning of meaningful log representations that could further improve the troubleshooting tasks critical for operation of IT systems.

References

1. Devlin, J., Chang, M.W., Lee, K., Toutanova, K.: Bert: pre-training of deep bidirectional transformers for language understanding. arXiv preprint arXiv:1810.04805 (2018)
2. Du, M., Li, F.: Spell: streaming parsing of system event logs. In: Proceedings of the 2016 IEEE 16th International Conference on Data Mining (ICDM), pp. 859–864 (2016)
3. Du, M., Li, F., Zheng, G., Srikumar, V.: DeepLog: anomaly detection and diagnosis from system logs through deep learning. In: Proceedings of the 2017 ACM SIGSAC Conference on Computer and Communications Security, pp. 1285–1298 (2017)
4. Fu, Q., Lou, J.G., Wang, Y., Li, J.: Execution anomaly detection in distributed systems through unstructured log analysis. In: Proceedings of the 2009 IEEE International Conference on Data Mining, pp. 149–158 (2009)
5. Hamooni, H., Debnath, B., Xu, J., Zhang, H., Jiang, G., Mueen, A.: LogMine: fast pattern recognition for log analytics. In: Proceedings of the 25th ACM International on Conference on Information and Knowledge Management, pp. 1573–1582 (2016)
6. He, P., Zhu, J., He, S., Li, J., Lyu, M.R.: An evaluation study on log parsing and its use in log mining. In: Proceedings of the 2016 46th Annual IEEE/IFIP International Conference on Dependable Systems and Networks (DSN), pp. 654–661 (2016)
7. He, P., Zhu, J., Zheng, Z., Lyu, M.R.: Drain: an online log parsing approach with fixed depth tree. In: Proceedings of the 2017 IEEE International Conference on Web Services (ICWS), pp. 33–40 (2017)
8. Jiang, Z.M., Hassan, A.E., Hamann, G., Flora, P.: An automated approach for abstracting execution logs to execution events. J. Softw. Maint. Evol.: Res. Pract. **20**, 249–267 (2008)
9. Liu, J., Zhu, J., He, S., He, P., Zheng, Z., Lyu, M.R.: Logzip: extracting hidden structures via iterative clustering for log compression. In: Proceedings of the 2019 34th IEEE/ACM International Conference on Automated Software Engineering (ASE), pp. 863–873. IEEE (2019)

10. Meng, W., et al.: LogAnomaly: unsupervised detection of sequential and quantitative anomalies in unstructured logs. In: Proceedings of the Twenty-Eighth International Joint Conference on Artificial Intelligence, IJCAI 2019. International Joint Conferences on Artificial Intelligence Organization, vol. 7, pp. 4739–4745 (2019)
11. Messaoudi, S., Panichella, A., Bianculli, D., Briand, L., Sasnauskas, R.: A search-based approach for accurate identification of log message formats. In: Proceedings of the 26th Conference on Program Comprehension, pp. 167–177 (2018)
12. Mizutani, M.: Incremental mining of system log format. In: Proceedings of the 2013 IEEE International Conference on Services Computing, pp. 595–602 (2013)
13. Nagappan, M., Vouk, M.A.: Abstracting log lines to log event types for mining software system logs. In: Proceedings of the 2010 7th IEEE Working Conference on Mining Software Repositories (MSR 2010), pp. 114–117 (2010)
14. Nandi, A., Mandal, A., Atreja, S., Dasgupta, G.B., Bhattacharya, S.: Anomaly detection using program control flow graph mining from execution logs. In: Proceedings of the 22nd ACM SIGKDD International Conference on Knowledge Discovery and Data Mining, pp. 215–224 (2016)
15. Nedelkoski, S., Cardoso, J., Kao, O.: Anomaly detection and classification using distributed tracing and deep learning. In: Proceedings of the 2019 19th IEEE/ACM International Symposium on Cluster, Cloud and Grid Computing (CCGRID), pp. 241–250 (2019)
16. Nedelkoski, S., Cardoso, J., Kao, O.: Anomaly detection from system tracing data using multimodal deep learning. In: Proceeding of the 2019 IEEE 12th International Conference on Cloud Computing (CLOUD), pp. 179–186 (2019)
17. Nedelkoski, S., Bogatinovski, J., Mandapati, A.K., Becker, S., Cardoso, J., Kao, O.: Multi-source distributed system data for AI-powered analytics. In: Brogi, A., Zimmermann, W., Kritikos, K. (eds.) ESOCC 2020. LNCS, vol. 12054, pp. 161–176. Springer, Cham (2020). https://doi.org/10.1007/978-3-030-44769-4_13
18. Shima, K.: Length matters: clustering system log messages using length of words. arXiv preprint arXiv:1611.03213 (2016)
19. Tang, L., Li, T., Perng, C.S.: LogSig: generating system events from raw textual logs. In: Proceedings of the 20th ACM International Conference on Information and Knowledge Management, pp. 785–794 (2011)
20. Taylor, W.L.: Cloze procedure: a new tool for measuring readability. J. Q. **30**, 415–433 (1953)
21. Vaswani, A., et al.: Attention is all you need. In: Proceedings of the Advances in Neural Information Processing Systems, pp. 5998–6008 (2017)
22. Xu, W., Huang, L., Fox, A., Patterson, D., Jordan, M.I.: Detecting large-scale system problems by mining console logs. In: Proceedings of the ACM SIGOPS 22nd Symposium on Operating Systems Principles, pp. 117–132 (2009)
23. Zhang, X., et al.: Robust log-based anomaly detection on unstable log data. In: Proceedings of the 2019 27th ACM Joint Meeting on European Software Engineering Conference and Symposium on the Foundations of Software Engineering, pp. 807–817 (2019)
24. Zhu, J., et al.: Tools and benchmarks for automated log parsing. In: Proceedings of the 2019 IEEE/ACM 41st International Conference on Software Engineering: Software Engineering in Practice (ICSE-SEIP), pp. 121–130. IEEE (2019)
25. Zhu, L., Laptev, N.: Deep and confident prediction for time series at uber. In: Proceedings of the 2017 IEEE International Conference on Data Mining Workshops (ICDMW), pp. 103–110 (2017)

Long-Term Pipeline Failure Prediction Using Nonparametric Survival Analysis

Dilusha Weeraddana[1][(✉)], Sudaraka MallawaArachchi[2], Tharindu Warnakula[2],
Zhidong Li[3], and Yang Wang[3]

[1] Data61-The Commonwealth Scientific and Industrial Research Organisation
(CSIRO), Eveleigh, Australia
`dilusha.weeraddana@data61.csiro.au`
[2] Monash University, Melbourne, Australia
[3] University of Technology Sydney, Sydney, Australia

Abstract. Australian water infrastructure is more than a hundred years
old, thus has begun to show its age through water main failures. Our
work concerns approximately half a million pipelines across major Australian cities that deliver water to houses and businesses, serving over
five million customers. Failures on these buried assets cause damage to
properties and water supply disruptions. We applied Machine Learning
techniques to find a cost-effective solution to the pipe failure problem
in these Australian cities, where on average 1500 of water main failures
occur each year. To achieve this objective, we construct a detailed picture and understanding of the behaviour of the water pipe network by
developing a Machine Learning model to assess and predict the failure
likelihood of water main breaking using historical failure records, descriptors of pipes and other environmental factors. Our results indicate that
our system incorporating a nonparametric survival analysis technique
called 'Random Survival Forest' outperforms several popular algorithms
and expert heuristics in long-term prediction. In addition, we construct
a statistical inference technique to quantify the uncertainty associated
with the long-term predictions.

Keywords: Advanced assets management · Machine learning · Data
mining · Nonparametric · Survival analysis · Random survival forest

1 Introduction

The degradation of urban water mains causes a major problem in urban engineering in Australia. The most common measures of pipeline breakage are the frequency of the water pipe breaks (breaks per 100 km per year) and the criticality
factor of the breakage. Pipeline failure rate varies widely, as it depends on various
factors, such as pipe material, pipe diameter and various other environmental and
operational conditions. The maintenance and renewal of water mains demand
high financial investments. Moreover, direct inspection of all water mains in

© Springer Nature Switzerland AG 2021
Y. Dong et al. (Eds.): ECML PKDD 2020, LNAI 12460, pp. 139–156, 2021.
https://doi.org/10.1007/978-3-030-67667-4_9

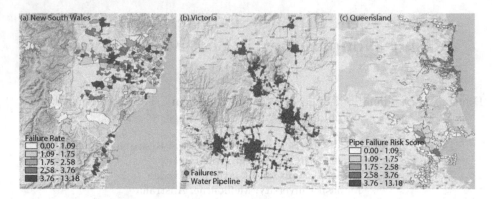

Fig. 1. Water pipeline failure statistics in three major regions of Australia, using three metrics; (a) pipeline failure rate (number of failures per year per 100 km), (b) failure count on the water pipeline network, (c) risk factor (failure likelihood × consequence of the failure) across the network.

a distribution system is extremely expensive. Therefore, a cost-effective break mitigation technique such as a prediction model that allows one to predict the water mains failure, would reduce the negative customer impact and the cost to serve. Consequently, this proactive maintenance model elaborates an optimized strategy for water mains maintenance and rehabilitation.

1.1 The Water Pipeline Failure Problem

This study concerns the failure analysis of the water pipeline network of three major cities located in three different states in Australia, namely: New South Wales (NSW), Victoria (VIC), and Queensland (QLD). The water network includes a total of 500,000 pipelines. The oldest pipes were laid in 1890 in Romsey, VIC and surrounding suburbs. The total length of this pipeline network is over 30,000 km. As depicted in Fig. 2 (a), a water main comprises of several pipes and each pipe comprises several pipe nodes buried in various ground levels.

Water pipe failures are mainly studied using three different metrics, namely: failure rate, failure count and the risk factor associated with the failure. Figure 1(a) shows the pipeline failure rates in major cities of NSW, Australia. The failure rate is the number of asset failures per 100 km per year. Higher failure rates are illustrated in darker red spots, and it clearly shows higher failure rates are not localized to one area, they are spread across the state. Breakages in the water main network in the region west of VIC is shown in Fig. 1(b). Figure 1(c) illustrates the risk distribution of pipeline network in south-east QLD. Across the entire region under our study, an average number of 1500 pipe failures occur each year, causing water supply disruptions and myriads of property and environmental damages. Figure 2 (b) shows the increasing trend in breakage of critical pipes (each water utility has its own method to identify the criticality of a water pipe depending on the risk associated with its breakage) in NSW from 2000 to 2017.

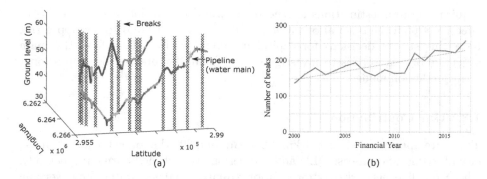

Fig. 2. (a) 3D schematic of a water pipeline which comprises of various individual pipes. X axis, Y axis, and Z axis denote latitude, longitude and the ground level of the water main, respectively. Vertical blue lines represent the breaks occurred in this water main since 2000, (b) Increment in number of critical pipe breaks over the recent years in Sydney, Australia.

1.2 Related Work

There has been a lot of work in recent years on pipe failure prediction in water infrastructure, ranging from physical models [1,7] to machine learning models [13,15,27], [12] and the combination of both [22].

Machine learning based pipe failure forecasting dates back to 1979 [21], where authors developed a forecasting technique to study how the number of breaks would change with time if the pipes were not replaced. In that study, authors used a Poisson model based on the age of the pipes. However, research carried out afterwards shows that the age is not the only factor that causes the pipeline failure. In fact, some of the very old pipes function more robust than their newly laid counterparts. Furthermore, the prediction of water main breaks has been studied widely using statistical based approaches, such as Poisson regression and Weibull models [2,23]. Most recently, tree-based Machine Learning techniques have been used to analyse water pipe breakages in Syracuse, USA [12], and QLD, Australia [14]. The former study shows that Gradient Boosting (GB) outperforms other methods when predicting high risk city blocks. The work reported in [25] uses a combination of Random Forest (RF) and linear regression to predict the long-term pipe failure likelihood for water and sewer pipes in QLD, Australia.

Although numerous research have been conducted on forecasting water pipeline failures, open questions still exist regarding the intricate relationship among the major factors causing pipe failure and their long-term effects on the life-time of a pipe. This may vary depending on the environment (weather, soil, ground level, pressure, etc.) that the pipe is laid in and the pipe maintenance approach of each water utility. Thus, prediction of the water main breaks becomes a complicated task due to their low failure rate and high cost of inspection, which have led to a sparse historical data.

Most of the research found in the literature predict short-term failure forecasting, which spans 1–3 years into the future [12]. However, water utilities require long-term estimations for the structural deterioration of water mains to effectively plan the renewal of water distribution systems and to develop a risk based investment decisions for capital interventions [24]. Therefore, the main objective of this work is to investigate pipe failure factors and develop a long-term prediction model.

1.3   Our Contribution

There are approximately 22850 km, 5900 km, and 2000 km of water mains in the city of Sydney, South-east QLD and the region West of Melbourne, respectively. These pipeline networks comprise approximately 500,000 water pipes serving major residential and industrial cities in Australia. We implement a machine learning based prediction model using the Random Survival Forest (RSF) to identify future pipe failure likelihoods for water main asset in these Australian cities. Firstly, we generate failure likelihood of each pipe using RSF, as it is fully nonparametric and does not impose a restrictive structure on data distribution or how the variables should be combined [26]. If the relationship between the independent variables and the dependent variable is complex with non linear interactions, then the RSF algorithm is capable of capturing these intricate relationships [9,19]. In our model, the predictions were validated by separating the data into training and testing samples. Afterwards, a derived list is generated and evaluated on the testing data. We further compare the results from RSF with a variety of other approaches, such as GB, RF and Weibull. Water authorities are often interested in obtaining a confidence interval for the predictions we produce. This is due to the fact that, pipe failure predictions suffer from various sources of error, such as the variations in weather conditions, new infrastructure developments, root clogs caused by near by trees, and many other sources, which are caused by the inherent stochastic and nonlinear characteristics of water pipe failures. In order to quantify the uncertainty in failure forecasting effectively, we have generated the uncertainty interval for the long-term prediction by treating RSF as quantile regression forests. As a result, for each point that is predicted with a RSF, we provide the perceived uncertainty of that prediction.

In the past, RSF has been employed in various medical related research [5,17]. However, to the best of our knowledge, this is the first model applied on pipe failure problem embracing the quantile regression forests [16] for uncertainty estimation, and proven on real-world datasets collected from multiple water authorities.

Our data analytical model provides a projection of the likelihood of future pipe failures. These likelihoods, along with the consequence of failures, are currently being used in current investment planning of each of these Australian water utilities, to make risk based investment decisions for capital interventions. Thus, our contributions help the water asset renewal programs to reduce the catastrophic consequence of water main failures and the cost to customers.

1.4   Preliminary

Survival and Hazard Functions. The survival function, $S(t)$ is a non-increasing function, which provides the probability that a subject will survive past time t [4,11].

$$S(t) = Pr(T > t) = \int_t^\infty f(u)du$$

(1)

Here, T is a continuous random variable with the probability density function: $f(u)$, or more generally, T represents the waiting time until the occurrence of an event. In our scenario, the survival function illustrates the probability that a particular pipe survives past a given time. The hazard function describes the event rate,

$$\lambda(t) = \lim_{\delta t \to 0} \frac{Pr(t < T \le t + \delta t | T > t)}{\delta t}$$

(2)

$$S(t) = \exp^{-\lambda(t)}$$

(3)

The Cumulative Hazard Function (CHF) provides the accumulated risk up to time t,

$$\mu(t) = \int_0^t \lambda(u)du$$

(4)

$\mu(t)$ can be seen as the sum of the risks accumulating from duration 0 to t. Thus, these functions are of intrinsically pivotal in forecasting about the condition of a pipe which has survived a certain time period.

Random Survival Forest. An extension of RF to the domain of survival analysis enhances its value greatly. In survival analysis, many different regression modeling strategies, such as Cox regression and Poisson regression, can be applied to predict the survival likelihoods. Extending the RF approach [3] to survival analysis provides an alternative way to build a robust asset failure prediction model. This technique safely omits the need to impose parametric or semi-parametric constraints on the underlying distributions and allows for an accurate prediction [9,17].

RSF consists of arbitrarily grown survival trees. Using independent bootstrap samples, each tree is grown by randomly selecting a subset of variables at each node and then splitting the node using the candidate variable that maximizes survival difference between daughter nodes. The tree is grown until saturation is reached due to the condition of each terminal node having no fewer than $d_0 > 0$ unique deaths (in our case, this referred to the number of pipe breakages). The output of each tree may be estimated as the CHF for each case, the estimator for which is the Nelson–Aalen estimator for the terminal node in which the case ends up [9],

$$\mu(t) = \sum_{t_j \leq t} \frac{d_j}{Y_j},$$

(5)

where t_j are the ordered pipe failure times for the terminal node. d_j and Y_j are the number of pipe failures and pipes at risk (number of pipes in the terminal node that are functioning) at time t_j in the terminal nodes. However, in our model, instead of the CHF, we derive an estimate of the survival probability for each terminal node using the Kaplan-Meier estimator [10] given by,

$$S(t) = \prod_{t_j \leq t} \left(1 - \frac{d_j}{Y_j}\right).$$

(6)

Given the CHF or survival estimate from a tree, an ensemble average is performed over the entire forest to produce the final prediction.

2   Data Analytic Model for Pipeline Failure Prediction

2.1   Data Extraction and Pre-processing

There are three main data sources used as the inputs to the analytical model:

- Network data: describes water main information such as asset number, installation date, material, size.
- Work order data: describes water main failure information such as asset number, failure date, location, and failure type (burst, fitting, leak).
- External data: includes information in addition to assets, such as weather data from the Bureau of Meteorology and census data from the Australian Bureau of Statistics, soil data, pressure data, pipe ground level data, etc.

The above data should be sufficiently accurate for the intended use, so a data quality review has been undertaken based on three key characteristics:

- Completeness: this is a statistical analysis that does not allow empty values.
- Validity: this is a statistical analysis that removes invalid values.
- Consistency: make sure that the data obtained from all the sources are consistent with each other.

The quality review demonstrates that the data is sufficient and accurate for further analysis. Accordingly, this process allows to establish a comprehensive data file with complete information for each asset that can be used as an important input for further analysis. Moreover, when information is gathered from multiple sources, prior to the adoption of advanced analytic techniques, it is essential to match the failure records with the network data and identify gaps in the datasets. In addition, environmental and demographic factors need to be matched with the network data. Specifically, failure records and information are assigned to the corresponding assets based on the work order number.

2.2 Factor Analysis

Once the data is pre-processed, the next step is to identify the factors that cause pipeline breakage and compare their relative impact on the network based on the water network information. Factor analysis measures the correlation between asset performance based on the comprehensive data and a large range of factors (including environmental, demographic and asset specific factors) [7,20]. While a significant amount of literature exists on the pipeline failure causes, this step is critical to discerning which of these causes would be the most important for each water utility. The asset performance is based on failure rate, which is the number of asset failures per 100 km per year. Both single factor analysis and multi-factor analysis have been performed to identify the possible driving factors. The asset performance is not usually related to only one factor, so it is essential to measure the correlation based on multiple factors.

For example, within operational factors, AC water mains were found to break more often than others in the regions of QLD as shown in Fig. 3 (b). It was also found that water mains with diameters less than 100 mm exhibit higher failure rates, compared to larger pipes (see Fig. 3 (a)). Moreover, a quantitative study on the ground level of water main and its impact on the pipe breakage is shown in Fig. 3 (c). It can be observed that the failure rate of pipes laid in the bottom 25% of ground level is twice higher than the pipes laid in the top 25% of ground level (above 75% of quantile).

To quantify the amount of pipe failure information stored in each of the features in isolation, we calculate the mutual information between the 'Pipe Failure' parameter and each feature (we have selected a basic set of asset specific features which are common to all three states). The data from all three states display a very similar dependence of the failures on the predictor variables. Therefore, the resulting information scores for the VIC dataset are presented in Fig. 3 (d). Pipe size (or diameter) shares the highest amount of mutual information with failures while pipe type has the least effect on failures. In general, all predictors by themselves display very low levels of mutual information indicating that by themselves, they do not predict failures sufficiently well. However, as we shall show later, the six features in unison will provide us with an excellent prediction model of pipe failures.

To this end, we also identified the potential advantages of analysing the factors causing pipe failures in different datasets across various Australian regions.

We have been working with a few water utilities and identifying the differences and the commonalities among these various datasets allow us to improve our knowledge in developing the prediction framework.

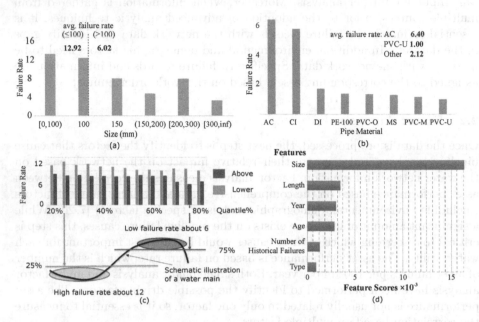

Fig. 3. Factor analysis examples: (a) Failure rate of water mains based on pipe size in pipelines located region west of Melbourne, (b) AC materials are more prone to break in QLD, (c) Factor analysis of the ground level of a water main and how it affects the pipe breakage in pipelines located region west of Sydney. (d) Feature importance scores for the VIC dataset, computed using the mutual entropy gain method.

2.3   Long-Term Failure Prediction

This phase involves predicting future water pipe failure probabilities. We framed this scenario as determining the likelihood of failure on each given pipe. The failure prediction is generated by training the RSF model on historical failure records and other factors, such as pipe material, pipe laid year, pipe diameter, etc. This trained model produces a survival probability score for each water main asset for each years into the future.

The RSF model utilized in this work uses data on the history of water pipeline network across major Australian cities. It specifically uses the failure history of pipes (the response) and their characteristics (the predictor variables). The response variable includes the minimum of the survival time: T_i, the right censoring time C_i and $\Delta_i = \Im\{T_i \leq C_i\}$ which is the censoring value indicating a pipe has failed ($\Delta_i = 1$) or was right-censored ($\Delta_i = 0$). The predictor variables $X_i = (X_i^1, ..., X_i^N)$ for respective pipe, i consists of both continuous variables, such as age, previous failures, as well as qualitative variables, such as pipe material and pipe size.

Prediction Uncertainty. We construct a simple, yet effective, statistical inference technique to quantify the uncertainty associated with the predictions generated by supervised learning ensembles. Here, we employ quantile regression forests in survival trees generated by the RSF model. The concept behind the Random survival quantile regression forests is, instead of recording the mean value of response variables in each tree leaf in the forest, record all observed responses in the leaf. The prediction can then be calculated as the mean of the response variables, as well as the full conditional distribution of response values for every x. Using the distribution, the prediction intervals for new instances can be generated by employing the appropriate percentiles of the distribution.

Following [8], the high-level description of the algorithm used in this work, along with the procedure for determining uncertainty, an be given as follows:

1. Ascertain the training year range and the prediction year range of pipe failure observations. A training data file is created on average for eight observation years of pipeline data. Each observation year contains information of all the pipes in the network, with an indication of whether a particular pipe has failed in that observation year or not. The observation year range for the training data is selected and restricted(e.g. 2005–2010).
2. N number of bootstrap samples are pulled from the training dataset by excluding on average 37% of the data, which is referred to as out-of-bag (OOB) data.
3. A survival tree is developed for each bootstrap sample. At each node of the tree, a p number of candidate variables are randomly selected. The node is split using the candidate variable that maximizes survival difference between daughter nodes.
4. Grow the tree to full size under the constraint that a terminal node should have no less than $d_0 > 0$ unique deaths.
5. Using OOB data, the prediction error for the ensemble survival is calculated.
6. Calculate survival probability for the predicting data range of observation years (e.g. 2011–2025) by recording all observed responses in the leaf, and obtaining conditional probability distribution of the response variable for every given set of predictor variables (x) of each pipe. Using the distribution, create prediction intervals for new instances by using the appropriate percentiles of the distribution to calculate the lower and upper bounds of the prediction uncertainty.

3   Case Study

We study the pipeline failure data from three major Australian states: VIC, NSW and QLD. Each of these three datasets were generated based on the results of observations made on pipelines in each observation year. As an example, the data statistics for selected laid year groups for VIC are presented in Table 1. This highlights the different dynamics associated in training and testing datasets for each laid year. For VIC and NSW, data spanning observations from 2000 to 2017 were available while for QLD, only data for observations from 2013 to 2017 were available. The key information recorded at these observations is represented as a boolean variable recording whether a failure was detected at the time. We also use auxiliary data regarding each of the observed pipes as input parameters to predict failures into the future. The full list of features used in our modelling is reported in the Fig. 3. We use the age of the pipe observed, the year in which it was laid in, the material that the pipe is made of, the number of previous failures and the size (diameter) of the pipe as predictor variables.

Table 1. Data statistics for selected set of laid year groups

LY	TrFC	TrFR	TeFC	TeFR	LY	TrFC	TrFR	TeFC	TeFR
2000	37	6.62	12	5.16	2006	23	3.86	9	3.63
2001	17	4.04	4	2.28	2007	20	5.55	10	6.67
2002	23	2.73	25	7.12	2008	16	3.18	5	2.39
2003	30	4.57	22	8.04	2009	8	1.37	13	5.35
2004	30	6.27	32	16.05	2010	2	0.51	21	12.98
2005	26	4.14	14	5.35					

LY=laid year, TrFC = training set (2005-2010) failure count, TrFR = training set failure rate, TeFC = testing set (2015-2017) failure count, TeFR = testing set failure rate

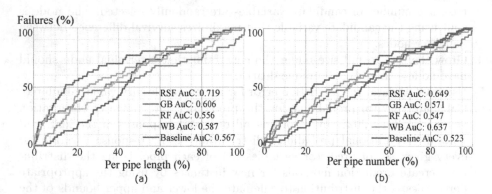

Fig. 4. ROC curves generated based on the predictions made by each technique for the state of VIC for the year 2017. Observation data collected in and before year 2010 was used for the training task. Therefore, the predictions illustrated here are made 7 years in advance. (a) depicts the ROC curves based on the total pipe length and (b) the number of pipes, respectively.

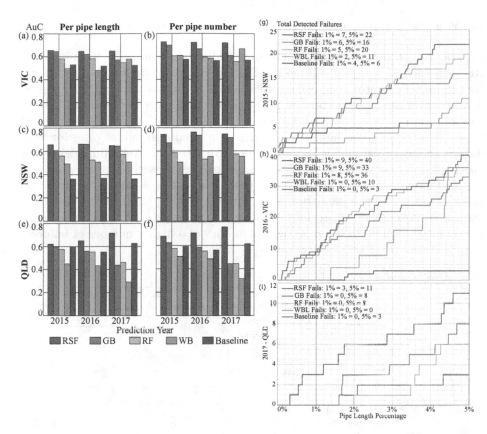

Fig. 5. Complete collection of bar plots depicting the AUC values generated for all prediction scenarios. (a) and (b) bar plots show the AUC values obtained for the VIC dataset for the (a) length based and (b) pipe number based ROC curves, respectively. Similarly, (c) and (d) plots correspond to the NSW dataset and (e) and (f) plots correspond to the QLD dataset. Inspection of the top 1% and 5% of ranked pipe length in (a) NSW - 2015, (h) VIC - 2016 and (i) QLD - 2017. In each case, the failures correctly identified using the predictions made by RSF (blue), GB (orange) and RF (yellow) techniques are shown therein. The two red vertical lines across each plot identifies the inspection points at 1% and 5% of total pipe length. The numerical failure count is indicated in the top left hand corner of each figure for convenience. (Color figure online)

3.1   Model Setting

Using this information, we perform a comprehensive comparison of performance between the Random Survival Forest technique and other widely used machine learning and statistical algorithms along with a baseline predictor. For VIC and NSW, we use the observations from 2000–2010 to train our machine learning models and only the data from the year 2013 for QLD. The probabilities of failure for pipes observed in the years 2015–2017 are then calculated and compared to the actual recorded observation. This provides all our methods a common benchmark to be compared against.

The algorithms we choose to measure RSF against are, GB technique [6], RF regression [3] and Weibull model [23]. GB is a machine learning technique that iteratively improves modelling using weak predictors [18]. In RF, independently drawn random sub-samples of the complete dataset are used to build an ensemble of regression trees. For all three algorithms (RSF, GB, RF), we use 100 trees when training the model. Additionally, we predicted failure rates for each dataset using a 2-parameter Weibull model. We fitted the Weibull model to the *age at first failure* distribution of all pipes in each dataset, and the computed parameters were used to estimate the probability of pipe failure by aging all pipes according to the prediction year. Further to this, the baseline predictor we use is the number of previous failures of a given pipe. We assign a higher probability of failure to the pipes with a history of a higher number of failures.

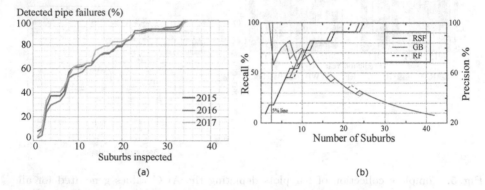

(a)   (b)

Fig. 6. (a) Suburb level model verification for the VIC pipe dataset. (b) Critical suburb prediction recall (blue) and precision (red) percentages for VIC suburbs in the year 2017. At 5% (out of a total of 42 suburbs), the recall percentage is at 18.18% and the precision is at 96.67% for all three techniques considered. The RSF is plotted as a solid line with the RF as a dashed line and the GB method response as a dotted line. This pattern of similar recall and precision curves across all three techniques and all years is evident especially at lower population levels. (Color figure online)

To easily compare the predictions, we plot the Receiver Operator Characteristic (ROC) curves of the predictions made by each predictor and compare them through the Area Under the Curve (AUC). Firstly, the prediction model was trained using the pipeline features and the failure data. Then the calibrated model was applied to predict the survival probability for each pipe from year 2015 to 2017. Then the pipes were ranked according to the survival probability of each pipe. Using the ranked list, actual failures from the lowest to the highest probability are accumulated (cumulative sum of failures). The percentage of detected failures is plotted against the percentage of inspected pipe lengths, and the percentage of inspected number of pipes. Predictors that reach higher true positive rates while maintaining low false positive rates are preferred for predicting purposes and correspondingly. We also note that we use two separate methods for generating the ROC. One method defines true positive and false positive rates based on the number of pipes predicted to fail correctly or falsely, while the other defines false positive rate as the total length of pipes incorrectly predicted to fail while retaining the usual definition of true positive rate. We term these pipe based ROC and length based ROC, respectively. While the pipe based ROC is the more natural definition of the two, Water Management authorities consider failures per unit length as an important parameter and the length based ROC accounts for this explicitly.

3.2   Experimental Results and Discussion

In order to quantify their prediction performance, we extensively studied the ROC curves generated by each machine learning technique under various scenarios. One such instance is shown in Fig. 4, where the complete set of ROC curves generated by each technique are plotted together. These curves are based on the predictions made for the year 2017 for the VIC pipe dataset. Pipe observations made in and before the year 2010 were used for training purposes and the predictions are made for a time period that is 7 years into the future (year 2017). The length based ROC curve for RSF clearly demonstrates a ≈5% prediction enhancement over the other two techniques, whereas the pipe number based ROC curve demonstrates a ~10% prediction enhancement.

The bar plots in Fig. 5 further demonstrate the superior prediction capabilities of the RSF technique compared to GB, RF and Weibull techniques. For the VIC pipe dataset, all predictions made using the RSF technique show better prediction results when compared with the GB and RF and Weibull. RSF outperforms both RF and GB by a considerable margin in the 2017 prediction. A similar observation can be made for both NSW and QLD pipe datasets; predictions made for a year further away from the last year in the training dataset (2010 for VIC and NSW, 2013 for QLD) tend to show better accuracy when predicted using the RSF technique. Furthermore, it is noted that there are some rare instances where the GB and RF techniques marginally outperform the predictions made by the RSF technique. This behavior is particularly observed for the NSW pipe dataset. This is because some divergent trends with respect to age are observed in the NSW records (failure rate for cast iron pipes decreases with the age for a subset of pipes).

To establish the effectiveness of the proposed techniques, it is important to know how many failures can be detected by inspecting the first few pipes within a group in the ranking order. In order to establish this, we studied the prediction data for the highest ranked 1% and 5% of pipe lengths in each year for each state. A sample of these observations is shown in Fig. 5(g)–(i). It is observed that using the RSF method has an advantage over using other methods in terms of detecting the most number of failures with a least amount of inspection effort.

We performed further analysis of predicted results using the RSF technique on the suburb level. For this study, the suburbs within a state are first sorted, based on the cumulative probability of all the pipes within the suburb for a given prediction year. This forms a suburb based dataset and the general procedure is then followed to obtain the suburb based ROC curve. Results for the VIC pipe dataset for the years 2015, 2016 and 2017 are shown in Fig. 6 (a). Based on the results, it is evident that by inspecting the pipes in the top 10 suburbs, it is possible to detect more than 60% of the total failures. Additionally, inspecting the pipes in only half of the total suburbs in the ranked order will result in detecting more than 80% of the total pipe failures for all years.

Extending our analysis of the prediction of pipe failures in suburbs, we turn our attention towards predicting critical suburbs within each of the three states analysed. It is extremely valuable for water management authorities to be able to restrict attention in specific years to monitoring only a select number of pipe systems located in specific suburbs. This comes with the benefits of reduced manpower and labour costs. We define critical suburbs within a specific year in our model to be suburbs that host a number pipe failures greater than the average for that year. Using this definition, we use our trained models to generate with aggregate failure probabilities for each suburb. Using these aggregate probability values and the actual number of failures occurring in each suburb in that year, we calculate the precision and the recall rates for the detection of critical suburbs. As shown in Fig. 6 (b), all three techniques we used as candidates demonstrated satisfactory recall (18.18%) and precision (96.67%) levels for suburbs in VIC for 2017. We also note that such similar behaviours were observed in the three techniques across years and geographic locations.

Our experimental results indicate that for most of the studied scenarios, the RSF technique outperforms other machine learning techniques, clearly highlighting its superior prediction capabilities in long-term predicting pipe failures. We also note that all three techniques perform better than the baseline as predicted by the historical number of previous failures in pipes. In general, the RSF technique tends to improve in its prediction accuracy or at least maintain the same accuracy as predictions are made further into the future. In contrast, we observe that the accuracy of the predictions made through GB and RF methods tend to deteriorate over time.

We finally observe that all the techniques used saturate at a maximum AUC level of around 0.65 to 0.75. While pipe failures may be extremely unpredictable, this to also due to the fact that our predictions are made further away into the future (5–7 years), and also due to the fewer number of features we use for predictions as compared to other similar studies [12].

In the suburb based analysis, we notice that by the aggregation of pipe failure probabilities predicted by RSF across suburbs, we were able to predict a significant proportion of pipe failures. In Fig. 6 (a), we clearly see that at 20%, more than 40% of the pipe failures were recovered. We also note that the curve for 2017 dominates that portion of the graph, again signalling the efficiency of the RSF technique for predicting failures further into the future. Our analysis of critical suburbs also reveals quite interesting facts regarding pipe failures and their distribution across suburbs. As Fig. 6 (b) clearly demonstrates, all three methods perform well in precisely recalling suburbs with greater-than-average numbers of failures in a year. This coupled with the results of Fig. 6 (a) also suggest that in a given year, the pipe failures are clustered in a few vulnerable suburbs within each state.

Prediction uncertainty for the long-term failure prediction also has been calculated. These long-term prediction curves show that actual failure rates align with the results generated by RSF and within the uncertainty interval. The mean failure prediction is ideal for modelling future behaviour of pipeline network benchmarking performance indices such as unplanned water interruption and water main breaks. In addition to the mean prediction, all the water utilities require uncertainty interval in order to evaluate the impact and cost of more targeted water network levels of service inform both short and long-term renewals budgets.

4 Discussion on RSF for Long-Term Pipe Failure Prediction

The empirical study we conducted here shows that, the longer we predict into the future, more degradation in the accuracy can be observed in RF and GB, whereas RSF remains quite consistent in the accuracy when we predict further into the future. This is because RSF explicitly takes, the time until the occurrence of a failure (Eq. 6), into account when calculating the CHF, making RSF to provide a robust prediction over a longer period of time. Secondly, RSF seeks a model that best explains the data and thus represents a suitable tool for exploratory analysis where prior information of the survival data is limited (consider the experimental results for QLD dataset, where we have only one year worth data for training and also the failure data is highly sparse). Thirdly, in case of multi-dimensional data, limitations of univariate regression approaches (i.e. Weibull method) such as unreliable estimation of regression coefficients or convergence problems do not apply to RSF. To the best of our knowledge, this is the first research conducted to explore the potential advantages of using RSF in pipe failure predictions along with the uncertainty estimation.

Currently, our predictive data analytic models are deployed in the city of Sydney, the region west of Melbourne and south-east QLD mainly for short-term prediction purposes. Each of these Australian water utilities are monitoring the number and the location of water main failures to validate our model. They also use our model in their internal financial modelling, risk distribution assessment planning and also to assist in the development of condition assessment programs. In addition to our previous work which have been deployed already, the study presents in this paper focuses on the development of a nonparametric survival analysis technique to determine which water main assets and suburbs are most likely to have water main failures in the next 5–7 years. Our results indicate that RSF opens up a new avenue for robust pipe failure prediction

5   Conclusions

The reliability of the water distribution network in any city is critical to delivering clean water supply to customers. Tailoring data science techniques to model the pipeline failure prediction provides accurate insights into water main networks. This will essentially assist water authorities to carry-out proactive pipeline maintenance. Therefore, we have presented a thorough survey of the landscape of nonparametric survival analysis as it pertains to predictions of survival rates and correspondingly decease rates of assets. We have used data from the water management authorities of three major Australian states to validate the survival analysis technique we propose, Random Survival Forest, to compare against other state-of-art machine learning techniques that have been proven effective the in similar applications. We perform a thorough analysis of the performance of the techniques in making predictions over multiple years. The results show that the Random Survival Forest (RSF) has consistently shown to outperform the other techniques, in long-term forecasting. To the best of our knowledge, this is the first research conducted to explore the potential advantages of using RSF in pipe failure predictions along with the uncertainty estimation. Ultimately, we believe this work, at the intersection of Machine Learning and Asset Management, will lead to more effective and proactive infrastructure maintenance in the water industry across the world.

Acknowledgement. We sincerely thank Australian water utilities: Sydney Water, UnityWater and Western Water for sharing data, expert domain knowledge and the valuable feedback.

References

1. Cronin, D.S., Pick, R.J.: Prediction of the failure pressure for complex corrosion defects. Int. J. Press. Vessels Pip. **79**(4), 279–287 (2002)
2. Asnaashari, A., McBean, E., Shahrour, I., Gharabaghi, B.: Prediction of watermain failure frequencies using multiple and poisson regression. Water Sci. Technol.: Water Supply **9**(1), 9–19 (2009)
3. Breiman, L.: Random forests. Mach. Learn. **45**(1), 5–32 (2001)
4. Cox, D.R.: Analysis of Survival Data. Routledge, Milton Park (2018)
5. Dietrich, S., et al.: Random survival forest in practice: a method for modelling complex metabolomics data in time to event analysis. Int. J. Epidemiol. **45**(5), 1406–1420 (2016)
6. Friedman, J.H.: Stochastic gradient boosting. Comput. Stat. Data Anal. **38**(4), 367–378 (2002)
7. Gould, S., Boulaire, F., Burn, S., Zhao, X.L., Kodikara, J.: Seasonal factors influencing the failure of buried water reticulation pipes. Water Sci. Technol. **63**(11), 2692–2699 (2011)
8. Ishwaran, H., Kogalur, U.B.: Random survival forests for R (2007)
9. Ishwaran, H., Kogalur, U.B., Blackstone, E.H., Lauer, M.S.: Random survival forests. Ann. Appl. Stat. **2**(3), 841–860 (2008). https://doi.org/10.1214/08-AOAS169

10. Kaplan, E.L., Meier, P.: Nonparametric estimation from incomplete observations. J. Am. Stat. Assoc. **53**(282), 457–481 (1958)
11. Klein, J.P., Moeschberger, M.L.: Survival Analysis: Techniques for Censored and Truncated Data. Springer, Heidelberg (2006). https://doi.org/10.1007/b97377
12. Kumar, A., et al.: Using machine learning to assess the risk of and prevent water main breaks. In: Proceedings of the 24th ACM SIGKDD International Conference on Knowledge Discovery and Data Mining, pp. 472–480 (2018)
13. Li, Z., et al.: Water pipe condition assessment: a hierarchical beta process approach for sparse incident data. Mach. Learn. **95**(1), 11–26 (2013). https://doi.org/10.1007/s10994-013-5386-z
14. Liang, B., et al.: Pipeline failure data analytics and prediction. In: OzWater, pp. 25–33. Australian Water Association (2018)
15. Luo, S., Chu, V.W., Zhou, J., Chen, F., Wong, R.K., Huang, W.: A multivariate clustering approach for infrastructure failure predictions. In: BigData Congress, pp. 274–281. IEEE Computer Society (2017)
16. Meinshausen, N.: Quantile regression forests. J. Mach. Learn. Res. **7**(Jun), 983–999 (2006)
17. Miao, F., Cai, Y.P., Zhang, Y.X., Li, Y., Zhang, Y.T.: Risk prediction of one-year mortality in patients with cardiac arrhythmias using random survival forest. Comput. Math. Methods Med. **2015** (2015)
18. Moisen, G.G., Freeman, E.A., Blackard, J.A., Frescino, T.S., Zimmermann, N.E., Edwards Jr., T.C.: Predicting tree species presence and basal area in Utah: a comparison of stochastic gradient boosting, generalized additive models, and tree-based methods. Ecol. Model. **199**(2), 176–187 (2006)
19. Nasejje, J.B., Mwambi, H.: Application of random survival forests in understanding the determinants of under-five child mortality in Uganda in the presence of covariates that satisfy the proportional and non-proportional hazards assumption. BMC Res. Notes **10**(1), 459 (2017)
20. Rajeev, P., Kodikara, J., Robert, D., Zeman, P., Rajani, B.: Factors contributing to large diameter water pipe failure. Water Asset Manag. Int. **10**(3), 9–14 (2014)
21. Shamir, U., Howard, C.D.: An analytic approach to scheduling pipe replacement. J.-Am. Water Works Assoc. **71**(5), 248–258 (1979)
22. Shi, L., Sun, L., Vidal Calleja, T., Miro, J.V.: Kernel-specific gaussian process for predicting pipe wall thickness maps. In: Australasian Conference on Robotics and Automation. AARA (2015)
23. Vanrenterghem-Raven, A., Eisenbeis, P., Juran, I., Christodoulou, S.: Statistical modeling of the structural degradation of an urban water distribution system: case study of New York city. In: World Water & Environmental Resources Congress, pp. 1–10 (2003)
24. Weeraddana, D., Hapuarachchi, H., Kumarapperuma, L., Khoa, N.L.D., Cai, C.: Long-term water pipe condition assessment: a semiparametric model using Gaussian process and survival analysis. In: Lauw, H.W., Wong, R.C.-W., Ntoulas, A., Lim, E.-P., Ng, S.-K., Pan, S.J. (eds.) PAKDD 2020. LNCS (LNAI), vol. 12085, pp. 487–499. Springer, Cham (2020). https://doi.org/10.1007/978-3-030-47436-2_37
25. Weeraddana, D., et al.: Utilizing machine learning to prevent water main breaks by understanding pipeline failure drivers. In: OzWater. Australian Water Association (2019)

26. Wey, A., Connett, J., Rudser, K.: Combining parametric, semi-parametric, and non-parametric survival models with stacked survival models. Biostatistics **16**(3), 537–549 (2015)

27. Zhang, B., et al.: Water pipe failure prediction: a machine learning approach enhanced by domain knowledge. In: Zhou, J., Chen, F. (eds.) Human and Machine Learning. HIS, pp. 363–383. Springer, Cham (2018). https://doi.org/10.1007/978-3-319-90403-0_18

Forecasting Error Pattern-Based Anomaly Detection in Multivariate Time Series

Seoyoung Park[1], Siho Han[2], and Simon S. Woo[2,3(✉)]

[1] Department of Statistics, Sungkyunkwan University, Seoul, South Korea
sera8522@g.skku.edu
[2] Department of Applied Data Science, Sungkyunkwan University,
Suwon, South Korea
siho.han@g.skku.edu
[3] Department of Computer Science and Engineering, Sungkyunkwan University,
Suwon, South Korea
swoo@g.skku.edu

Abstract. The advent of Industry 4.0, partly characterized by the development of cyber-physical systems (CPSs), naturally entails the need for reliable security schemes. In particular, accurate detection of anomalies is of paramount importance, as even a small number of anomalous instances can trigger a catastrophic failure, often leading to a cascading one, throughout the CPS due to its interconnectivity. In this work, we aim to contribute to the body of literature on the application of anomaly detection techniques in CPSs. We propose novel Functional Data Analysis (FDA) and Autoencoder-based approaches for anomaly detection in the Secure Water Treatment (SWaT) dataset, which realistically represents a scaled-down industrial water treatment plant. We demonstrate that our methods can capture the underlying forecasting error patterns of the SWaT dataset generated by Mixture Density Networks (MDNs). We evaluate our detection performances using the F_1 score and show that our methods empirically outperform the baseline approaches—cumulative sum (CUSUM) and static thresholding. We also provide a comparative analysis of our methods to discuss their abilities as well as limitations.

Keywords: Anomaly detection · Forecasting error patterns · Cyber-physical systems

1 Introduction

An essential constituent of the Industry 4.0 trend [30], a cyber-physical system (CPS) is a decentralized integration of physical components characterized by its autonomy and interconnectivity [42]. As its name indicates, a CPS refers to the close linkage of physical hardware in smart grids and industrial control systems

---

S. Park and S. Han—Co-first authors.

© Springer Nature Switzerland AG 2021
Y. Dong et al. (Eds.): ECML PKDD 2020, LNAI 12460, pp. 157–172, 2021.
https://doi.org/10.1007/978-3-030-67667-4_10

(ICSs), such as water treatment plants, with software elements. The increasing complexity of CPSs due to rapid advancements in computation and communication technology naturally entails the need for reliable security schemes [3], of which an accurate detection of anomalous instances is indispensable for the prevention of cascading failure across the whole system [44]. The occurrence of anomalies is ascribed to a variety of causes, such as cyber-attacks on communication networks and computing resources [19,28], as well as physical attacks on other areas of vulnerability, including sensors and actuators [1]; technical failures such as hardware issues, software bugs, operator errors, and server misconfiguration may also cause the CPS to malfunction, resulting in unexpected spikes or point anomalies in a given time series [32,48]. In any case, even a small number of anomalous events can be harmful to the system as a whole, posing a real-world challenge in the Fourth Industrial Revolution era. Moreover, the mitigation of false positives due to mediocre anomaly detection performance will be costly, because the process would require the main infrastructures to stop running for manual inspections by human experts. It is therefore crucial to correctly distinguish minor glitches from perilous anomalies, such that the false alarm rate is minimal.

In this work, we propose novel functional data analysis (FDA) and Autoencoder-based approaches to detect anomalies in the physical properties data collected by iTrust, Center for Research in Cyber Security, Singapore University of Technology and Design (SUTD), from the Secure Water Treatment (SWaT) testbed [27], which represents a scaled-down industrial water treatment plant. More specifically, we focus on leveraging the forecasting error data generated from the SWaT dataset using Mixture Density Networks (MDNs) [5], as analyzing the underlying patterns of forecasting error has shown to be effective when detecting anomalies [11,41]. While simple rule-based techniques using the Out-Of-Limit (OOL) approach may suffice to detect point anomalies, more sophisticated methods are required to detect contextual anomalies [18,22,43]. The latter type of anomalies are defined only in a specific context with contextual and behavioral attributes: in time series, temporal information determines the order of the samples within the data [7]. For instance, the forecasting error time series for the actuator MV101 in the SWaT dataset, which represents a motorized valve that controls the water flow to the raw water tank [13], plotted in Fig. 1, contains a long sequence of anomalous samples for more than 11 h (approximately between 7.1×10^5 s and 7.5×10^5 s) that are close to the value 0. These anomalies would not be correctly identified if a simple OOL technique were to be used due to the values being significantly lower than the cutoff threshold, resulting in an extremely high false alarm rate, as they comprise a considerable portion of anomalies in the MV101 data. The challenges arising from such properties justify our choice of forecasting error patterns as anomaly indicators, rather than simply the magnitudes of the error.

Our proposed methods aim to capture forecasting error patterns, given the smoothness of data achieved by several preprocessing steps, to effectively detect point and contextual anomalies. In particular, we formulate a semi-supervised

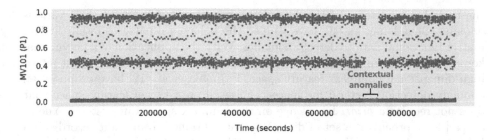

Fig. 1. Plot of the forecasting error time series representing the actuator MV101 of the process P1, where anomalous instances are marked in red. A long sequence of contextual anomalies can be observed between approximately 7.1×10^5 s and 7.5×10^5 s, as indicated on the graph.

anomaly detection problem as a representation of challenging real-world scenarios, where anomalies are rare compared to normal data, hence the use of only this latter data for training. In addition, the intermediate stages of our proposed methods can be served as dimension and noise reduction techniques. Through extensive experiments on the SWaT dataset, we empirically demonstrate that our methods significantly outperform baseline methods, namely Cumulative Sum (CUSUM) [29] and static thresholding [11]. Also, we evaluate the relative performance of our methods to weigh their pros and cons. Our contributions are summarized as follows:

- We propose two novel approaches based on FDA and Autoencoder to detect anomalies in multivariate forecasting error data generated from the SWaT dataset using MDNs.
- We devise a new algorithm for anomaly detection, which applies dynamic thresholding to identify point and contextual anomalies based on the tangent error.
- We empirically demonstrate that our methods significantly outperform baseline anomaly detection methods—CUSUM and static thresholding—when evaluated on the SWaT dataset.

2   Related Work

Due to potential threats of cyber and physical attacks against CPSs, a sizeable body of literature has sought to devise anomaly detection techniques for CPS security and reliability. In this section, we provide a high-level overview of the previous work that is directly relevant to our proposed methods.

Considering the partly discrete nature of data collected from CPSs, FDA can facilitate data analysis by converting point values into functional data [34]. The latter type of data can be broadly categorized into 2 types, analytic and non-analytic. Typical examples of the former include polynomial, exponential, logarithmic, and trigonometric functions, while those of the latter include absolute value and piecewise polynomial functions [35]. Allowing further analysis of the underlying patterns of the data, FDA has been applied to data of various domains. Shaadan et al. [40] have studied the behavioral patterns in particulate matter functional data to detect anomalies. With the application of a polynomial kernel function to implicitly transform telemetry data at each time point into a high-dimensional nonlinear feature space, Fujimaki et al. [12] have proposed an anomaly detection method for spacecraft based on a kernel feature space that does not require *a priori* knowledge on the complex spacecraft systems. Yue et al. [49] have processed a set of discrete points obtained from audit records on a behavior session using Fourier transform (FT) to detect intrusions, enabling the detection of anomalies without having to know the actual flaws of the intruded system or to observe the specific actions exploiting those flaws. Febrero et al. [9] have analyzed outlier detection in nitrogen oxide emission functional data using the Fraiman and Muniz depth (FMD), the h-modal depth (MD), and the random projection depth (RPD), of which the main idea is to determine a robust estimate C, such that $Pr(D_n(x_i) \leq C) = 0.01$, $i = 1, 2, \ldots, n$, where $D_n(x_i)$ denotes the functional depth of the ith functional curve x_i. Meanwhile, anomalies in CPSs are rare in practice that their detection often relies on semi-supervised or unsupervised learning schemes [7]. In this work, we apply FT to tackle a semi-supervised anomaly detection problem, using only the normal data of the SWaT dataset as our training set, to mimic real-world scenarios with high scarcity of anomalous samples.

Unlike discriminative models, generative models can effectively handle the class imbalance issue by learning solely from normal data to detect samples with deviant behaviors that result in significant reconstruction errors of the output relative to the original input [39, 45]. Due to the scarcity of data representing anomalies, as well as the difficulty in modeling realistic scenarios to generate anomalies in CPSs, neural network-based approaches also often adopt semi-supervised or unsupervised schemes in the context of anomaly detection [6]. Accordingly, Autoencoders [36] are widely used for the identification of anomalous instances in multivariate time series [25, 37]. An Autoencoder encodes the input data to a lower-dimensional space from which the latent variables are decoded back to the original input space. Trained only on normal data, the Autoencoder is then fed with unseen data to produce its reconstruction, and the sample for which the reconstruction error, that is, the distance between the input and the reconstructed values, exceeds a certain threshold is deemed anomalous. An Autoencoder thus attempts to learn a compressed representation of the original input while preserving the most relevant features based on its encoder-decoder structure. However, the latent space to which the original input is mapped is not continuous, since Autoencoders cannot model data distributions. To this end, Variational Autoencoders (VAEs) [21] have been proposed for the detection of anomalies in multivariate time series [16, 50]. Due to their ability to learn the parameters of normal data distributions, VAEs have shown to outperform vanilla Autoencoders in terms of anomaly detection using reconstruction probability [2]. Other variants of Autoencoders, such as Deep Autoencoding Gaussian Mixture Model (DAGMM) [51] and Long Short-Term

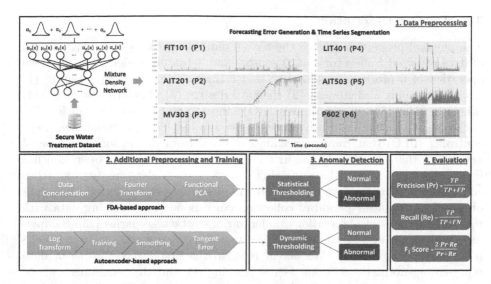

Fig. 2. Overview of our approach for the detection of anomalies in the SWaT dataset. (1) The data preprocessing step includes forecasting error generation using MDNs and time series segmentation. The preprocessed data via this initial step is then used for both FDA and Autoencoder-based approaches. (2) For the former approach, additional preprocessing steps—data concatenation, FT, and FPCA—are taken before the detection phase. For the latter approach, data is transformed into a logarithmic scale before the training of the Autoencoder; the tangent error is then calculated from the reconstruction error smoothed via moving average. (3) For the detection of anomalies, statistical and dynamic thresholds are applied to the FDA and Autoencoder-based approaches, respectively, and (4) the performance is evaluated based on the F_1 score.

Memory (LSTM) Encoder-Decoder [17,26], as well as other deep learning-based methods, such as Recurrent Neural Networks (RNNs) [10,14] and Generative Adversarial Networks (GANs) [4,15,23,24], have also shown compelling results for anomaly detection. In this work, we use a 1D convolutional Autoencoder, mainly due to the representative power of convolution layers, to encode the temporal patterns of only the normal data from the multivariate forecasting error time series generated from the SWaT dataset. We also extend the calculation of ordinary reconstruction errors to that of tangent errors, such that we can leverage the fluctuating forecasting error patterns to our advantage for the detection of anomalies.

3 Proposed Methods

In this section, we explain each step of our proposed approaches in detail. A high-level overview is presented in Fig. 2.

3.1   Dataset Description

The physical properties data of the SWaT dataset[1] consists of 11 days' worth of data collected from a total of 51 sensors and actuators under the continuous operation of 6 processes, referred to as P1 through P6, each representing (1) raw water supply and storage, (2) chemical dosing, (3) ultrafiltration (UF), (4) dechlorination, (5) reverse osmosis (RO), and (6) RO permeate transfer, UF backwash, and cleaning. Each process is connected to a programmable logic controller (PLC) that converts readings received from the sensors into control actions for the actuators. In the first 7 days, data under normal operation has been collected, while in the remaining 4 days, data undergoing attacks (thus containing both normal and abnormal samples) has been collected. The data obtained in the latter period of collection corresponds to streaming data with real-time updates of observations under various attack scenarios [13]. Samples have been logged every second from 8:28:14 p.m. on December 22, 2015 to 2:59:58 p.m. on January 2, 2016. We herein refer to the first 7 days' worth of data representing normal conditions as the training set and the remaining 4 days' worth of data as the test set, in which we aim to detect anomalies. A summarized description of the dataset is provided in Table 1.

Table 1. Summary of the physical properties data of the SWaT dataset.

Data	Date of collection	Total instances	Label	Total instances
Training	Dec. 22, 2015 20:28:14 - Dec. 28, 2015 09:59:58	480,705	0 (normal)	873,667
Test	Dec. 28, 2015 10:01:34 - Jan. 02, 2016 14:59:58	449,824	1 (anomaly)	56,862

In our work, we use the whole dataset, containing data representing all 6 processes, as well as its subset, containing data representing only 4 processes, excluding P2 and P6. We herein refer to the former as P_{all} and the latter as $P_{1,3,4,5}$. The reason for this feature selection is ascribed to the peculiar patterns observed in the forecasting error data corresponding to the processes P2 and P6, as shown in Fig. 3. In addition, since we surmised that there exists a considerable correlation between sensors and actuators belonging to the same process, a more specialized feature selection approach was avoided; that is, we dealt with each process as a whole (multivariate time series), instead of treating each sensor or actuator data from the processes individually (univariate time series).

3.2   Data Preprocessing

Applied to both our FDA and Autoencoder-based methods, the preprocessing techniques used in our work are (1) forecasting error generation using MDNs

---

[1] https://itrust.sutd.edu.sg/itrust-labs_datasets/.

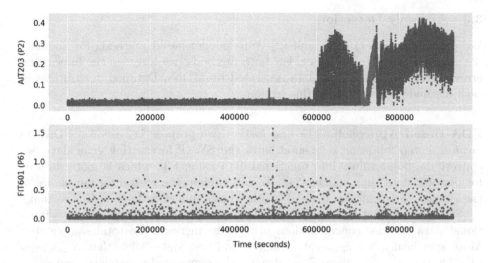

Fig. 3. Forecasting error plots for the sensors AIT203 and FIT601 of processes P2 and P6, respectively, where anomalous instances are marked in red. (Color figure online)

and (2) time series segmentation, which facilitate the analysis of the underlying patterns of multivariate time series. Detailed intuitions behind adopting these techniques are explained in this section.

Forecasting Error Generation. Combining neural networks and mixture density models, MDNs [5] can effectively model conditional density functions of target data given some input data. We leverage this attribute to forecast the normal behaviors of the SWaT sensor and actuator data by training an MDN solely with normal data collected in the first 7 days and testing it with anomalous data collected in the remaining 4 days. Data corresponding to the past 90 s is used to train the model, which then makes predictions corresponding to the next 10 s. To account for the relative importance of specific parts of the input data for the prediction of future behaviors, we incorporate attention mechanisms [46] that direct the attention of the model more to the sample values that are closely located to the predicted output in terms of time, that is, the sample at the last of the 90 s. By calculating the difference between the values in the original and predicted sequences, we obtain the forecasting error data, which is used throughout the experiments in our work.

Time Series Segmentation. We segment a multivariate time series of dimension $D = 51$ into non-overlapping windows. Since the inherent patterns of time windows vary from one another, we choose to experiment with different window sizes and investigate the effect of each window size on the anomaly detection performance.

3.3   Anomaly Detection

We propose novel (1) FDA and (2) Autoencoder-based methods for anomaly detection in the SWaT dataset. For both methods, we leverage the forecasting error patterns of the SWaT data generated by MDNs. Detailed descriptions of each method are provided in this section.

FDA-Based Approach. The first method we propose is based on FDA. As an initial step following segmentation of the SWaT forecasting error data, we convert the point values into functional data using FT, which is more suitable for periodic data than splines [34]. We will explain the reason for our choice of the Fourier basis function instead of other basis functions later in this section.

Since a sufficient amount of data is required for its conversion into functional data, we use concatenations of data to increase the total sample size. More specifically, we aggregate training and test data, such that $S_{concat} := \bigcup_{j=0}^{k}(S_{train} \cup s_{test,j})$, where S_{train} denotes the segmented training set and $s_{test,j}$ denotes the jth window of the segmented test set S_{test}. All subsets of S_{concat} are then converted into k sets of functional data through FT. We apply functional principal component analysis (FPCA) to achieve dimension reduction, such that pattern recognition is possible without compromising important information on the original data. Lastly, we apply statistical thresholding to isolate anomalies using the interquartile range (IQR) approach, where samples lying below the 25th percentile and above the 75th percentile are considered anomalous.

With the basis set $\{1, cos(\cdot), sin(\cdot)\}$, the Fourier series $f(x)$ in the space $L^2((-\pi, \pi])$ is defined as follows:

$$f(x) = a_0 + \sum_{n=1}^{\infty} (a_n cos(nx) + b_n sin(nx)), \quad x \in (-\pi, \pi]. \tag{1}$$

Since $L^2((-\pi, \pi])$ is much more flexible than the space $\mathbb{C}^{\infty}$, the Fourier basis function has the edge over other functions for FDA. Also, the calculation of Fourier coefficients is simpler with the basis set $\{1, cos(\cdot), sin(\cdot)\}$, which is completely orthogonal, and the computing time is much faster than when using other functional forms. Using FT, we expect the resulting functional form of abnormal samples will show a high spike, caused by the difference in forecasting error patterns of normal and anomalous instances, making it distinguishable from that of normal samples. We use functions in R's `fda` package [33] to apply FT and FPCA.

Autoencoder-Based Approach. The second method we propose is based on an Autoencoder. Our encoder and decoder respectively consist of five and six 1D convolution layers, of which the last one is added for dimension matching between the input and the output. Accordingly, all convolution layers except the last one are followed by a rectified linear unit (ReLU) [31] activation layers, while zero-padding is applied to every output of a convolution layer. The last two convolution layers of the encoder and the first two convolution layers of the decoder are followed by max-pooling [47] and upsampling, respectively.

We apply log transformation to the segmented forecasting error data, such that the distribution becomes more uniform. We also add Gaussian noise to the input for generalization purposes. To account for the distances between the neighboring components of the original input and the reconstructed output of the log-transformed forecasting error data, we use the Soft-Dynamic Time Warping (Soft-DTW) [8] loss function, which is differentiable as opposed to DTW. Unlike mean squared error (MSE), which simply compares the distance between two samples at a given time point, Soft-DTW searches for an optimal alignment between two time series, that is, the log-transformed input X and its reconstructed output $\hat{X}$ in our case. More formally, we aim to minimize the following loss function for all n samples:

$$\mathcal{L}(X, \hat{X}) = -\log \sum_{i=0}^{n} e^{-\|X_i - \hat{X}_i\|^2} - \log \sum_{i=0}^{n-1} e^{-\|X_i - \hat{X}_{i+1}\|^2} - \log \sum_{i=0}^{n-1} e^{-\|X_{i+1} - \hat{X}_i\|^2}.$$

(2)

The distances taken into account by the loss function $\mathcal{L}(X, \hat{X})$ in Eq. 2 is illustrated in Fig. 4. After training our model with the Adam optimizer [20], we use Fast-DTW, which reduces the quadratic time complexity of DTW to a linear one [38], to calculate the reconstruction error ϵ, to which smoothing via moving average (MA) is applied. We then calculate the tangent error ϵ_{tan}, using values of ϵ that are 30 time windows apart ($\delta = 30$), as follows:

$$\epsilon_{tan} = \frac{\epsilon_{MA}^t - \epsilon_{MA}^{t-\Delta}}{\Delta},$$

(3)

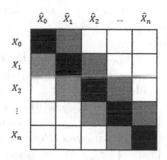

Fig. 4. Alignment matrix of X and $\hat{X}$. The pathways colored in black, red and green represents the distances $\|X_i - \hat{X}_i\|^2$, $\|X_i - \hat{X}_{i+1}\|^2$ and $\|X_{i+1} - \hat{X}_i\|^2$ from Eq. 2, respectively.

where $\Delta = \delta \cdot W = 30 \cdot W$, and ϵ_{MA}^t and $\epsilon_{MA}^{t-\Delta}$ denote the respective reconstruction errors at time points t and $t - \Delta$ obtained after MA smoothing. We refer to the tangent error calculated for the training and test reconstruction data as $E_{tan,train}$ and $E_{tan,test}$. From these measures, we aim to discover error patterns that have significantly changed compared to Δ seconds ago, hence a metric that solely considers the rate of change among the tangent error values and not their magnitudes as such, which can be useful for the detection of contextual anomalies.

Lastly, we present our algorithm for the detection of anomalies in the test set in Algorithm 1, where $\mu_{tan,train}$ and $\sigma_{tan,train}$ denote the mean and standard deviation of $E_{tan,train}$, respectively. The latter statistics are utilized to define an initial threshold of τ_0, which, when exceeded, marks the start of a range of anomalies or an anomaly section. The main idea of our algorithm is to dynamically update the latter threshold based on the change in tangent error of the test set, $E_{tan,test}$. The end of an anomaly section is marked by another quantity τ_{end}, which is updated based on a previously encountered maximum tangent error, denoted by $anomalyPeak$. Consequently, all samples within the anomaly section bounded by τ_0 and τ_{end} are considered anomalous and labeled as '1'.

Algorithm 1: Tangent error-based dynamic thresholding

Inputs : $E_{tan,test}$ of size l

Initialize: A set of predictions $pred(E_{tan,test})$

A threshold $\tau_0 = \mu_{tan,train} + 5 \cdot \sigma_{tan,train}$

A constant $\alpha = 0.5$

```
1  for i = 1 to l do
2      previousPred ← pred(E_tan,test[i − 1])
3      if previousPred == 0 then          // Previous sample was normal
4          if E_tan,test[i] > τ_0 then         // Check for start of anomaly
5              pred(E_tan,test[i]) ← 1                   // Mark as anomalous
6              anomalyPeak ← E_tan,test[i]      // Initialize anomaly peak
7              τ_end ← −α · anomalyPeak   // Initialize end-of-anomaly
                    threshold
8      else                                // Previous sample was anomalous
9          if E_tan,test[i] < τ_end then        // Check for end of anomaly
10             pred(E_tan,test[i]) ← 0                    // Mark as normal
11         else                         // Check for continuance of anomaly
12             pred(E_tan,test[i]) ← 1                // Mark as anomalous
13             if E_tan,test[i] > anomalyPeak then
14                 anomalyPeak ← E_tan,test[i]    // Update anomaly peak
15                 τ_end ← −α · anomalyPeak   // Update end-of-anomaly
                        threshold
16         end
17     end
18 end
```

---

4   Experimental Results and Discussion

In this section, we report our experimental results regarding the anomaly detection performance and compare them to those of baseline approaches. We also comparatively assess our methods and discuss their limitations.

4.1   Comparison with Baselines

We compare the anomaly detection performances of our methods with those of two baseline approaches—CUSUM [29] and static thresholding [11]. The former also segments a given time series into fixed-size windows and calculates the

Table 2. Best overall anomaly detection performances of our methods and baseline approaches. Our FDA and Autoencoder-based methods outperform the baseline approaches, CUSUM and static thresholding (denoted by STATIC). Overall, the Autoencoder-based approach achieved the highest F_1 score (marked in bold) for the window size $W = 15$ and processes P_{all}.

Evaluation metric	CUSUM	STATIC	FDA (Ours)	AE (Ours)
Precision	0.5423	0.4897	0.8113	0.9792
Recall	0.6888	0.5372	0.7021	0.6869
F_1 Score	0.6068	0.5124	0.7528	**0.8074**

cumulative sum of p-norm for each window; the resulting value serves as the anomaly score for its corresponding window, which is deemed anomalous if the score exceeds a predefined threshold. The latter simply uses the p-norm as the anomaly score. In our work, we use $p = 4$ (4-norm) for the evaluation of both CUSUM and static thresholding.

The overall detection performance results are summarized in Table 2, where we can observe that both our methods (denoted by FDA and AE) significantly outperform the baseline methods: the F_1 score of FDA is 24.06% (0.1460) and 46.92% (0.2404) higher than those of CUSUM and static thresholding (denoted by STATIC), and that of AE is 33.06% (0.2006) and 57.57% (0.2950) higher than those of CUSUM and STATIC. Presented in Table 2, the best overall performances of FDA and AE were achieved for the window size $W = 30$ and processes $P_{1,3,4,5}$ for the former, and $W = 15$ and P_{all} for the latter. The superior performances of our methods are mainly attributed to their ability to detect long sequences of contextual anomalies in the SWaT dataset. Consequently, accurate predictions of these anomalies result in a significant increase of the number of true positives (TP), which is directly reflected in the increase of our evaluation metrics, precision (Pr) and recall (Re), defined as $Pr = \frac{TP}{TP+FP}$ and $Re = \frac{TP}{TP+FN}$, where FP and FN denote false positives and false negatives, respectively. Since the F_1 score is a harmonious mean of Pr and Re, defined as $F_1 score = \frac{2 \cdot Pr \cdot Re}{Pr+Re}$, our methods achieve high F_1 scores compared to CUSUM and STATIC. We can also observe in Table 2 that Pr of FDA and AE are particularly high: that of FDA is 49.60% (0.2690) and 65.67% (0.3216) higher than those of CUSUM and STATIC, and that of AE is 80.56% (0.4369) and 99.96% (0.4895) higher than those of CUSUM and STATIC. Since a high Pr implies a low FP, our methods clearly demonstrate their ability to correctly identify most anomalies as shown in Fig. 5, minimizing the potential cost of high FP, one of which is the need for the main infrastructures of the SWaT processes to stop running for manual inspections by human experts.

4.2   Comparative Analysis of Our Methods

Meanwhile, the assessment of each of our methods based on only their highest F_1 scores may render our analysis hasty. Thus, we would rather attempt to provide

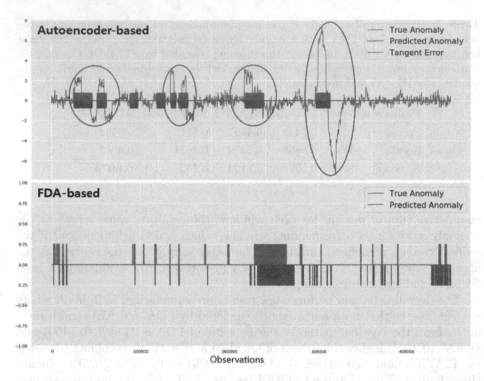

Fig. 5. Examples of anomaly prediction results for our Autoencoder (top) and FDA-based (bottom) methods. For the former, noticeable tangent error fluctuations are circled in red. Some portions have been omitted due to space limitations.

a comparative analysis of the two approaches in terms of their performances with respect to varying window sizes W and different subsets of the SWaT dataset involved, that is, $P_{1,3,4,5}$ and P_{all}.

The anomaly detection performances of our methods evaluated on window sizes $W = \{15, 30, 60, 120, 240\}$, and processes $P_{1,3,4,5}$ and P_{all} are shown in Table 3, where, in the last column, we can observe that the F_1 score of AE quickly drops with increasing W. On the other hand, the change in F_1 score for FDA with respect to W is rather minimal, resulting in a higher overall mean (0.7061) and lower standard deviation (0.0332) compared to those of AE (0.6060 and 0.1434, respectively). The most intuitive explanation for this performance drop for AE can be given based on the reduction of dataset size when W increases. For instance, the dataset size following segmentation with window size $W = 240$ will be 16 times smaller than that following segmentation with window size $W = 15$, resulting in a lack of data from which the model can extract relevant features; further, the encoded data may not result in effective latent representation. On the other hand, FDA includes a data enlargement process through aggregation of the training and test data (S_{concat}), as described in Sect. 3.3.

Table 3. Anomaly detection performances of our methods (denoted by FDA and AE) for all window sizes W, and processes $P_{1,3,4,5}$ and P_{all}. The higher F_1 score among those of FDA and AE from their evaluation on each of the dataset used is marked in bold.

Dataset	Precision		Recall		F_1 score	
(W_P)	FDA	AE	FDA	AE	FDA	AE
$15\_P_{1,3,4,5}$	0.8075	0.9127	0.6926	0.6888	0.7456	**0.7851**
15_P_{all}	0.7105	0.9792	0.7136	0.6869	0.7120	**0.8074**
$30\_P_{1,3,4,5}$	0.8113	0.6381	0.7021	0.7235	**0.7528**	0.6781
30_P_{all}	0.6717	0.7731	0.7291	0.7195	0.6992	**0.7453**
$60\_P_{1,3,4,5}$	0.7418	0.4921	0.7247	0.7857	**0.7332**	0.6052
60_P_{all}	0.6515	0.5574	0.6777	0.7676	**0.6643**	0.6458
$120\_P_{1,3,4,5}$	0.7107	0.3682	0.7136	0.7981	**0.7121**	0.5039
120_P_{all}	0.6057	0.3361	0.681	0.8023	**0.6411**	0.4737
$240\_P_{1,3,4,5}$	0.6897	0.2808	0.7408	0.8475	**0.7143**	0.4218
240_P_{all}	0.6975	0.2687	0.6755	0.7383	**0.6863**	0.3940

Also noteworthy, the effect of the involved processes differs in the two methods. While the exclusion of P2 and P6 contributes to the increase in F_1 score for FDA for all window sizes W, it is not always the case for AE: for relatively small window sizes $W = \{15, 30, 60\}$, higher F_1 scores are achieved for P_{all} than for $P_{1,3,4,5}$. This is attributed to our proposed algorithm, which accounts for the change in tangent error values. For instance, we can observe in Fig. 5 that anomalous instances are well characterized by the tangent error metric, which largely fluctuates with the occurrence of anomalies.

5 Conclusion and Future Work

We present two novel approaches based on FDA and Autoencoder to detect anomalies in multivariate time series. We leverage the underlying patterns of forecasting error data instead of the error values per se and propose promising options for the detection of contextual anomalies. Our experimental results show that the proposed methods significantly outperform baseline approaches in terms of the F_1 score when evaluated on a popular CPS dataset, of which only the normal data is used for training. Based on the relative performance of the proposed methods, we can formulate research problems to be tackled in the future, one of which is the improvement of thresholding techniques to better cope with inferior time series forecasting performance of deep learning models due to a lack of samples representing anomalies in a real-world scenario. Also, we plan to extend the application scope of our methods to other CPS datasets, such as smart grid data, and deploy our algorithms in a testbed environment to validate their performance. We believe that our work contributes to the reliability and security of CPSs, in which anomalous instances are scarce, in that the proposed methods effectively identify anomalies by exploiting solely the data representing the normal state of a CPS.

Acknowledgments. This research was supported by the Energy Cloud R&D program through the National Research Foundation (NRF) of Korea, funded by the Ministry of Science and ICT (MSIT) (No. 2019M3F2A1072217), and the NRF grant, funded by MSIT (No. 2017R1C1B5076474 and No. 2020R1C1C1006004). We would like to thank Youngrok Choi for his earlier contributions to this project.

References

1. Alguliyev, R., Imamverdiyev, Y., Sukhostat, L.: Cyber-physical systems and their security issues. Comput. Ind. **100**, 212–223 (2018)
2. An, J., Cho, S.: Variational autoencoder based anomaly detection using reconstruction probability. Spec. Lect. IE **2**(1), 1–18 (2015)
3. Baheti, R., Gill, H.: Cyber-physical systems. Impact Control Technol. **12**(1), 161–166 (2011)
4. Belenko, V., Chernenko, V., Kalinin, M., Krundyshev, V.: Evaluation of GAN applicability for intrusion detection in self-organizing networks of cyber physical systems. In: 2018 International Russian Automation Conference (RusAutoCon), pp. 1–7. IEEE (2018)
5. Bishop, C.M.: Mixture density networks (1994)
6. Chandola, V., Banerjee, A., Kumar, V.: Outlier detection: a survey. ACM Comput. Surv. **14**, 15 (2007)
7. Chandola, V., Banerjee, A., Kumar, V.: Anomaly detection: a survey. ACM Comput. Surv. (CSUR) **41**(3), 1–58 (2009)
8. Cuturi, M., Blondel, M.: Soft-DTW: a differentiable loss function for time-series. In: Proceedings of the 34th International Conference on Machine Learning, vol. 70, pp. 894–903. JMLR.org (2017)
9. Febrero, M., Galeano, P., González-Manteiga, W.: Outlier detection in functional data by depth measures, with application to identify abnormal nox levels. Environmetrics: Official J. Int. Environmetrics Soc. **19**(4), 331–345 (2008)
10. Filonov, P., Kitashov, F., Lavrentyev, A.: RNN-based early cyber-attack detection for the Tennessee Eastman process. arXiv preprint arXiv:1709.02232 (2017)
11. Filonov, P., Lavrentyev, A., Vorontsov, A.: Multivariate industrial time series with cyber-attack simulation: fault detection using an LSTM-based predictive data model. arXiv preprint arXiv:1612.06676 (2016)
12. Fujimaki, R., Yairi, T., Machida, K.: An approach to spacecraft anomaly detection problem using kernel feature space. In: Proceedings of the Eleventh ACM SIGKDD International Conference on Knowledge Discovery in Data Mining, pp. 401–410 (2005)
13. Goh, J., Adepu, S., Junejo, K.N., Mathur, A.: A dataset to support research in the design of secure water treatment systems. In: Havarneanu, G., Setola, R., Nassopoulos, H., Wolthusen, S. (eds.) CRITIS 2016. LNCS, vol. 10242, pp. 88–99. Springer, Cham (2017). https://doi.org/10.1007/978-3-319-71368-7_8
14. Goh, J., Adepu, S., Tan, M., Lee, Z.S.: Anomaly detection in cyber physical systems using recurrent neural networks. In: 2017 IEEE 18th International Symposium on High Assurance Systems Engineering (HASE), pp. 140–145. IEEE (2017)

15. Goodfellow, I., et al.: Generative adversarial nets. In: Advances in Neural Information Processing Systems, pp. 2672–2680 (2014)

16. Guo, Y., Liao, W., Wang, Q., Yu, L., Ji, T., Li, P.: Multidimensional time series anomaly detection: a GRU-based gaussian mixture variational autoencoder approach. In: Asian Conference on Machine Learning, pp. 97–112 (2018)

17. Habler, E., Shabtai, A.: Using LSTM encoder-decoder algorithm for detecting anomalous ADS-B messages. Comput. Secur. **78**, 155–173 (2018)

18. Hayes, M.A., Capretz, M.A.: Contextual anomaly detection in big sensor data. In: 2014 IEEE International Congress on Big Data, pp. 64–71. IEEE (2014)

19. Huang, K., Zhou, C., Tian, Y.C., Yang, S., Qin, Y.: Assessing the physical impact of cyberattacks on industrial cyber-physical systems. IEEE Trans. Ind. Electron. **65**(10), 8153–8162 (2018)

20. Kingma, D.P., Ba, J.: Adam: a method for stochastic optimization. arXiv preprint arXiv:1412.6980 (2014)

21. Kingma, D.P., Welling, M.: Auto-encoding variational bayes. arXiv preprint arXiv:1312.6114 (2013)

22. Kosek, A.M.: Contextual anomaly detection for cyber-physical security in smart grids based on an artificial neural network model. In: 2016 Joint Workshop on Cyber-Physical Security and Resilience in Smart Grids (CPSR-SG), pp. 1–6. IEEE (2016)

23. Li, D., Chen, D., Goh, J., Ng, S.K.: Anomaly detection with generative adversarial networks for multivariate time series. arXiv preprint arXiv:1809.04758 (2018)

24. Li, D., Chen, D., Jin, B., Shi, L., Goh, J., Ng, S.-K.: MAD-GAN: multivariate anomaly detection for time series data with generative adversarial networks. In: Tetko, I.V., Kůrková, V., Karpov, P., Theis, F. (eds.) ICANN 2019. LNCS, vol. 11730, pp. 703–716. Springer, Cham (2019). https://doi.org/10.1007/978-3-030-30490-4_56

25. Macas, M., Wu, C.: An unsupervised framework for anomaly detection in a water treatment system. In: 2019 18th IEEE International Conference On Machine Learning And Applications (ICMLA), pp. 1298–1305. IEEE (2019)

26. Malhotra, P., Ramakrishnan, A., Anand, G., Vig, L., Agarwal, P., Shroff, G.: LSTM-based encoder-decoder for multi-sensor anomaly detection. arXiv preprint arXiv:1607.00148 (2016)

27. Mathur, A.P., Tippenhauer, N.O.: Swat: a water treatment testbed for research and training on ICS security. In: 2016 International Workshop on Cyber-Physical Systems for Smart Water Networks (CySWater), pp. 31–36. IEEE (2016)

28. Mo, Y., Sinopoli, B.: Integrity attacks on cyber-physical systems. In: Proceedings of the 1st International Conference on High Confidence Networked Systems, pp. 47–54 (2012)

29. Morgenstern, V., Upadhyaya, B., Benedetti, M.: Signal anomaly detection using modified CUSUM method. In: Proceedings of the 27th IEEE Conference on Decision and Control, pp. 2340–2341. IEEE (1988)

30. Mosterman, P.J., Zander, J.: Industry 4.0 as a cyber-physical system study. Softw. Syst. Model. **15**(1), 17–29 (2016)

31. Nair, V., Hinton, G.E.: Rectified linear units improve restricted Boltzmann machines. In: Proceedings of the 27th International Conference on Machine Learning (ICML 2010), pp. 807–814 (2010)

32. Narayanan, S.N., Mittal, S., Joshi, A.: Obd_securealert: an anomaly detection system for vehicles. In: 2016 IEEE International Conference on Smart Computing (SMARTCOMP), pp. 1–6. IEEE (2016)

33. Ramsay, J.O., Hooker, G., Graves, S.: R package "fda" (version 2.4.8). https:// CRAN.R-project.org/package=fda. Accessed 15 Jan 2020
34. Ramsay, J., Silverman, B.W.: Functional Data Analysis, 1st edn. Springer, New York (2005). https://doi.org/10.1007/b98888
35. Ramsay, J.O.: Functional data analysis. Encycl. Stat. Sci. 4 (2004)
36. Rumelhart, D.E., Hinton, G.E., Williams, R.J.: Learning internal representations by error propagation. Technical report. California Univ San Diego La Jolla Inst for Cognitive Science (1985)
37. Sakurada, M., Yairi, T.: Anomaly detection using autoencoders with nonlinear dimensionality reduction. In: Proceedings of the MLSDA 2014 2nd Workshop on Machine Learning for Sensory Data Analysis, pp. 4–11 (2014)
38. Salvador, S., Chan, P.: Toward accurate dynamic time warping in linear time and space. Intell. Data Anal. 11(5), 561–580 (2007)
39. Schneider, P., Böttinger, K.: High-performance unsupervised anomaly detection for cyber-physical system networks. In: Proceedings of the 2018 Workshop on Cyber-Physical Systems Security and PrivaCy, CPS-SPC 2018, pp. 1–12. Association for Computing Machinery, New York (2018). https://doi.org/10.1145/3264888. 3264890
40. Shaadan, N., Jemain, A.A., Latif, M.T., Deni, S.M.: Anomaly detection and assessment of PM10 functional data at several locations in the Klang valley, Malaysia. Atmos. Pollut. Res. 6(2), 365–375 (2015)
41. Shalyga, D., Filonov, P., Lavrentyev, A.: Anomaly detection for water treatment system based on neural network with automatic architecture optimization. arXiv preprint arXiv:1807.07282 (2018)
42. Shi, J., Wan, J., Yan, H., Suo, H.: A survey of cyber-physical systems. In: 2011 International Conference on Wireless Communications and Signal Processing (WCSP), pp. 1–6. IEEE (2011)
43. Song, X., Wu, M., Jermaine, C., Ranka, S.: Conditional anomaly detection. IEEE Trans. Knowl. Data Eng. 19(5), 631–645 (2007)
44. Stouffer, K., Falco, J., Scarfone, K.: Guide to industrial control systems (ICS) security. NIST Spec. Publ. 800(82), 16 (2011)
45. Taormina, R., Galelli, S.: Real-time detection of cyber-physical attacks on water distribution systems using deep learning. In: World Environmental and Water Resources Congress, vol. 2017, pp. 469–479 (2017)
46. Vaswani, A., et al.: Attention is all you need. In: Advances in Neural Information Processing Systems, pp. 5998–6008 (2017)
47. Weng, J., Ahuja, N., Huang, T.S.: Cresceptron: a self-organizing neural network which grows adaptively. In: Proceedings of the 1992 IJCNN International Joint Conference on Neural Networks, vol. 1, pp. 576–581. IEEE (1992)
48. Wu, L., Kaiser, G.: An autonomic reliability improvement system for cyber-physical systems. In: 2012 IEEE 14th International Symposium on High-Assurance Systems Engineering, pp. 56–61. IEEE (2012)
49. Yue, B., Zhao, Y., Xu, Z., Fu, H., Ma, F.: An anomaly intrusion detection method using Fourier transform. J. Electron. 21(2), 135–139 (2004)
50. Zhang, C., Chen, Y.: Time series anomaly detection with variational autoencoders. arXiv preprint arXiv:1907.01702 (2019)
51. Zong, B., et al.: Deep autoencoding Gaussian mixture model for unsupervised anomaly detection (2018)

Applied Data Science: Web Mining

Neural User Embedding from Browsing Events

Mingxiao An[1] and Sundong Kim[2(✉)]

[1] Carnegie Mellon University, Pittsburgh, USA
mingxiaa@andrew.cmu.edu
[2] Data Science Group, Institute for Basic Science, Daejeon, South Korea
sundong@ibs.re.kr

Abstract. The deep understanding of online users on the basis of their behavior data is critical to providing personalized services to them. However, the existing methods for learning user representations are usually based on supervised frameworks such as demographic prediction and product recommendation. In addition, these methods highly rely on labeled data to learn user-representation models, and the user representations learned using these methods can only be used in specific tasks. Motivated by the success of pretrained word embeddings in many natural language processing (NLP) tasks, we propose a simple but effective neural user-embedding approach to learn the deep representations of online users by using their unlabeled behavior data. Once the users are encoded to low-dimensional dense embedding vectors, these hidden user vectors can be used as additional user features in various user-involved tasks, such as demographic prediction, to enrich user representation. In our neural user embedding (NEU) approach, the behavior events are represented in two ways. The first one is the ID-based event embedding, which is based on the IDs of these events, and the second one is the text-based event embedding, which is based on the textual content of these events. Furthermore, we conduct experiments on a real-world web browsing dataset. The results show that our approach can learn informative user embeddings by using the unlabeled browsing-behavior data and that these user embeddings can facilitate many tasks that involve user modeling such as user-age prediction and -gender prediction.

Keywords: User embedding · Web browsing · Demographic prediction

1 Introduction

The Internet has accumulated enormous amount of user-behavior data such as the data related to web browsing [11], news reading [25], advertisement clicking [29], and product purchasing [2], which are generated by hundreds of millions of online users. The deep understanding of users based on their online-behavior data is critical to providing personalized services, such as customized online

© Springer Nature Switzerland AG 2021
Y. Dong et al. (Eds.): ECML PKDD 2020, LNAI 12460, pp. 175–191, 2021.
https://doi.org/10.1007/978-3-030-67667-4_11

advertising [29] and personalized news recommendation [25], to them. Therefore, learning accurate and informative user representations using the massive user-behavior data is important in many practical applications.

One of the conventional methods of learning user representations is based on supervised-learning frameworks, such as user profiling [28], personalized recommendation [25], and product-rating prediction [2]. For example, Zhang et al. [28] proposed to use long short-term memory (LSTM) [10] to learn the representations of social-media users on the basis of the microblogging messages posted by them, for predicting their ages. Wang et al. [25] proposed to learn user representations on the basis of the news articles clicked by the users by using knowledge-aware convolutional neural networks (CNNs) and attention networks [1] for news recommendation. Lu et al. [15] extracted hidden user features from user product reviews by using recurrent neural networks (RNNs) and multiple attention networks for product-rating prediction. However, these methods rely on a large amount of labeled data to learn user-representation models. Not only annotating sufficient samples is expensive and time-consuming but also the user representations learned using these methods are highly restricted to certain purposes only, thereby restricting their generalization to relevant tasks. For example, the user representations learned from the user-age prediction task provides negligible assistance to the user-gender prediction task [12,26]. Consequently, we must introduce more generalizable user representations, which completely excavate the underlying properties of the massive behavioral data.

Notably, highly generalizable representation learning is also one of the central problems in NLP, where inspirations can be brought from the recent success on pretrained word embeddings [4,16,20]. These word embeddings are usually pretrained on a large-scale unlabeled corpus, and they can be applied to many NLP tasks as initial word representations or as additional word features [4,20]. In addition, many studies have proven that these pretrained word embeddings can boost the performance of many important NLP tasks [4,20]. For example, the state-of-the-art performance can be achieved in machine reading comprehension, semantic-role labeling, and named-entity recognition by incorporating as additional word features the word embeddings that are pretrained using the ELMo model [20]. These word embeddings are usually trained on a large-scale unlabeled corpus based on some linguistic heuristics and assumptions, e.g.., "You shall know a word by the company it keeps." For instance, Mikolov et al. [16] proposed a CBOW model to pretrain word embeddings by predicting a target word on the basis of its surrounding words in a sentence. Peters et al. [20] proposed the ELMo model to pretrain contextualized word embeddings based on a language model by predicting the next word in a sentence according to the previous words in the sentence. However, these word-embedding methods could not be directly applied to learn user embeddings, since online user behavior includes diverse interactions between users and events in multiple sessions and simply concatenating texts is known to be suboptimal [27].

In this study, we propose a simple but effective neural user embedding (NEU) approach to learn the deep representations of online users on the basis of unlabeled behavior data generated by the users. In our approach, online users are encoded to low-dimensional dense embedding vectors, which can capture the rich hidden information of online users and can be applied as additional user features to boost the performance of various user-modeling tasks, such as demographic prediction and personalized recommendation. To learn these user embeddings from the user-behavior data, we propose an event-prediction framework to predict the behavior events that these users may have by analyzing their embedding vectors. Our event-prediction framework contains two modules to represent the behavior events. The first one is ID-based event embedding, wherein each event is mapped to a low-dimensional dense vector on the basis of the IDs of these events. The second one is the text-based event embedding, wherein we first extract the texts in these events and then use a text encoder to learn the semantic representations of these events. Furthermore, we conduct extensive experiments on a real-world web browsing dataset crawled using a commercial search engine, named Bing[1] and we also perform two user demographic-prediction tasks, namely, user-age and -gender prediction. The experimental results show that the user embeddings learned using the unlabeled web browsing behavior data can encode the rich latent information of online users and can effectively improve the performance of existing query-based age and gender prediction models.

The major contributions of this paper are three fold as follows:

1. We propose a *NEU* approach to learn user embeddings using unlabeled user-behavior data; these user embeddings can be used to capture rich user information and can enhance various user-involved applications by acting as additional user features.
2. We propose a user-behavior event-prediction framework to learn user representations. Our framework can exploit both event IDs and semantic information of events.
3. We evaluate our approach on a real-world user-behavior dataset and two demographic-prediction tasks.

2   Related Work

Here, we introduce several representative user-modeling methods in different user-involved applications. The first scenario is of user profiling, which aims to predict user attributes, such as age, gender, profession, and interests, on the basis of user-generated data such as blogs and social-media messages [13]. User-profiling methods rely on learning accurate user-feature representations by using user-generated data to predict user attributes [3,21,27,28]. For example, Rosenthal and McKeown [21] used many handcrafted features to represent blog users for predicting their ages. In the recent years, many deep-learning methods have been proposed to learn hidden user representations for user profiling. For example, Zhang et al. [28] used LSTM to learn the representations of users by using social-media logs, for predicting their demographics. Farnadi et al. [5] proposed a multimodal fusion model to learn user representations by using texts, images,

---

[1] https://www.bing.com.

and user relations to predict the ages and genders of Facebook users. Wu et al. [27] used hierarchical attention network to extract user representation from search queries. Chen et al. [3] applied heterogeneous graph attention networks for semi-supervised user profiling from JD.com.

The second scenario is of recommender system and product-rating prediction. Many popular recommender-system methods and product-rating prediction methods involve the learning of both user and item representations [2,15,25]. For example, Wang et al. [25] proposed to learn user representations on the basis of news articles clicked by these users using CNN and attention network for news recommendation. Lu et al. [15] proposed the use of RNNs and multiple attention networks to learn user representations in order to perform product recommendation on the basis of analyzing both user-item ratings via matrix factorization and the user-generated reviews. Chen et al. [2] also proposed to learn user representations for performing product-rating prediction on the basis of the reviews posted by users, by using CNNs and attention networks.

Although these methods can be used to effectively learn user representations for user profiling, recommender systems, and product-rating prediction, there are several drawbacks as follows. First, the user representations learned using these methods are designed for a specific task and usually cannot be generalized to other tasks [12,26]. For example, the user representations learned using the age-prediction task usually have limited informativeness for the gender-prediction task. Therefore, these methods can only encode latent user features in specific dimensions and cannot capture the global information of users. Second, these methods usually rely on labeled data to learn user representations. In many scenarios, the process of annotating sufficient amount of labeled data to learn accurate user representations is expensive and time consuming. Different from these methods, in our approach, we learn deep user representations by using large-scale unlabeled user-behavior data. The user representations learned using our approach can encode the global information of users and can be applied to various user-modeling tasks such as age and gender prediction as additional user features to improve their performance.

Network-embedding methods are also related to this work, as we can regard both users and behavior events as nodes in a graph and user-behavior records as the edges connecting the user nodes and behavior-event nodes. Subsequently, network-embedding methods can be used to learn the vectors of users from the graph. For example, DeepWalk [19] and Node2Vec [9] applied the skip-gram technique [16] on vertex sequences that were generated via truncated random walk on the graph. LINE [23] preserved both the first- and second-order proximities in its objective function. However, these popular network-representation methods had two major differences with our approach. First, they were designed for homogeneous graphs. However, the graph in the problem of user embeddings is bipartite. Although BiNE [6] can be applied to bipartite graphs, it relies on the relations between the same kinds of nodes, which is not available in our task. Second, these methods usually could not incorporate the textual information of nodes. Although Tu et al. [24] considered textual data in their CANE model, the model required all the nodes to have relevant texts; however, in our task, the textual data of users are not always available. Therefore, the CANE model is difficult to be applied in our user-embedding task.

3   Our Approach: Neural User Embedding *(NEU)*

Here, we present our *NEU* approach to learn neural user embeddings from the user-behavior data. In our approach, each user is mapped to a low-dimensional dense embedding vector to capture the latent characteristics of the user. To learn these user embeddings from the user-behavior data, we assume that we can predict the behavior events that these users may have, by analyzing their user-embedding vectors. We explore two approaches to represent behavior events. The first approach, denoted by *NEU-ID*, is the ID-based event embedding, wherein each event is mapped to a low-dimensional dense vector on the basis of the IDs of these events. The second approach, denoted by *NEU-Text*, is the text-based event embedding, wherein we extract the texts in the events and use a text encoder to encode the textual content into vector representations. Next, we introduce both the approaches and a model-training method.

3.1   *NEU-ID* Model for User Embedding

The framework of our *NEU-ID* approach is depicted in Fig. 1a. In our *NEU-ID* approach, each user u is mapped to a low-dimensional dense vector $\mathbf{u} \in \mathcal{R}^D$ by using a user-embedding matrix $\mathbf{U} \in \mathcal{R}^{N_U \times D}$ according to the user IDs, where D denotes the user-embedding dimension and N_U the number of users. In addition, each behavior event e is also mapped to a low-dimensional dense vector $\mathbf{e} \in \mathcal{R}^D$ by using an event-embedding matrix $\mathbf{E} \in \mathcal{R}^{N_E \times D}$ according to the event IDs, where N_E denotes the number of events. In our approach, we assume that the user and the events share the same embedding dimension. Both the user-embedding matrix and event-embedding matrix are randomly initialized and tuned in the model-training stage. Subsequently, we predict the probability of a user u having a behavior event e on the basis of the embeddings of both this user and this event, as follows:

$$p_u^e = p(e|u) = \frac{\exp(\boldsymbol{u}^\top \boldsymbol{e})}{\sum_{e' \in \mathcal{E}} \exp(\boldsymbol{u}^\top \boldsymbol{e'})}. \tag{1}$$

In our approach, we want to maximize the probabilities of all events behaved by users which are recorded in the large-scale user-behavior data. We denote the set of all users as $\mathcal{U}$ and the set of all events behaved by user u as $\mathcal{E}(u)$. Accordingly, the likelihood of all the user behaved events is formulated as follows:

$$\prod_{u \in \mathcal{U}} \prod_{e \in \mathcal{E}(u)} p_u^e = \prod_{u \in \mathcal{U}} \prod_{e \in \mathcal{E}(u)} \frac{\exp(\boldsymbol{u}^\top \boldsymbol{e})}{\sum_{e' \in \mathcal{E}} \exp(\boldsymbol{u}^\top \boldsymbol{e'})}. \tag{2}$$

In our approach, we jointly tune the user embeddings and event embeddings to maximize the likelihood in Eq. (2). Therefore, the user embeddings learned using

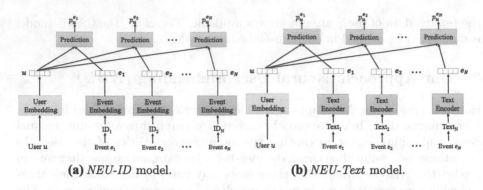

(a) *NEU-ID* model.

(b) *NEU-Text* model.

Fig. 1. Framework of our *NEU-ID* and *NEU-Text* model.

our *NEU-ID* approach can predict the events that a user may have. Accordingly, the hidden characteristics and patterns of users based on their behavior events can be effectively encoded to their user embeddings.

3.2 *NEU-Text* Model for User Embedding

In many user behavior events such as web browsing and advertisement clicking, there exists rich textual information such as the title and contents in the webpage and the keywords in advertisements. Exploiting the semantic meaning of the texts in behavior events can help learn more accurate event representations, which are, in turn, beneficial for learning user embeddings. Let us consider two webpages that may have different titles but considerably similar textual content, for example, *"Tesla Model X for Sale"* and *"Tesla buyers can get a tax credit."* Although browsing both these two webpages involve different behavior event IDs, both the browsing events are closely related in the semantic space and, therefore, may indicate the same user interest, i.e., the interest towards the price of the Tesla car. In addition, exploiting the textual content in behavior events can process the new events. Notably, new events do not have IDs, but their representations can be learned using their text.

Therefore, in our *NEU-Text* approach, we utilize the event text to learn event representations in our framework, as depicted in Fig. 1b. The framework of our *NEU-Text* model is considerably similar to that of the *NEU-ID* model, except that in the former the event representation is learned using event texts by employing a text encoder, rather than using the event IDs. In addition, different text encoders can be applied to our *NEU-Text*

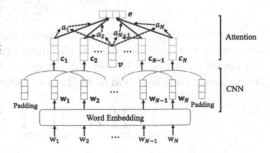

Fig. 2. Architecture of the text encoder.

model. In this study, we introduce a neural-network-based text encoder, whose architecture is depicted in Fig. 2.

Next, we briefly introduce our neural-network-based text encoder. As depicted in Fig. 2, there exist three major layers in the text encoder. The first one is the word-embedding layer. It aims to convert words in a text to a low-dimensional dense vector. We denote the word sequence in a text t as $[w_1, w_2, ..., w_N]$, where N represents the length of the text. In the word-embedding layer, this word sequence is transformed to a vector sequence $[\mathbf{w}_1, \mathbf{w}_2, ..., \mathbf{w}_N]$ by using a word-embedding matrix $\mathbf{W} \in \mathcal{R}^{V \times D_W}$, where V denotes vocabulary size and D_W the word-embedding dimension.

The second layer in the text encoder is a CNN, which is used to capture the local contexts of words to learn contextual word representations. We denote by $\mathbf{c}_i$ the contextual word representation of the i-th word in a text learned using the CNN, and it is computed as follows:

$$\mathbf{c}_i = \mathrm{ReLU}(\mathbf{C} \times \mathbf{w}_{[i-M:i+M]} + \mathbf{b}), \tag{3}$$

where $\mathbf{w}_{[i-M:i+M]}$ denotes the concatenation of the word embeddings between $i-M$ and $i+M$. In addition, $\mathbf{C}$ and $\mathbf{b}$ denote the parameters of the convolutional filters in the CNN, and $2M+1$ is the window size. ReLU is the activation function used [7]. The output of this layer is a sequence of contextual word representations $[\mathbf{c}_1, \mathbf{c}_2, ..., \mathbf{c}_N]$.

The third layer is an attention network [1]. Different words usually have different informativeness for event representation. For example, the title of a webpage may be *"Tesla Model 3 Deliveries In China To Begin In March 2019."* The words "Tesla" and "Deliveries" are more informative than "Begin" for representing the webpage. Therefore, we used the attention mechanism [1] to select important words in order to learn informative text-based event representations. The attention weight of the i-th word in text t is formulated as follows:

$$a_i = \tanh(\mathbf{v} \times \mathbf{c}_i + v), \tag{4}$$

$$\alpha_i = \frac{\exp(a_i)}{\sum_{j=1}^{N} \exp(a_j)}, \tag{5}$$

where $\mathbf{v}$ and v denote the parameters of the attention network. The final representation of an event, based on the text thereof, is the summation of the contextual word representations weighted by their attention weights as follows:

$$\mathbf{e} = \sum_{i=1}^{N} \alpha_i \mathbf{c}_i. \tag{6}$$

In our *NEU-Text* model, both the user embeddings and text encoder are learned using the data by maximizing the likelihood of the behavior events, as shown in Eq. (2).

3.3 Model Training

As previously mentioned in Sect. 3.1 and 3.2, the objective of our event-prediction framework for learning user embeddings is to maximize the likelihood of behavior events on the basis of user and event representations. The objective function of our models is the log-likelihood of behavior events, and it is formulated as follows:

$$\sum_{u \in \mathcal{U}} \sum_{e \in \mathcal{E}(u)} \log \frac{\exp(\boldsymbol{u}^\top \boldsymbol{e})}{\sum_{e' \in \mathcal{E}} \exp(\boldsymbol{u}^\top \boldsymbol{e}')}. \tag{7}$$

However, because the number of behavior events is considerably large, it is significantly costly to compute the denominator part in Eq. (7). However, inspired by [17], we counter this problem by employing negative sampling. For each positive user-event pair (u, e) that actually exists in the user-behavior data, we randomly sample K negative events $e_i^- \in \mathcal{E}, i = 1, 2, \cdots, K$. Subsequently, the objective function can be simplified as follows:

$$\log \frac{\exp(\boldsymbol{u}^\top \boldsymbol{e})}{\sum_{e' \in \mathcal{E}} \exp(\boldsymbol{u}^\top \boldsymbol{e}')} \approx \log \sigma(\boldsymbol{u}^\top \boldsymbol{e}) + \sum_{i=1}^{K} \mathbb{E}_{e_i^- \sim P(e)} \log \sigma(-\boldsymbol{u}^\top \boldsymbol{e}_i^-), \tag{8}$$

where σ denotes the sigmoid function and $P(e)$ the probability distribution of negative events. By following the work in [17], $P(e)$ is defined as follows:

$$P(e) = \frac{f(e)^{0.75}}{\sum_{e' \in \mathcal{E}} f(e')^{0.75}}, \tag{9}$$

where $f(e)$ denotes the frequency of event e.

In our *NEU* approach, both the *NEU-ID* model and *NEU-Text* models were separately trained. Because the IDs of events and the textual content in events may contain complementary information for modeling users, in our *NEU* approach, the user embeddings learned using both the models are concatenated together as the final representations of online users. These user embeddings can encode useful global information of users, and they can be used in many tasks as additional user features to improve their performance.

4 Experiments

In our experiments, we trained user embeddings using real-world web browsing data. In addition, we verified the effectiveness of these user embeddings by applying them to search-query based age and gender prediction. As depicted in Fig. 3, we utilize the user-embedding vectors trained using browsing data to boost the query-based classification. In the present service, browsing data are not used as user features. Therefore, we designed the experiments in this section to estimate the best way to introduce browsing data as additional user features to the present service.

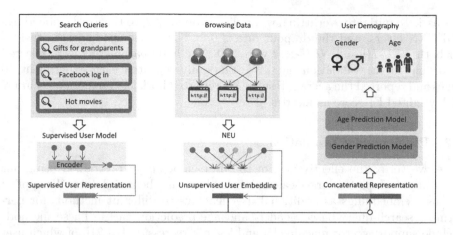

Fig. 3. Overall framework of age and gender prediction. Our NEU embedding is attached to task-specific representation, resulting in improved performance.

4.1 Datasets and Experimental Settings

Datasets. We created a real-world web browsing behavior dataset by crawling the web browsing records of 25,000 anonymous users on Bing, from February 1, 2018 to July 31, 2018. We used the webpage titles in the web browsing records as the event representation. The average number of browsing events per user was 584. In addition, we collected the search queries generated by these users during the same period, along with their gender and age-group tags, to build two datasets for performing search-query-based age- and gender-prediction tasks. The two datasets are denoted by *Gender* and *Age*, respectively. The average number of search queries per user was 211. There are 13,496 (53.98%) male users and 11,504 (46.02%) female users and their age groups are summarized in Table 1. In both *Gender* and *Age* datasets, we randomly sampled 20,000 users as the training set and remaining as the test set. In addition, we randomly sampled 10% users from the training set for validation.

Experimental Settings. In our experiments, two sets of word embeddings were pretrained on all the collected search queries and webpage titles. The embedding dimension was 200. In the training of our *NEU-ID* model, we filtered out the webpages visited by fewer than five distinct users. Consequently, 93,345 unique events and 3,130,070 user-event pairs were left. However, in the training of our *NEU-Text* model, we used all the 2,994,625 available events and 14,500,108 user-event pairs. In the text encoder, the window size of the CNN was 3 and the number of filters 200. We used Adam [14] with learning rate 0.001, and the batch size was set to 500. The number of negative events

Table 1. Age distribution of users.

Age range	Ratio
$[0, 18]$	0.94%
$[18, 24]$	6.81%
$[25, 34]$	14.20%
$[35, 49]$	29.08%
$[50, 64]$	30.56%
$[64, \infty)$	18.41%

K was 4. To mitigate overfitting, we applied dropout [22] to the word embeddings and CNN layers, and the dropout rate is 0.2. The user-embedding dimensions for both *NEU-ID* and *NEU-Text* were 200. The hyperparameters were selected according to the validation data. We randomly repeated each experiment 10 times and reported the average results. We reported both the prediction accuracy and weighted F1-score as metrics.

4.2 Performance Evaluation

Here, we verify the effectiveness of the user embeddings pretrained using our *NEU* approach on the large-scale unlabeled browsing-behavior data. We applied the user embeddings as additional user features to different methods for performing search-query-based gender- and age-prediction tasks. These methods include support vector machine [8] and logistic regression [18,21], of which user search queries are transformed to TF-IDF feature vectors as input, LSTM [10] and CNN, of which user queries are concatenated into a long document then applied as input, LSTM+Att, CNN+Att, LSTM+HieAtt and CNN+HieAtt, of which word-level attention or hierarchical attentions are used to make a final user representation instead of using a global max-pooling layer.

The experimental results of both tasks are summarized in Table 2. From the results, we can see that after incorporating as additional user features the user embeddings pretrained using our *NEU* approach, the predictive power of *all* the classifiers significantly improved for both tasks. For example, after incorporating the pretrained user embeddings, the age-prediction F-score of the CNN+HieAtt method increased from 38.33% to 46.90%. In addition, different methods achieved significant performance improvements after incorporating our user embeddings. Therefore, these results validate that the user embeddings pretrained using our *NEU* approach on the large-scale unlabeled browsing-behavior data contain useful latent information of online users, and that they can improve the performance of various tasks that involve user representations.

Table 2. Gender- and age-prediction performances of different methods both with and without pretrained user embeddings. Notably, U.E. denotes the user embeddings pretrained by our approach.

	Gender prediction				Age prediction			
	Accuracy		F-score		Accuracy		F-score	
	Without	With U.E.	Without	With U.E.	Without	With U.E.	Without	With U.E.
SVM	62.87 ± 0.28	72.98 ± 0.49	61.47 ± 0.50	72.92 ± 0.49	36.41 ± 0.45	42.66 ± 0.54	34.85 ± 0.57	41.70 ± 0.49
LR	62.92 ± 0.37	73.18 ± 0.65	62.05 ± 0.37	73.11 ± 0.64	39.26 ± 0.28	45.45 ± 0.23	36.33 ± 0.34	43.81 ± 0.58
LSTM	65.57 ± 0.55	75.3 3 ± 0.34	63.92 ± 0.92	75.29 ± 0.36	40.83 ± 0.96	47.7 7± 0.66	35.36 ± 1.21	45.97 ± 0.56
CNN	65.55 ± 0.37	75.40 ± 0.33	64.42 ± 0.57	75.34 ± 0.34	40.35 ± 0.62	47.77 ± 0.66	34.51 ± 0.88	45.46 ± 0.64
LSTM+Att	65.91 ± 0.40	75.21 ± 0.34	64.74 ± 0.80	75.13 ± 0.36	40.75 ± 0.60	47.84 ± 0.54	35.70 ± 0.60	45.95 ± 0.56
CNN+Att	66.14 ± 0.50	75.47 ± 0.33	64.61 ± 0.79	75.40 ± 0.31	41.30 ± 0.61	47.66 ± 0.70	36.01 ± 0.89	46.07 ± 0.63
LSTM+HieAtt	66.58 ± 0.39	75.37 ± 0.30	65.58 ± 0.57	75.32 ± 0.31	42.36 ± 0.54	48.19 ± 0.53	37.69 ± 0.83	46.41 ± 0.49
CNN+HieAtt	66.83 ± 0.61	75.72 ± 0.36	65.98 ± 0.71	75.67 ± 0.38	42.66 ± 0.43	48.61 ± 0.56	38.33 ± 0.61	46.90 ± 0.62

4.3   Model Effectiveness

Here, we explore the effectiveness of our *NEU-ID* and *NEU-Text* models in learning user embeddings from user-behavior data. The experimental results are presented in Table 3. The baseline method used in this experiment is CNN+HieAtt. The experimental settings are the same as those in Sect. 4.2.

According to Table 3, both the user embeddings pretrained by our *NEU-ID* model and those pretrained using our *NEU-Text* model can effectively improve both gender- and age-prediction performances, thereby showing that the user embeddings learned using both event IDs and event texts are effective. Interestingly, user embeddings learned using *NEU-Text* performs very well on gender prediction and user embeddings learned using *NEU-ID* performs very well on age prediction. Although the reason for this phenomenon is not clear, our results validate that the user embeddings pretrained by both the models using user-behavior data contain complementary information, and that combining them is more powerful for representing online users than using them separately.

Table 3. Effectiveness by having both *NEU-ID* and *NEU-Text* models.

	Gender		Age	
	Accuracy	*F-score*	*Accuracy*	*F-score*
w/o **U.E.**	66.83 ± 0.61	65.98 ± 0.71	42.66 ± 0.43	38.33 ± 0.61
NEU-ID	73.32 ± 0.39	73.19 ± 0.41	48.10 ± 0.47	46.19 ± 0.84
NEU-Text	75.60 ± 0.44	75.51 ± 0.45	46.73 ± 0.61	44.75 + 0.55
Both	**75.72 ± 0.36**	**75.67 ± 0.38**	**48.61 ± 0.56**	**46.90 ± 0.62**

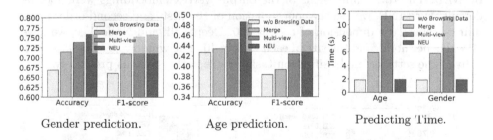

Gender prediction.   Age prediction.   Predicting Time.

Fig. 4. Comparison between our *NEU* approach with baseline methods that use browsing data as direct input.

4.4   Comparison with Direct Input

Here, we compare our proposed *NEU* model with the following two models that utilize the browsing data as direct input: Merge and Multi-view. In the Merge model, browsing titles are considered additional textual information. We use the CNN+HieAtt model explained in Sect. 4.2 to parse the textual information. In the Multi-view model, the browsing data are passed through the CNN+HieAtt model with different parameters, and the concatenation of both the channel outputs is presented to the classifier. The results are depicted in Fig. 4.

According to Fig. 4, our *NEU* model performed better than the other methods that use the browsing data as direct input. Other than the textual information, the user vectors trained using the *NEU* model hold the potential relationship among users. Therefore, the unsupervised embedding method can achieve better result. In addition, the Multi-view model can also effectively improve the performance, and it outperforms the Merge model in terms of both age and gender prediction. This indicates that browsing titles are considerably different from search queries, and that the former should be considered another type of text. In addition, the performance boost provided by the Multi-view model validates that age and gender information can be mined using viewed page titles.

In addition, replacing the direct input by user vectors can dramatically decrease the predicting time according to Fig. 4. Although the initial training *NEU* takes some time, the marginal cost of incorporating *NEU* to a new task is little. Because user representations can be used in many demographic prediction tasks, more time can be saved upon increasing the number of tasks in practice.

4.5 Comparison with Network Embedding

Here, we compare our model with network-embedding methods such as DeepWalk [19], Node2Vec [9], LINE [23], and BiNE [6]. This comparison is necessary, as the vertex embeddings generated using network-embedding methods are also generally used in age and gender prediction. Therefore, we can also try to use them as additional features, and then we can compare the result obtained by *NEU*. The total dimensions of the output vertex embeddings were 400, by default. For LINE, we use the concatenation of the first- and second-order proximities, where the dimension of each is 200. DeepWalk and Node2Vec, we have 10 walks per node, and the size of each walk is 40. The window size for skip-gram is five. The settings of classification tasks are kept the same as previously.

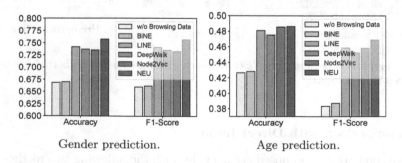

Gender prediction.   Age prediction.

Fig. 5. Comparison between our model and network-embedding methods.

According to Fig. 5, all network-embedding methods enhance the age- and gender-classification performances, thereby indicating that network-embedding methods are effective in training the user embeddings. However,the results obtained using network-embedding methods are not as high as that achieved using our *NEU* method, especially for gender classification. The primary reason should be the lack of textual context. As discussed in Sect. 4.3, textual information is important for gender prediction. Therefore, our approach can gain higher accuracy and F1-score significantly, as it uses textual information, especially in the gender-prediction task. These results validate the importance of combining both the ID- and text-based event representations once again.

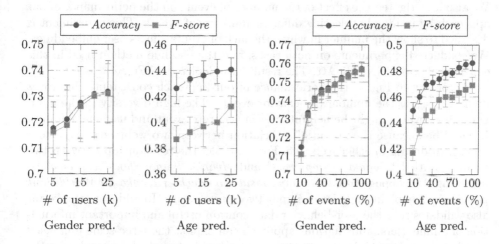

Fig. 6. Effect of the number of users.   **Fig. 7.** Effect of the number of events.

4.6 Effect of the Number of Users

In this experiment, we investigated the effect of the number of users on the performance of *NEU*, by introducing additional users during the embedding training. We conducted the demographic prediction base on 5,000 users including 1,000 test users, and we varied the number of users for training from 5,000 to 25,000. Each set of training users contained all the users from the previous set.

According to Fig. 6, the performance of *NEU* continued to increase upon introducing additional users to the embedding training. Although the additional users were not present in the classification task, their browsing behaviors helped build better representations for all users and behaviors. For example, if a few users had visited a online shopping mall titled *"Love necklace handmade jewelry for her"*, then the text information may be insufficient for the model to produce an accurate representation for this event. However, when this event occurs more number of times upon introducing additional users, the event representation can becomes finer than before. Accordingly, the user embeddings built using more accurate event representation could be more informative and less noisy. This result shows that our method can be used in semi-supervised scenarios wherein we have massive amount of unlabeled user-behavior data but small amount of labeled users. We think introducing additional users will be helpful especially for the cases wherein the labeled data are noisy or insufficient, as the rich information mined from an unlabeled user can be the key to overcoming the restrictions of end-to-end supervised models.

4.7   Effect of the Number of Events

We also investigated the effect of the number of events on the performance of our approach by randomly selecting different numbers of events. This experiment is designed to show the manner in which the lack of events affects user embeddings. We conducted experiments on both tasks, and the baseline method used in this experiment was CNN+HieAtt. The results are depicted in Fig. 7.

According to Fig. 7, the performance of our approach consistently improves upon increasing the number of behavior events. The high diversity in the number of user events might be helpful to form accurate event and user embeddings. It could be because the potential association between two webpages may occur upon introducing another webpage. For example, the relationship between two webpages titled *"Donald Trump News"* and *"Ivanka Trump Shoes"* can be clarified using the webpage titled *"Trump considered daughter Ivanka for World Bank post,"* which explain the relationship between two people. In addition, this result also validates that the user-behavior data contain useful and important information of online users, and that our approach can exploit the large-scale unlabeled user-behavior data to learn accurate user embeddings, which can enhance the performance of many different tasks by acting as additional user features.

4.8   Qualitative Analysis of the User-Embedding Results

For performing qualitative studies, we visualize the user-embedding results generated using the *NEU* model and compare the websites that the users browsed. The t-SNE results achieved using the learned user embeddings are depicted in Fig. 8. Each point denotes each user in our dataset. From the result, we could observe some groups of users in the embedding space and confirm that each user group had a similar interest by observing their browsing histories. In Table 4, we present the browsing titles/histories of the users from four different groups. From the results, we can validate the representation power of our *NEU* model.

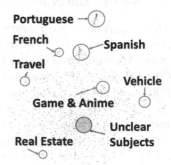

Fig. 8. User groups observed from their embeddings.

Table 4. Browsing histories of users in the same group.

Cluster	User ID	Browsing titles
Game	317	Epic Games' Fortnite
		Razer Cortex: Game Booster
		Battle Pass Season 4
	958	Xbox Games: Immerse Yourself in all the Action | Xbox
		Twitch Prime Pack
		Xbox One Accessories | Xbox
Vehicle	873	Used 2015 Nissan Rogue for sale in Knowville, TN 37922: Sport Utility Details - Autotrader
		Used 2001 Dodge Dakota SLT for sale in Alcoa, TN 37701: Truck Details - Autotrader
		CarMax - Browse used cars and new cars online
	51	Subaru Cars, Sedans, SUVs | Subaru of America
		Motorcycles for Sale on CycleTrader.com: New & Used Motorcycles
		2016 Victory High-Ball Base, Athens OH - Cycletrader.com
Real estate	653	Arcata Real Estate - Arcata CA Homes for Sale | Zillow
		Every American should collect "Federal Rent Checks" - Money Morning
		Table Lamps for Bedroom, Living Room and More | Lamp Plus
	168	Real Estate | Homes for Sale - 0 Homes | Zillow
		15325 SE 155th P1 UNIT F2, Renton, WA 98058 | MLS #1261468 | Zillow
		Unitus Mortgage: Refinance Home Loans
Travel	488	Western Caribbean Cruises & Vacations | Celebrity Cruises
		Tickets | Santana - Divination Tour 2018 - Calgary, AB at Ticketmaster
		7 Foods that Help Fight Arthritis Pain | ActiveBeat
	607	Air Canada Vacations
		YYZ to MID Flights | Expedia
		The 10 Best Nashville Tours, Tickets, Excursions & Activities 2018 | Viator

5   Conclusion

We proposed an *NEU* approach to learn the deep representations of online users by using their large-scale unlabeled behavioral data. In our approach, online users were encoded to low-dimensional dense vectors. In addition, we proposed an event-prediction framework to learn the user embeddings by predicting the behavior events that these users may have, on the basis of their embedding vectors. We explored two methods to represent events, one was based on event IDs and other on the textual content in events. The experiments on real-world datasets validated the effectiveness of our approach.

Acknowledgments. The main part of this work was done when the two authors were interns at Microsoft Research Asia. The authors would like to thank Fangzhao Wu, Zheng Liu, and Xing Xie (Microsoft Research Asia) for their support and discussions. This work was partially supported by the Institute for Basic Science (IBS-R029-C2).

References

1. Bahdanau, D., Cho, K., Bengio, Y.: Neural machine translation by jointly learning to align and translate. In: ICLR (2015)
2. Chen, C., Zhang, M., Liu, Y., Ma, S.: Neural attentional rating regression with review-level explanations. In: WWW, pp. 1583–1592 (2018)
3. Chen, W., et al.: Semi-supervised user profiling with heterogeneous graph attention networks. In: IJCAI, pp. 2116–2122 (2019)
4. Collobert, R., Weston, J., Bottou, L., Karlen, M., Kavukcuoglu, K., Kuksa, P.: Natural language processing (almost) from scratch. J. Mach. Learn. Res. **12**, 2493–2537 (2011)
5. Farnadi, G., Tang, J., De Cock, M., Moens, M.F.: User profiling through deep multimodal fusion. In: WSDM, pp. 171–179 (2018)
6. Gao, M., Chen, L., He, X., Zhou, A.: BiNE: bipartite network embedding. In: SIGIR, pp. 715–724 (2018)
7. Glorot, X., Bordes, A., Bengio, Y.: Deep sparse rectifier neural networks. In: AISTATS, pp. 315–323 (2011)
8. Goel, S., Hofman, J.M., Sirer, M.I.: Who does what on the web: A large-scale study of browsing behavior. In: ICWSM, pp. 120–137 (2012)
9. Grover, A., Leskovec, J.: node2vec: calable feature learning for networks. In: KDD, pp. 855–864 (2016)
10. Hochreiter, S., Schmidhuber, J.: Long short-term memory. Neural Comput. **9**(8), 1735–1780 (1997)
11. Hu, J., Zeng, H.J., Li, H., Niu, C., Chen, Z.: Demographic prediction based on user's browsing behavior. In: WWW, pp. 151–160 (2007)
12. Kim, R., Kim, H., Lee, J., Kang, J.: Predicting multiple demographic attributes with task specific embedding transformation and attention network. In: SDM, pp. 765–773 (2020)
13. Kim, S.M., Xu, Q., Qu, L., Wan, S., Paris, C.: Demographic inference on twitter using recursive neural networks. In: ACL, pp. 471–477 (2017)
14. Kingma, D.P., Ba, J.: Adam: A method for stochastic optimization. arXiv preprint arXiv:1412.6980 (2014)
15. Lu, Y., Dong, R., Smyth, B.: Coevolutionary recommendation model: mutual learning between ratings and reviews. In: WWW, pp. 773–782 (2018)
16. Mikolov, T., Chen, K., Corrado, G., Dean, J.: Efficient estimation of word representations in vector space. In: ICLR Workshop (2013)
17. Mikolov, T., Sutskever, I., Chen, K., Corrado, G., Dean, J.: Distributed representations of words and phrases and their compositionality. In: NeurIPS (2013)

18. Nguyen, D., Gravel, R., Trieschnigg, D., Meder, T.: "How old do you think I am?" A study of language and age in Twitter. In: ICWSM, pp. 439–448 (2013)

19. Perozzi, B., Al-Rfou, R., Skiena, S.: Deepwalk: Online learning of social representations. In: KDD, pp. 701–710 (2014)

20. Peters, M., et al.: Deep contextualized word representations. In: NAACL, pp. 2227–2237 (2018)

21. Rosenthal, S., McKeown, K.: Age prediction in blogs: a study of style, content, and online behavior in pre-and post-social media generations. In: ACL (2011)

22. Srivastava, N., Hinton, G.E., Krizhevsky, A., Sutskever, I., Salakhutdinov, R.: Dropout: a simple way to prevent neural networks from overfitting. J. Mach. Learn. Res. **15**(1), 1929–1958 (2014)

23. Tang, J., Qu, M., Wang, M., Zhang, M., Yan, J., Mei, Q.: LINE: Large-scale information network embedding. In: WWW, pp. 1067–1077 (2015)

24. Tu, C., Liu, H., Liu, Z., Sun, M.: CANE: context-aware network embedding for relation modeling. In: ACL, pp. 1722–1731 (2017)

25. Wang, H., Zhang, F., Xie, X., Guo, M.: DKN: deep knowledge-aware network for news recommendation. In: WWW, pp. 1835–1844 (2018)

26. Wang, P., Guo, J., Lan, Y., Xu, J., Cheng, X.: Multi-task representation learning for demographic prediction. In: ECIR, pp. 88–99 (2016)

27. Wu, C., Wu, F., Liu, J., He, S., Huang, Y., Xie, X.: Neural demographic prediction using search query. In: WSDM, pp. 654–662 (2019)

28. Zhong, D., Li, S., Wang, H., Zhou, G.: User classification with multiple textual perspectives. In: COLING, pp. 2112–2121 (2016)

29. Zhou, G., et al.: Deep interest network for click-through rate prediction. In: KDD (2018)

6VecLM: Language Modeling in Vector Space for IPv6 Target Generation

Tianyu Cui[1,2], Gang Xiong[1,2(✉)], Gaopeng Gou[1,2], Junzheng Shi[1,2], and Wei Xia[1,2]

[1] Institute of Information Engineering, Chinese Academy of Sciences, Beijing, China
{cuitianyu,xionggang,gougaopeng,shijunzheng,xiawei}@iie.ac.cn
[2] School of Cyber Security, University of Chinese Academy of Sciences, Beijing, China

Abstract. Fast IPv6 scanning is challenging in the field of network measurement as it requires exploring the whole IPv6 address space but limited by current computational power. Researchers propose to obtain possible active target candidate sets to probe by algorithmically analyzing the active seed sets. However, IPv6 addresses lack semantic information and contain numerous addressing schemes, leading to the difficulty of designing effective algorithms. In this paper, we introduce our approach 6VecLM to explore achieving such target generation algorithms. The architecture can map addresses into a vector space to interpret semantic relationships and uses a Transformer network to build IPv6 language models for predicting address sequence. Experiments indicate that our approach can perform semantic classification on address space. By adding a new generation approach, our model possesses a controllable word innovation capability compared to conventional language models. The work outperformed the state-of-the-art target generation algorithms on two active address datasets by reaching more quality candidate sets.

Keywords: IPv6 target generation · Deep learning · Data mining · Network measurement · Natural language processing

1 Introduction

Host discovery has always been a vital research method in the field of network measurement. By exploiting the ability of modern hardware and connectivity, tools like Zmap [9] and Masscan [13] have been able to complete the exploration of the global IPv4 address space, which has fundamentally enhanced the ability of researchers to conduct wide-ranging assessments of Internet services.

However, as has long been recognized, IPv6's much larger address space [7] renders exhaustive probing completely infeasible. A recently proposed solution is to design a target generation algorithm [11,19,25] to generate a candidate set that may be active. Systems are required to analyze the potential distribution characteristics of the active address set and infer the target clustering area. The

© Springer Nature Switzerland AG 2021
Y. Dong et al. (Eds.): ECML PKDD 2020, LNAI 12460, pp. 192–207, 2021.
https://doi.org/10.1007/978-3-030-67667-4_12

design of effective analysis algorithms directly determines the ability of model learning and the quality of generated candidate sets.

While prior work has obtained a preliminary understanding of active addresses distribution [12, 21], the results commonly lack interpretability because IPv6 address consisting entirely of digits misses semantics [3], conditioning our inability to infer active addresses using sequence relationships. The reason mainly comes from numerous customizable IPv6 addressing schemes [20, 24]. The complexity of address composition causes difficulty in algorithmic inferences.

The representative target generation algorithms include Entropy/IP [11] and 6Gen [19], which are designed based on human observation and assumptions on network data. Human intervention may result in the algorithm overly dependent on experience and lose adaptability to the data set. The question of how to push the candidate set generated by the algorithm from quantity to quality remains.

To address these problems, we consider a new approach employing deep learning to facilitate effective IPv6 target generation. Word embedding [16] and language modeling [2, 6, 14, 17] are a critical component of systems that require modeling long-term dependency, with successful applications such as summarization and machine translation. By word-to-vector space mapping, word vectors expose the semantic relationships between various words. Language models can estimate the probability distribution of a sequence of words by supervised learning. Based on these principles, we propose to construct an IPv6 vector space with a certain degree of semantic relationship. Through learning the semantic, a language model can autonomously infer the components of active addresses to generate more effective candidate results.

Conventional language models are used to model deterministic sequence dependency. The predicted sequence results are basically consistent with the original data set. In the target generation work, language models are required an innovation to satisfy creative sequence generation.

In this paper, we develop a new concrete instantiation of the target generation algorithm 6VecLM through deep learning, which includes two mechanisms IPv62Vec and Transformer-IPv6. IPv62Vec maps the entire active address space to a semantic vector space, where addresses with similar sequences will be classified into the same cluster. Semantic address vectors will be learned by Transformer-IPv6 to implement IPv6 language modeling. By modeling with a Transformer network [26], our work can comprehensively consider multiple sequence relationships and generate creative and semantically similar sequences to the data set. To serve the generative task, we decide to employ a new generation approach based on cosine similarity and softmax temperature [18] to substitute the probability prediction in language models. Through choosing various sampling strategies, the model can generate expected and creative host targets.

Contributions: Our contributions can be summarized as follows:

1) We explored the construction of IPv6's semantic space for the first time. IPv62Vec can effectively cluster the active address space into several classes.
2) We designed a new target generation algorithm Transformer-IPv6 for language modeling in the vector space. The new generation approach we used can render the language model obtaining creative sequences.

3) Experiments show that our approach outperformed conventional language models and state-of-the-art target generation algorithms on multiple metrics.

Roadmap. Section 2 summarizes the prior researches related to our work. Sect. 3 introduces the background of the IPv6 target generation. Sect. 4 and Sect. 5 highlights the overall design of IPv62Vec and Transformer-IPv6 components in 6VecLM. Sect. 6 shows the evaluation results and Sect. 7 concludes the paper.

2   Related Work

Prior work on IPv6 target generation falls into two broad categories: (1) analyzing known addresses similarity to understand allocation patterns and (2) designing algorithms that generate candidate targets to scan. In addition, we will introduce (3) the related applications exploring semantic relationships.

2.1   Address Similarity Learning

To measure behavioral similarity among network hosts, Coull et al. [4] proposed semantically meaningful metrics for common data types found within network data and compare its performance to a metric that ignores such information to underscore the utility. Ring et al. [23] designed IP2Vec to learn the similarity of IP addresses. They used the meta-information about traffic as the context of the address to train the Word2Vec [16] model. experiments demonstrate the effectiveness of clustering IP Addresses within a botnet data set. Our work is also based on Word2Vec to implement address similarity learning. However, active host discovery is the problem where meta-information is often lacking. We only rely on the active address set to discover new active hosts in this paper.

In the prior work of IPv6 active address set analysis, Planka et al. [21] first explored the potential patterns of IPv6 active addresses in time and space. They used Multi-Resolution Aggregate plots to quantify the correlation of each portion of an address to grouping addresses together into dense address space regions. Gasser et al. [12] employed entropy clustering to classify the hitlist into different addressing schemes. These efforts indicate that researchers have found a certain pattern hidden in the active IPv6 address sets, which provides a basis for the feasibility of IPv6 target generation.

2.2   Target Generation Algorithm

Ullrich et al. [25] used a recursive algorithm for the first attempt to address generation. They iteratively searched for the largest match between each bit of the address and the current address range until the undetermined bits were left, which is used to generate a range of addresses to be scanned. Murdock et al. [19] introduced 6Gen, which generates the densest address range cluster by combining the closest Hamming distance addresses in each iteration. Foremski et al. [11] used Entropy/IP for efficient address generation. They used a Bayesian network to model the statistical dependence between the values of different defined

	Human-readable Text Format	Commonly Used Address Format
• Fixed IID	2001:0db8:0106:0001:**0000:0000:0000:0003**	2001:db8:106:1:**:3**
• Low 64-bit Subnet	2001:0db8:0100:0015:**0000:0000:000a:0005**	2001:db8:100:15:**:a:5**
• SLAAC EUI64	2001:0db8:0000:4144:**f816:3ef f:f e57:0e6d**	2001:db8:0:4144:**f816:3ef f:f e57:e6d**
• SLAAC Privacy	2001:0db8:f bd0:0021:**7c61:2880:3148:36e1**	2001:db8:f bd0:21:**7c61:2880:3148:36e1**

Fig. 1. Sample IPv6 addresses in presentation format with the low 64 bits shown bold.

segments. This learned statistical model can then generate target addresses for scanning. Different from the previous approaches, our work tries to focus on the semantics of IPv6 for the first time to achieve the target generation algorithm through neural networks.

2.3 Word Embedding and Language Modeling

In order to explore the semantic relationship between words, Mikolov et al. [16] proposed Word2Vec to learn high-quality distributed vector representations and prove the availability in measuring syntactic and semantic word similarities. With the development of word embeddings, Bengio et al. [2] first employed neural networks to learn the joint probability function of sequences for substituting statistical language modeling, which subsequently led to deep learning gaining many successful experiences on language models [6,14,17].

Recently, Vaswani et al. [26] proposed a completely self-attention-based network architecture Transformer. The model achieves state-of-the-art performance on the WMT 2014 English-to-German and English-to-French translation task. More work [1,5,8,22] relies on the advantages of this model to achieve breakthroughs in applications. Our work is also based on the Transformer network. We modified the model to implement semantic discovery in the vector space.

3 Preliminary

While prior work of IPv6 analysis has obtained preliminary insights, active host discovery in the large and missing semantic IPv6 address space is still a huge challenge. In this section, we provide basic information on IPv6 and highlight our consideration of IPv6 target generation.

3.1 IPv6 Addressing Background

To explain IPv6 knowledge and the domain-specific terms we used in this paper, we provide a brief background on IPv6 addressing. We refer the reader to RFC 2460 [7] for a detailed description of the protocol.

An IPv6 address consists of a global network identifier, subnet prefix, and an interface identifier [3]. It is composed of 128-bit binary digits, which are usually represented in human-readable text format, using 8 groups of 4 hexadecimal digits and separating them by colons, as shown in Fig. 1. Each of the hexadecimal digits is called a nybble. IPv6 addresses usually use "::" to replace groups of consecutive zero values and omit the first zero value in each group.

However, IPv6 addresses are not simply composed of meaningless digits. There are many IPv6 addressing schemes and network operators are reminded to treat interface identifiers as semantically opaque [3]. Administrators have the option to use various standards to customize the address types. In addition, some IPv6 addresses have SLAAC [24] address format that the 64-bit IID usually embeds the MAC address according to the EUI-64 standard [24] or uses completely pseudo-random [20]. Consider the sample addresses in Fig. 1. In increasing order of complexity, these addresses appear to be: (1) an address with fixed IID value (::3). (2) an address with a structured value in the low 64 bits (perhaps a subnet distinguished by: a). (3) a SLAAC address with EUI-64 Ethernet-MAC-based IID (ff:fe flag). (4) a SLAAC privacy address with a pseudorandom IID.

3.2 Target Generation Consideration

IPv6 Address Space. Active hosts are scattered in the sparse IPv6 space because of excessive address reserves. Limited by computational power, a brute-force approach to probe the entire network space of IPv6 is almost impossible. The distribution of active addresses is also difficult to extract. Therefore, we consider constructing a vector space with good interpretability, where the distance between vectors can be defined as the relationship between addresses. However, IPv6's large address range means that even hexadecimal IPv6 addresses have 32 nybbles. It is difficult to build a high-quality representation of an address vector in a high-dimensional space. Our work will focus on address space representation through model learning and utilize dimensionality reduction techniques to obtain active address clustering areas.

IPv6 Semantic. It may be difficult to effectively train the learning model when analyzing the structure of the address set due to the opaque semantics of IPv6 addresses and the existence of multiple addressing schemes. In order to design an effective target generation algorithm, we believe that reasonably mining the semantic information of address composition is particularly critical. We define the IPv6 semantics by building sequences of address words. The context of the address word sequences can be learned through a model to generate the address vectors with semantic discrimination, which contributes to learning address word sequence relationships to speculate address composition by language modeling.

4   IPv62Vec

In this section, we will introduce our first component in 6VecLM, IPv6 vector space mapping technology IPv62Vec. We outline the underlying ideas of the work including word building, sample generation, and model training.

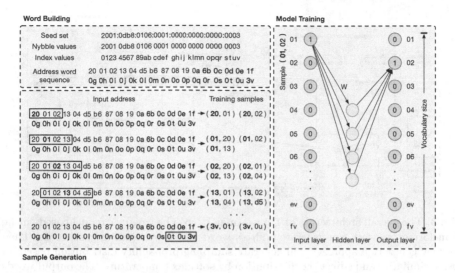

Fig. 2. The overall architecture of IPv62Vec. An address word sequence is composed of the nybble and index values of an address from the seed set. The training samples are generated from the corresponding combinations of input words (highlighted in red color) and context words. The neural network is trained with samples and outputs the vector representation W of the input word. (Color figure online)

4.1 Word Building

Constructing effective semantic information requires to define a new semantic representation of the address. In Fig. 2, we re-represent each nybble of the hex address to create address words. We define the value V_i of the i-th nybble in an address, where $V \in \{0, 1, ..., f\}$. The index i is defined as S_i, where $S \in \{0, 1, ..., v\}$. The i-th address word in a new representation is composed of the nybble value and index value as $V_i S_i$ (e.g. the 11th nybble value 2 is represented as the address word 2a). All address words built from the seed set is defined as vocabulary. Our purpose is to distinguish the nybble values at different indexes. We consider that the same nybble values usually have different degrees of semantic importance according to their position in an address. Differentiating work contributes to discovering the semantic information of key positions (e.g. The 23rd-26th nybbles of the SLAAC EUI-64 address–fffe).

4.2 Sample Generation

After determining the address words, we follow the word selection process of Mikolov et al. [16] to select input words and context to generate training samples. As shown in Fig. 2, we perform a word selection operation on each address word sequence in the seed set. When a certain word of the sequence is selected as an input word, words from the surrounding window of the input word are chosen as context words for building training samples. The window size is 5.

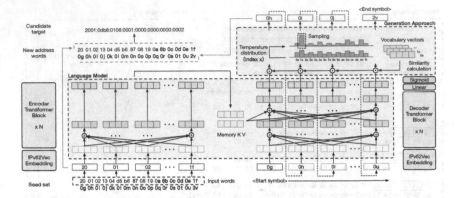

Fig. 3. The overall architecture of Transformer-IPv6. The language model based on the Transformer network can generate address word vector by learning past words. Then the new generation approach will generate sampling probability distribution based on cosine similarity and adjust the distribution by softmax temperature. The output words are recombined with the input to generate a new candidate target.

4.3   Model Training

Since neural networks cannot be fed with words, each word is represented as a One-Hot vector and the length of this vector is equal to the size of the vocabulary. The number of input and output neurons of the neural network is equal to the size of the vocabulary. Further, the output layer uses a softmax classifier and indicates the probabilities that a particular word appears in a specific context. The neural network is fed with the input word and tries to predict the probability of the context word. The output layer of the neural network indicates how likely each word of the vocabulary may be found in the context of the input word. After training, the final hidden layer result is the vector representation of the input word. In this paper, we used 100 neurons in our hidden layer.

5   Transformer-IPv6

After obtaining the IPv6 address vector, the next step is how to apply this semantic vector space to complete network-related tasks. In this section, we will employ the address vector to realize the IPv6 target generation through building the second component in 6VecLM, our language model Transformer-IPv6.

5.1   Language Modeling

Language models assign a probability distribution over sequences $t_{0:L}$ by factoring out the joint probability as follows, where L is the sequence length:

$$P(t_{0:L}) = P(t_0) \prod_{i=1}^{L} P(t_i|t_{0:i-1}) \qquad (1)$$

To model the conditional probability $P(t_i|t_{0:i-1})$, we train a Transformer network to process the address word sequence $t_{0:i-1}$. Transformer is a self-attention-based deep network. We consider focusing on the semantic importance of address word in the sequence for modeling the probability of the next word given the preceding word, until obtaining an entire address word sequence.

The architecture of Transformer-IPv6 is shown in Fig. 3. The input words are converted from addresses in the seed set according to the word building method in Sect. 4. The vector representation, which is determined by the pre-trained IPv62Vec, of the first 16 words of the sequence is inputted in the model to predict the last 16 words. The model then stacks n layers of Transformer encoder block to encode latent vector as a memory. Following Vaswani et al. [26], the Transformer encoder block contains a multi-head self-attention sub-layer followed by a feed-forward network of two fully connected sub-layers. A residual connection and layer normalization are followed each of the two sub-layers. The attention mechanism can help addresses consider critical parts of the sequence when predicting words, while the multi-head attention mechanism observes more address word combinations by training multiple attentions:

$$\text{Attention}(Q, K, V) = \text{softmax}(\frac{QK^T}{\sqrt{d_k}})V \tag{2}$$

$$\begin{aligned} \text{MultiHead}(Q, K, V) &= \text{Concat}(\text{head}_1, ..., \text{head}_h)W^O \\ \text{where head}_i &= \text{Attention}(QW_i^Q, KW_i^K, VW_i^V) \end{aligned} \tag{3}$$

where Q, K, V is the output of the upper layer. $d_q, d_k, d_v, d_{\text{model}}$ are the dimensions of the matrix W_i^Q, W_i^K, W_i^V, and the model input. The linear projections are parameter matrices $W_i^Q \in \mathbb{R}^{d_{\text{model}} \times d_q}$, $W_i^K \in \mathbb{R}^{d_{\text{model}} \times d_k}$, $W_i^V \in \mathbb{R}^{d_{\text{model}} \times d_v}$, $W_i^O \in \mathbb{R}^{d_{\text{model}} \times hd_v}$.

The last 16 words use the mask method [26] to select the current input of the Transformer decoder and ensure that the model's predictions are only conditioned on past words. Transformer decoder block inserts a second multi-head self-attention sub-layer, which performs attention weights computation while keeping encoder memory as attention input K and V.

The model finally predicts the next address word vector through a linear layer and a sigmoid activation function until completing an entire address generation process. Our model uses Transformer block layers $n = 6$, attention head numbers $h = 10$, parameter matrix dimension $d_q = d_k = d_v = 10$, and model input dimension $d_{\text{model}} = 100$.

5.2 Generation Approach

In order to complete the target generation task in the vector space, We expect the generated word vector y_{pred} to have a high semantic similarity to the target word vector y_{true}. Therefore, our model uses the cosine distance as the loss function L:

$$\cos(\theta) = \frac{\sum_{i=1}^{n} y_{\text{true}}^{(i)} \cdot y_{\text{pred}}^{(i)}}{\sqrt{\sum_{i=1}^{n} (y_{\text{true}}^{(i)})^2} \cdot \sqrt{\sum_{i=1}^{n} (y_{\text{pred}}^{(i)})^2}} \tag{4}$$

$$L = 1 - \cos(\theta)$$

Unlike conventional language models that directly model word probabilities, our approach predicts word vectors to preserve the semantic information of the vector space. Since our training samples are address vectors with semantic relationships obtained by IPv62Vec, minimizing the cosine distance can obtain prediction targets with similar context structure to the seed address. This approach aims to choose the closest address word in the vector space, which contributes to discovering the active addresses cluster area.

After generating the address word vector in each epoch, we calculate the cosine similarity between the predicted word vector and each word vector containing the current index in the vocabulary, which is used as a basis for sampling the predicted words. We employ the softmax function to convert the cosine similarity $\cos(\theta)$ to the word sampling probability $P(i)$:

$$P(i) = \frac{e^{\cos(\theta)_i}}{\sum_{j=1}^{C} e^{\cos(\theta)_j}} \ , \ i = 1, ..., C \tag{5}$$

Where C is the number of words with the current index in the vocabulary.

To build an effective generative model, we consider two word sampling strategies: greedy sampling and random sampling. The greedy sampling selects the word with the highest sampling probability in each epoch, while the generated address is always similar to the training set address and raises a high repetition rate. Random sampling ignores the sampling probability and always randomly selected words, while the generated address has high randomness and excessively loses semantics, thus leading to a low activity rate. To seek a balance between maintaining semantics and creativity, we use softmax temperature [18] to readjust the probability distribution:

$$Pr(i) = \frac{e^{\log P(i)^{1/t}}}{\sum_{j=1}^{C} e^{\log P(j)^{1/t}}} \ , \ i = 1, ..., C \tag{6}$$

Where temperature t is a hyperparameter. A high temperature t leads close sampling probability of each word, thus the sampled address is more random. While a low temperature t enhances the difference of original sampling probability, which results in a strong ordering of the generated address.

6   Evaluation

In this section, we evaluate the performance of our approach. We will introduce the data set and evaluation method used in the experiment and show the effectiveness of our approach on the active address set.

Table 1. The detail of the two active address datasets we used in the paper.

Dataset	Seeds	Period	Collection method
IPv6 Hitlist	100,000	January 9, 2020	Public
CERN IPv6 2018	90,010	March 2018–July 2018	Passive measurement

6.1 Dataset

Our experimental datasets are mainly from two parts, a daily updated public dataset IPv6 Hitlist and a measurement dataset CERN IPv6 2018. Table 1 summarizes the datasets used in this paper. The public dataset IPv6 Hitlist is from the data scanning the IPv6 public list for daily active addresses, which is provided by Gasser et al. [12]. In addition, we passively collected address sets under the China Education and Research Network from March to July 2018. We continued to scan and track the IPs that are still keeping active as our measurement dataset CERN IPv6 2018.

6.2 Evaluation Method

Scanning Method. To evaluate the activity of the generated address, we use the Zmapv6 tool [9] to perform ICMPv6, TCP/80, TCP/443, UDP/53, UDP/443 scans on the generated address. When the query sent by any scanning method gets a response, we will determine the address as active. Noting the difference in activity between hosts at different times, we maintain continuous scanning of the host for 3 days to ensure the accuracy of our method.

Evaluation Metric. Since IPv6 target generation is different from text generation tasks, we need to define a new evaluation metric for the address generative model. In the case of a given seed set, $N_{candidate}$ represents the number of the generated candidate set, N_{hit} represents the number of generated active addresses, N_{gen} represents the generated address that is active and not in the seed set. Then the active hit rate r_{hit} and active generation rate r_{gen} of the model can be computed as

$$r_{hit} = \frac{N_{hit}}{N_{candidate}} \times 100\% \qquad r_{gen} = \frac{N_{gen}}{N_{candidate}} \times 100\% \qquad (7)$$

r_{hit} can represent the model's learning ability to learn from the seed set. r_{gen} highlights the model's generation ability to generate new active addresses.

6.3 IPv6 Vector Space

To illustrate the effectiveness of our approach IPv62Vec, we use the active address set IPv6 Hitlist and construct training samples as described in Sect. 4. After training the model, we extract the hidden layer parameters of the model to build the mapping relationship between the address word and the word vector.

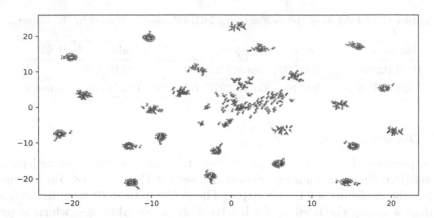

Fig. 4. The address word distribution by using t-SNE visualization. The words with a similar context in an address are clustered in the vector space.

Word Vector Space. Since address word vectors have high dimensions, we use t-SNE technology [15] to reduce the dimensionality of word vectors to facilitate display. Figure 4 shows the semantic relationship between address words in word vector space. Address words with similar contexts perform a tight cluster in vector space. We found that most address words are clustered according to their index attributes, which indicates that different nybble values with the same index possess similar contexts. While the long distance between address words with the same nybble value indicates that they keep different contexts with different indexes, which confirms our intention on word building. Nybble value 0 is an exception to this proposition because address words with nybble value 0 at some index perform a certain degree of clusters, such as index value 4–5, g–l. We surmise that consecutive zeros in the address are the reason for this situation. In addition, the address words with index values 0–7 are close, which indicates that the network prefixes of addresses often have similar structures.

Address Vector Space. The address vectors are determined by combining the address word vectors contained in each address in the address set. In addition, we use a One-Hot vector to represent the characters 0-f and construct the address vector according to the active addresses composition as a baseline for comparison. We use t-SNE technology to reduce the dimension of the address vectors and employ DBSCAN [10] to complete the unsupervised clustering for display. Figure 5 shows the address distribution in the vector space. IPv62Vec successfully divided addresses into several classes. Addresses under the same class perform a high similarity. The One-Hot address vector cannot mine the addresses similarity due to the lack of semantic information.

IPv62Vec only relies on the address sequences to perform effective address similarity learning. The network prefix, subnet identifier, and interface identifier in an address cannot be determined due to the opaque sequence. IPv62Vec extracts the potential network features, thus performing an effective address clus-

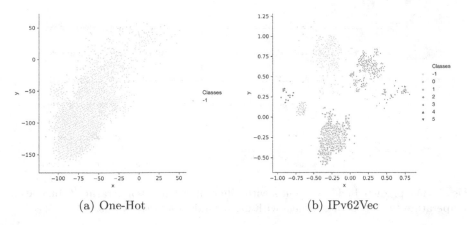

(a) One-Hot   (b) IPv62Vec

Fig. 5. Comparison of address vector distribution between Ont-Hot Encoding and IPv62Vec. IPv62Vec classified the address set to 6 categories by learning the address similarity on the data set IPv6 Hitlist.

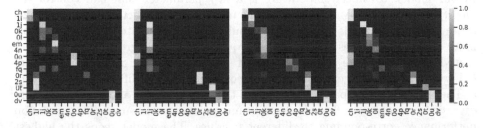

Fig. 6. An example of 4 attention heads on address word sequence in the decoder self-attention in layer 5 of 6. Transformer-IPv6 can conclude a comprehensive target by considering multiple attention results.

ter. By extracting features on IPv6 addresses, we consider that the approach will also be feasible for other network tasks, such as encrypted traffic classification.

6.4   Address Attention

After performing the address vector space mapping, we use Transformer-IPv6 to learn the address vector. To illustrate the effectiveness of Transformer-IPv6, we performed attention visualization work to show the model performance. Figure 6 shows the model's focus on IPv6 addresses. In IPv6 address words, each word has a clear object of attention in the entire address, which enables the associated information in the address to be effectively mined. For example, 1i, 1j, 0k, and 0l may respectively have strong correlations in Fig. 6. In addition, the multi-head attention mechanism guarantees the diversity of address attention, which renders the final address word output of the model to integrate multiple possibilities.

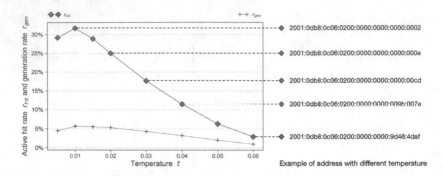

Fig. 7. The predicted address results with different softmax temperature t. Increasing temperature t will generate targets with more random address words.

6.5   Temperature

In our model, softmax temperature is a key parameter that can control the quality of the addresses generated by the model. When selecting a high temperature t, the model tends to sample randomly and the generated address contains more creative sequences. The model is required to sample greedily and the generated address is more close to the seed set when keeping a low temperature t. Figure 7 shows the generation results corresponding to different temperature t. The increase of temperature promotes the generation of address words more diversified. In order to find the equilibrium point, we measured the generation performance corresponding to different t values. The model keeps the highest active hit rate r_{hit} and active generation rate r_{gen} when $t = 0.01$. We recommend t value between 0–0.05 to ensure the model performance.

6.6   Evaluation Results

Baselines. The baselines in our experiments for comparison mainly contain: (1) conventional language model. RNN [17], LSTM [14] and GCNN [6] are the prior paradigms that have shown significant gains in language modeling. In addition, we added IPv62Vec and our generation approach to the conventional language model for adapting the model to target generation tasks. (2) target generation algorithm. Entropy/IP [11] and 6Gen [19] are the state-of-the-art address generation tools that can also efficiently generate active IPv6 targets. We employed the open-source code of Entropy/IP and implemented 6Gen according to the algorithm described by the authors to build baselines.

Experimental Results. Table 2 demonstrates the performance of all the compared models based on the public data set IPv6 Hitlist. The results show 6VecLM outperforms all the baselines, which confirms the advantage of the IPv62Vec and Transformer-IPv6 mechanism. Entropy/IP and 6Gen performs poorly compared to our approach due to lacking IPv6 semantics and adaptability to data

Table 2. The experimental results by comparing with conventional language models and target generation algorithms Entropy/IP and 6Gen. Results show that 6VecLM reached the best performance in our experiments.

Category	Model	$N_{candidate}$	N_{hit}	N_{gen}	r_{hit}	r_{gen}
Conventional	RNN [17]	34,604	995	851	2.88%	2.46%
Language	LSTM [14]	34,636	727	564	2.10%	1.63%
Model	GCNN [6]	34,817	787	649	2.26%	1.86%
Target generation	Entropy/IP [11]	69,167	8,321	2,540	12.03%	3.67%
Algorithm	6Gen [19]	67,712	4,612	1,638	6.81%	2.42%
Adding	RNN [17]	44,242	12,133	2,409	27.42%	5.44%
IPv62Vec and	LSTM [14]	61,950	10,640	2,019	17.18%	3.26%
Generation approach	GCNN [6]	52,046	11,360	2,146	21.83%	4.12%
Our approach	6VecLM	46,461	**15,406**	**2,883**	**33.16%**	**6.21%**

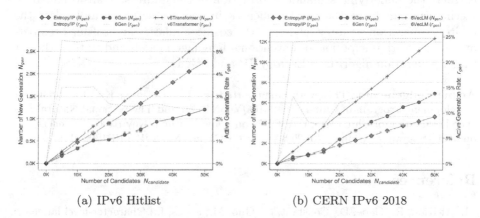

(a) IPv6 Hitlist   (b) CERN IPv6 2018

Fig. 8. The experimental results by comparing with Entropy/IP and 6Gen on the two datasets. N_{gen} and r_{gen} are evaluated under the different candidate set $N_{candidate}$.

sets. By adding generation approach and IPv62Vec mechanism, the conventional language models can reach a not bad performance. While our model is more competent to the target generation task due to the multiple address attention mechanism in Transformer-IPv6.

6.7 Generating Ability

In order to evaluate the model's generating ability, in Fig. 8, we tested the number of new active generation N_{gen} and the active generation rate r_{gen} under the different number of generated candidates on the two address sets. The performance of Entropy/IP and 6Gen is slightly different under the two data sets. Because of the algorithm's ability to adapt to different data sets, the data set CERN IPv6 2018 may have denser address areas than IPv6 Hitlist, which is more conducive to 6Gen algorithm. However, experiment results indicate that our approach reaches a better performance than the other two algorithms. 6VecLM can find 1.23–2.66 times and 1.78–2.31 times more hits than Entropy/IP and 6Gen. The model has a stable and good performance under different data sets. We expect that future target generation algorithms will be capable of generating more valid targets under a limited size of the candidate set to ensure predicting high-quality candidate sets.

7   Conclusion

In this work, we explored the basic challenge of generating promising IPv6 addresses to scan. We presented 6VecLM, an approach to map addresses to a vector space and implement an IPv6 language model that can generate addresses. The address vector generated by IPv62Vec mechanism in 6VecLM effectively extracts the underlying semantic information of the address. Transformer-IPv6 mechanism can learn the word sequences in the vector space and select the address generation strategy relying on the cosine similarity and softmax temperature. The work is superior to conventional language models and state-of-the-art target generation algorithms Entropy/IP and 6Gen.

Acknowledgements. This work is supported by The National Key Research and Development Program of China (No. 2016QY05X1000) and The National Natural Science Foundation of China (No. U1636217) and The National Key Research and Development Program of China (No. 2020YFE0200500 and No. 2018YFB1800200).

References

1. Al-Rfou, R., Choe, D., Constant, N., Guo, M., Jones, L.: Character-level language modeling with deeper self-attention. In: Proceedings of the AAAI Conference on Artificial Intelligence, vol. 33, pp. 3159–3166 (2019)
2. Bengio, Y., Ducharme, R., Vincent, P., Jauvin, C.: A neural probabilistic language model. J. Mach. Learn. Res. **3**, 1137–1155 (2003)
3. Carpenter, B., Jiang, S.: Significance of IPv6 interface identifiers. Internet Engineering Task Force, pp. 1–10 (2014)
4. Coull, S.E., Monrose, F., Bailey, M.: On measuring the similarity of network hosts: Pitfalls, new metrics, and empirical analyses. In: NDSS (2011)
5. Dai, Z., Yang, Z., Yang, Y., Carbonell, J., Le, Q.V., Salakhutdinov, R.: Transformer-xl: Attentive language models beyond a fixed-length context. arXiv preprint arXiv:1901.02860 (2019)
6. Dauphin, Y.N., Fan, A., Auli, M., Grangier, D.: Language modeling with gated convolutional networks. In: Proceedings of the 34th International Conference on Machine Learning, vol. 70, pp. 933–941. JMLR. org (2017)
7. Deering, S., Hinden, R.: Internet protocol, version 6 (IPv6) specification (1998). RFC2460 (2016)
8. Devlin, J., Chang, M.W., Lee, K., Toutanova, K.: Bert: Pre-training of deep bidirectional transformers for language understanding. arXiv preprint arXiv:1810.04805 (2018)

9. Durumeric, Z., Wustrow, E., Halderman, J.A.: ZMap: fast internet-wide scanning and its security applications. In: Presented as part of the 22nd USENIX Security Symposium (USENIX Security 13), pp. 605–620 (2013)

10. Ester, M., Kriegel, H.P., Sander, J., Xu, X., et al.: A density-based algorithm for discovering clusters in large spatial databases with noise. In: KDD, vol. 96, pp. 226–231 (1996)

11. Foremski, P., Plonka, D., Berger, A.: Entropy/IP: uncovering structure in IPv6 addresses. In: Proceedings of the 2016 Internet Measurement Conference, pp. 167–181 (2016)

12. Gasser, O., et al.: Clusters in the expanse: understanding and unbiasing IPv6 hitlists. In: 2018 Proceedings of the Internet Measurement Conference, pp. 2018, 364–378 (2018)

13. Graham, R.D.: Masscan: Mass ip port scanner (2014). https://github.com/robertdavidgraham/masscan

14. Grave, E., Joulin, A., Usunier, N.: Improving neural language models with a continuous cache. arXiv preprint arXiv:1612.04426 (2016)

15. Maaten, L.V.D., Hinton, G.: Visualizing data using t-SNE. J. Mach. Learn. Res. **9**, 2579–2605 (2008)

16. Mikolov, T., Chen, K., Corrado, G., Dean, J.: Efficient estimation of word representations in vector space. arXiv preprint arXiv:1301.3781 (2013)

17. Mikolov, T., Karafiát, M., Burget, L., Černocký, J., Khudanpur, S.: Recurrent neural network based language model. In: Eleventh Annual Conference of the International Speech Communication Association (2010)

18. Müller, R., Kornblith, S., Hinton, G.E.: When does label smoothing help? In: Advances in Neural Information Processing Systems, pp. 4696–4705 (2019)

19. Murdock, A., Li, F., Bramsen, P., Durumeric, Z., Paxson, V.: Target generation for internet-wide IPv6 scanning. In: Proceedings of the 2017 Internet Measurement Conference, pp. 242–253 (2017)

20. Narten, T., Draves, R., Krishnan, S.: Privacy extensions for stateless address autoconfiguration in IPv6. Technical report, RFC 3041, January 2001

21. Plonka, D., Berger, A.: Temporal and spatial classification of active IPv6 addresses. In: Proceedings of the 2015 Internet Measurement Conference, pp. 509–522 (2015)

22. Radford, A., Narasimhan, K., Salimans, T., Sutskever, I.: Improving language understanding by generative pre-training (2018). https://s3-us-west-2.amazonaws.com/openai-assets/researchcovers/languageunsupervised/languageunderstanding paper.pdf

23. Ring, M., Dallmann, A., Landes, D., Hotho, A.: IP2vec: Learning similarities between ip addresses. In: 2017 IEEE International Conference on Data Mining Workshops (ICDMW), pp. 657–666. IEEE (2017)

24. Thomson, S., Narten, T., Jinmei, T., et al.: IPv6 stateless address autoconfiguration. Technical report, RFC 2462, December 1998

25. Ullrich, J., Kieseberg, P., Krombholz, K., Weippl, E.: On reconnaissance with ipv6: a pattern-based scanning approach. In: 2015 10th International Conference on Availability, Reliability and Security, pp. 186–192. IEEE (2015)

26. Vaswani, A., et al.: Attention is all you need. In: Advances in Neural Information Processing Systems, pp. 5998–6008 (2017)

Calibrating User Response Predictions in Online Advertising

Chao Deng[✉], Hao Wang, Qing Tan, Jian Xu, and Kun Gai

Alibaba Group, Beijing, China
{fengyang.dc,wh111044,qing.tan,xiyu.xj,jingshi.gk}@alibaba-inc.com

Abstract. Predicting user response probability such as click-through rate (CTR) and conversion rate (CVR) accurately is essential to online advertising systems. To obtain accurate probability, calibration is usually used to transform predicted probabilities to posterior probabilities. Due to the sparsity and latency of the user response behaviors such as clicks and conversions, traditional calibration methods may not work well in real-world online advertising systems. In this paper, we present a comprehensive calibration solution for online advertising. More specifically, we propose a calibration algorithm to exploit implicit properties of predicted probabilities to reduce negative impacts of the data sparsity problem. To deal with the latency problem in calibrating delayed responses, e.g., conversions, we propose an estimation model to leverage post-click information to approximate the real delayed user responses. We also notice that existing metrics are insufficient to evaluate the calibration performance. Therefore, we present new metrics to measure the calibration performance. Experimental evaluations on both real-world datasets and online advertising systems show that our proposed solution outperforms existing calibration methods and brings significant business values.

Keywords: Online advertising · Calibration · Click-through rate prediction · Conversion rate prediction

1 Introduction

Online advertising is a multi-billion dollars industry with an annual revenue of 107 billion US dollars for the full year of 2018 in the United States only [27]. Compared to traditional advertising industry such as TV, online advertising provides services that tie advertisers' payment directly to measurable user responses such as clicks and conversions. Therefore, predicting user response probability accurately has become one of the essential problems in online advertising [5,12,20]. The most common tasks are click-through rate (CTR) prediction and conversion rate (CVR) prediction.

Predicting user response probability is usually treated as a supervised learning problem. A unique challenge is that the supervision labels are binary observations. For example, in CTR prediction, the observation is that a user either

© Springer Nature Switzerland AG 2021
Y. Dong et al. (Eds.): ECML PKDD 2020, LNAI 12460, pp. 208–223, 2021.
https://doi.org/10.1007/978-3-030-67667-4_13

clicks or not clicks an ad and there is no ground-truth of the underlying click probability. Therefore, most existing work for user response prediction strives to learn binary classifiers and the optimization objectives are based on classification performance such as Area-Under-Curve (AUC) of Precision-Recall (PR) and/or Receiver Operating Characteristic (ROC) curves [6]. Even if some classifiers are modeled to output the user response probability estimations directly, there are still many factors accounting for the discrepancy between *predicted probabilities* and *posterior probabilities*. These factors include inaccurate modeling assumption, deficiencies in the learning algorithm [10,13], hidden features being not available at training and/or serving time [20], data up/down sampling [12,15], etc. While much research effort has been endeavored to address these factors, *calibration* provides a complementary and alternative approach to resolve the discrepancies by transforming predicted probabilities to posterior probabilities directly [10,15,21]. There are two additional benefits associated with calibration from the perspective of advertising system designs. First, calibration is helpful for a loosely coupled system design which separates the concerns of optimization in the auction and the machine learning machinery [20]. Second, calibration is a light-weight solution to cope with the real-time changes in the online environment whenever the user response prediction models are not able to capture the changes in a timely manner.

For these reasons, in online advertising systems, calibration is usually designed as a module to transform predicted probabilities to posterior probabilities. Figure 1 shows the architecture of a common online advertising system. When an ad request arrives, a set of candidate ads are selected by an AD SELECTION module. Then the predicted probabilities of these ads are produced by a PREDICTION module. These predicted probabilities are calibrated by the CALIBRATION module to posterior probabilities, which are important input for the following RANKING module, where an auction mechanism determines which ad will be shown. Finally, the top ranked ad is shown to the user, and user behaviors are tracked. The tracked behavior data are used for prediction model training and calibration function learning.

The online advertising applications pose at least the following two unique challenges to calibrating user response predictions:

- **Sparsity**. User response behaviors are usually very rare. For example, the CTR in certain scenarios may be less than 1% [32]. The number of conversions can be even smaller. According to our experience from an e-commerce advertising platform, the CVR of some electronic product ads is less than 0.1%. The data sparsity problem makes it difficult to estimate the underlying probabilities from the observations.
- **Latency**. User response behaviors may have substantial delays. For example, it may take several days for a user to convert (e.g., place an order) after she clicks an ad. If only short-term responses are considered , the underlying probability will be underestimated. On the other hand, calibration would be stalled if we wait for a long time to collect response data for calibration.

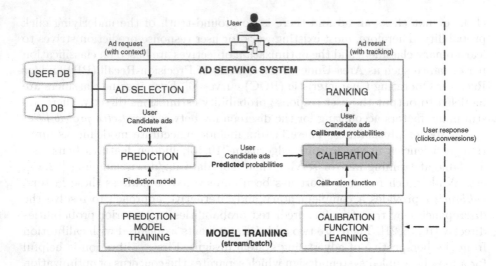

Fig. 1. An illustration of a common advertising system.

In this paper, we present a comprehensive calibration solution for online advertising. More specifically, to cope with the sparsity problem, we propose a simple yet effective calibration algorithm. This algorithm exploits the property that the predicted probabilities can rank samples well (with high AUC) and adds smoothness constraint to ensure that the calibrated probabilities keep the same order with the original predicted ones. To tackle the latency challenge, we propose an estimation model to leverage post-click information to approximate the real delayed responses.

The key contributions of the paper can be summarized as follows:

- We propose the *Smoothed Isotonic Regression* (SIR) algorithm for user response prediction calibration. The algorithm learns a monotonically increasing function to transform predicted probabilities to posterior probabilities and effectively handles data sparsity.
- We propose the *Post-Click Conversion Estimation Model* (PCCEM) for delayed response prediction calibration. The model leverages short-term post-click behaviors for conversion approximation and effectively solves the delayed response problem.
- We present new metrics to measure the calibration performance. Experimental evaluations on two real-world datasets and online advertising systems demonstrate the effectiveness of our calibration solution.

2   Related Work

There has been extensive research on user response prediction [4,8,17–19,28,34], and calibration methods have been introduced as part of the prediction solution

[3,5,9,12,15,20]. However, the importance of calibration is usually underrated, and there is no special study on calibration in online advertising to the best of our knowledge.

In a more general paradigm, calibration can be regarded as a process to produce a function to transform predicted probabilities to posterior probabilities. Existing calibration methods could be divided into parametric and nonparametric ones. Platt's method [25] is a traditional parametric method, which tries to fit a sigmoid calibration function [14]. Beta calibration [14] added more flexibility by assuming that the scores are beta distributed. These methods may fail when their parametric assumptions are not met.

The most popular nonparametric method is Isotonic Regression [23,26,31]. This method tries to find a monotonically increasing function to minimize the squared error between the calibrated probabilities and user response values. A commonly used algorithm for isotonic regression is the pair-adjacent-violator (PAV) [1] algorithm. On sparse datasets, the spiking problem [24] makes this method sensitive to the samples with maximum and minimum predicted probabilities. Another commonly used calibration method is binning method [29,30], of which the main idea is to divide samples into bins and calibrate a predicted probability to the posterior probability of the bin it belongs to. One limitation of this method is that the number of bins needs to be set properly. The BBQ method [22] was then proposed to consider different number of bins and use their weighted average to yield more robust calibrations. However, It is hard to calculate accurate posterior probability of each bin on sparse dataset and binning based methods may not work well.

3   User Response Prediction Calibration

In this section, we first define the problem of calibration and give a brief overview of our calibration solution. Then we introduce the Smoothed Isotonic Regression (SIR) algorithm for a general calibration solution and the Post-Click Conversion Estimation Model (PCCEM) for solving the delayed response problem.

3.1   Problem Definition and Solution Overview

Calibration was defined as a measure: a binary classifier is perfectly calibrated if for a sample of examples with predicted probability p, the expected proportion of positives is close to p [2,14,21]. However, calibration has been recently used to denote the process of obtaining the posterior probabilities [20]. It is beneficial to define the calibration problem more precisely.

Let $\mathcal{X} \subseteq [0,1]$ be the predicted probability space from a prediction model and $\mathcal{Y} = \{0,1\}$ be the user response space where 0 denotes negative response and 1 denotes positive response. Let random variable X denote the predicted probability and Y denote the response value. We define the conditional expectation of Y given $X = x$ as

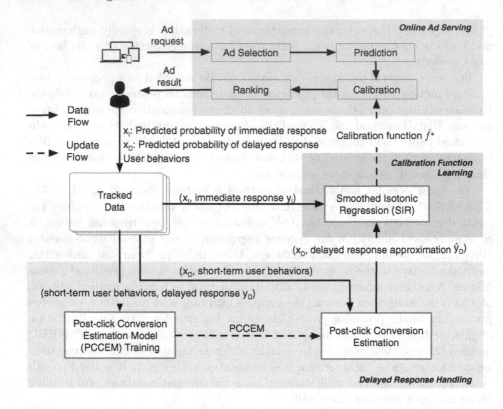

Fig. 2. Calibration solution overview.

$$E[Y|X = x] = \sum_{y \in \mathcal{Y}} y \lim_{\epsilon \to 0^+} P(Y = y | |X - x| \le \epsilon)$$

$$= \lim_{\epsilon \to 0^+} P(Y = 1 | |X - x| \le \epsilon) \tag{1}$$

We are particularly interested in the error function

$$J(X, Y) = \int_{\mathcal{X}} (E[Y|X = x] - x)^2 dx \tag{2}$$

If $J(X, Y) = 0$, the prediction model is said to be perfectly calibrated. Otherwise, let $f : \mathcal{X} \to [0, 1]$ denote a function from $\mathcal{X}$ to $[0, 1]$. The goal of calibration is to find the function

$$f^* = \arg\min_f \int_{\mathcal{X}} (E[Y|X = x] - f(x))^2 dx \tag{3}$$

In this paper, we slightly abuse the terminology so that we define *calibration* as the process of finding the optimal function f^*.

Algorithm 1. Smoothed Isotonic Regression

Input: training set $T = \{(x_i, y_i)|i \leq N\}$; bin size n;
Output: mapping function $f(x)$;

Phase 1 – Binning

1: Sort T according to x_i, get list $L = [(x_i, y_i)]$, where $\bigvee\limits_{i \leq j} x_i \leq x_j$
2: Initialize empty list BL
3: Number of bins $K = \lfloor \frac{N}{n} \rfloor$
4: **for** $k = 0$ to $K - 1$ **do**
5:　　$S = \{i|nk \leq i < n(k+1)\}$
6:　　$l_k = \min\limits_{i \in S} x_i$, $u_k = \max\limits_{i \in S} x_i$, $v_k = \frac{\sum_{i \in S} y_i}{|S|}$, $c_k = |S|$
7:　　Append bin (l_k, u_k, v_k, c_k) to BL
8: **end for**

Phase 2 – Pair Adjacent Violator for Bins

9: Initialize empty list IBL
10: **for** $i = 0$ to $|BL| - 1$ **do**
11:　　Initialize $l = l_i$, $u = u_i$, $v = v_i$, $c = c_i$
12:　　**if** IBL is empty **then**
13:　　　$IBL = [(l, u, v, c)]$
14:　　　**continue**
15:　　**end if**
16:　　Choose bin $t = (l_t, u_t, v_t, c_t)$ at the end of IBL
17:　　**while** $v \leq v_t$ **do**
18:　　　$l = l_t$, $v = \frac{v \times c + v_t \times c_t}{c + c_t}$, $c = c + c_t$
19:　　　Remove t from IL
20:　　　Choose bin $t = (l_t, u_t, v_t, c_t)$ at the end of IBL
21:　　**end while**
22:　　Append new bin (l, u, v, c) to IBL
23: **end for**

Phase 3 – Interpolation

24: Initialize empty list ML
25: **for** $i = 0$ to $|IBL| - 2$ **do**
26:　　$m_i = \frac{l_i + u_i}{2}$, $m_{i+1} = \frac{l_{i+1} + u_{i+1}}{2}$
27:　　$a = \frac{v_{i+1} - v_i}{m_{i+1} - m_i}$, $b = v_i - a m_i$
28:　　Append bin (m_i, m_j, a, b) to ML
29: **end for**
30: $f(x) = a_i x + b_i$　　if　$(l_i, u_i, a_i, b_i) \in ML, l_i < x \leq u_i$
31: **return** $f(x)$

However, finding f^* is not a trivial task. First, it is impossible to calculate $E[Y|X = x]$ directly because we can only observe limited samples drawn from the joint distribution of (X, Y). The best thing one can do is to find an approximate function $\hat{f}^*$ based on these observed samples. Second, in real world applications, the environment may change over time and the prediction model may not capture these changes in a timely manner, so that the joint distribution of (X, Y) may change over time as well. Therefore the calibration function $\hat{f}^*$ is not static and should be updated timely. Third, another unique challenge brought by online advertising is that user responses can be delayed [4]. Such delays hinder $\hat{f}^*$ to be updated in time.

We propose a generic calibration solution to tackle all these challenges. Figure 2 shows the architecture of this solution. The Smoothed Isotonic Regression (SIR) module receives samples and updates the calibration function $\hat{f}^*$ for the online calibration module. For immediate response (click) prediction calibration, the calibration function can be learned with predicted probability x_I and immediate response y_I directly in the SIR module. On the other hand, for delayed response (conversion) prediction calibration, a post-click conversion estimation mechanism is designed to leverage the post-click user behaviors to approximate the delayed response. More specifically, the Post-Click Conversion Estimation Model (PCCEM) Training module collects short-term user behaviors and delayed response y_D to learn the PCCEM, which is used in the Post-Click Conversion Estimation module to produce the approximated delayed response $\hat{y}_D$. The benefit of this design is that the SIR module can also receive calibration learning samples $(x_D, \hat{y}_D)$ for delayed response prediction in a timely manner.

3.2   Smoothed Isotonic Regression (SIR)

On the one hand, the joint distribution of (X, Y) usually changes over time in real-world applications. Hence, we believe that nonparametric methods would be preferable to those based on some distribution assumption when designing the calibration function $\hat{f}^*$. On the other hand, recent advances of the prediction models that optimize objectives based on ranking performance such as AUC [9,16,20,34] provide more and more accurate rankings. This property could be useful while learning calibration function. Therefore, we propose Smoothed Isotonic Regression (SIR), a practical nonparametric method.

The details of SIR is presented in Algorithm 1. The inputs of SIR are training set T and bin size n. First, a binning strategy is used to produce a sorted list of bins BL (Phase 1). Second, Isotonic Regression is applied to ensure that the posterior probability of each bin in BL is monotonically increasing. We adopt the pair-adjacent-violator (PAV) algorithm due to its computational efficiency. However, the vanilla PAV algorithm needs to be modified to be applicable to bins. For two adjacent bins, if the monotonicity is violated, they are pooled together to generate a new bin (Phase 2). Finally, an interpolation strategy is used to derive a monotonic and smoothed function. We note that SIR does not put any constraint on the choice of interpolation strategy. For simplicity, we only present the linear interpolation strategy (Phase 3).

3.3   PCCEM Based Calibration for Delayed Response

The SIR algorithm proposed in Sect. 3.2 is a general algorithm suitable for various calibration tasks. However, there is another challenge for certain calibration

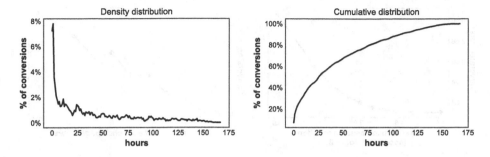

Fig. 3. Distribution of conversion within different time delays.

tasks: when user responses have substantial delays. For example, a conversion may happen several days later after the click. Figure 3 shows a case of such conversion delays for an advertiser on an e-commerce website. As we can see from the cumulative distribution, only a small portion (8%) of the conversions happen within an hour after the click and about 17% conversions have delays for more than 100 h. The delays of the responses result in difficulty for updating the calibration function: if we only use conversions in a short period of time after the click, the conversion rate may be substantially underestimated after calibration. However, to collect all the conversions, we may have to wait for a long time, e.g., a couple of days, which is undesirable for the calibration function to be updated timely. To deal with this problem, we introduce the Post-Click Conversion Estimation Model (PCCEM) to leverage short-term post-click user behaviors for conversion approximation. Then we use these approximated conversions to update calibration functions.

Before detailing PCCEM, we provide an intuitive example as follows. Suppose a user clicks an ad on the first day, and places an order five days later. Although we can not observe the conversion until five days later, there can be plenty of post-click user behaviors that can help us predict how likely the user will convert. For example, the user may spend a long time on the landing page and add the item to shopping cart, etc. These post-click behaviors usually happen in a short period of time, e.g., within a few minutes after the click. Strong evidence shows that these user behaviors are very good conversion predictors. Figure 4 illustrates one such example: both the landing page session duration and the number of page views have positive correlation with conversion rate.

The PCCEM is built on top of the short-term user behaviors to produce a *post-click score* which quantifies the probability of the final conversion. The model is fitted with a dataset with post-click information as features and real conversions as labels, capturing patterns in the post-click behaviors that are correlated with the final conversion. It is worth noting that the conventional conversion rate prediction model used in ad auctions is unable to leverage such information since the conversion rates are predicted and used *before* the ad impressions and clicks.

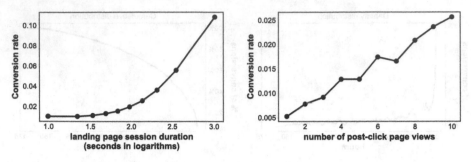

Fig. 4. The relationship between post-click information within one hour after the click and average conversion rate.

Algorithm 2. PCCEM based Calibration for Delayed Response

1: **for** long-term period $T = 1, 2, 3, \ldots$ **do**
2:     Generate training set S by using post-click information as features and real responses as labels
3:     Fit PCCEM m_T based on S
4:     **for** each short-term period t in T **do**
5:         **for** each request i in time period t **do**
6:             Collect post-click information
7:             Use m_T to produce user response score $\hat{y}_i$
8:             Generate a sample $(x_i, \hat{y}_i)$
9:         **end for**
10:     Generate sample set D of recent samples
11:     Update f by using SIR based on D
12:     **end for**
13: **end for**

With PCCEM, the update procedure of calibration solution for delayed response consists of two parts: PCCEM is used to generate post-click score for each click. These scores are used to update the calibration function. The algorithm is shown in Algorithm 2.

4　Metrics

In this section, we first review traditional metrics and discuss their defects. Then, we propose new metrics which can better quantify the calibration performance.

Predicted click over click (PCOC)[1][9,12] is the most commonly used quantitative metric for measuring calibration performance, which is calculated as the ratio of the average calibrated probability and the posterior probability on the whole dataset. The posterior probability is underestimated if PCOC < 1 and overestimated if PCOC > 1. The less PCOC deviates from 1, the better the

---

[1] In the literature this metric is called *calibration*. We use a different name here to avoid confusion.

calibration is. However, PCOC is insufficient to evaluate the calibration performance. The following example shows the the defect of PCOC.

EXAMPLE 1 (Defect of PCOC). *Suppose we have 20,000 samples, and half of them have calibrated probability 0.2 whose posterior probability is 0.4 (underestimated). The other samples have calibrated probability 0.8 and their posterior probability is 0.6 (overestimated). However, the PCOC of these 20,000 samples is $\frac{0.2*10000+0.8*10000}{0.4*10000+0.6*10000} = 1$. These samples have well-calibrated probability values according to PCOC, but they really don't.*

The misleading result roots from the fact that PCOC does not consider the distribution of calibrated probabilities. If we know the joint distribution of (X, Y), we can calculate PCOC for each given x as $\frac{E[Y|X=x]}{x}$. However, we can only get limited samples drawn from the joint distribution of (X, Y). An alternative is to aggregate samples with similar calibrated probabilities to approximately calculate $E[Y|X = x]$ and evaluate the error on different x. Based on this idea, we present a new metric *calibration-N* (Cal-N). First, the calibrated probabilities are divided into N bins with equal frequency and PCOC for each bin is calculated. Then, the error of the i−th bin can be defined as

$$error_i = \begin{cases} PCOC_i - 1 & PCOC_i \geq 1 \\ \frac{1}{PCOC_i} - 1 & PCOC_i < 1 \end{cases} \tag{4}$$

Note that when $PCOC_i < 1$, we use its reciprocal so that overestimation and underestimation are equally treated. Finally, we use the root mean square to accumulate these N errors. To put it formally, Cal-N is defined as

$$\text{Cal-}N = \sqrt{\frac{\sum_{i=1}^{N} error_i^2}{N}} \tag{5}$$

The lower the value is, the better the predicted probabilities are calibrated. Compared with PCOC, Cal-N can accumulate the calibration error across different calibrated probability subspaces. Consider Example 1 again, the Cal-N ($N = 2$) of the 20,000 samples is 2.4, which significantly differs from 0, which means these samples don't have well-calibrated probability values.

In online advertising, a *campaign* is the minimum entity for an advertiser to setup a marketing strategy, which includes budget, target ad audience, creatives and bid price, etc. Therefore, we are concerned with calibration performance of each advertising campaign. Thus we also propose a domain-specific metric *grouped calibration-N* (GC-N), which is the weighted average Cal-N of m campaigns.

$$\text{GC-}N = \frac{\sum_{j=1}^{m} w_j \text{Cal-}N_j}{\sum_{j=1}^{m} w_j} \tag{6}$$

where Cal-N_j and w_j are Cal-N and importance weight of campaign j respectively. In our experiments, w_j can be the number of samples in the j-th campaign. The lower GC-N is, the better the calibrated result is for each campaign.

It is worth mentioning that *log-loss* is commonly used to compare the prediction performance in some binary supervised learning problem literatures [11,18]. A smaller log-loss means better probabilities. However, the absolute value of log-loss is not a good indicator of how well the predictions are calibrated. For example, suppose we get samples with 0/1 labels drawn from a binomial distribution with $p = 0.5$, the log-loss on these samples is not 0 if we predict the ground truth probability 0.5 for each sample. In real world applications, it is important to know the performance of perfectly calibrated probabilities because this could help us to measure the available performance optimizations for a certain problem.

5   Experimental Evaluation

In this section, we conduct experiments on three real-world online advertising datasets. First, we report the experiment results on CTR calibration, comparing SIR with state-of-the-art methods. For delayed response prediction calibration problem, we report the experiment results on CVR calibration. Our solution is also deployed for online A/B test.

5.1   Evaluation of SIR

Dataset. The experiments are conducted on two datasets. *Dataset A* is from a world-leading advertising platform[2]. This dataset comprises roughly 50 million impressions randomly sampled from the ad serving log from July 1 to July 25, 2018. Each impression has its predicted CTR and a label indicating whether the user clicks on the ad. *Dataset B* is a public dataset from iPinYou[3]. Details of this dataset are introduced in [32]. Since there is no predicted CTR in this dataset, we first use the training set to construct a prediction model with GBDT [7], then produce predicted probabilities on test set. The test set augmented with predicted CTR is used for our experiment on calibration.

To make the experimental setup similar to real-world application scenarios, for both datasets, we update and evaluate calibration functions of different methods by an hourly sliding window: for each hour, the calibration function of each campaign is updated on the set of samples in the past 24 h (training set) and evaluated on the set of samples in next hour (test set). Then, we aggregate all the hourly results.

---

[2] Dataset A is available at https://tianchi.aliyun.com/dataset/dataDetail?dataId=40792.

[3] iPinYou dataset is available at http://data.computational-advertising.org.

Comparative Experiment. To validate the effectiveness of our method, we compare the performance of Smoothed Isotonic Regression (SIR) against the state-of-the-art methods. The methods of the comparative experiment are as follows:

(1) **BBQ** [22]: the state-of-the-art binning based method. This method considers different number of bins and uses their weighted average to yield more robust calibration results. The parameters of BBQ are set as the same as [22].

(2) **IR** [3]: the most popular nonparameteric method. We implement this method by the PAV algorithm [1].

(3) **Beta calibration** [14]: the state-of-the-art parametric method. This method assumes that the predicted probabilities and user response values are beta distributed.

(4) **SIR**: our proposed method in this paper. We set bin size n as 1,000.

Table 1 shows GC-N of these methods on the test sets with various N. Usually a larger N will help us evaluate the performance in a more detailed way (Recall that GC-N reduces to PCOC with $N = 1$). As we can see, SIR can decrease GC-N 10.4% on average on dataset A and 29.9% on average on dataset B respectively. SIR outperforms BBQ because SIR leverages the property that the predicted probabilities can rank samples well. BBQ assumes that all bins are independent with each other, so it fails to exploit the ranking relationships between different bins. SIR outperforms IR because the binning phase in SIR reduces the effect of the spiking problem [24] of the PAV algorithm. This problem makes PAV algorithm sensitive to the samples with maximum predicted probabilities and positive labels. This would make IR performs instability on sparse dataset. As we can see, the performance of IR is bad on dataset A. Beta calibration has the closet performance to SIR, which decreases GC-N 9.9% on average on dataset A and 27.7% on average on dataset B respectively. Since beta calibration has pre-defined parametric function curve, the performance is less affected by data sparsity. But it's also this parametric assumption that limits its accuracy of fitting, while SIR is more adaptive to various of distributions due to no distribution assumption.

5.2 Evaluation of PCCEM Based Calibration

Dataset. The dataset in this experiment is from a world-leading advertising platform[4]. This dataset comprises roughly 7 million clicks randomly sampled from the ad serving log from July 1 to July 21, 2018. Each record in the dataset contains information related to a click, including pre-click and post-click features. The pre-click features include a lot of user behavior information before click such as number of views/purchases. The post-click features includes landing page session duration after click, number of views/purchases after click in an hour, number of add items into cart/favorites in an hour. Data between July 15th to July 21th are used as test set.

---

[4] This dataset is available at https://tianchi.aliyun.com/dataset/dataDetail?dataId= 40796.

Table 1. GC-N of different methods on two test sets. For a sufficient comparison, we use $N = 3, 4, 5$ to calculate GC-N. The best value of them on test sets is highlighted.

Dataset	Method	$N = 3$	$N = 4$	$N = 5$
A	No calibration	0.63	0.67	0.72
	(1) BBQ	0.57 (-9.5%)	0.63 (-6.0%)	3.05 ($+323.6\%$)
	(2) IR	0.82 ($+30.0\%$)	0.89 ($+32.8\%$)	0.98 ($+36.1\%$)
	(3) Beta calibration	0.56 (-11.1%)	0.60 (-10.4%)	0.66 (-8.3%)
	(4) SIR	**0.56 (-11.1%)**	**0.60 (-10.4%)**	**0.65 (-9.7%)**
B	No calibration	0.40	0.45	0.46
	(1) BBQ	0.71 ($+77.5\%$)	0.81 ($+80.0\%$)	0.84 ($+82.6\%$)
	(2) IR	0.31 (-22.5%)	0.35 (-22.2%)	0.40 (-13.0%)
	(3) Beta calibration	0.27 (-32.5%)	0.32 (-28.9%)	0.36 (-21.7%)
	(4) SIR	**0.27 (-32.5%)**	**0.30 (-33.3%)**	**0.35 (-23.9%)**

Comparative Experiment. In this section, we compare the conversion calibration performance of different methods. To make it fair, all the methods use our proposed SIR algorithm as the calibration algorithm. We consider the following methods:

(1) **Short-term Calibration (STC):** For each click, we only use the short-term conversions that are within one hour after the clicks to update the calibration function.
(2) **Long-term Calibration (LTC):** For each click, we wait for 7 days to get the true conversions. In this case, the calibration function can only be updated with real conversions in the past 7 days and the clicks no later than 7 days ago.
(3) **PCCEM based Calibration (PCCEM):** We update PCCEM on a daily basis: each day whenever the new conversion data are available, PCCEM is updated. For each hour, when the new short-term behavior data are available, post-click scores are produced by the latest PCCEM, then the calibration function is updated based on these scores.

Table 2 shows the experiment results for the three methods. As we can see, STC has the worst performance as the short-term conversions underestimates the conversion rates by a large margin. Method LTC makes calibration function updating stalled so it may not capture the relationship between predicted conversion rate and real conversion in time. Method PCCEM outperforms both methods, effectively improving calibration performance with delayed response.

5.3   Online Evaluation

To investigate whether our proposed approach can help improve the business performance, we also conducted an online A/B test experiment on a world-leading advertising platform. The experiment lasted for seven days with the setup that the control bucket uses the predicted probabilities given by a deep

Table 2. GC-N of three calibration methods for predicted conversion rate calibration.

Method	$N = 3$	$N = 4$	$N = 5$
No calibration	4.46	4.47	4.49
(1) STC	5.67 (+27.1%)	5.91 (+32.2%)	7.11 (+58.4%)
(2) LTC	1.57 (−64.8%)	1.72 (−61.5%)	1.81 (−59.7%)
(3) PCCEM	**0.50 (−88.8%)**	**0.58 −87.0%)**	**0.71 (−84.2%)**

Table 3. Business result of A/B test.

Business metrics	RPM	CTR	ROI
Improvement (%)	+3.86%	+8.93%	+5.07%

learning-based prediction model and the test bucket uses the further calibrated probabilities produced by our proposed approach. Each bucket was assigned 10% of all the online traffic which was in the magnitude of tens of millions. Generally speaking, an advertising platform strives to provide values to the advertisers, the users, and the platform itself. A better response prediction is expected to contribute to all these three values. We use *return on investment* (ROI)[5] to indicate the advertisers' benefit, CTR to indicate the user experience and *revenue per mille* (RPM) to quantify the platform's gain. Results are shown in Table 3, we can observe that our calibration solution can increase RPM by 3.86%, CTR by 8.93%, and ROI by 5.07%.

6   Conclusion and Future Work

In this paper, we introduced a calibration solution for user response prediction in online advertising, including the SIR algorithm for data sparsity problem and the PCCEM for delayed response problem. We also proposed new metrics to evaluate the effectiveness of calibration. Experiment results on real-world datasets have proven that the calibration solution can lead to significantly better results both in terms of technical measurements and business performance.

There is an interesting direction for our future work. In many applications, the distribution of the observed samples can be different from the distribution of the ones whose response probabilities need to be predicted and calibrated [33]. Therefore, it is beneficial to design an unbiased calibration algorithm in this case.

References

1. Ayer, M., Brunk, H.D., Ewing, G.M., Reid, W.T., Silverman, E., et al.: An empirical distribution function for sampling with incomplete information. Ann. Math. Stat. **26**(4), 641–647 (1955)

---

[5] The *return* is the value of the conversions and the *investment* is the cost charged by the advertising platform.

2. Bella, A., Ferri, C., Hernández-orallo, J., Ramírez-quintana, M.J.: Calibration of machine learning models
3. Borisov, A., Kiseleva, J., Markov, I., de Rijke, M.: Calibration: A simple way to improve click models. In: Proceedings of the 27th ACM International Conference on Information and Knowledge Management, pp. 1503–1506. ACM (2018)
4. Chapelle, O.: Modeling delayed feedback in display advertising. In: Proceedings of the 20th ACM SIGKDD International Conference on Knowledge Discovery and Data Mining, KDD 2014, pp. 1097–1105. ACM, New York (2014). https://doi.org/10.1145/2623330.2623634
5. Chappelle, O., Manavoglu, E., Rosales, R.: Simple and scalable response prediction for display advertising. ACM Trans. Intell. Syst. Technol. 2(3), Article 1 (2015). https://doi.org/10.1145/0000000.0000000, http://arxiv.org/abs/1502.07526
6. Fawcett, T.: An introduction to roc analysis. Pattern Recognit. Lett. **27**(8), 861–874 (2006)
7. Friedman, J.H.: Greedy function approximation: a gradient boosting machine. Ann. Stat. **29**, 1189–1232 (2001)
8. Gentile, C., Li, S., Kar, P., Karatzoglou, A., Etrue, E., Zappella, G.: On context-dependent clustering of bandits. arXiv preprint arXiv:1608.03544 (2016)
9. Graepel, T., Candela, J.Q., Borchert, T., Herbrich, R.: Web-scale Bayesian click-through rate prediction for sponsored search advertising in microsoft's bing search engine. In: Proceedings of the 27th international conference on machine learning (ICML-10), pp. 13—20 (2010)
10. Guo, C., Pleiss, G., Sun, Y., Weinberger, K.Q.: On calibration of modern neural networks. In: Proceedings of the 34th International Conference on Machine Learning, ICML 2017, Sydney, NSW, Australia, 6–11 August 2017, pp. 1321–1330 (2017). http://proceedings.mlr.press/v70/guo17a.html
11. Guo, H., Tang, R., Ye, Y., Li, Z., He, X.: DeepFM: a factorization-machine based neural network for CTR prediction. In: IJCAI International Joint Conference on Artificial Intelligence, pp. 1725–1731 (2017). https://doi.org/10.1145/2988450.2988454
12. He, X., et al.: Practical lessons from predicting clicks on ads at Facebook. In: Proceedings of the Eighth International Workshop on Data Mining for Online Advertising, pp. 1–9. ACM (2014)
13. King, G., Zeng, L.: Logistic regression in rare events data. Polit. Anal. **9**(2), 137–163 (2001)
14. Kull, M., Silva Filho, T., Flach, P.: Beta calibration: a well-founded and easily implemented improvement on logistic calibration for binary classifiers. In: Artificial Intelligence and Statistics, pp. 623–631 (2017)
15. Lee, K.c., Orten, B., Dasdan, A., Li, W.: Estimating conversion rate in display advertising from past performance data. In: Proceedings of the 18th ACM SIGKDD International Conference on Knowledge Discovery and Data Mining, pp. 768–776. ACM (2012)
16. Li, C., Lu, Y., Mei, Q., Wang, D., Pandey, S.: Click-through prediction for advertising in twitter timeline. In: Proceedings of the 21th ACM SIGKDD International Conference on Knowledge Discovery and Data Mining, pp. 1959–1968. ACM (2015)
17. Li, S., Karatzoglou, A., Gentile, C.: Collaborative filtering bandits. In: Proceedings of the 39th International ACM SIGIR conference on Research and Development in Information Retrieval, pp. 539–548. ACM (2016)

18. Liu, Q., Yu, F., Wu, S., Wang, L.: A convolutional click prediction model. In: Proceedings of the 24th ACM International on Conference on Information and Knowledge Management - CIKM 2015, pp. 1743–1746 (2015). https://doi.org/10.1145/2806416.2806603, http://dl.acm.org/citation.cfm?doid=2806416.2806603

19. Lu, Q., Pan, S., Wang, L., Pan, J., Wan, F., Yang, H.: A practical framework of conversion rate prediction for online display advertising. In: Proceedings of the ADKDD 2017, p. 9. ACM (2017)

20. Mcmahan, H.B., et al.: Ad click prediction: a view from the trenches. In: In Proceedings of the ACM SIGKDD International Conference on Knowledge Discovery and Data Mining, KDD, pp. 1222–1230 (2013)

21. Menon, A.K., Jiang, X.J., Vembu, S., Elkan, C., Ohno-Machado, L.: Predicting accurate probabilities with a ranking loss. In: Proceedings of the. International Conference on Machine Learning. International Conference on Machine Learning, vol. 2012, p. 703. NIH Public Access (2012)

22. Naeini, M.P., Cooper, G.F., Hauskrecht, M.: Obtaining well calibrated probabilities using Bayesian binning. In: AAAI, pp. 2901–2907 (2015)

23. Niculescu-Mizil, A., Caruana, R.: Predicting good probabilities with supervised learning. In: Proceedings of the 22nd International Conference on Machine Learning, pp. 625–632. ACM (2005)

24. Pal, J.K.: Spiking problem in monotone regression: penalized residual sum of squares. Stat. Prob. Lett. **78**(12), 1548–1556 (2008)

25. Platt, J., et al.: Probabilistic outputs for support vector machines and comparisons to regularized likelihood methods. Adv. Large Margin Classif. **10**(3), 61–74 (1999)

26. Robertson, T., Robertson, T.: Order restricted statistical inference. Technical report (1988)

27. Statista: Online advertising revenue in the united states from 2000 to 2018 (2019). https://www.statista.com/statistics/183816/us-online-advertising-revenue-since-2000/. Accessed 02 Apr 2020

28. Yang, H., Lu, Q., Qiu, A.X., Han, C.: Large scale CVR prediction through dynamic transfer learning of global and local features. In: Workshop on Big Data, Streams and Heterogeneous Source Mining: Algorithms, Systems, Programming Models and Applications, pp. 103–119 (2016)

29. Zadrozny, B., Elkan, C.: Learning and making decisions when costs and probabilities are both unknown. In: Proceedings of the Seventh ACM Sigkdd International Conference on Knowledge Discovery and Data Mining, pp. 204–213. ACM (2001)

30. Zadrozny, B., Elkan, C.: Obtaining calibrated probability estimates from decision trees and Naive Bayesian classifiers. In: ICML, vol. 1, pp. 609–616 (2001)

31. Zadrozny, B., Elkan, C.: Transforming classifier scores into accurate multiclass probability estimates. In: Proceedings of the eighth ACM SIGKDD International Conference on Knowledge Discovery and Data Mining, pp. 694–699. ACM (2002)

32. Zhang, W., Yuan, S., Wang, J., Shen, X.: Real-time bidding benchmarking with ipinyou dataset. arXiv preprint arXiv:1407.7073 (2014)

33. Zhang, W., Zhou, T., Wang, J., Xu, J.: Bid-aware gradient descent for unbiased learning with censored data in display advertising. In: Proceedings of the 22nd ACM SIGKDD International Conference on Knowledge Discovery and Data Mining, pp. 665–674. ACM (2016)

34. Zhou, G., et al.: Deep interest network for click-through rate prediction. In: Proceedings of the 24th ACM SIGKDD International Conference on Knowledge Discovery and Data Mining, pp. 1059–1068. ACM (2018)

An Advert Creation System for 3D Product Placements

Ivan Bacher[1](✉), Hossein Javidnia[1], Soumyabrata Dev[1,2], Rahul Agrahari[1], Murhaf Hossari[1], Matthew Nicholson[1], Clare Conran[1], Jian Tang[3], Peng Song[3], David Corrigan[3], and François Pitié[1,4]

[1] ADAPT SFI Research Centre, Trinity College Dublin, Dublin, Ireland
ivan.bacher@adaptcentre.ie
[2] School of Computer Science, University College Dublin, Dublin, Ireland
[3] Huawei Ireland Research Center, Dublin, Ireland
[4] Department of Electronic and Electrical Engineering, Trinity College Dublin, Dublin, Ireland

Abstract. Over the past decade, the evolution of video-sharing platforms has attracted a significant amount of investments on contextual advertising. The common contextual advertising platforms utilize the information provided by users to integrate 2D visual ads into videos. The existing platforms face many technical challenges such as ad integration with respect to occluding objects and 3D ad placement. This paper presents a Video Advertisement Placement & Integration (Adverts) framework, which is capable of perceiving the 3D geometry of the scene and camera motion to blend 3D virtual objects in videos and create the illusion of reality. The proposed framework contains several modules such as monocular depth estimation, object segmentation, background-foreground separation, alpha matting and camera tracking. Our experiments conducted using Adverts framework indicates the significant potential of this system in contextual ad integration, and pushing the limits of advertising industry using mixed reality technologies.

Keywords: Advertisement · Augmented reality · Deep learning

1 Introduction

With the popularity of 4G networks and the decline in data traffic tariffs, the video content industry has maintained a relatively high growth rate. It is expected that the overall market size in 2021 will approach 211 billion RMB. An increase of 351% as compared to 2018 [1]. Such growing video demand and the increase of user generated videos creates additional challenges for advertisement and marketing agencies. The agencies need to devise innovative strategies

I. Bacher, H. Javidnia and S. Dev—Authors contributed equally.

The ADAPT Centre for Digital Content Technology is funded under the SFI Research Centres Programme (Grant 13/RC/2106) and is co-funded under the European Regional Development Fund.

© Springer Nature Switzerland AG 2021
Y. Dong et al. (Eds.): ECML PKDD 2020, LNAI 12460, pp. 224–239, 2021.
https://doi.org/10.1007/978-3-030-67667-4_14

to attract the attention of end-users. Traditionally, advertisements were added into existing videos as overlay, pre-roll, mid-roll or post-roll. These approaches are disruptive to the user's experience for online streaming applications.

In this paper, we solve the problem of disruptive user experience by creating a 3D-advertisement creation system. Therefore, this work describes a proof-of-concept prototype system that enables users to seamlessly insert a 3D object in any user-generated video. The user can select a 3D object from the library of 3D objects with in the proof-of-concept prototype, which can then be placed on any planar surface within a video scene. Our system can automatically analyze different depth layers in a video sequence and seamlessly integrate new 3D objects with proper occlusion handling.

1.1 Related Work

In the literature, there are several works in the area of advertisements in images and video streams. However, most of the existing work focus on the identification of logos and advertisement billboards in videos. Covell *et al.* in [8] used audio and video features to accurately identify the sections in the video that contain the ads. This assists them to replace the existing ads with user-specific adverts in redistributed television materials. Hussain *et al.* proposed a novel framework in [17] that understands the general sentiments of adverts using a large image- and video- datasets. Recently, Nautiyal *et al.* [28] used pre-trained deep learning models to identify existing 2D adverts in video streams, and seamlessly replace them with new adverts. Using large-scale annotated datasets of billboards [10, 11], it provides them with an end-to-end framework for video editors to perform 2D advert placements. Their system assists in detecting frames in a video that contains a billboard [15], localizes the billboard in the detected frame [12], and subsequently replace the existing billboard with a new 2D advertisement. In this paper, we generalize this problem of product placements into any user-generated videos, and artificially augment 3D adverts into the existing scenes.

1.2 Contributions and Organization of the Paper

Our contribution in this paper is two fold: (a) we propose a proof-of-concept prototype system that enables 3D computer graphic advertisement objects to be inserted seamlessly into video streams; and (b) we thereby establish a new paradigm in product placements for marketing agencies. Our proposed system will greatly assist video editors and content producers reduce the time it takes to dynamically generate augmented videos.

The remainder of the paper is arranged as follows: Sect. 2 briefly describe the technology behind our cloud-based advertisement creation system. Section 3 presents information regarding the design and development of the proof-of-concept prototype system. Section 4 describes the various use cases and associated applications for the developed prototype. Section 5 concludes the paper.

2   Technology

This section describes the methods and technologies employed in the Adverts framework including the monocular depth estimation module, camera tracking, interactive segmentation and background matting.

2.1   Monocular Depth Estimation

Monocular depth estimation is used to understand the 3D geometry of the scene and anchor the 3D plane on which the object will be placed. The classical depth estimation approaches heavily rely on multi-view geometry [4,9,20,33,38] such as stereo image [31,32]. These methods acquire depth information by utilising visual cues and different camera parameters which are not often available in offline monocular videos. The idea of using the monocular image to capture depth information could potentially solve the memory requirement issue of the conventional methods, but it is computationally difficult to capture the global properties of a scene such as texture variation or defocus information. The recent advancement of Convolutional Neural Networks (CNN) and publicly available datasets have significantly improved the performance of monocular depth estimation [5,13,14,22,37].

Several deep learning based monocular depth estimation networks are studied and evaluated in this research [13,16,23,24,34]. Among these, the network proposed by Hu *et al.* [16] illustrated a superior performance in terms of accuracy and computational time compared to others. More importantly, this model showed a better generalization in depth scales due to the multi-scale feature fusion module integrated in the architecture. Figure 1 presents a sample of the monocular depth estimation followed by a localised plane in the scene. The orientation of the plane is obtained by calculating the normals from the depth information. The model by Hu *et al.* [16] is employed as the first module in Adverts framework.

Fig. 1. Monocular depth estimation on a real-world scene. From left to right: input image; estimated depth map, localised plane using normal estimation.

2.2   Camera Tracking

One of the very essential components of any augmented reality platform is tracking the camera motion to seamlessly integrate 3D object into the scene. Online augmented reality tools often utilise accelerometer, GPS, and solid state compass to track the camera motion in real-time. Such information is not available in offline scenarios. Adverts framework takes advantage of the traditional Structure from Motion (SfM) pipeline. Initially the user identifies a certain number of keyframes with manually matched feature points. Further, SIFT features [26] are detected and matched between the selected Keyframes followed by an optimization applied to refined the 3D projected points. The next step involves automatic feature matching between keyframes and non-keyframes. The matched features from each non-keyframes are triangulated and reconstructed using the previous keyframes $[R|T]$. The final step of the camera tracking process involves a large scale sparse bundle adjustment [2] with least square optimization applied to refine each non-keyframe's $[R|T]$. Figure 2 illustrates the pipeline implemented to track the camera motion in Adverts framework. The camera projection matrix obtained for each frame is later used to project the 3D objects to camera space.

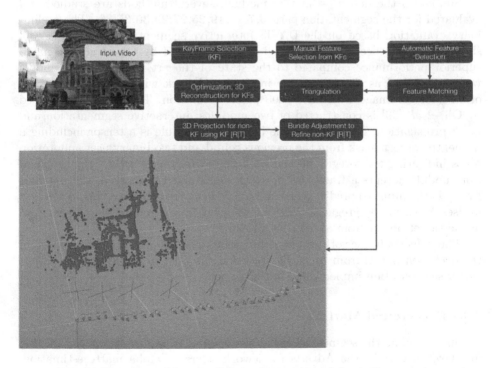

Fig. 2. Camera tracking pipeline implemented in Adverts framework.

2.3   Interactive Segmentation

Determining the occluding object in an augmented reality application highly depends on the location of the 3D object and structural accuracy of the estimated depth map. By differentiating different layers of depth information, one can consistently integrate a virtual object in the scene. However, the quality of the integration result depends on the following measures:

1. How accurate the general structure of depth map is?
2. How temporally consistent the depth of the occluding object is across the entire video?
3. How much detail is preserved in the depth structure of the occluding object?

To compensate for the flaws of the estimated depth maps such as inaccurate depth boundaries, Adverts framework takes advantage of object segmentation in videos to produce binary masks of the occluding object. The generated masks are later used to estimate the alpha matte which is explained in the next section. This module allows users to interactively select the occluding object and decide which part of the scene is causing the occlusion by providing a broader control over tracking the occluding object across the entire video.

Similar to the depth estimation module, several methods are studied and evaluated for the segmentation part [3,7,18,19,25,27,29,30,35,36]. The preliminary evaluation based on the DAVIS interactive segmentation benchmark [6] showed that the deep learning based model proposed by Oh *et al.* [29] has a superior performance compared to the state of the art methods. This model was also ranked as the fastest one in the benchmark with the inference time of $0.2s$ for an image with 800×600 pixels resolution. The network proposed by Oh *et al.* [29] is constructed of two modules: interactive segmentation and mask propagation. The input to the interactive module is a tensor including a frame, the object mask from the previous round and two binary user annotation maps indicating the foreground and background regions. Further, the propagation module accepts a frame, its previous mask and the mask of the previous frame as the input to predict a new mask for the current frame. This model also utilises a Feature Aggregation Module designed to accumulate the information of the target object from all user interactions.

The Adverts framework employs the model from [29] to interactively obtain the occlusion masks from users. Figure 3 demonstrate an example of the interactive segmentation implemented in this paper.

2.4   Background Matting

To further refine the segmentation masks acquired from Sect. 2.3 and achieving fine level of details, the Adverts framework refers to alpha matte estimation. This is performed to calculate the opacity value of each blended pixel in the foreground object.

Fig. 3. Illustration of occlusion mask. User input, segmentation, propagation.

Generally, the composition image I_i is represented as a linear combination of the background B_i and foreground F_i colors [39]:

$$I_i = \alpha_i F_i + (1 - \alpha_i)B_i \tag{1}$$

where $\alpha_i \in [0, 1]$ denotes the opacity or alpha matte of the foreground at pixel i. Often users provide guidance in a form of a trimap to solve this problem. Trimap assigns a label to every pixel as foreground $\alpha = 1$, background $\alpha = 0$ or unknown opacity. The goal of the alpha matting algorithms is to estimate the opacity value of the unknown regions by utilising the pixel values from known regions. To achieve this goal, we investigated the effect of known background information in the matting process. This is done by introducing a Background-Aware Generative Adversarial Network to estimate alpha channels. Unlike the conventional methods, this architecture is designed to accept a 7 channel volume, where the first 3 channels contain the RGB image, the second 3 channels contain the RGB background information and the last channel contains the trimap. The preliminary experiments using the trained model indicates a significant improvement in the accuracy of the alpha mattes compared to the state of the art. The full details of this module including the background reconstruction and matting blocks are available as a preprint article on arXiv [21].

3   System Design

In this section we describe the design and development, as well as the main technologies used for building the proof-of-concept prototype system that this work presents. The system can be split into two main components: user interface and back-end. Figure 4 illustrates the main structure of the system.

3.1   User Interface

The user interface was implemented as a web application using modern web based technologies. This choice is supported by the fact that web based technologies only need a browser to run, thus making them cross platform compatible. The main web based technologies utilised in the user interface are: Aurelia, Three.js, Async, Bootstrap, Fontawesome, as well as state of the art web APIs

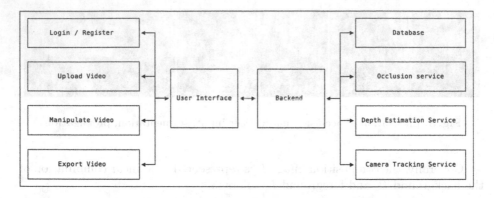

Fig. 4. System design of our proposed Adverts framework.

such as the Broadcast Channel API, the Canvas API, the Fetch API, and the Web Storage API[1].

Aurelia can be described as a core building block of the user interface and is a JavaScript client framework for web, mobile, and desktops. Three.js is an open source JavaScript library for 3D graphics on the web that supports WebGL. Async is a JavaScript library which provides functions for working with asynchronous code. Bootstrap is an open source CSS framework and Fontawesome is an icon toolkit. The Broadcast Channel API allows bi-directional communication between browsing contexts (windows, tabs, frames, or iframes) and workers on the same origin. The Canvas API can be used for animation, game graphics, data visualisation, photo manipulation, and real-time video processing. The Fetch API provides an interface for fetching resources. The Web Storage API allows browsers to persistently store key/value pairs.

Figure 5 depicts the video-editor of the user interface, which can be split into three main components. Component A presents the user with a visual output of the manipulated video. It consists of a canvas element to which frames of the selected video, as well as 3D objects, occlusion masks, and depth estimation frames are drawn. If we examine Fig. 5, we can see that a 3D model of a yellow duck is superimposed onto the current selected video. Figure 6 illustrates the design of the rendering canvas for component A of the video-editor page. The rendering canvas consists of one main canvas, and several hidden canvasses. The hidden canvasses are created in memory and each is responsible for rendering a specific video layer. The layers are composed of video frames of the selected video, depth estimation frames, 3D scene frames, background subtraction frames, alpha matting frames, and foreground reconstruction frames. The canvasses are updated ever iteration of the main rendering loop, where pixels are then extracted and merged into the main canvas.

---

[1] Aurelia, Three.js, Async, Bootstrap, Fontawesome, Whammy, Broadcast Channel API, Canvas API, Fetch API, Storage API.

Fig. 5. User interface of our Adverts framework.

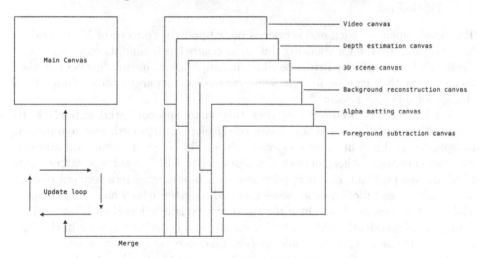

Fig. 6. Design of canvas layers in Adverts framework.

Component B consists of a side-menu where the user is presented with a series of options. These options include opening the plane inspector, model inspector, light inspector, camera tracking inspector, mask inspector, as well as export and debug options. The plane inspector is used to add a plane to 3D scene superimposed onto the current frame. The position of the frame is calculated using depth estimation information. This is crucial, as the plane acts as a anchor point for a 3D model. Users can use the plane inspector to change to positions, size, and rotation of the inserted plane. Further, the user also has the option to hide/show to plane. The model inspector allows users to select one of the included 3D models, and add it to the plane within the superimposed 3D scene. Similar to the plane inspector, users can change the position, rotation, and size of the 3D model. Figure 5 shows the model inspector option currently selected. The light inspector allows users to manipulate lighting settings with the 3D scene to mimic those of the selected video. Users can switch lights on and off, as well as change the color, intensity, and position of various lights (ambient lights, spot lights, etc.). The camera tracking inspector allows users to add tracking points to the video on several key frames and start the tracking process in the back-end. The mask inspector allows users to add an occlusion mask layer to the selected video for a range of frames. The user can select an object within the video that a mask should be created for.

Component C includes options that allow users to play, pause, fast forward, and reverse the selected video. Further, an overview timeline and a detailed timeline are also included in this component. The overview timeline shows the user in which frame they are currently located with the video. The detailed timeline shows the user which tracks have been added to the selected video.

3.2   Back-End

The development of back-end service is built upon the concept of Microservices Architecture. Reason for choosing this architecture lies behind its core concept of the Single Responsibility Principle which in simple terms means "gather together those things that change for the same reason and separate those things that change for different reasons."

A microservices architecture takes this same approach and extends it to the loosely coupled services which can be developed, deployed, and maintained independently. Each of these services is responsible for a discrete task and can communicate with other services through simple APIs to solve a larger complex business problem. In our application, each module was independent of each other only those functionalities were grouped together which has a dependency with the previous module. Once the microservice is developed each one can be deployed independently which offer improved fault isolation whereby in the case of an error in one service the whole application doesn't stop functioning. Another benefit which this architecture brings to the table is the freedom to choose technology stack(programming language, databases, cache etc) which is best suited for the service instead of using the one-size-fits-all approach.

The blue shaded region in Fig. 7 is the back-end service where three major microservices work as a backbone to the whole application for providing core services. The three major services are: API Gateway, Video Upload, Video Processing.

API Gateway. This service is the first point of contact to any request coming from outside of the world. The API Gateway is responsible for authenticating the request and then navigating the request to the requested service. This service can also be called a proxy server as this is the interface for all the requests coming

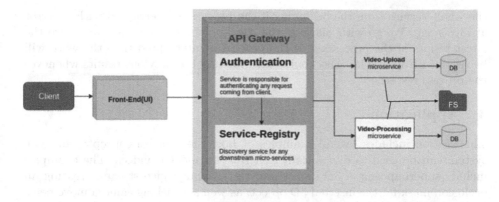

Fig. 7. Back-end Microservices Architecture in Adverts framework.

from the outside world and prevent direct exposure of the delicate services. The API Gateway accommodate the User Authentication and Service Registry Module

Authentication. For authenticating user request, JWT Authentication mechanism is being used which is one of the recommended standard authentication processes for microservices architecture. JWT is open industry-standard RFC 7519 for representing claims securely between two parties. It provides a compact and self-contained way for securely transmitting information between parties as a JSON object. This information can be verified and trusted because it is digitally signed.[jwt.io].

Service Registry. Service Registry is the discovery application for the microservices who want to use the API gateway for authentication and as a proxy server. We have used Eureka Service Registry developed and maintained by Netflix for their microservices which is robust and fast and known for its efficiency.

Video Upload Service. The input video upload is supported by video-upload services where the uploaded video is pre-processed i.e. the frames are extracted from the video and the frames are then stored in four different resolution (original, 480p, 720p, 1080p) along with storing the video property information in the database.

Video Processing Service. This is the service where the core functionalities are written and support the application. This service is majorly divided into three different components: Depth Estimation, Occlusion Detection, and Camera Tracking.

All these three components are dependent, hence they have been grouped together to support each other in functionality. Since the whole video processing is of heavy computation and time-consuming it does not make sense to restart the whole service from the beginning if the process is interrupted with an error at any stage. We go with marking the checkpoints at each stage and save the latest results of the processing. If the process is interrupted then the work will be resumed from the last checkpoint with loading the previous results whenever the process restarts.

4   Application

This section includes several examples of how the proof-of-concept prototype system can be used to dynamically generate augmented videos. The examples include superimposing a 3D object into the existing video stream, creating an occlusion mask for the inserted 3D object, as well as tracking camera movement. Figure 8 illustrates these processes.

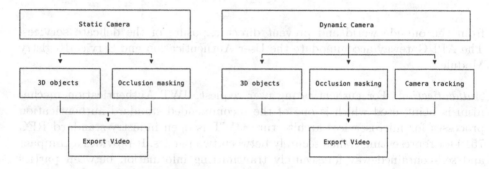

Fig. 8. Static and dynamic camera workflows.

4.1   Superimposing 3D Objects

Superimposing a 3D object onto a video stream consists of two main steps: 1) adding plane to the video and 2) adding a 3D model to the superimposed plane. Figure 9 shows an example of both steps. To add a plane to the video, the user can select the plane inspector menu option (more details in Sect. 3.1). Once this option is selected, the user can move the mouse cursor over the video, which will display a temporary plane at the current position of the cursor. The user can then click the left mouse button to permanently add the plane to the video. The position of the plane is calculated from the position of the mouse cursor, as well as from depth information obtained from the depth estimation module. Additionally, the user also has the option to manually change the position, rotation, and size of the superimposed plane, after it was added to the video. To add a 3D object to the superimposed plane, the user can select the model inspector menu option (more details in Sect. 3.1). Once the menu option is selected, the user can choose a 3D object to add to the video from the 3D object library. The user has the option to manually change the position, rotation, and scale of the 3D object.

Fig. 9. Illustration of superimposing plane and 3D object onto video.

4.2 Occlusion Masks

To create an occlusion mask for a superimposed 3D object, the user must first determine the occluding area with the video. Once determined, the user can select this area and a segmentation mask is created using the interactive segmentation module described in Sect. 2.3. The user can then choose to modify the segmentation mask, or propagate it over a range of video frames. After verifying if mask propagation was successful, the user can start the process of creating the occlusion masks for the range of selected frames. Once complete, the interface can be used to export the newly created video. Figure 10 depicts the final result of adding a 3D object and an occlusion mask to the video.

Fig. 10. Superimposing 3D object onto video with occlusion mask.

4.3 Camera Tracking

To track camera motion over a series of frames, the user must first manually add tracking feature points to the video. The user can add these points by using the mouse cursor to place a marker over the corresponding object within the video. Each frame that the user adds these points to becomes a keyframe. Once the user has completed adding the markers, the camera tracking process can be started. More information regarding this process can be found in Sect. 2. An additional window in the user interface is automatically opened, once the tracking process has completed. This window displays a 3D view of the video scene, in which matched feature points and debug cameras are drawn. Figure 11 B shows an example of this window. The red circles correspond to the matched feature points that the camera tracking algorithm returns. This additional window gives the user an overview of how the cameras position and orientation can change over a series of frames. Further, it can also be used to help with the placement of 3D objects and lights.

Fig. 11. Camera tracking - A) User interface B) 3D view of tracking information

5   Conclusion

In this work we presented a proof-of-concept prototype system that enables 3D computer graphic advertisement objects to be inserted seamlessly into video streams. Our goal was to establish a new paradigm in product placements for marketing agencies and facilitate the process of dynamically generating augmented videos. We show that the proof of concept prototype can be used to inserted 3D objects into existing video streams, create occlusion masks for these objects, as well as track camera movement.

References

1. China: short video market revenue 2016–2021. Accessed 15 Mar 2020. https://www.statista.com/statistics/874562/china-short-video-market-size/
2. Agarwal, S., Snavely, N., Seitz, S.M., Szeliski, R.: Bundle adjustment in the large. In: Daniilidis, K., Maragos, P., Paragios, N. (eds.) ECCV 2010. LNCS, vol. 6312, pp. 29–42. Springer, Heidelberg (2010). https://doi.org/10.1007/978-3-642-15552-9_3

3. Aksoy, Y., Oh, T.H., Paris, S., Pollefeys, M., Matusik, W.: Semantic soft segmentation. ACM Trans. Graph. (TOG) **37**(4), 1–13 (2018)
4. Basha, T., Avidan, S., Hornung, A., Matusik, W.: Structure and motion from scene registration. In: 2012 IEEE Conference on Computer Vision and Pattern Recognition, pp. 1426–1433. IEEE (2012)
5. Bazrafkan, S., Javidnia, H., Lemley, J., Corcoran, P.: Semiparallel deep neural network hybrid architecture: first application on depth from monocular camera. J. Electron. Imaging **27**(4), 043041 (2018)
6. Caelles, S., et al.: The 2018 Davis challenge on video object segmentation. arXiv preprint arXiv:1803.00557 (2018)
7. Chen, Y., Pont-Tuset, J., Montes, A., Van Gool, L.: Blazingly fast video object segmentation with pixel-wise metric learning. In: Proceedings of the IEEE Conference on Computer Vision and Pattern Recognition, pp. 1189–1198 (2018)
8. Covell, M., Baluja, S., Fink, M.: Advertisement detection and replacement using acoustic and visual repetition. In: IEEE Workshop on Multimedia Signal Processing, pp. 461–466. IEEE (2006)
9. Dai, Y., Li, H., He, M.: Projective multiview structure and motion from element-wise factorization. IEEE Trans. Pattern Anal. Mach. Intell. **35**(9), 2238–2251 (2013)
10. Dev, S., et al.: The ALOS dataset for advert localization in outdoor scenes. In: 2019 Eleventh International Conference on Quality of Multimedia Experience (QoMEX), pp. 1–3. IEEE (2019)
11. Dev, S., et al.: The CASE dataset of candidate spaces for advert implantation. In: 2019 16th International Conference on Machine Vision Applications (MVA), pp. 1–4. IEEE (2019)
12. Dev, S., et al.: Localizing adverts in outdoor scenes. In: Proceedings of IEEE International Conference on Multimedia & Expo Workshops (ICMEW), pp. 591–594. IEEE (2019)
13. Fu, H., Gong, M., Wang, C., Batmanghelich, K., Tao, D.: Deep ordinal regression network for monocular depth estimation. In: Proceedings of the IEEE Conference on Computer Vision and Pattern Recognition, pp. 2002–2011 (2018)
14. Godard, C., Mac Aodha, O., Brostow, G.J.: Unsupervised monocular depth estimation with left-right consistency. In: Proceedings of the IEEE Conference on Computer Vision and Pattern Recognition, pp. 270–279 (2017)
15. Hossari, M., et al.: ADNet: a deep network for detecting adverts. arXiv preprint arXiv:1811.04115 (2018)
16. Hu, J., Ozay, M., Zhang, Y., Okatani, T.: Revisiting single image depth estimation: toward higher resolution maps with accurate object boundaries. In: 2019 IEEE Winter Conference on Applications of Computer Vision (WACV), pp. 1043–1051. IEEE (2019)
17. Hussain, Z., et al.: Automatic understanding of image and video advertisements. In: Proceedings of the IEEE Conference on Computer Vision and Pattern Recognition, pp. 1705–1715 (2017)
18. Jain, S.D., Xiong, B., Grauman, K.: FusionSeg: learning to combine motion and appearance for fully automatic segmentation of generic objects in videos. In: 2017 IEEE Conference on Computer Vision and Pattern Recognition (CVPR), pp. 2117–2126. IEEE (2017)
19. Jang, W.D., Kim, C.S.: Interactive image segmentation via backpropagating refinement scheme. In: Proceedings of the IEEE Conference on Computer Vision and Pattern Recognition, pp. 5297–5306 (2019)

20. Javidnia, H., Corcoran, P.: Accurate depth map estimation from small motions. In: Proceedings of the IEEE International Conference on Computer Vision Workshops, pp. 2453–2461 (2017)
21. Javidnia, H., Pitié, F.: Background matting. arXiv preprint arXiv:2002.04433 (2020)
22. Kuznietsov, Y., Stuckler, J., Leibe, B.: Semi-supervised deep learning for monocular depth map prediction. In: Proceedings of the IEEE Conference on Computer Vision and Pattern Recognition, pp. 6647–6655 (2017)
23. Lasinger, K., Ranftl, R., Schindler, K., Koltun, V.: Towards robust monocular depth estimation: Mixing datasets for zero-shot cross-dataset transfer. arXiv preprint arXiv:1907.01341 (2019)
24. Li, Z., Snavely, N.: MegaDepth: learning single-view depth prediction from internet photos. In: Proceedings of the IEEE Conference on Computer Vision and Pattern Recognition, pp. 2041–2050 (2018)
25. Lim, L.A., Keles, H.Y.: Learning multi-scale features for foreground segmentation. Pattern Anal. Appl. **23**, 1369–1380 (2018)
26. Lowe, D.G.: Object recognition from local scale-invariant features. In: Proceedings of the Seventh IEEE International Conference on Computer Vision, vol. 2, pp. 1150–1157. IEEE (1999)
27. Maninis, K.K.: Video object segmentation without temporal information. IEEE Trans. Pattern Anal. Mach. Intell. **41**(6), 1515–1530 (2018)
28. Nautiyal, A., et al.: An advert creation system for next-gen publicity. In: Brefeld, U., et al. (eds.) ECML PKDD 2018. LNCS (LNAI), vol. 11053, pp. 663–667. Springer, Cham (2019). https://doi.org/10.1007/978-3-030-10997-4_47
29. Oh, S.W., Lee, J.Y., Xu, N., Kim, S.J.: Fast user-guided video object segmentation by interaction-and-propagation networks. In: Proceedings of the IEEE Conference on Computer Vision and Pattern Recognition, pp. 5247–5256 (2019)
30. Papazoglou, A., Ferrari, V.: Fast object segmentation in unconstrained video. In: Proceedings of the IEEE International Conference on Computer Vision, pp. 1777–1784 (2013)
31. Scharstein, D., Pal, C.: Learning conditional random fields for stereo. In: 2007 IEEE Conference on Computer Vision and Pattern Recognition, pp. 1–8. IEEE (2007)
32. Scharstein, D., Szeliski, R.: A taxonomy and evaluation of dense two-frame stereo correspondence algorithms. Int. J. Comput. Vis. **47**(1–3), 7–42 (2002). https://doi.org/10.1023/A:1014573219977
33. Schonberger, J.L., Frahm, J.M.: Structure-from-motion revisited. In: Proceedings of the IEEE Conference on Computer Vision and Pattern Recognition, pp. 4104–4113 (2016)
34. Tosi, F., Aleotti, F., Poggi, M., Mattoccia, S.: Learning monocular depth estimation infusing traditional stereo knowledge. In: Proceedings of the IEEE Conference on Computer Vision and Pattern Recognition, pp. 9799–9809 (2019)
35. Wang, Q., Zhang, L., Bertinetto, L., Hu, W., Torr, P.H.: Fast online object tracking and segmentation: a unifying approach. In: Proceedings of the IEEE Conference on Computer Vision and Pattern Recognition, pp. 1328–1338 (2019)
36. Wug Oh, S., Lee, J.Y., Sunkavalli, K., Joo Kim, S.: Fast video object segmentation by reference-guided mask propagation. In: Proceedings of the IEEE Conference on Computer Vision and Pattern Recognition, pp. 7376–7385 (2018)

37. Xu, D., Wang, W., Tang, H., Liu, H., Sebe, N., Ricci, E.: Structured attention guided convolutional neural fields for monocular depth estimation. In: Proceedings of the IEEE Conference on Computer Vision and Pattern Recognition, pp. 3917–3925 (2018)
38. Yu, F., Gallup, D.: 3D reconstruction from accidental motion. In: Proceedings of the IEEE Conference on Computer Vision and Pattern Recognition, pp. 3986–3993 (2014)
39. Chuang, Y.-Y., Curless, B., Salesin, D.H., Szeliski, R.: A Bayesian approach to digital matting. In: 2001 IEEE Conference on Computer Vision and Pattern Recognition (CVPR), vol. 2, pp. 264–271 (2001). https://doi.org/10.1109/CVPR.2001.990970

Estimating Precisions for Multiple Binary Classifiers Under Limited Samples

Rahul Tripathi[✉], Srinivasan Jagannathan, and Balaji Dhamodharaswamy

Amazon, Seattle, USA
{rahtripa,sjaganna,dhbalaji}@amazon.com

Abstract. Machine learning classifiers often require regular tracking of performance measures such as precision, recall, F1-score, *etc.*, for model improvement and diagnostics. The population over which accuracy metrics are evaluated can be too large for a full ground-truth assessment and so only small random samples are chosen for estimation. Ground-truthing often requires human review, which is expensive. Moreover, in some business applications, it may be preferable to minimize human contact with the data in order to improve privacy safeguards. Thus, sampling methods that can provide estimates with low margin of error, high confidence, and small sample size are highly desirable. With an ensemble of multiple binary classifiers, choosing the right sampling method with these desired properties and small size for the collective sample becomes even more important. We propose a sampling method to estimate the precisions of multiple binary classifiers that exploits the overlaps between their prediction sets. We provide theoretical guarantees that our estimators are unbiased and empirically demonstrate that the precision metrics estimated from our sampling technique are as good (in terms of variance and confidence interval) as those obtained from a uniform random sample.

We applied our sampling technique to performance evaluation of an ensemble of binary classifiers. The reduction in sample size depends on the extent of overlap between the predicted positive set of the ensemble and that of the individual classifiers. Since we do not have a closed form solution for quantifying the impact of the overlap, we relied on simulations to investigate how the overlap between an ensemble (parent) and component (child) classifier affects the overall sample size. We found that for every combination of parent and child intersection ratio we tested on, there were significant savings in sample size. Moreover, across all these simulations, we found a mean reduction of 33% in the sample size needed from a child. Our simulations also confirm that the precision metrics estimated from the samples generated using our sampling technique have accuracy comparable to those estimated from uniform random sampling.

Keywords: Model precision · Crowd-sourcing · Sampling

© Springer Nature Switzerland AG 2021
Y. Dong et al. (Eds.): ECML PKDD 2020, LNAI 12460, pp. 240–256, 2021.
https://doi.org/10.1007/978-3-030-67667-4_15

1   Introduction

Machine learning (ML) models rely on the assumption that the target data distribution is close (in statistical sense) to the training data distribution. While the latter is static when the models are being developed and trained, in many applications, the target data distribution may vary over time due to the dynamic nature of production workloads that are classified by the models. In order to continuously evaluate ML models, accuracy metrics such as *precision* and *recall* need to be measured on a regular basis. For a binary classifier that classifies any instance into either $\mathcal{P}$ (positives) or $\mathcal{N}$ (negatives), precision is defined as the fraction of instances predicted as positive that are in fact positive whereas recall is defined as the fraction of positive instances that are correctly predicted as positive. More precisely, if TP, FP, FN denotes True-Positive, False-Positive, and False-Negative instances respectively based on the classification decisions by the model, then precision = TP/(TP + FP) and recall = TP/(TP + FN).

One of the main bottlenecks in tracking these model performance metrics is the need for labeling of the target data used in the evaluation. The label assignment process, called *annotation* or *ground-truthing*, in ML applications is often done manually, which is not scalable. In particular, the cost of annotating a dataset increases significantly with the size of the dataset. Additionally, in some applications, it is preferable to minimize the exposure of data to manual reviewers, for example, to improve privacy safeguards and increase security.

Quite often, an ML application is composed of an ensemble of multiple classifiers. As a result, for model performance diagnostics and tracking, it becomes important to evaluate accuracy metrics of not only the ensemble but also each individual ML classifiers. Therefore, a challenging problem is how to estimate the performance metrics (e.g., precision) of multiple (binary) classifiers with low error, high confidence, and minimal ground truth cost. In this paper, we focus only on the precision performance metric, however, our techniques can be generalized to other measures.

There are two main approaches to estimating the precision of a classifier: *simple random sampling* and *stratified sampling*. These sampling approaches select a small, but statistically relevant, number of instances, called a *sample* from the underlying population (i.e., the predicted positive set of a classifier). Based on the ground-truth assignment of labels to instances in the sample, the precision is estimated using the formula, discussed earlier, but applied to the sample instead of the population.

In simple random sampling, one chooses a uniformly random sample from the population. The main parameter here is the sample size, which as explained in Sect. 3.2, depends on the desired level of accuracy and confidence. Simple random sampling is quite effective in that it yields an unbiased estimator for the precision. However, it can result in a larger sample size than possible with a stratified sampling.

Stratified sampling divides the population into k disjoint strata or bins, for some fixed k. It requires two important considerations: (a) stratification method - how the bins/strata are constructed and (b) allocation method - how the sample size is split across all the bins. It is expected that stratification results in near homogenous bins, i.e., bins containing high concentration of instances with same ground-truth labels, and therefore it lowers the variance of the precision within each bin. By giving different weights to bins and taking a weighted average of the precision estimation from each bin, we can get an unbiased estimator for the precision of the classifier. Also, if the variance in each bin is low, the resulting estimator will also have low variance over the population.

In this work, we use the observation that if a random sample for one classifier overlaps with the prediction set of another, then we can reuse the common instances so that only a smaller sample size is needed for the other classifier. Large-scale production systems often consist of multiple binary classifiers whose individual predictions contribute to the final decision of an ensemble composed of individual classifiers. This observation is particularly useful in such systems since the classifiers are expected to have overlaps in their prediction sets. We give theoretical justification and share experimental findings to show that the new sampling scheme, based on this observation, reaches the same accuracy at a significantly reduced sample size.

We describe our algorithms for estimating the precisions of multiple binary classifiers in Sect. 4. We address the case of an ensemble model and its constituent binary classifiers (Sect. 4.1). We present both theoretical and experimental results to demonstrate that our solution achieves the desired objectives: low error, high confidence, and low ground truth sample size compared to the baseline (Sects. 4.1 and 5). Generalization of our method to other accuracy metrics (e.g., recall) is explained in Sect. 6. Finally, we conclude with a summary of the main results (Sect. 7).

2   Related Work

Bennett et al. [1] adapts stratified sampling techniques to present an online sampling algorithm to evaluate the precision of a classifier. They experimentally demonstrate that their algorithm achieves an average reduction of 20% in sample size compared to simple random sampling and other types of stratified sampling to get the same level of accuracy and confidence. Similarly, Kumar [6] proposes strategies based on stratified sampling to estimate the accuracy of a classifier. They also experimentally show that their methods are more precise compared to simple random sampling for accuracy estimation under constrained annotation resources. In Kataria et al. [4], an iterative stratified sampling strategy is presented that continuously learns a stratification strategy and provides improved accuracy estimates as more labeled data is available. However, for more thanone classifier, it is unclear whether these stratified sampling based methods applied individually to the constituent classifiers would give a similar saving on the size of the collective sample set. Our proposed sampling algorithm relies on uniform random sampling and achieves significant saving (e.g., on average ≈33% and ≈85% average reduction in sample size for any individual classifier in two different experimental settings) compared to the baseline of simple random sampling when applied individually on multiple classifiers for estimating their precisions.

For multiple classifiers, unsupervised methods for estimating classifier accuracies, ranking them, and constructing a more accurate ensemble classifier based solely on classifier outputs over a large unlabeled test data are presented in [3,7,8]. However, these methods rely on assumptions such as conditional independence of classifiers or certain constraints on classifier errors, which limits their practical applicability in many situations. Our work makes no assumption regarding the classifiers.

3 Preliminaries

3.1 Notation

We consider binary classifiers that map instances from some universe Ω to either *positives* ($\mathcal{P}$) or *negatives* ($\mathcal{N}$) label. The *predicted positive set* (*predicted negative set*) of a classifier is the set of all instances that it maps to $\mathcal{P}$ (resp., $\mathcal{N}$). Let sequence $S \subseteq \Omega$ be an ordered multi-set in Ω and denote its length by $|S|$, which includes duplicity. A subsequence of a sequence S contains a subset of elements and preserves their ordering in S. The notation $A - B$ denotes the set difference between any two sets A and B in Ω. For any sequence S and set A, we denote $S \cap A$ to denote the subsequence of S that contains all and only those elements that are in A. If S and T are sequences, then $S + T$ denotes the sequence obtained by appending T to S to the right of S. If a is any instance and S is a sequence, then $\text{count}(a, S)$ denotes the number of occurrences of a in S.

3.2 Sample Size to Estimate Precision

The precision of a classifier C for positives $\mathcal{P}$ can be estimated by uniformly sampling instances from the predicted positive set of C. Given a sample with sufficient number of such instances, we can label each instance to determine the number of True-Positives (TPs) and False-Positives (FPs) in the collection. A point estimate $\hat{p}$ for the precision p is: $\hat{p} = \text{TP}/(\text{TP} + \text{FP})$.

To determine a $(1 - \delta)$-confidence-interval with $\pm\epsilon$ additive margin of error, the sample size needed is given by $\epsilon \geq z_{1-\delta} \times s/\sqrt{n}$, where s is the standard deviation of each random instance, n is the number of samples, and for any $0 < \alpha < 1$, z_α is the α'th quantile[1] of the standard normal distribution. Since a

---

[1] z_α is a factor such that a normal r.v. $N(\mu, \sigma)$ lies inside the interval $\mu \pm z_\alpha\sigma$ with probability α.

sample instance being in $\mathcal{P}$ is a Bernoulli trial with success probability equal to precision p, its variance is $s^2 = p(1-p)$. Plugging into the earlier equation gives

$$\epsilon \geq z_{1-\delta} \times \sqrt{\frac{p(1-p)}{n}}. \tag{1}$$

Thus, for $\epsilon = 0.03$, $\delta = 0.05$, and the maximum variance assumption ($p = 1/2$), the sample size estimate is 1068. If one is willing to make stronger assumptions, e.g. precision is guaranteed to be at least some threshold p_0, then the sample size estimate can be considerably reduced (e.g., 385 if $p_0 \geq 90\%$ and 278 if $p_0 \geq 93\%$). We denote the sample size needed to estimate precision within $\pm\epsilon$ additive error and $1-\delta$ confidence by $n_{\epsilon,\delta}$.

4   Optimized Precision Estimation by Recycling Samples

Given a collection of binary classifiers, estimating the precision for each one requires generating samples and assigning a label to each instance. The label assignment (*annotation*), is typically a manual, laborious process whose cost is proportional to the size of a sample. The baseline approach to estimate the precisions of a collection of k binary classifiers requires labeling individual sample sets for each of the classifiers. Anchoring on one of the classifiers (called *parent* in the follow up discussion), we consider the remaining classifiers as its children. Any classifier whose predicted positive set is presumed to have significant overlap with those of the remaining classifiers is a good choice for the parent. For example, an ML system may be composed of multiple binary classifiers with an ensemble of them as the authoritative classifier. In this setting, the ensemble could be considered a parent classifier because we expect the predicted positive set of the ensemble to overlap with that of each individual classifier. In this section, we explain how samples from a parent classifier can be recycled to generate subsamples of each child classifier, and thereby reduce the combined sample size.

4.1   Classifiers with Overlapping Predicted Positive Sets

Suppose we have a parent classifier P with predicted positive set, denoted A_P, and a sample S_P generated from A_P. Given a child classifier C with predicted positive set, denoted A_C, we exploit the overlap between A_P and A_C to generate a sample S_C using S_P. We show the sample S_C retains the statistical property needed for an unbiased estimator of the precision of C, provided S_P possessed the same. Thus, it results in a smaller sample size to estimate the precisions of both C and P compared to the baseline. We demonstrate empirically that our estimates are within the desired margin of error and acceptable confidence.

Algorithm 1, called RecycleSamplesForPrecision, takes the input (a) the predicted positive sets A_P and A_C, (b) the size n_C of the sample needed to estimate the precision of C, (c) the sample S_P, and (d) an option UniformSample or UniformShuffle. It generates a sample S_C and estimates precision $\hat{p}_C$ for C. In Line 1, S^+ equals the subsequence (with repetitions) of all elements in S_P that belong to A_C. In Line 2, a call to a function UniformSample is made, which generates a uniform sample with replacement from a population. The function takes three arguments (a) the population to sample from, (b) the size of the sample, (c) and whether (or not) to sample with replacement. The subsequence S^- is a uniformly generated from $A_C - A_P$ and its size is required to satisfy: $|S^-|/|S^+| = |A_C - A_P|/|A_P \cap A_C|$. This is needed because any uniformly generated sample S from A_C of size $|S^-| + |S^+|$ is expected to contain instances from the disjoint sets $A_P \cap A_C$ and $A_C - A_P$ in proportion to their sizes. S^{remain} in Line 3 includes uniformly generated instances from A_C to backfill any shortage from just S^+ and S^- combined.

Any single instance in $S^+ + S^-$ is not uniformly distributed over A_C. To see this, if e_1 is the first instance and e_l is the last instance in this subsequence, then e_1 is likely to come from S^+ and e_l from S^-. Therefore, in such a case, e_1's distribution is over $A_P \cap A_C$ whereas e_l's over $A_C - A_P$, and so they are not uniform over A_C. The function MixSequence, defined in Algorithm 2, ensures that $S^+ + S^-$ is a uniformly random sample from A_C, and so the estimator $\hat{p}_C$ in Line 8 is an unbiased estimator of the precision of C.

Algorithm 1: RecycleSamplesForPrecision

Data: Classifiers C and P with predicted positives sets A_C and A_P, respectively; sample size n_C for C; a sequence S_P of uniformly random instances from A_P; and a parameter option $\in$ [UniformSample, UniformShuffle].

Result: Estimated precision $\hat{p}_C$ of C and a sequence S_C of uniformly random instances from A_C of size n_C.

1 $S^+ \leftarrow S_P \cap A_C$.
2 $S^- \leftarrow$ UniformSample($A_C - A_P$, $|A_C - A_P| \times |S^+|/|A_P \cap A_C|$, replace=True)
3 $S^{\mathrm{remain}} \leftarrow$ UniformSample(A_C, $\max(0, n_C - (|S^+| + |S^-|)$, replace=True)
4 $S_C \leftarrow$ MixSequence($S^+ + S^-$, option) $+ S^{\mathrm{remain}}$
5 **if** $|S_C| > n_C$ **then**
6 $\quad |\quad S_C \leftarrow S_C[0 : n_C]$
7 **end**
8 $\hat{p}_C \leftarrow$ fraction of positives instances in S_C
9 **return** $\hat{p}_C$, S_C

Figure 1 shows the child-parent relationship and the sets involved in generating the final sample S_C. We transform the sequence $S^+ + S^-$ into a sequence S of uniformly random instances in Algorithm 2. Two possible ways are considered: (a) uniform sampling and (b) uniform shuffling. The former is nothing but sampling with replacement and the latter is without replacement. In both options, we show that the new sequence consists of uniformly random instances from A_C, and so the average $\hat{p}_C$ is unbiased. We distinguish between these two

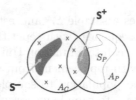

Fig. 1. A parent classifier P overlapping with a child classifier C on their predicted positive sets A_P and A_C, respectively. The right circle represents A_P and the left A_C. A random sample S_P, shown as a closed curve, intersects with A_C as shown in the shaded region. This shaded region represents S^+, the shaded closed curve in $A_C - A_P$ represents S^-, and the sprinkled tick-marks represent S^{remain} in the description of Algorithm 1.

Algorithm 2: MixSequence

Data: A sequence S and a parameter
     option $\in$ [UniformSample, UniformShuffle].
Result: Rearranged sequence S, where option determines the rearrangement
     method.
1 **if** option *equals* UniformSample **then**
2 | $S \leftarrow$ UniformSample(S, $|S|$, replace=True)
3 **else if** option *equals* UniformShuffle **then**
4 | $S \leftarrow$ UniformShuffle(S, $|S|$)
5 **return** S

---

options because only uniform sampling would result in total independence of instances. This observation is useful in situations where C itself is an ensemble of other classifiers. In such a situation, we may want to start with a uniformly *independent* sample S_C of C (i.e., sample with replacement), then recursively apply Algorithm 2 by treating C as a parent classifier and its constituent classifiers as children of C.

In Lemma 1, we show that with uniform sampling as the option in the function MixSequence, the original sequence $S^+ + S^-$ is transformed into a sequence composed of independent and uniformly distributed instances from A_C.

Lemma 1. *Let* $S =_{df}$ *MixSequence*($S^+ + S^-$, option = UniformSample) *in Algorithm 1. Then* S *is a sequence of i.i.d. uniformly random instances from* A_C.

Proof. Let s_i denote the i'th random instance in S. Since each s_i is chosen uniformly from $S^+ + S^-$ with replacement, s_i's are independent and identically distributed. We now show that, for any $a \in A_C$ and any i, it holds that $\Pr[s_i = a] = 1/|A_C|$. Fix some $a \in A_P \cap A_C$. Let $S^+ + S^- =_{df} Y_1, Y_2, \dots, Y_{|S|}$. Then

$$\Pr[s_i = a] = \sum_{\ell > 0} \Pr[s_i = a \mid |S^+| = \ell] \times \Pr[|S^+| = \ell]$$

$$= \sum_{\ell > 0} \sum_{k=1}^{|S|} \frac{1}{|S|} \Pr[Y_k = a \mid |S^+| = \ell] \times \Pr[|S^+| = \ell]$$

$$= \sum_{\ell > 0} \frac{|A_P \cap A_C|}{|A_C| \ell} \sum_{k=1}^{|S^+|} \Pr[Y_k = a \mid |S^+| = \ell] \times \Pr[|S^+| = \ell]$$

$$= \frac{|A_P \cap A_C|}{|A_C|} \sum_{\ell > 0} \frac{1}{\ell} \times \sum_{k=1}^{\ell} \frac{1}{|A_P \cap A_C|} \times \Pr[|S^+| = \ell] = \frac{1}{|A_C|}.$$

Here, the second equality uses the fact that s_i equals Y_k (for any k) with probability $1/|S|$. The third equality uses $|S| = |S^+| + |S^-| = \frac{|A_C| \times |S^+|}{|A_P \cap A_C|}$, $|S^+| = \ell$, and the fact that since $a \in A_P \cap A_C$, the terms are zero for $k \in [|S^+| + 1, |S|]$. The fourth equality uses the fact that each element of $A_P \cap A_C$ is equally likely to be the k'th element of S^+, therefore $\Pr[Y_k = a \mid |S^+| = \ell]$ is equal to $\frac{1}{|A_P \cap A_C|}$. Additionally, $\sum_{k=1}^{\ell} \Pr[|S^+| = \ell]$ equals $l \times \Pr[|S^+| = \ell]$.

For the case where $a \in A_C - A_P$, the analysis is analogous, with minor differences. Instead of k varying over $[1, |S^+|]$ in the third and the fourth equality, we now have k vary over $[|S^+| + 1, |S|]$, and the term $\frac{1}{|A_P \cap A_C|}$ is replaced by $\frac{1}{|A_C - A_P|}$ inside the second summation in the fourth equality. □

In Lemma 2, we show that with uniform shuffling as the option in the function MixSequence, the original sequence $S^+ + S^-$ is transformed into a sequence composed of uniformly distributed instances from A_C, but the instances are not independent.

Both Lemmas 1 and 2 appear identical in that they generate a uniform random sample. However, the main distinction is in the *independence* of the resulting sequence: uniform sampling results in an independent sequence whereas uniform shuffle in Lemma 2 does not imply independence. Nevertheless, we explain below that Lemma 2 gives rise to a stratified sampling procedure. We empirically show that the mean, the std dev., and the 95% confidence interval of precision errors are comparable to that of the simple random sample (see Table 1 and Fig. 2(b) and 2(c)).

Lemma 2. Let $S =_{df}$ MixSequence($S^+ + S^-$, option = UniformShuffle) in Algorithm 1. Then S is a sequence of identically and uniformly distributed instances from A_C.

Proof. Let s_i denote the i'th random instance in S. Fix an element a of $A_P \cap A_C$. (A similar argument will apply if $a \in A_C - A_P$.) We will show that, for any $1 \le i \le |S|$, $\Pr[s_i = a] = 1/|A_C|$, and so the lemma would follow. By the law of total probability,

$$\Pr[s_i = a] = \sum_{S^+, S^-} \Pr[s_i = a \mid S^+, S^-] \times \Pr[S^+, S^-]$$

$$= \sum_{S^+, S^-} \frac{\text{count}(a, S^+ + S^-)}{|S|} \times \Pr[S^+, S^-]$$

$$= \sum_{S^+, S^-} \frac{|A_P \cap A_C|}{|A_C|} \times \frac{\text{count}(a, S^+)}{|S^+|} \times \Pr[S^+, S^-]$$

$$= \frac{|A_P \cap A_C|}{|A_C|} \sum_{S^+} \frac{\text{count}(a, S^+)}{|S^+|} \times \sum_{S^-} \Pr[S^+, S^-]$$

$$= \frac{|A_P \cap A_C|}{|A_C|} \sum_{S^+} \frac{\text{count}(a, S^+)}{|S^+|} \times \Pr[S^+]$$

$$= \frac{|A_P \cap A_C|}{|A_C|} E[X_a], \tag{2}$$

where X_k, for any $k \in A_P \cap A_C$, is a random variable that equals the fraction of times k occurs in S^+ when each element in S (and so S^+) is chosen uniformly at random. Here, the second equality follows since S is a uniformly random shuffle of $S^+ + S^-$. In the third equality, we use the fact that $a \in A_P \cap A_C$ implies $a \in S^+$, and so $\text{count}(a, S^+ + S^-) = \text{count}(a, S^+)$. We also use $|S| = \frac{|A_C| \times |S^+|}{|A_P \cap A_C|}$ there. Note that $\sum_{k \in A_P \cap A_C} X_k = 1$ and, by symmetry, $E[X_k] = E[X_{k'}]$ for any $k, k' \in A_P \cap A_C$. Hence, by the linearity of expectation, $E[X_a] = 1/|A_P \cap A_C|$. It follows from Eq. (2) that $\Pr[s_i = a] = 1/|A_C|$. ☐

Lemma 2 shows that the resulting sequence is composed of uniformly distributed instances over A_C. This shows that the estimator $\hat{p}_C$ in Algorithm 1 with option = UniformShuffle is unbiased. Note that Algorithm 1 with option = UniformShuffle is just a special case of *stratified sampling* with *proportional allocation* [2] involving the two strata $A_C - A_P$ and $A_C \cap A_P$. This is true because we maintained the ratio of $|S^+|$ to $|S^-|$ as that of $|A_C \cap A_P|$ to $|A_C - A_P|$. Since $A_C \cap A_P$ is expected to be more homogeneous than A_C and likewise for $A_C - A_P$, the stratification should reduce the variance of $\hat{p}_C$.

Lemma 3 expresses the amount of saving in sample size in terms of various probability events. As evident from Lemma 3, the saving in the sample size depends on the extent of the overlap of $A_P \cap A_C$ relative to A_P and to A_C. We refer to the ratios $|A_P \cap A_C|/|A_P|$ as PIR (parent intersection ratio) and $|A_P \cap A_C|/|A_C|$ as CIR (child intersection ratio).

Lemma 3. *Let $S =_{df}$ MixSequence($S^+ + S^-$, option), $n_P =_{df} |S_P|$, and $X =_{df}$ $|S^+|$ in Algorithm 1. Let Savings denotes the number of sample instances saved by Algorithm 1 relative to the baseline (simple random sampling) of sample size n_C. Then the following statements hold:*

(a) $X \mid n_P \sim B(n_P, \frac{|A_P \cap A_C|}{|A_P|})$.

(b) *Savings = X when $\frac{X}{n_C} \leq \frac{|A_P \cap A_C|}{|A_C|}$.*

(c) Savings $= |S[0 : n_C] \cap S^+|$ when $\frac{X}{n_C} > \frac{|A_P \cap A_C|}{|A_C|}$.

Here, $B(n, p)$ denotes the binomial distribution with parameters n and p.

Proof. Part (a) follows because each instance in S^+ arises because of the successful Bernoulli trial of choosing an element in $A_P \cap A_C$ uniformly and independently from A_P. Hence, the distribution of $X = |S^+|$ given n_P is binomial with number of trials n_P and success probability $\frac{|A_P \cap A_C|}{|A_P|}$. For Part (b), we note that if $\frac{X}{n_C} \leq \frac{|A_P \cap A_C|}{|A_C|}$, then $|S| = |S^+| + |S^-| \leq n_C$. Hence, in this case, we can reuse all of S^+ and so Savings equals X. The condition in Part (c) implies that $|S| > n_C$ and so S^{remain} would equal the empty set. Therefore, any saving we get would be due to only those instances in S^+ that also occur in the first n_C instances in S. □

Using Lemma 3, we can describe the distribution of savings as a function of PIR, CIR, n_P, and n_C.

5   Experiments and Results

5.1   Metrics for Comparison

We consider the below metrics for comparing Algorithm 1 against simple random sampling. Our simulations involve multiple trials in which each trial requires a distinct seed for randomly selecting parameters of the simulation. For trial i,

- %**Savings**: this is the percentage savings in the sample size achieved by our algorithm against a simple random sample. Formally, if $s_{i,a}$ denotes the sample size required by our algorithm and $s_{i,r}$ denotes the sample size required by a simple random sample in trial i, then this equals $\frac{s_{i,r} - s_{i,a}}{s_{i,r}} \times 100$.
- %**PrecisionError**: this is the percentage absolute deviation of the point estimate from actual precision. Formally, if $\hat{p}_i$ and p_i denote the estimated and the actual precisions in trial i, respectively, then this equals $\frac{|\hat{p}_i - p_i|}{p_i} \times 100$.
- %**MaxCIError**: this is percentage width of the 95% confidence interval relative to the actual precision. Formally, if $[l_i, u_i]$ denotes the 95% confidence interval and p_i is the actual precision in trial i, then this equals $\frac{\max\{|l_i - p_i|, |u_i - p_i|\}}{p_i} \times 100$.

5.2   Simulations

We compare the sample sizes required by Algorithm 1 against that of simple random samples for an ensemble of three classifiers whose properties and performance are randomly chosen. We consider a majority vote ensemble model (denoted MVE) as parent of three models (denoted ML1, ML2, ML3), which are its children. We compare three different sampling algorithms for precisions: (1) **Simple Random Sample (SRS):** each of MVE, ML1, ML2, and ML3 requires a separate sample of size 1100. As noted in Sect. 3.2, 1100 samples are sufficient to estimate precision within ±0.03 additive error and 95% confidence. The collective sample size in this case is at most 4400 and can be lower if sample instances repeat across the collective samples. (2) **RecycleSampleForPrecision(RSFP)-Shuffle** This is Algorithm 1 with *option*=UniformShuffle. Here, MVE requires 1100 samples, but ML1, ML2, and ML3 have reduced sampling requirements because of overlap between predicted positive sets of MVE and ML1, MVE and ML2, and MVE and ML3. (3) **RecycleSampleForPrecision(RSFP)-Sample:** This is Algorithm 1 with *option*=UniformSample.

We choose the parameters of our simulation as follows: (1) *population size of positives*: set to one million; (2) *number of trials*: 200 (a separate random seed is used in each trial); (3) *class ratio of positives to the size of the entire population*: randomly chosen from $[0.01, 0.5]$ range; (4) *precision of model ML1*: randomly chosen from $[0.70, 1.0]$ range; (5) *recall of model ML1*: randomly chosen from $[0.1, 1.0]$ range; and (6) *precisions of models ML2 and ML3*: for each, randomly chosen between 0.5 and that of ML1.

These random choices range over almost all permissible values of these parameters. Thus, our simulations were designed to cover arbitrary model performance characteristics, and empirically illustrate the validity of our algorithms. The data within each trial is generated independently as follows. First, the models ML1, ML2, and ML3 are simulated by independently assigning each one scores between 0 and 1 using a truncated exponential distribution with shape parameter 0.5. In order to de-correlate the scores across models, we divide the scores into blocks of size 0.03 and randomly shuffle the scores of ML2 and ML3 within each block. Next, for any fixed choice of (a) the class ratio of positives to the population size, (b) precisions of the models, and (c) their recalls, we determine the right decision threshold for each of the models so that their predicted positive sets satisfy the precision and recall constraints. Next, fixing one of the models, say ML1, we uniformly assign true-positive labels over the predicted positive set and randomly assign the remaining false-negative labels over the predicted negative set of ML1. Here, we considered different variations of random assignment of false-negative labels: uniform selection and exponentially decaying selection. Once the scores and the thresholds of ML1, ML2, and ML3 are determined, MVE is also uniquely defined. Together with the label assignments, we use the data within each trial to estimate precisions using these sampling algorithms.

Figure 2 reports the metrics for comparison between our algorithms (RFSP-Sample and RFSP-Shuffle) and the SRS (baseline) algorithm in the simulation. In all trials and for each model, we compare %Savings in sample size, %PrecisionError of the point estimate from actual precision, and %MaxCIError of the 95% confidence interval relative to the actual precision. All three algorithms use a simple random sample of size 1100 for MVE. Thereafter, the baseline algorithm draws independent random samples of size 1100 for each of three ML models whereas RFSP-Sample and RFSP-Shuffle require the same sample size but different option setting. Since sampling for MVE is done similarly in all three algorithms, we report comparison results only for ML1, ML2, and ML3.

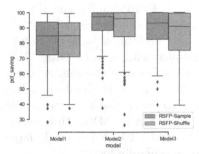 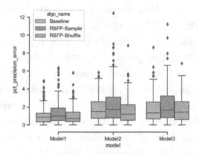

(a) %Saving in sample size is compared against simple random sampling for all three models. High Savings correspond to reduced sample size required to estimate precision.

(b) %PrecisionError of the *point estimate* from the actual is compared between Baseline and our algorithms. A low precision error corresponds to a tight point estimate.

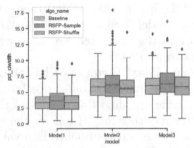

(c) %MaxCIError in the width of the *confidence interval* relative to the actual is compared between Baseline and our algorithms. A low width corresponds to narrow confidence interval around the actual.

Fig. 2. The boxplots are based on 200 trials with random selection of parameters of the experiment. (a) Both RSFP-Sample and RSFP-Shuffle show significant savings in sample size compared to simple random sampling for all the models. The median %Savings reach above 85% in all cases because of high overlap between predictive positives sets. Notice also that %Savings is low (<40%) in certain trials in which this overlap is small. (b) The %PrecisionError of all three algorithms are low (e.g., the third quartile is around 3% or less), which suggests that the *point estimates* from RSFP-Sample and RSFP-Shuffle are tight. The outliers correspond to those trials where the estimate deviates too far from actual, which is possible in up to 5% of trials. (c) The %MaxCIError of RSFP-Sample and RSFP-Shuffle are also close to the Baseline for all three models, which suggests that the samples generated from RSFP-Sample and RSFP-Shuffle produce as narrow *confidence interval* as the simple random sampling. The outliers correspond to the trials in which one of the (upper or lower) limits of the generated confidence interval deviates too far from the actual, which is possible in a small % of trials.

Figure 2(a) shows that both RSFP-Sample and RSFP-Shuffle can lead to significant %Savings in sample size compared to SRS. The savings is >85% in at least half of the trials. The higher savings occur because of large overlap in the predicted positives sets of the majority ensemble with each of ML1, ML2, and ML3 in various trials. The overlap is possibly because the model scores were somewhat positively correlated with each other during the simulation. The simple random sample for any one model, say ML1, may also save on sample size because of possible overlap with the random sample for MVE. However, if the size of predicted positive sets of both ensemble and ML1 are extremely large compared to the sample size (1100), the overlap is generally low.

In Fig. 2(b), we can see that %PrecisionError of all the algorithms are all close to each other for each of the models (ML1, ML2, and ML3) and that RSFP-Shuffle and baseline show slightly lower error than RSFP-Sample. For example, the median, quartliles, and the inter-quartlile ranges for all three models are lower for both RSFP-Shuffle and baseline than for RSFP-Sample. This shows that both RSFP-Shuffle and SRS allow to produce an equally tight point estimate of the precision with RSFP-Sample a little behind these two in accuracy. The extreme outlier instance for Model 2 corresponds to a trial run in which the actual precision was 0.539 and the estimated precision was 0.472, and so the %PrecisionError turned out to be 12.43%.

In Fig. 2(c), we notice that %MaxCIError of each of the algorithms are again close to each other. This shows that the confidence intervals produced from RSFP-Sample and RSFP-Shuffle are almost as narrow as those from the simple random sampling. The extreme outlier for ML2 is for the case when the actual precision was 0.539 and the estimated confidence interval from RSFP-Sample was [0.442, 0.501], and so %MaxCIError turned out to be 17.92%.

5.3 Savings in Sample Size as a Function of PIR and CIR

We ran a simulation of Algorithm 1 to evaluate the amount of savings for different PIR and CIR values. Here, we present results for *option*=UniformShuffle because, as reported in Fig. 2 (see Sect. 5.2), the random shuffle provides more accurate precision estimates. Specifically, we ran 200 trials each for 361 combinations of PIR and CIR (values in the range [0.05, 0.95] in increments of 0.05). In each trial of the simulations we randomly chose $|A_P \cap A_C|$ in the range [10K, 100K] and performed a random selection of samples in S_P. In Table 1, we report the percentage saving in sample sizes (mean as well as 95% confidence interval) for a representative subset of PIR and CIR values we used in the simulations. We found a mean savings of 33.68%, and that in 95% of the simulation runs, the savings varied between 4.43% and 82.98%. In Fig. 3, we present a surface plot of the mean savings in the simulation runs as a function of PIR and CIR. Our simulations confirm that significant amount of savings is achieved as the amount of overlap between parent and child classifiers increases. Furthermore, the error in precision due to our sampling method is extremely low (both mean and std. dev. of %PrecisionError $\leq 1.51\%$) across all choices considered.

5.4 Practical Application of Algorithm 1

We applied our sampling algorithm to evaluate precision of binary classifiers for the offensive content detection problem, studied in [5]. We start with the labeled dataset considered in that work, take random subsets of it to create a train set with 4.5M texts and a distinct test set with 2M texts. The ratio of positives and negatives is kept 1:1 in both train and test sets. We implemented three binary classifiers, namely Bi-LSTM, CNN, and LogReg, described in [9], and trained

Table 1. Percentage saving in sample size as a function of Parent-Intersection-Ratio and Child-Intersection-Ratio is shown. For each combination of PIR and CIR below, we report over 200 trials the mean and the 95% confidence interval of %Saving in sample size and the mean and the std. dev. of %PrecisionError when applying Algorithm 1. As seen below, the savings increase with increasing values for PIR and CIR, i.e., with the amount of overlap between parent and child classifiers. Also, the mean, the std. dev, and the 95% confidence interval of %precision errors from our algorithm are 1.367%, 1.269%, and [0.0379, 4.697]%, resp., which match closely with those from *simple random sampling* that are 1.368%, 1.267%, and [0.0385, 4.678]%, resp.

PIR	CIR	Mean % Saving	2.5^{th}% Saving	97.5^{th}% Saving	mean% Precision error	std. dev% Precision error
0.05	0.05	4.72	3.64	5.38	1.50	1.46
0.05	0.25	4.91	3.39	6.42	1.39	1.27
0.05	0.45	4.95	3.23	6.54	1.36	1.31
0.05	0.65	4.92	3.06	6.58	1.43	1.31
0.05	0.85	4.91	2.79	6.70	1.30	1.16
0.25	0.05	4.93	3.65	6.11	1.51	1.33
0.25	0.25	24.21	20.60	25.57	1.45	1.32
0.25	0.45	24.73	19.73	27.79	1.20	1.12
0.25	0.65	24.61	18.70	27.73	1.44	1.36
0.25	0.85	24.61	18.70	27.75	1.26	1.28
0.45	0.05	4.89	3.55	6.29	1.37	1.31
0.45	0.25	24.58	20.63	26.72	1.45	1.28
0.45	0.45	43.85	36.15	46.03	1.31	1.17
0.45	0.65	44.45	33.75	48.56	1.46	1.28
0.45	0.85	44.47	32.82	48.74	1.37	1.42
0.65	0.05	4.83	3.28	6.20	1.29	1.11
0.65	0.25	24.34	18.44	26.88	1.34	1.19
0.65	0.45	44.17	33.15	47.19	1.50	1.50
0.65	0.65	63.68	48.02	67.07	1.37	1.16
0.65	0.85	64.39	49.58	69.21	1.41	1.39
0.85	0.05	4.82	3.19	6.20	1.50	1.43
0.85	0.25	24.29	18.98	26.95	1.43	1.40
0.85	0.45	44.00	32.28	47.36	1.39	1.34
0.85	0.65	64.14	48.59	67.95	1.33	1.24
0.85	0.85	84.04	55.49	88.83	1.30	1.32

them on our train dataset. We also created a majority vote ensemble (MVE) of them. As in Sect. 5.3, we present results for *option*=UniformShuffle only.

We fed the Reddit test dataset as input to the classifiers discussed above, and applied Algorithm 1 on the predicted positive sets to sample for precision. We ran 1000 trials to generate samples and calculated %PrecisionError and %Savings from the samples in each trial. In Table 2, we observe that our algorithm can save >88% in the number of samples needed to estimate precision, while obtaining very low errors in the precision estimates derived from the smaller sample sizes. Note that there is no saving for MVE since in this case our algorithm defaults to a simple random sample. Moreover, the percentage precision errors of our algorithm closely matches that of simple random sample for each model.

6   Generalizing to Other Performance Measures

The focus of this paper has been on the precision metric. However, as previously stated, our approach can be generalized to other metrics such as recall. In this section, we illustrate how we can generalize to recall calculations.

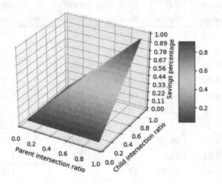

Fig. 3. A surface plot of %Savings as a function of parent intersection ratio (PIR) and child intersection ratio (CIR). Notice that %Savings increase as both PIR and CIR increase with %Savings can reach as high as 90%.

Table 2. The sizes of predicted positive sets, and the min/max of %PrecisionError and %Savings over 1000 trials for each of the component models in MVE (majority vote ensemble) are reported here. Note that there are no metrics for MVE alone, as we would use simple random sample for MVE—the parent classifier.

Model	Predicted positive set size	min/max % precision error (Algorithm 1)	min/max % precision error (simple random sample)	min/max % savings
BiLSTM	776653	0.03/0.25%	0.03/0.34%	88.96/94.73%
CNN	848576	0.04/0.50%	0.04/0.41%	97.52/97.90%
LogReg	884272	0.01/0.56%	0.01/0.65%	90.22/91.72%

Suppose C_1 and C_2 are classifiers with their precisions p_1 and p_2, recalls r_1 and r_2, predicted positive sets A_1 and A_2, and predicted negative sets B_1 and B_2, respectively. Then the number of true-positives for C_1 and C_2 are given by $p_1|A_1|$ and $p_2|A_2|$, respectively. Since the total size of positives in the population (say N) is fixed and independent of the classifiers, it follows that $\frac{r_1}{r_2} = \frac{p_1|A_1|/N}{p_2|A_2|/N} = \frac{p_1|A_1|}{p_2|A_2|}$. Generally, sizes of A_1, A_2, B_1, and B_2 are known. Therefore, if tight estimates $\hat{p}_1$ on p_1, $\hat{p}_2$ on p_2, and $\hat{r}_1$ on r_1 are known, then we can obtain a tight estimate $\hat{r}_2$ on r_2. Also, since recall equals TP/(TP + FN) and false omission rate (FOR) equals FN/(TN + FN), we get recall $r_1 = \frac{p_1|A_1|}{p_1|A_1|+f_1|B_1|}$ where f_1 is the FOR of C_1. It follows that the recall estimation problem for multiple classifiers is reducible to obtaining tight estimates on their precisions and a tight estimate on the FOR of a *single* classifier, say C_1.

Sampling for estimating the FOR and for estimating the precision of a classifier C_1 are over disjoint populations B_1 and A_1, respectively. So, applying Algorithm 1 to estimate FOR (by treating B_1 and A_1 as child-parent relationship) will not result in reduced sample size. The overlap between B_1 (predicted negative set of C_1) and predicted positive sets of other classifiers are also expected to be weak. Therefore, it is unlikely that Algorithm 1 will help. In this case, estimating FOR for classifier C_1 under limited annotations should ideally be done using stratified sampling approaches suggested in [1,4,6]. In other words, a combination of Algorithm 1 for estimating the precisions of multiple classifiers and stratified sampling method for estimating the FOR of a *single* classifier would suffice to estimate the recalls of multiple classifiers to achieve an overall reduction in number of samples.

7   Conclusion

We presented a sampling algorithm RecycleSamplesForPrecision to estimate precisions of multiple binary classifiers with minimal sample size. Our algorithm makes use of two properties: (a) the predicted positive sets of classifiers quite often have significant overlaps and (b) if a random sample for estimating precision of one classifier overlaps with the predicted positive set of another classifier, then we can reuse the common instances to reduce the sample size. We showed that our algorithm results in uniformly distributed random samples. We ran experiments with an ensemble of three classifiers (with randomly assigned accuracy metrics) and observed (in Fig. 2) that, for each individual classifier in the ensemble, (a) the mean %savings is >80% and (b) the distribution of %PrecisionError and %MaxCIError are all close to the baseline (simple random sample). In particular, our algorithm with option=UniformShuffle gives a slightly tighter estimate compared to option=UniformSample. Next, focusing only on RecycleSamplesForPrecision with option=UniformShuffle, we ran experiments over a wide range of possible ratios of intersections for parent and child classifiers, and observed consistent savings in samples sizes across all these scenarios, where the amount of savings increases with the amount of intersection ratios. Over all the runs of this experiment (see Table 1), we observe (a) a mean

%savings of ≈33% and (b) the mean, the std. dev., and the 95% confidence interval of %PrecisionError are 1.367%, 1.269%, and [0.0379, 4.697]%, respectively, which are comparable to those of simple random sampling.

References

1. Bennett, P.N., Carvalho, V.R.: Online stratified sampling: evaluating classifiers at web-scale. In: Proceedings of the 19th ACM CIKM, pp. 1581–1584. ACM (2010)
2. Cochran, W.G.: Sampling Techniques. Wiley, Hoboken (2007)
3. Jaffe, A., Nadler, B., Kluger, Y.: Estimating the accuracies of multiple classifiers without labeled data. In: Artificial Intelligence and Statistics, pp. 407–415 (2015)
4. Katariya, N., Iyer, A., Sarawagi, S.: Active evaluation of classifiers on large datasets. In: IEEE 12th International Conference on Data Mining, pp. 329–338. IEEE (2012)
5. Khatri, C., Hedayatnia, B., Goel, R., Venkatesh, A., Gabriel, R., Mandal, A.: Detecting offensive content in open-domain conversations using two stage semi-supervision. CoRR abs/1811.12900 (2018)
6. Kumar, A., Raj, B.: Classifier risk estimation under limited labeling resources. In: Phung, D., Tseng, V.S., Webb, G.I., Ho, B., Ganji, M., Rashidi, L. (eds.) PAKDD 2018. LNCS (LNAI), vol. 10937, pp. 3–15. Springer, Cham (2018). https://doi.org/10.1007/978-3-319-93034-3_1
7. Parisi, F., Strino, F., Nadler, B., Kluger, Y.: Ranking and combining multiple predictors without labeled data. Proc. Natl. Acad. Sci. 111(4), 1253–1258 (2014)
8. Platanios, E.A., Blum, A., Mitchell, T.: Estimating accuracy from unlabeled data. In: Proceedings of the Thirteenth Conference on Uncertainty in Artificial Intelligence (UAI), pp. 682–691 (2015)
9. Tripathi, R., Dhamodharaswamy, B., Jagannathan, S., Nandi, A.: Detecting sensitive content in spoken languages. In: Proceedings of the 6th IEEE International Conference on Data Science and Advanced Analytics (DSAA) (2019)

Applied Data Science: Transportation

Automation of Leasing Vehicle Return Assessment Using Deep Learning Models

Mohsan Jameel[1(✉)], Mofassir ul Islam Arif[1], Andre Hintsches[2],
and Lars Schmidt-Thieme[1]

[1] Information Systems and Machine Learning Lab, University of Hildesheim,
Hildesheim, Germany
{mohsan.jameel,mofassir,schmidt-thieme}@ismll.uni-hildesheim.de
[2] Volkswagen Financial Services AG, Braunschweig, Germany
Andre.Hintsches@vwfs.com

Abstract. The vehicle damage assessment includes classifying damage
and estimating its repair cost and is an essential process in vehicle leasing and insurance industries. It contributes heavily to the actual cost the
customer has to pay. The standard practices follow manual identification
of damages and cost estimation of repairs, resulting in noisy images of
the damaged parts, inconsistent categorization of damage types, and high
variance in repair costs estimation between two appraisers.

We employ explainable machine learning to highlight how the standard ML models and their training protocols fail when dealing with a
dataset acquired without a standard procedure. In this paper, we present
a multi-task image regression model for the leasing vehicle return assessment that leverages the car configuration to reduce the cost of repair
assessment. Our solution achieves a 50% error reduction in the repair
cost estimates. Furthermore, we present remedies base on hierarchical
taxonomy and cost-sensitive loss to improve the damage classification
accuracy.

Keywords: Image classification · Computer vision · Cost-sensitive ·
Deep learning · Explainable machine learning

1 Introduction

Leasing vehicles such as luxury cars, cooperate vehicle fleets etc., is an attractive
option for many customers as it provides a cost-effective alternative to buying
those vehicles. It is estimated that the market share of the leasing vehicle industry will grow more than USD 300 billion by 2021 [1]. The vehicle is used by
the customer for a contracted period of time. At the end of the contract, an
appraiser inspects the vehicle for damages and generates a report using the pictures of damages and their associated repair cost. Traditional methods rely on
manual identification of damage and cost estimation of repairs, which results

---

M. Jameel and M. I. Arif—Both authors contributed equally to this research.

© Springer Nature Switzerland AG 2021
Y. Dong et al. (Eds.): ECML PKDD 2020, LNAI 12460, pp. 259–274, 2021.
https://doi.org/10.1007/978-3-030-67667-4_16

in noisy images of the damaged parts, inconsistent categorization of damage types, and high variance in repair costs estimation between two appraisers. The high variance in the cost of repair means that either the customer or the leasing company were overburdened by the disproportionate estimates.

In recent years, the enhancement in the modeling capacity of deep learning models for image analysis have made the automation efforts feasible in many fields such as medical image diagnostics, roadside sign recognition, autonomous driving, predictive maintenance, etc. Damage assessment of leased vehicles presents another challenging application with huge potential to reap benefits of advancement in the area deep learning and computer vision. The damage assessment comprises two main components, 1) identification and classification of the damage type and 2) predicting the cost of repair for that particular damage. The two components are related, as the accurate classification leads to accurate cost estimates. Although, there are many off-the-shelf deep learning solutions for object detection and classification, however, tuning them to an industrial setting brings its own challenges.

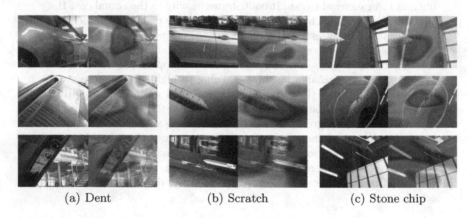

(a) Dent   (b) Scratch   (c) Stone chip

Fig. 1. In each sub figure, the image on the left is the original and on the right is GradCAM generated overlay. The region of the images used by the model for decision making are highlighted using GradCAM. The magenta color region surrounded by violet is the focus region used for decision making. (Color figure online)

In this paper, we used the data collected by one of the leading vehicle leasing company in Europe. The data collected by the traditional approach was highly noisy, unstructured and labels were inconsistently categorized as in contrast to benchmark datasets available for the research purposes. To showcase the problem, we trained an *Inception_v3* model using transfer learning techniques and fine-tuned it on the label images from the company data. The classification model was able to predict the correct damage labels with a nominal 50% accuracy. The strength of these models comes from extracting useful representations from images, these representations are then used for decision making in a classification setting. To further investigate the reason for low accuracy, we employed

GradCAM [10] to explain the decision made by the model, which is presented in Fig. 1. Although there are some instances for which the model was able to base its decision on the correct damage representation, however, due to the high level of noise and incorrect images, a large majority of the decisions are based on incorrect representations such as classifying a 'Scratch' based on the floor of the workshop. Another main issue was the incorrect categorization of the damages, for example, damages with similar visual representation were given two different labels. Lastly, the cost of repair estimates has a high variance between different observations for similar damage types.

In this paper, we tackle these problems using a combination of well-established pre-processing techniques and explainable machine learning to identify and rectify the problems in the automation process. Firstly, we properly annotated the images using bounding boxes that help in capturing a proper representation of the damage types and remove noisy images. The problem of inconsistent damage labels was tackled by defining the hierarchical class taxonomy. Secondly, to better utilize the cost of repair information in the damage classification, we defined a cost-sensitive classification loss. And Lastly, we define a cost regression model that uses both images and vehicle meta-features to predict the cost of repair.

To recap, our contributions are:

- We used a data-driven pre-processing procedure for adapting an industrial dataset to a machine learning problem and used explainable machine learning to define better damage categorization.
- We define a cost-sensitive classification loss as the classification error has an associated penalty in terms of cost estimation.
- We present a cost regression model that leverages both car information as well as damage images to reduce the variance in cost estimation.

2   Related Work

The detection and classification of damage from the picture and assessing the cost is the main task of the leasing vehicle return assessment process. The assessment of damage is not unique to the leasing vehicle return assessment process. It is a core component of the insurance claim process such as vehicle and housing damage claims. However, there are limited research studies conducted in this particular area. In this direction, Patil et al. [8] created a small dataset of damage cars through web crawling. They used some standard CNN models to extract image features and feed it to an SVM classifier for predictions. The dataset is limited to only dent and broken glass/light damage types and did not include any cost estimations. Li et al. [5] conducted a study on detecting the fraud in a car insurance claim and generated a damage dataset by crawling the Internet for the damage images. They used an object detector to identify damage parts and build a system to check for fraudulent claims. Although both the studies target detection of the damages but they still fall short of providing or discussing a complete solution for damage assessment. Previous studies were limited in their scope of exploring other types of damages that are frequent in the real world dataset. On the contrary they focused on dent and scratch, which are easily distinguishable due to distant features. In our case study, we worked with 14 different damage types, which occur frequently in real-world applications.

There are some literature available on related applications on damage detection. Maeda et al. [6] conducted a study on detection of road damage such as cracks. The data was collected using mobile device, which consists of 8 different types of damages, and used variety of object detection models to build an automated solution. Similarly, the assessment of damages to a building after disaster was studied by [7]. There is a commercial interest in the automation of the damage insurance claim, which is evident from the fact that there are number of startups working in this area such as Ant Financial and Tractable.ai to name a few.

3   Methodology

In this section, we will formulate the leasing vehicle assessment process as a multi-task machine learning problem and present the cost-sensitive loss for damage classification.

3.1   Problem Formulation

The leasing vehicle return assessment process consists of two main tasks, i.e classify a damage type and estimate its cost of repair. Generally, a multi-task learning [11] setup best suits this type of problem. Let $\mathcal{X} = \{\mathcal{X}^v, \mathcal{X}^p\}$ define a set of input space, where $\mathcal{X}^v \in \mathbb{R}^V$ is a set of vehicle features such as model, make, color, body part, etc., and $\mathcal{X}^p \in \mathbb{R}^{H \times W}$ is a set of associated pictures/images to a capture visual representation of specific damage. The task-specific output space $\mathcal{Y} = \{\mathcal{Y}^d, \mathcal{Y}^c\}$, where $\mathcal{Y}^d \in \mathbb{R}^D$ represents a set of damages and $\mathcal{Y}^c \in \mathbb{R}$ represents the cost of repair. The dataset set $\mathcal{D} = \{\mathbf{x}_i, y_i^d, y_i^c\}_{i=1}^N$ consists of N observations. To learn a joint model for two tasks, we have two sets of model parameters, a set of model parameters θ^s that is shared between tasks and task specific model parameters θ^d and θ^c. We want to learn a mapping function for each task, which can be defined as,

$$\hat{y}_d(\mathbf{x}, \theta^s, \theta^d) : \mathcal{X} \rightarrow \mathcal{Y}^d \tag{1}$$

$$\hat{y}_c(\mathbf{x}, \theta^s, \theta^c) : \mathcal{X} \rightarrow \mathcal{Y}^c \tag{2}$$

We also have a specific loss for each task i.e. a cross-entropy loss $\hat{\mathcal{L}}_d(\cdot, \cdot)$ for damage classification and squared loss $\hat{\mathcal{L}}_c(\cdot, \cdot)$ for cost of repair assessment. The multi-task objective function thus becomes:

$$\underset{\theta^s, \theta^d, \theta^c}{\arg\min} \, \alpha^c \hat{\mathcal{L}}_c \left(y^c, \hat{y}_c(\mathbf{x}, \theta^s, \theta^c)\right) + \alpha^d \hat{\mathcal{L}}_d \left(y^d, \hat{y}_d(\mathbf{x}, \theta^s, \theta^d)\right) \tag{3}$$

where $\hat{\mathcal{L}}^j(y^j, \hat{y}_j(\mathbf{x}, \theta^s, \theta^j)) = \frac{1}{N} \sum_{(\mathbf{x}, y^j) \in \mathcal{D}} \mathcal{L}_j(y^j, \hat{y}_j(\mathbf{x}, \theta^s, \theta^j)), j = \{d, c\}$. The task-specific weights $\alpha^c \in \mathbb{R}^+$ and $\alpha^d \in \mathbb{R}^+$ are hyperparameters, which are used to control the weight of a specific task in the overall loss. However since we are dealing with an industrial dataset, the images are noisy and collected without a machine learning application in mind. Therefore, directly using a machine learning model on this dataset does not yield the desired results. With this in mind, we propose to solve the classification and regression problem separately.

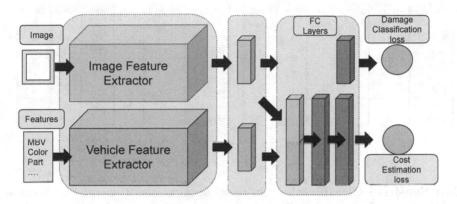

Fig. 2. The cost regression model for vehicle leasing return assessment process. The upper part of the diagram corresponds to the image classification model and lower part corresponds to a cost regression model.

3.2 Damage Classification

Damage classification is an important part of cost of repair estimation since the type of damage directly impacts the cost. We have used *Inception_v3* and *Resnet20* as image feature extractors and since they are complex models we have used transfer learning to initialize their weights pre-trained on ImageNet. Transfer learning has shown to be an effective method to retrain a model with limited data. For training the models, we propose to use two variations of the multi-class classification loss function $\mathcal{L}_d(\cdot, \cdot)$, the standard cross entropy loss and cost-sensitive classification loss.

Cross Entropy Loss: The cross entropy loss is used as a proxy loss for a misclassification rate, defined in Eq. (4).

$$\mathcal{L}_d\left(y^d, \hat{y}_d(\mathbf{x}, \theta^s, \theta^d)\right) = \begin{cases} 1, & \text{if } y^d \neq \hat{y}_d(\mathbf{x}, \theta^s, \theta^d). \\ 0, & \text{otherwise.} \end{cases} \tag{4}$$

In this loss function, if a model prediction does not match the target label, it incurs an error. The error is always one, irrespective of the incorrect label selected by the model. This loss is widely used and best suited for situations in which the penalty for all misclassifications is equal. The cost matrix for misclassification is given in Table 1. For example, if the model misclassifies a scratch as a dent or stone-chip, the penalty of the mistake is the same.

Table 1. (left) Cost matrix for misclassification, and (right) Cost matrix based on average cost difference between pairs of damages

Damage Class	Scratch	Dent	Stone -chip	Wear	Burnt -Hole
Scratch	0	1	1	1	1
Dent	1	0	1	1	1
Stone -chip	1	1	0	1	1
Wear	1	1	1	0	1
Burnt -Hole	1	1	1	1	0

Damage Class	Scratch	Dent	Stone -chip	Wear	Burnt -Hole
Scratch	0	49	327	28	34
Dent	49	0	377	78	84
Stone -chip	327	377	0	298	292
Wear	28	78	298	0	5
Burnt -Hole	34	84	292	5	0

Cost-Sensitive Classification: In many applications, the cost for misclassification is not the same for all types of mistakes, for example, customer churn prediction. In our problem, we are given the cost of repair for each instance of the damage. The cost of repair of two different damage types could vary significantly. For this purpose we created a cost matrix by recording for each pair of damage the difference between their average cost of repair, a subset of the cost matrix is shown in Table 1. Again taking the same example as before, now if a scratch is misclassified as a dent, it will incur a penalty of 49. On the other-hand, misclassifying a scratch as a stone-chip will result in a penalty of 327. Therefore, we used this information in the loss function and define a cost-sensitive loss given in Eq. (5).

$$\mathcal{L}_d \left(y^d, \hat{y}_d(\mathbf{x}, \theta^s, \theta^d)\right) = \begin{cases} c_{y^d, \hat{y}_d(\mathbf{x},\theta^s,\theta^d)}, & \text{if } y^d \neq \hat{y}_d(\mathbf{x}, \theta^s, \theta^d). \\ 0, & \text{otherwise.} \end{cases} \tag{5}$$

where $c_{.,.}$ is an element of the cost matrix $\mathbf{C} \in \mathbb{R}^{D \times D}$.

3.3 Cost Regression

The task of predicting the cost of repair can be categorized as a regression problem, which is defined in Eq. (2). There are many state-of-the-art machine learning models, such as Gradient Boosted Decision Trees (XGB) [3] and Random Forest (RF) [2], which have shown to perform exceptionally good on the regression task for vector data. However, in our problem, we were given a mix of vector data and images, more specifically, damage labels are encoded in images. To include both, the vector data such as car information and pictures of damage, we use a deep neural network. We used a CNN based feature extractor i.e. *Inception_v3* and *Resnet*, to learn the latent representation of the images. The latent representations of images are concatenated with car features and become the input to the fully connected feed-forward neural network, as shown in Fig. 2. We used mean squared error (MSE) loss $\mathcal{L}_c(y^c, \hat{y}_c(\mathbf{x}, \theta^s, \theta^c) = (y^c - \hat{y}_c(\mathbf{x}, \theta^s, \theta^c)^2$ to train the model.

$$\hat{\mathcal{L}}_c\left(y^c, \hat{y}_c(\mathbf{x}, \theta^s, \theta^c)\right) = \frac{1}{N} \sum_{(\mathbf{x}, y^c) \in \mathcal{D}} (y^c - \hat{y}_c(\mathbf{x}, \theta^s, \theta^c))^2 \tag{6}$$

4 Experiments

This section talks about the dataset, the steps taken to make is compatible with a machine learning setting and lays out the results for our classification and cost regression.

Table 2. Statistics of leasing vehicle return dataset

Name	Reports	Images with Cost	Damage Types	Models (mbv)	Colors	Parts	repair actions
Count	39,000	342,029	35	51	165	166	21

Table 3. Statistics of dataset after annotation phase

Name	Damage classes	Sampled Images	Annotated Images	Total Crops
Count	14 + 1	48,000	17,083	25,228

4.1 Dataset

The dataset used in this paper was collected by one of the leading vehicle leasing company in Europe. It is made up of 40,000 reports that have been generated manually by appraisers at the end of a leasing contract. The appraiser inspects the car for damages, identify the damages, photograph them, and provides an estimate for the cost of those repairs. There are 342,029 photograph images of damages and each image has a corresponding body part, damage type and the estimate for the cost of repair. Overall, there are 166 meta-level body parts and 35 damage types available in the collected dataset. Apart from the damage specific information, we also have detailed meta-features about the vehicle out of which the more relevant features are model, make and color. The information of the car model was available at a very fine grain level i.e. interior configurations, and variation in the trim levels. We combined these models in high-level groups represented by MBV, which are based on the model rather than the variants of the same model, for example, the same car model with different trim levels is treated as one model. The color of the car also plays an important role in the repair cost estimation, as metallic or exotic colors cost more than the standard colors. Table 2 provides an overview of the number of reports in the dataset and the final number of these features.

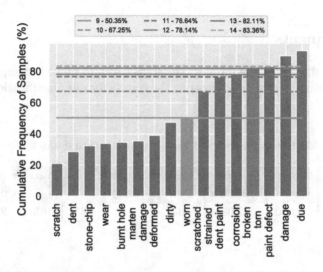

Fig. 3. The frequency of samples for top-14 damage types, represented as a cumulative frequency plot. It shows that first 9 classes listed from left to right covers around 50% of the data, whereas, top 14 classes covers 83% data. This shows only 14 classes out of 35 total classes constitute majority of the data.

The dataset consists of 35 damage types, however, there are two main problems with these damage types. Firstly, the damage types can be categorized into optical and non-optical damages. Optical damage has a visual appearance and can be captured through pictures, for example, 'dent', 'scratch' and 'stone-chips' etc. Whereas, non-optical damage cannot be captured or defined using visual features, for example 'smell of a bad odor', 'missing item' or 'play in a component'. Therefore, the non-optical damage types cannot be included in the classification task. Secondly, the damage types suffer from a typical long-tail distribution, some of the damage types did not have enough samples. To overcome these two problems, we picked 14 most frequent optical damage classes, which are shown in Fig. 3 as a cumulative frequency bar plot. We can see that ≈84% of the dataset can be covered using only the top 14 classes.

4.2   Exploratory Data Analysis

In this section, we perform an exploratory data analysis to understand the useful relationship between different features. We used Kernel Density estimate of cost and different features and plotted them in Fig. 4[1]. To presents more meaningful information in these plots, we used a single car model (mbv) to represent the relationship between the color, damage, part, and the cost. In Fig. 4(a) shows the cost of repair of a particular body part is higher than other, which verifies that different body parts require a different type of repairs. Figure 4(b) shows the cost relationship with color and again some colors have a higher cost of repair. It is also to be noted that it might also depend on the extent of the damage i.e. a small scratch might cost less to repair than a bigger scratch. Lastly, Fig. 4(c) shows the relationship between the cost and damage, which is similar to the color relationship. This can be caused by the extent of that damage but it is highlighted that the final cost for damage is also impacted by the variance in the opinion of an appraiser. We also wanted to see how the different parts and colors were related to the damages, to see if particular damage is always related to a certain part/color. Figure 5(a) shows that damage and color do not hold a strong correlation as is expected. Conversely, we can see in Fig. 5(b) that the damage and part appear to have a strong correlation. A 'stone chip' frequently appears at the curved lining, where the paint is weakest. From this analysis, we are able to infer that the model, color, part and damage under consideration have an impact on the final cost and therefore need to be included in the model as auxiliary information.

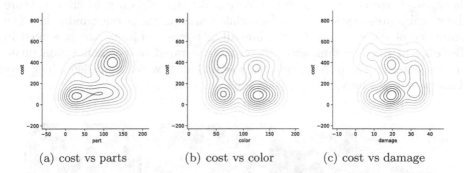

(a) cost vs parts   (b) cost vs color   (c) cost vs damage

Fig. 4. The plots represents Kernel Density estimate between cost and different car features present in the dataset.

4.3   Data Cleaning and Annotation

The task of image classification relies heavily on the quality of the images being trained on. The damage images in the reports are taken without a standard

---

[1] The values of the cost of repair is always greater than 0, however, because of the kernel density function some contours appear to be below zero values.

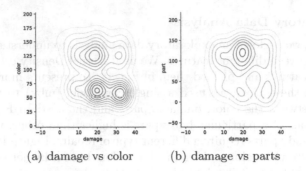

(a) damage vs color   (b) damage vs parts

Fig. 5. The plots represents Kernel Density estimate between damage and other car features present in the dataset.

acquisition procedure and therefore vary significantly. Variance in lighting conditions, distance from the damage, noisy backgrounds, and even dirty car parts make the task of learning useful representations more challenging. In order to learn a useful classifier for the damages, we annotated the dataset using bounding boxes. We annotated the images with bounding boxes and marked those images as 'dirty', which have a noisy background, dirty car, poor lighting, high reflections, and blurrey images. We randomly sampled 3500 images from each damage class to be annotated but because of the high level of noise, only 17, 083 were annotated, while the rest were marked as 'dirty'. Furthermore, we created crops of images using the bounding boxes, which resulted in 25, 000 crops of damages. These crops are useful to learn a damage classifier, as crops capture the visual representation of damages while reducing the background noise. The summary of the statistics for the bounding box annotations are presented in Table 3 and Fig. 7. Examples of the crops generated by the annotation phase are presented in Fig. 6. An extra class was included, which we called a 'negative damage class' to provide negative examples for training.

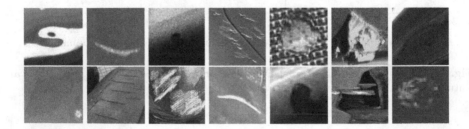

Fig. 6. Crops of damages

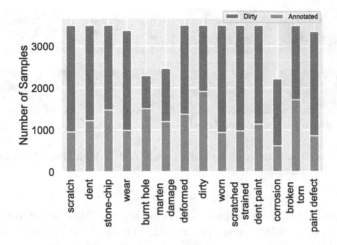

Fig. 7. The number of samples for annotation vs actual clean images annotated.

4.4 Damage Classification

In this section, we perform the experiments for the classification task. For this purpose, we used 25,228 crop images dataset, which consists of color images, sized 225 × 225, with 14 damage classes and a 'negative damage class'. The data was split into 90% train-set and 10% test-set, such that test-set contains equal samples from each damage class. We perform 10 experiment runs, and for each run creating a new train/test split. We used *Inception_v3* [12] and *Resnet20* [4] pretrained on the ImageNet dataset [9]. The training of these models was done using SGD with momentum $\mu = 0.99$ and the learning rate η was searched in the grid $\eta = \{0.001, 0.01, 0.05, 0.1\}$.

Cross Entropy Loss. In the first set of experiments, we trained the classifier using standard cross entropy loss given in Eq. (4). The results presented in Fig. 10(a) show the classification accuracy on varying the number of damage classes. It is evident from the results, as we increase the number of classes, the complexity of the problem increases and the accuracy drops. The first column for 3 damage classes consists of 'Scratch', 'Dent' and 'Chip-Stone', which have very distinctive damage patterns, therefore, both the models were able to achieve very good results. However, once we start to increase the number of classes, the accuracy starts to degrade. The most significant drop in accuracy was observed at 10 classes and more.

In order to investigate the performance degradation, we used an explainable machine learning approach called GradCAM [10], which provides a method to visualize the gradients of the image per pixel and gain insight on the regions in an image used by the model for its decision to assign a particular class. The GradCAM analysis on a few similar classes is shown in Fig. 8. At a cursory glance, it becomes evident that the images for 'Scratch' and 'Scratched' classes,

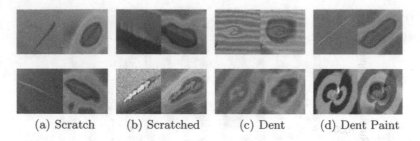

(a) Scratch   (b) Scratched   (c) Dent   (d) Dent Paint

Fig. 8. Analysis of the damage crops using GradCAM. It highlights that some damage types are visually similar.

and 'Dent' and 'Dent Paint' appear to be causing very similar activations in the model. This is caused by the similar manifestation of the damages on the car i.e 'Dent' and 'Dent Paint' are both visually similar. This will lead to confusion between these classes and lead to poor classification accuracy.

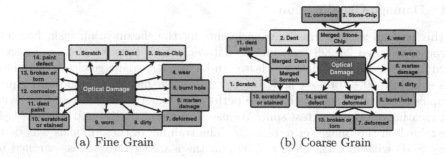

(a) Fine Grain   (b) Coarse Grain

Fig. 9. Damage class taxonomy, a) original taxonomy and b) proposed taxonomy

In addition to GradCAM, we analyzed the confusion matrix on our test set to identify the confusing cases. A certain pair of classes are being confused with each other, which was evident from the confusion matrix, for example 'Dent Paint' and 'Dent', and 'Corrosion' and 'Stone chip' are frequently confused. This problem highlights that the degradation in the performance of a machine learning model is not necessarily caused by the training or model choice, but it stems from the non-standard categorization of the damage labels. To rectify the problem of non-standard categorization of the damage labels, we proposed to group similar classes based on their visual representations. We defined a hierarchy taxonomy of the damage labels, which we referred to as 'Coarse Grain' (CG) taxonomy Fig. 9(b), whereas, the original class taxonomy is referred as 'Fine Grain' (FG) taxonomy Fig. 9(a). The classification accuracy for the CG taxonomy is presented in Fig. 10(b). It is observed that both *Inception_v3* and *Resnet20* model perform at par with each other. To compare the results of FG and CG taxonomy, we have to compare 9 classes results in Fig. 10(a) with Fig. 10(b), and it becomes clear that despite increasing the confusing samples by keeping the number of classes same, there is no degradation in the accuracy.

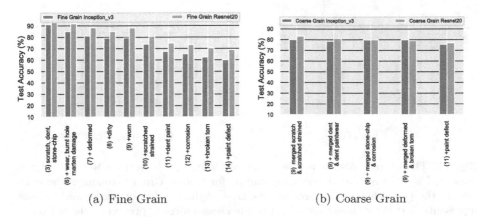

(a) Fine Grain   (b) Coarse Grain

Fig. 10. The classification accuracy on the test-set was presented a) for Fine Grain taxonomy and b) for Coarse Grain taxonomy. The numbers in the () on the x-axis represents the case of the number of classes and '+' sign represents that these classes are added to the classes already present in the left bar.

Cost Sensitive Classification: In the second set of experiments, we trained the damage classifier using cost sensitive loss given in Eq. (5). We used the same training protocol as in the previous section, the only change was the evaluation metric, which is changed from accuracy to cost-sensitive cost define similar to Eq. (5). The results are presented in Fig. 11(a) and Fig. 11(b) for FG and CG taxonomies respectively. The models trained on cost-sensitive loss had a lower misclassification error as compared to the one which was trained on the misclassification rate. It is also evident from the results if the problem is well defined, for example in the case of classification of 3 damage types, the misclassification error is very low, therefore, the performance of both the methods is equal.

4.5   Cost Regression

In this section, we perform the experiments for the prediction of the cost of repair. We used the same dataset as explained in the damage classification section, however, now the target is to predict the cost of repair. We used the car features given in Table 2 with 14 damage classes to predict the cost of repair. The data was split using a three-fold validation strategy, where two folds are used for training and one for testing. The state-of-the-art models such as RF and XGB were trained on this data excluding the images and using the appraiser assigned damage type. We also build a custom Feed Forward neural network (FNN), which consists of two fully connected layers with Relu activation function and

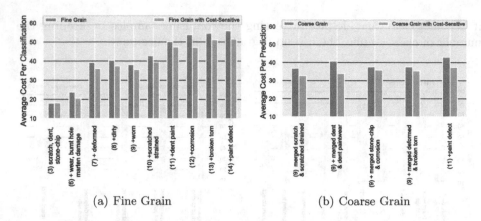

(a) Fine Grain   (b) Coarse Grain

Fig. 11. The average cost of misclassification (lower the better) on the test-set was presented a) for Fine Grain taxonomy and b) for Coarse Grain taxonomy. The numbers in the () on the x-axis represents the case of the number of classes and '+' sign represents that these classes are added to the classes already present in the left bar.

Table 4. The results of cost regression task.

Model	Features	RMSE
Average model	∅	237.43 ± 0.73
Linear regression	Parts, mbv, color, damage, action	106.41 ± 3.81
Random Forest (RF) [2]	Parts, mbv, color, damage, action	85.84 ± 3.99
XGboost (XGB) [3]	Parts, mbv, color, damage, action	84.77 ± 1.78
FNN (our)	Parts, mbv, color, damage, action	**82.3 ± 2.8**
FNN + Image (our)	Image, parts, mbv, color, action	**83.6 ± 0.73**

dropouts. We performed extensive grid search to find the optimal number of nodes $\{32, 64, 128, 256\}$, dropout rates $\{0.1, 0.3, 0.5, 0.7, 0.9\}$ and learning rate $\{0.01, 0.05, 0.1, 0.5\}$. Lastly, we combined inputs to FNN with the image latent features learned in classification task, this helps to remove the dependence on the true damage labels provided by an appraiser at inference. The RMSE scores of different models are summarized in Table 4. The regression models were able to achieve comparable RMSE score. However, it can be seen that FNN with image feature does not require information about the true damage labels, which it infers from the image feature. Lastly, Fig. 12 shows a comparison between the natural variance in the dataset as compared to the error made by the models. The mean cost variance of the dataset is higher than the model prediction errors, which means if a customer goes for a repair, the estimate of the appraiser has a variance of approximately ±172. Whereas, the model was able to significantly reduce the variance to approximately ±80.

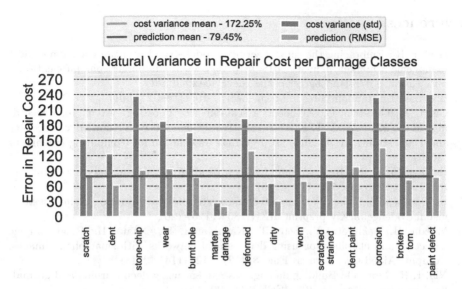

Fig. 12. The comparison of natural variance in the dataset with the prediction made by the model.

5 Conclusion

In this paper, we have presented the challenges encountered when translating the gains made in the field of machine learning to a real-world application and the necessary steps to overcome those challenges. Translating the gain to a propriety dataset requires a data-driven approach to transform the dataset into one that lends itself to machine learning problems. We show how explainable machine learning can be employed to understand the factors causing the machine learning models to under-perform and design a strategy to be applied to similar datasets. This work has also shown a novel application of cost-sensitive loss functions to a new use-case, where widely used cross entropy loss does not capture the important aspects of the task at hand. We experimentally show the gains made by leveraging cross domain knowledge i.e. using bounding boxes to improve classification accuracy. Lastly, we developed a cost regression solution, which leverages latent features from both images and vehicle feature to improve the regression task. We were able to significantly reduce variance in the cost estimation as compare to the manual estimations by appraisers.

Acknowledgment. This work is co-funded by the industry project "Data-driven Mobility Services" of ISMLL and Volkswagen Financial Services (https://www.ismll. uni-hildesheim.de/projekte/dna_en.html).

References

1. Berger, R.: Embracing the caras-a-service model - the European leasing and fleet management market, January 2018. https://www.rolandberger.com/publications/publication_pdf/roland_berger_car_as_a_service_final.pdf
2. Breiman, L.: Random forests. Mach. Learn. **45**(1), 5–32 (2001)
3. Chen, T., Guestrin, C.: XGboost: a scalable tree boosting system. In: Proceedings of the 22nd ACM SIGKDD International Conference on Knowledge Discovery and Data Mining, KDD 2016, pp. 785–794. Association for Computing Machinery (2016)
4. He, K., Zhang, X., Ren, S., Sun, J.: Deep residual learning for image recognition. In: CVPR 2016, pp. 770–778 (2016)
5. Li, P., Shen, B., Dong, W.: An anti-fraud system for car insurance claim based on visual evidence. arXiv preprint arXiv:1804.11207 (2018)
6. Maeda, H., Sekimoto, Y., Seto, T., Kashiyama, T., Omata, H.: Road damage detection and classification using deep neural networks with smartphone images. Comput.-Aided Civ. Infrastr. Eng. **33**(12), 1127–1141 (2018)
7. Nia, K.R., Mori, G.: Building damage assessment using deep learning and ground-level image data, pp. 95–102. IEEE, May 2017
8. Patil, K., Kulkarni, M., Sriraman, A., Karande, S.: Deep learning based car damage classification. In: 2017 16th IEEE International Conference on Machine Learning and Applications (ICMLA), pp. 50–54, December 2017
9. Russakovsky, O., et al.: ImageNet large scale visual recognition challenge. Int. J. Comput. Vis. **115**(3), 211–252 (2015). https://doi.org/10.1007/s11263-015-0816-y
10. Selvaraju, R.R., Cogswell, M., Das, A., Vedantam, R., Parikh, D., Batra, D.: Grad-CAM: visual explanations from deep networks via gradient-based localization. In: 2017 IEEE International Conference on Computer Vision (ICCV), pp. 618–626 (2017)
11. Sener, O., Koltun, V.: Multi-task learning as multi-objective optimization. In: Bengio, S., Wallach, H., Larochelle, H., Grauman, K., Cesa-Bianchi, N., Garnett, R. (eds.) Advances in Neural Information Processing Systems, vol. 31, pp. 527–538. Curran Associates, Inc. (2018)
12. Szegedy, C., Vanhoucke, V., Ioffe, S., Shlens, J., Wojna, Z.: Rethinking the inception architecture for computer vision. In: Proceedings of IEEE Conference on Computer Vision and Pattern Recognition (2016)

A Route-Affecting Region Based Approach for Feature Extraction in Transportation Route Planning

Fandel Lin[1], Hsun-Ping Hsieh[1,2(✉)], and Jie-Yu Fang[1]

[1] Institute of Computer and Communication Engineering,
National Cheng Kung University, Tainan, Taiwan
{q36084028,hphsieh,q36074170}@mail.ncku.edu.tw
[2] Department of Electrical Engineering, National Cheng Kung University,
Tainan, Taiwan

Abstract. Traffic deployment is highly correlated with the quality of life. Current research for passenger flow estimation in transportation route planning focuses on origin-destination matrices (OD) analysis; however, we claim that urban functions and geographical environments around passing area and stations should also be considered because they affect the demand of public transportation. For the route-based demand prediction task, we therefore define route affecting region (RAR) to model the influential region of routes. Based on the proposed RAR, we further proposed route-based feature extraction approaches along with adopting several regression models to do high accurate inference. Given heterogeneous features and faced with the competitive and transfer effects of existing routes, our proposed RAR-based feature engineering methods are effective for handling and combining dynamic and static data which are high-correlated with passenger volumes. The experiments on bus-ticket data of Tainan and Chicago, with public transit network structures different from each other, show the adaptability and better performance of our proposed RAR-based approach compared to traditional OD-based feature extraction strategies.

Keywords: Feature engineering · Origin-destination matrix · Feature extraction · Traffic demand · Deep Neural Network (DNN)

1 Introduction

Traffic deployment is highly correlated with the quality of life [1]. Governments or traffic management authorities employ new transportation services such as bus or MRT routes to serve residents. For residents, new services bring convenience

---

This work was partially supported by Ministry of Science and Technology (MOST) of Taiwan under grants 108-2221-E-006-142 and 108-2636-E-006-013. Meanwhile, we are grateful to Tainan City Government for providing the bus ticket data.

Y. Dong et al. (Eds.): ECML PKDD 2020, LNAI 12460, pp. 275–290, 2021.
https://doi.org/10.1007/978-3-030-67667-4_17

and reduce pollution. For traffic management authorities, the number of passengers traveling by their deployed services is an important indicator. However, constructing unwanted and redundant routes or stations can lead to environmental damage and resource waste. Besides, according to our interview with civil servants in the bureau of transportation, they pointed out that the current procedure in planning new routes turns out to be lengthy due to many stakeholders or delegates involved in them. In addition, the overwhelming number of requests from the public makes it difficult to decide where to construct new routes and stations. Therefore, an effective evaluation of potential Passenger Flow (PF) for the requested route in a timely manner turns out to be crucial and helpful for traffic management authority (user).

A plethora of frameworks utilizing either deep learning methods or statistical analyses have been developed to solve such PF inference in designing new public transportation routes. However, their works often rely on surveying data in building origin-destination matrices (OD). In this work, we claim that the geographical environment and urban functions of the trajectories between two stations could also affect the demand for deploying public transportation. Meanwhile, none of the previous research investigated and solved the problem of inferring route-based PF utilizing route-based feature extraction strategies. Therefore, we propose route-affecting region (RAR) to model the influential region of routes; besides, we investigate the correlations between various features in RAR and PF in this work.

To show the adaptability of the proposed RAR-based feature extraction approaches in the PF inference problem, we take bus-ticket data for the realistic public transit networks of Tainan and Chicago, which represent radial and square road structure respectively, for evaluation. Several regression models including DeepNeuralNetwork (DNN) for Regression, Support Vector Regression, Linear Regression, and XGBoost are adopted along with our proposed RAR-based feature extraction methods for inferencing PF. Evaluation results show that RAR-based strategy outperforms traditional OD-based strategy for at most 39%. Moreover, the DNN for Regression with RAR-based scenario surpasses all other OD-based settings for 33% to 87%, which demonstrates the effectiveness of our newly proposed route-based feature extraction strategies.

In particular, we made the following contributions:

- We introduce route-affecting region (RAR) to model the influential region of routes and thereafter propose several route-based feature engineering methods for potential PF inference in transportation route planning.
- Given heterogeneous features and faced with the competitive and transfer effects of existing routes, the proposed RAR and feature engineering methods are effective for handling dynamic and static data together.
- Experiments on two cities, with public transit network structures different from each other, show the adaptability of the proposed RAR-based approaches compared to traditional OD-based feature extraction strategies.

The system flow for the PF inference framework along with RAR-based feature engineering strategy is illustrated in Fig. 1. This framework with the pro-

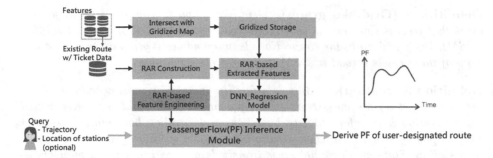

Fig. 1. System Flow for the PF inference framework.

posed RAR can be adopted for several kinds of urban transportation system, such as subway or bus, and be utilized in any cities where ticket data is available for the government or transportation management authority.

2   Background

We look into previous researches that focus on inferencing the PF of designed transportation routes in order to decide which features are needed to be considered and extracted for further inference. Some works in designing new transportation routes focused on reducing transportation time through route adjustment or shift [2–6]. There are also some works that optimized route planning with distance, time, transference considered [7,8]. Some works studied the problem of predicting arrival time [9–11] or future PF [12–14] based on regression analysis. There are also some researches that inferred PF utilizing machine learning or deep learning techniques [15–20]. However, most of the above works focused on dealing with existing routes utilizing OD-based feature extraction strategies.

By focusing on designing new transportation routes [2–6] and analyzing the considered and extracted features of formulation for previous works, we generalize six kinds of relevant urban features in inferring the PF for new routes deployed in transportation networks. To be more specific, surrounding POI, human mobility, road network structure (network structure and trajectory length), relationship with existing routes (transference and competition), population structure, time information (waiting time and journey interval) are considered as the input features of the inference model. However, most importantly, all the previous works utilized these features on an OD-based analysis.

3   Preliminary

Definition 1 (Grid). *We divide the city into disjointed grids (0.1 km * 0.1 km) [21] and store all features that are correlated with PF (e.g. population in this grid, whether existing routes passed this grid, etc.) into corresponding grid.*

Definition 2 (Grid-like graph). *Grid-like graph is composed of disjointed grids that records connections as original road network based on OpenStreetMap (OSM). Each grid stores the connections between adjacent grids in its eight directions if there exists a road in OSM that connects each other.*

Definition 3 (Station). *Station is a facility or area for passenger to regularly get-into or get-off the mass transit transportation. (Note that the mass transit transportation here refers to city bus, light rails, trolley bus, etc.) Passenger need to pay by smart card when getting-into or/and getting-off the mass transit at a station. Station in original mass transit data or as input given by users as system is a point with latitude and longitude; but is labelled on a grid that the point located at in grid-like graph.*

Definition 4 (Trajectory). *Trajectory is the path that certain mass transit takes. Trajectory in original mass transit data or as input given by users as system is a series of road junctions; but turns into a series of connecting grids in the graph-like grids for further PF inference and route recommendation.*

Definition 5 (Route). *Route is a set of combination of trajectory and stations. Note that same series of trajectories with different set of stations does refer to different route. Route is a series of connecting grids and several grids labelled as stations in the grid-like graph; however, since we divide the city into disjointed grids with a meticulous size, the actual route in real world (OSM) can be easily reproduced given a sequence of grids. Therefore, though some re-projections from grid to actual road network are needed, no other superfluous process needed to handle in post-processing.*

Definition 6 (Passenger Flow (PF)). *Since the price is fixed fare for Chicago and Tainan bus transit system, and most of mass transit transportation system in other cities, the passenger flow along the route here refers to the total passengers who passed any point along the route. To be more specific, PF is counted once someone pay by smart card when getting-into or getting-off the mass transit at a station of a route.*

Definition 7 (Origin-destination (OD) matrix). *An origin-destination (OD) matrix is essential for efficient traffic control and management [22], which has been utilized in modeling congestion and estimating travel time by specifying the travel demands between two nodes in network. The input data is usually based on survey data from the region of origin and destination [23].*

4 PF Inference

4.1 Problem Definition

The PF inference problem is defined as follows. Given a base map with urban heterogeneous features and a set of trajectories for the designated route with its stations labeled from users, our goal is to infer the passenger flow PF for that route.

The workflow of RAR-based approach for PF inference is displayed in Fig. 2, which mainly consists of three components. In data preprocessing, we divide the city into disjointed grids (e.g., 0.1 km * 0.1 km) [21], and all features are fetched and stored in grids for further extraction. The second component is training models. The feature set for each existing route is extracted and integrated as the training data along with its corresponding ticket data, which is associated with the timestamp and PF for each route. We treat various features as inputs and PF values as the predictive label. We tried several machine learning methods as training models, and the DNN for regression gets the most promising result in our evaluation. In the third component, the pre-trained model can be utilized for the query route given by the user to infer PF value.

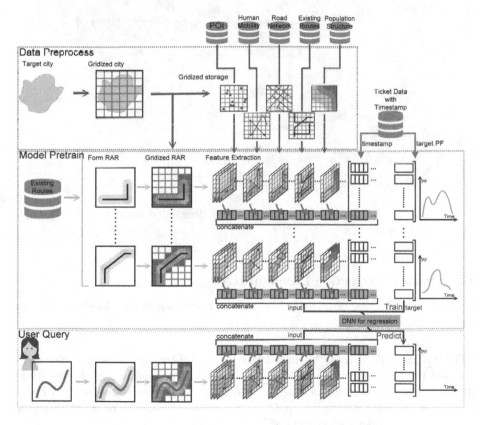

Fig. 2. Workflow of RAR-based approach for PF inference.

4.2   Route Affecting Region (RAR)

Our idea is that the demand for public transportation is not only based on the origin and destination, but also geographical environment and urban functions of nearby areas. Thus, we proposed RAR or considering these PF-related features. A route can comprise multiple segments that contain successive points close to each other. Then we can draw a circle for each point, where we consider each point as the center of a circle, and then RAR formed by a set of circles. Based on Design Manual for Urban Sidewalks [24], the walking tolerance for pedestrians is 400 to 800 m; we thereafter extract corresponding features within RAR. The green area in Fig. 3 is an example of RAR of given route q_s to q_d, with a radius of 0.4 km.

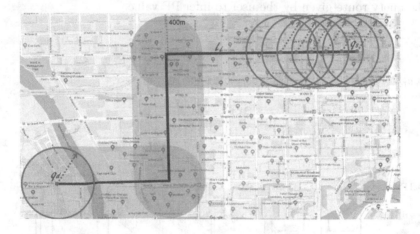

Fig. 3. A RAR example of user-designated route.

4.3   Feature Extraction Based on RAR

To infer the PF value of a trajectory correctly, we consider six kinds of relevant urban features in RAR:

POI-Related Features. Various POIs (Point-Of-Interest, specific point location such as transportation hubs or entertainment venues) and their density in RAR indicate the function of a region, which are highly correlated to the PF of a route. For example, a high PF might be associated with route to many shopping centers. We consider two aspects of POI features:

POI Density. The density of POI indicates the popularity of a certain activity type in RAR, which also refers to the function of a region. As the example mentioned above, a high density of certain types of POI such as shopping centers and schools can result in high PF value.

POI Entropy. The POI entropy shows the diversity of purpose when people visit the nearby area of a route and is based on Information Theory [25]:

$$Entropy(l_i) = -\sum_{\gamma}^{\gamma \in \Gamma} (\frac{N_\gamma(l_i, r)}{N(l_i, r)} \times \log \frac{N_\gamma(l_i, r)}{N(l_i, r)}) \tag{1}$$

Where Γ indicates set of POI, and γ refers to certain type of POI. Besides, N (l_i, r) displays the total number of POI in RAR of trajectory l_i based on radius r, $N_\gamma(l_i, r)$ displays the number of type-γ POI in RAR of trajectory l_i.

Human Mobility. The total number of pick-up and drop-off records of Taxi (in Chicago) or Bike (in Tainan) trip records that occur in RAR are accumulated as the leaving and incoming flow respectively. Records taking place in same RAR are viewed as transition flow. Dividing these values by number of all taxi records, several floating-point numbers are derived.

Road Network Structure. Based on grid-like graph, the degree and closeness centrality in RAR are calculated as floating-point numbers. Degree centrality identifies the total number of reachable vertices for all intersections in RAR, and closeness centrality shows the average distance between one intersection to another in RAR.

Competition and Transference with Existing Routes. To quantify competitive relationship and since the road network structure is already reconstructed into grid-like graph, grids that holds designated routes and each existing route can be labelled respectively. Then a simple algorithm is run to calculate the number of grids that are labelled as both designated route and certain existing one, or grids that are labelled as designated route but nearby grids in its RAR are labelled as certain existing route. Through this process, the numbers of overlap grids and nearby grids between given route and each existing route are derived; meanwhile, if certain existing route is transferable, the grids (except for overlap and nearby ones) of that existing route are viewed as extended grids. Dividing these values by total grids of corresponding existing route, several floating-point numbers are derived, representing overlap/ nearby/ extended region between given and each existing route respectively.

Population Structure. People in RAR of different ages and genders have different intentions for taking public transportation. Consequently, based on population pyramid (if available) for each village, Population in certain RAR of different ages and genders are derived as several floating-point numbers.

Time Information and Granularity. Seasons and holidays can influence the passenger flow of public transportation. We adopt one-hot encoding to record the time information for each ticket record.

4.4   Inference Model Construction

We adopt and modify multiple machine learning methods including Support Vector Regression (SVR), Linear Regression (LR), XGBoost, and DNN for Regression, to derive the PF for the designated route respectively. The input data includes all the features extracted based on the RAR of the given route, while the output is the inferred PF value of that route. The architecture of DNN for Regression is a feed-forward neural network with many levels of non-linearities. Meanwhile, all our input features are rescaled to 0 to 1, the type of the hidden units for 4 dense layers is ReLU and the output unit is linear.

As the features been considered include relationship with existing routes and other relevant factors as described above, we believe that the model is trained/adapted to infer the PF of an arbitrary route (either new one or existing one) with the features that are extracted based on RAR of the route.

Descriptions of the features for inference model are provided in Table 1 Due to the nature of sources provided by authorities of cities, there exists slight differences. For instance, population pyramid for each village (the smallest administrative district; the average area of each village in Tainan City is about 2.91 square kilometers, the value turns to 0.25 square kilometers for the city centre) is only available for Tainan dataset.

Table 1. Descriptions of the independent variables (features) used in our models.

Feature	Description
Timestamp	84 variables for the time of day and day of week (each 2-h interval)
POI-density	10 variables for each type of POI (category based on Foursquare[a])
POI-entropy	1 variable for the entropy of POI
Human mobility	3 variables for each type of flow (transition, incoming, leaving)
Road network	2 variables for each type of centrality (degree, closeness)
Existing routes	15 variables for top-5 related existing routes (3 values for each)
Population structure	1 variable for total population (Chicago) 6 variables for different ages and genders (Tainan)

[a]https://developer.foursquare.com/docs/resources/categories

5   Evaluation

5.1   Datasets

We use bus-ticket data on two different types (radial and square structure) of public transit networks from Tainan City Government and Chicago Transit

Authority (CTA[2]). The data for Tainan lists ticket id, route id, timestamps, and the starting and ending stations; on the other hand, the data for Chicago lists route id, timestamps, and number of passengers. The datasets contain 14,336,226 and 231,196,847 ticket records respectively and hold at least 100 routes and thousands of stations in service. The public transit networks for both cities are shown in Fig. 4.

The urban spaces of Tainan and Chicago are divided into 505,296 and 330,335 disjointed grids (0.1 km * 0.1 km) based on EPSG: 3857, which is a variant of the Mercator projection and acts as the standard for web mapping applications. Since the unit for this projection is meter, we are able to divide the urban spaces into disjointed grids based on meters precisely considering the ellipsoidal datum when generating grid-like graph and pre-processing relevant features in the Geographic Information System. Meanwhile, only 94,282 and 91,320 grids (vertices) would be considered in route recommendation, which are reduced by about three times compared to the original road network structure. On the other hand, static features including POI and road network structures are extracted from GoogleMap and OpenStreetMap. The population are fetc.hed from respective agencies. Finally, we take bike trips and taxi trips that list pick-up and drop-off location as human mobility. Details of both cities are presented in Table 2.

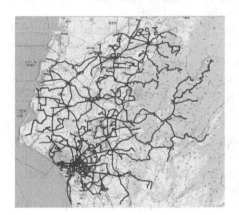 

Fig. 4. Public transit networks for Tainan (left) and Chicago (right) on the same scale.

5.2 Evaluation Setting

To demonstrate the effectiveness of our proposed route-based feature extraction strategies, we hold two kinds of experimental scenarios to be compared. The first one is feature extraction based on our proposed route-affecting region (*RAR-based*), and the second one is to extract features of regions from simply origin-destination stations (*OD-based*) traditionally. We then developed four

---

[2] http://www.transitchicago.com/

Table 2. Size figures for our input instances.

Instance	Dataset	Tainan	Chicago
Bus data	Existing routes	104	139
	Existing stations	6.575	11,592
	Ticket records	14,336,226	231,196,847
	Period	01/2017–12/2017	11/2017–10/2018
Gridized map	Grids (0.1 km * 0.1 km)	505,296	330,335
	Grids labelled with road	94,282	91,320
Other features	POI	8,734	21,889
	Bike trips (for human mobility)	139,478	N/A
	Taxi trips (for human mobility)	N/A	68,461,612
	Road nodes	237,866	390,509
	Road edges	414,409	560,810
	Census blocks (for population)	14,730	46,293

methods for inference: (a) *Support Vector Regression (SVR)*, (b) *Linear Regression (LR)*, (c) *XGBoost*, and (d) *DNN for Regression (DNN-Reg)*, along with two baseline methods: (e) *Median value* and (f) *Average value* using the median and average value of PF in all training routes respectively.

Meanwhile, we conduct other experiments to investigate the importance of different features and the relationship between RAR-based strategy and each feature set; Table 3 shows the components of feature sets used in the evaluation. Apart from time information, which is included in all feature sets, each of the six urban relevant features is selected respectively from *set I* to *set V*; meanwhile, we split the POI-related features into two parts–density and entropy–to see the influence between them. *Set VII* refers to the static location-based features, including POI-related ones and population structure; similarly, *set VIII* represents the static transportation-based features consist of road network and competition/ transference with existing routes. Furthermore, *set IX* combines all the static features, and human mobility is later considered in *set X*.

The evaluation is based on the leave-one-out method, where we iteratively leave one route PF data out of the complete data, and then use the rest of data to train the model and infer the value of the left out route based on the features extracted from the RAR of the route. Then we compare the inference value with the ground-truth, which is the total number of passengers that had taken the route. We choose the normalized root-mean-square error (RMSE) as the metric.

5.3   Evaluation of Feature Selection

We summarize the preliminary evaluation by illustrating the normalized RMSE of PF results for all comparative and baseline methods, the latter are depicted as horizontal lines since they are not affected by feature sets. Considering the walking tolerance for pedestrians (0.4 km to 0.8 km), Fig. 5 (for Tainan dataset)

Table 3. Strategy sets to be used in preliminary evaluation.

Research	Feature set									
	I	II	III	IV	V	VI	VII	VIII	IX	X
Timestamp	√	√	√	√	√	√	√	√	√	√
POI-density	√						√		√	√
POI-entropy						√	√		√	√
Human mobility		√								√
Road network					√			√	√	√
Existing route				√				√	√	√
Population structure			√					√	√	√

and Fig. 6 (for Chicago dataset) demonstrate the normalized RMSE of PF results on *RAR-based* scenario with ten different feature sets for RAR range from 0.4 km to 0.8 km.

Based on preliminary evaluation, we first focused on the Tainan dataset and identified that for a single urban relevant feature, the feature *sets I* and *VI* show better results than the others. This indicates that POI-related features turn out to have much more impact among all features. Furthermore, for static features, location-based ones (*set VII*) outperform transportation-based ones (*set VIII*) for 13% and 10% while utilizing XGBoost and DNN for Regression respectively; however, the result for Linear Regression turns out to be −583% from transportation-based to location-based features in normalized RMSE.

Next, we moved to the Chicago dataset, in general, feature *sets I* and *V* show better results than the others while focusing on a single urban relevant feature. In other words, POI-related features still turn out to have much more impact among all features. Furthermore, for static features, although location-based ones (*set VII*) outperform transportation-based ones (*set VIII*) for 7% to 16% while utilizing DNN for Regression, the results for SVR, Linear Regression and XGBoost turn out to be −3% to −15% from transportation-based to location-based features in normalized RMSE. Besides, we can observe that the light of the performance of *set V* for Chicago is better than Tainan.

We conclude that acting as a public transit focusing on metropolis; the square route structure in Chicago could construct a tighter connection for passengers to transfer between one another (e.g. a single station is shared by multiple routes, as Fig. 4 demonstrates). However, the relationship between existing routes is not useful since the radial routes structure for Tainan is to commute from suburbs or mountain areas to the city centre.

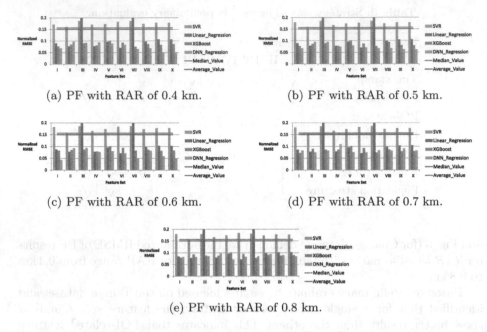

(a) PF with RAR of 0.4 km.

(b) PF with RAR of 0.5 km.

(c) PF with RAR of 0.6 km.

(d) PF with RAR of 0.7 km.

(e) PF with RAR of 0.8 km.

Fig. 5. Normalized RMSE of PF for different methods and feature sets on RAR-based scenario in Tainan dataset.

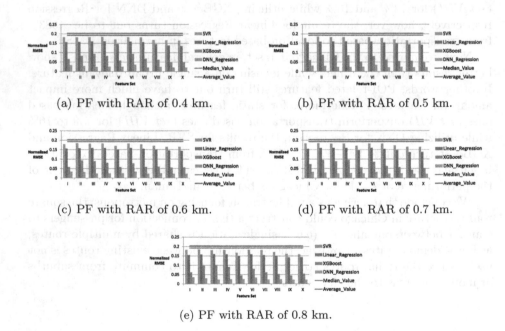

(a) PF with RAR of 0.4 km.

(b) PF with RAR of 0.5 km.

(c) PF with RAR of 0.6 km.

(d) PF with RAR of 0.7 km.

(e) PF with RAR of 0.8 km.

Fig. 6. Normalized RMSE of PF for different methods and feature sets on RAR-based scenario in Chicago dataset.

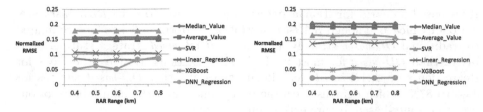

Fig. 7. Normalized RMSE of PF results for different methods with RAR range from 0.4 km to 0.8 km on RAR-based scenario in Tainan (left) and Chicago (right) dataset.

5.4 Evaluation of RAR Setting

For this experiment, we extract features by varying different RAR-ranges, showing how RAR-settings can influence the effectiveness.

Based on feature *set X*, which means to take all the urban relevant features into consideration, Fig. 7 depicts the normalized RMSE of PF results for all methods with RAR range from 0.4 km to 0.8 km (according to the walking tolerance for pedestrians) on *RAR-based* scenario.

We first notice that the performance of DNN for regression varies dramatically considering the RAR range in Tainan dataset. The experimental result shows that there exists a not small increase of RMSE for feature set I between RAR range of 0.6 km and 0.7 km when DNN for regression is utilized. Such phenomenon only takes place in Tainan dataset. We conclude that the reason is, when the RAR range is set too large, the POIs that being considered could not represent the characteristics that route actually covers. Evaluation results show that 0.4 to 0.6 km of RAR range ends up in outperforming other RAR ranges in Tainan dataset.

5.5 Evaluation of Feature Extraction Strategies

Finally, for this experiment, we extract features under *RAR-based* and *OD-based* scenario and treat each feature set as the input data for each method with RAR/Radius of 0.4 km.

We compared the results for RAR-based and OD-based scenario under a RAR/Radius of 0.4 km, which was the best radius as the preliminary evaluation reveals. Table 4 show the normalized RMSE of PF results for different methods, scenarios and feature sets with a RAR/Radius of 0.4 km for both two datasets, where the best performance in each feature set is labelled in bold; while the best performance in each dataset is labelled in red. Although the advantage of the proposed RAR strategy is not significant for each single urban relevant feature (feature *sets I* to *VI*), the DNN for Regression on RAR-based scenario consistently gains the best results when considering multiple heterogeneous features (*sets VII* to *X*).

Accordingly, overall performance results for both datasets under RAR/Range of 0.4 km are shown in Fig. 8, where DNN for Regression gains the best normalized RMSE and outperforms other comparative methods for at least 41% in

RAR-based scenario for Tainan dataset and at least 57% in *RAR-based* scenario for Chicago. Though *RAR-based* strategy performs −6% to 39% compared to traditional *OD-based* strategy; in general, *RAR-based* strategy still performs better than *OD-based* for most method and conditions. Besides, the DNN for Regression with *RAR-based* scenario surpasses all other *OD-based* settings for 33% to 87%, which demonstrates the effectiveness and the adaptability of our proposed route-based feature extraction strategies.

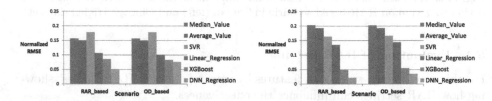

Fig. 8. Normalized RMSE of PF results for different methods and scenarios under a RAR/Radius of 0.4 km in Tainan (left) and Chicago (right) dataset.

Table 4. Strategy sets to be used in preliminary evaluation.

City	Method	Scenario	I	II	III	IV	V	VI	VII	VIII	IX	X
— Experimental settings			Feature set									
Tainan	Median value		0.1564	0.1564	0.1564	0.1564	0.1564	0.1564	0.1564	0.1564	0.1564	0.1564
	Average value		0.1501	0.1501	0.1501	0.1501	0.1501	0.1501	0.1501	0.1501	0.1501	0.1501
	SVR	OD-based	0.1716	0.1660	0.1852	0.1718	0.1738	0.1837	0.1850	0.1735	0.1787	0.1785
		RAR-based	0.1712	0.1610	0.1858	0.1699	0.1738	0.1826	0.1858	0.1735	0.1784	0.1783
	LR	OD-based	0.0880	0.0773	0.1951	**0.0770**	0.0961	0.0705	0.2289	0.0949	0.1006	0.1006
		RAR-based	0.0895	**0.0771**	0.4770	0.0773	0.0962	**0.0703**	0.2458	0.0971	0.1062	0.1063
	XGBoost	OD-based	0.0748	0.0802	0.0909	0.0841	0.1013	0.0887	0.0767	0.0905	0.0844	0.0830
		RAR-based	0.0762	0.0846	0.0903	0.0812	0.1013	0.0931	0.0750	0.0860	0.0826	0.0859
	DNN-Reg	OD-based	**0.0656**	0.0900	**0.0797**	0.1055	0.0922	0.0747	0.0732	0.0828	0.0723	0.0760
		RAR-based	0.0682	0.0972	0.0923	0.0941	**0.0887**	0.0835	**0.0654**	**0.0683**	**0.0514**	0.0509
Chicago	Median value		0.2031	0.2031	0.2031	0.2031	0.2031	0.2031	0.2031	0.2031	0.2031	0.2031
	Average value		0.1927	0.1927	0.1927	0.1927	0.1927	0.1927	0.1927	0.1927	0.1927	0.1927
	SVR	OD-based	0.1803	0.1768	0.1826	0.1836	0.1704	0.1886	0.1816	0.1691	0.1714	0.1663
		RAR-based	0.1823	0.1816	0.1827	0.1837	0.1704	0.1823	0.1813	0.1672	0.1665	0.1641
	LR	OD-based	0.1609	0.1620	0.1625	0.1632	0.1484	0.1631	0.1619	0.1502	0.1446	0.1442
		RAR-based	0.1563	0.1568	0.1591	0.1598	0.1451	0.1589	0.1560	0.1448	0.1380	0.1355
	XGBoost	OD-based	0.0765	0.0858	0.1028	0.0854	0.0583	0.1010	0.0719	0.0642	0.0456	0.0545
		RAR-based	0.0659	0.0831	0.0932	0.0830	0.0583	0.0936	0.0686	0.0622	0.0483	0.0492
	DNN-Reg	OD-based	0.0452	0.0393	0.0625	0.0736	0.0545	0.0911	0.0430	0.0538	0.0355	0.0355
		RAR-based	**0.0338**	**0.0290**	**0.0238**	**0.0254**	**0.0254**	**0.0366**	**0.0212**	**0.0253**	**0.0216**	0.0216

6   Conclusion

This work proposes a route-based approach for potential passenger flow inference in transportation route planning. We define route-affecting region (RAR) to model the influential region of routes and adopt several regression models along with our proposed RAR-based feature extraction strategies. Given heterogeneous features and faced with the competitive and transfer effects of existing routes, our proposed RAR and feature engineering methods are effective for handling dynamic and static data. Experimental results on bus-ticket data of Tainan and Chicago, with radial and square road structures, show the adaptability of our proposed RAR-based approaches compared to traditional OD-based feature extraction strategies. Although RAR-based strategy performs -6% to 39% compared to traditional OD-based strategy based on utilized methods, the DNN for Regression with RAR-based scenario surpasses all other OD-based settings for 33% to 87%.

References

1. Steg, L., Gifford, R.: Sustainable transportation and quality of life. J. Transp. Geogr. **13**(1), 59–69 (2005)
2. Cancela, H., Mauttone, A., Urquhart, M.E.: Mathematical programming formulations for transit network design. Transp. Res. Part B: Methodol. **77**, 17–37 (2015)
3. Mauttone, A., Urquhart, M.E.: A route set construction algorithm for the transit network design problem. Comput. Oper. Res. **36**, 2440–2449 (2009)
4. Pternea, M., Kepaptsoglou, K., Karlaftis, M.G.: Sustainable urban transit network design. Transp. Res. Part A: Policy Pract. **77**, 276–291 (2015)
5. Quadrifoglio, L., Li, X.: A methodology to derive the critical demand density for designing and operating feeder transit services. Transp. Res. Part B: Methodol. **43**(10), 922–935 (2009)
6. Szeto, W.Y., Wu, Y.: A simultaneous bus route design and frequency setting problem for Tin Shui Wai, Hong Kong. Eur. J. Oper. Res. **209**(2), 141–155 (2011)
7. Guihaire, V., Hao, J.-K.: Transit network timetabling and vehicle assignment for regulating authorities. Comput. Ind. Eng. **59**(1), 16–23 (2010)
8. Yan, Y., Meng, Q., Wang, S., Guo, X.: Robust optimization model of schedule design for a fixed bus route. Transp. Res. Part C: Emerg. Technol. **25**, 113–121 (2012)
9. Chien, S.I.-J., Ding, Y., Wei, C.: Dynamic bus arrival time prediction with artificial neural networks. J. Transp. Eng. **128**(5), 429–438 (2002)
10. Lin, Y., Yang, X., Zou, N., Jia, L.: Real-time bus arrival time prediction: case study for Jinan, China. J. Transp. Eng. **139**, 1133–1140 (2013)
11. Cheng, S., Liu, B., Zhai, B.: Bus arrival time prediction model based on APC data. In: the 6th Advanced Forum on Transportation of China (2010). https://doi.org/10.1049/cp.2010.1123
12. Arabghalizi, T., Labrinidis, A.: How full will my next bus be? A framework to predict bus crowding levels. In: Proceedings of the 8th International Workshop on Urban Computing. ACM, Anchorage (2019). https://doi.org/10.13140/RG.2.2.12969.75368
13. Yap, M., Cats, O., Arem, B.: Crowding valuation in urban tram and bus transportation based on smart card data. Transp.: Transp. Sci. (2018). https://doi.org/10.1080/23249935.2018.1537319
14. Wei, Y., Chen, M.-C.: Forecasting the short-term metro passenger flow with empirical mode decomposition and neural networks. Transp. Res. Part C **21**, 148–162 (2012)

15. Chen, Q., Li, C., Guo, W.: Railway passenger volume forecast based on IPSO-BP neural network. In: International Conference on Information Technology and Computer Science (2009). https://doi.org/10.1109/ITCS.2009.187

16. Lin, F., Hsieh, H.-P.: An intelligent and interactive route planning maker for deploying new transportation services. In: The proceedings of the 26th ACM SIGSPATIAL International Conference on Advances in Geographic Information Systems, pp. 620–621 (2018)

17. Mo, Y., Su, Y.: Neural networks based real-time transit passenger volume prediction. In: The 2nd International Conference on Power Electronics and Intelligent Transportation System (PEITS) (2009). https://doi.org/10.1109/PEITS.2009.5406782

18. Nam, K., Schaefer, T.: Forecasting international airline passenger traffic using neural networks. Logist. Transp. Rev. **31**(3), 239 (1995)

19. Sun, Y., Leng, B., Guan, W.: A novel wavelet-SVM short-time passenger flow prediction in Beijing subway system. Neurocomputing **166**, 109–121 (2015)

20. Tsai, T.-H., Lee, C.-K., Wei, C.-H.: Neural network based temporal feature models for short-term railway passenger demand forecasting. Expert Syst. Appl. **36**(2), 3728–3736 (2009)

21. Silman, L.A., Barzily, Z., Passy, U.: Planning the route system for urban buses. Comput. Oper. Res. **1**(2), 201–211 (1974)

22. Yang, H., Zhou, J.: Optimal traffic counting locations for origin-destination matrix estimation. Transp. Res. Part B: Methodol. **32**(2), 109–126 (1998)

23. Peterson, A.: The origin-destination matrix estimation problem - analysis and computations. Doctoral dissertation, Department of Science and Technology. Linkoping University, Sweden (2007). urn:nbn:se:liu:diva-8859

24. Su H.-M., Kuan, C.-C.: Planning and design guidelines. In: Design Manual for Urban Sidewalks, vol. 4, no. 1, pp. 1–4 (2003)

25. Cover, T.M., Thomas, J.A.: Entropy, relative entropy and mutual information. Elem. Inf. Theory **2**(1), 12–13 (1991)

Real-Time Lane Configuration with Coordinated Reinforcement Learning

Udesh Gunarathna[✉], Hairuo Xie, Egemen Tanin, Shanika Karunasekara, and Renata Borovica-Gajic

School of Computing and Information Systems, The University of Melbourne, Melbourne, VIC, Australia
pgunarathna@student.unimelb.edu.au,
{xieh,etanin,karus,renata.borovica}@unimelb.edu.au

Abstract. Changing lane configuration of roads, based on traffic patterns, is a proven solution for improving traffic throughput. Traditional lane-direction configuration solutions assume pre-known traffic patterns, hence are not suitable for real-world applications as they are not able to adapt to changing traffic conditions. We propose a dynamic lane configuration solution for improving traffic flow using a two-layer, multi-agent architecture, named Coordinated Learning-based Lane Allocation (CLLA). At the bottom-layer, a set of reinforcement learning agents find a suitable configuration of lane-directions around individual road intersections. The lane-direction changes proposed by the reinforcement learning agents are then coordinated by the upper level agents to reduce the negative impact of the changes on other parts of the road network. CLLA is the first work that allows city-wide lane configuration while adapting to changing traffic conditions. Our experimental results show that CLLA can reduce the average travel time in congested road networks by 20% compared to an uncoordinated reinforcement learning approach.

Keywords: Reinforcement learning · Spatial database · Graphs

1 Introduction

The goal of traffic optimization is to improve traffic flows in road networks. Traditional solutions normally assume that the structure of road networks is static regardless of how the traffic changes in real-time [6]. A less-common way to optimize traffic is by changing road network configurations at real time. We focus on dynamic lane-direction changes, which can help balance the usage of traffic lanes in many circumstances, e.g. when the traffic lanes in one direction become congested while the traffic lanes in the opposite direction are underused [11,20].

The impact of dynamic lane-direction configurations can be shown in the following example (Fig. 1). In Fig. 1a, there are 4 north-bound lanes and 4 south-bound lanes. Traffic is congested in the north-bound lanes. Figure 1b shows the dramatic change of traffic flow after lane-direction changes are applied, where

© Springer Nature Switzerland AG 2021
Y. Dong et al. (Eds.): ECML PKDD 2020, LNAI 12460, pp. 291–307, 2021.
https://doi.org/10.1007/978-3-030-67667-4_18

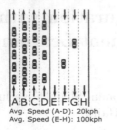
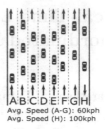

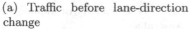

(a) Traffic before lane-direction change

(b) Traffic after lane-direction change

Fig. 1. The impact of lane-direction change on traffic flow. There are 20 vehicles moving in the north-bound direction and 2 vehicles moving in the south-bound direction.

the direction of E, F and G is reversed. The north-bound vehicles are distributed into the additional lanes, resulting in a higher average speed of the vehicles. At the same time, the number of south-bound lanes is reduced to 1. Due to the low number of south-bound vehicles, the average speed of south-bound traffic is not affected. The lane-direction change helps improve the overall traffic efficiency in this case. There is no existing approach for applying such lane-direction changes at the network level at real-time, which can help improve traffic efficiency of a whole city. We aim to scale this to city-wide areas. The emergence of connected autonomous vehicles (CAVs) [14] can make such large-scale dynamic lane-direction changes a common practice in the future. Compared to human-driven vehicles, CAVs are more capable of responding to a given command in a timely manner [4]. CAVs can also provide detailed traffic telemetry data to a central traffic management system in real time, which is important to dynamic traffic optimization.

In order to optimize the flow of the whole network, one needs to consider the impact of possible lane-direction changes on all the other traffic lanes. In many circumstances, one cannot simply allocate more traffic lanes at a road segment for a specific direction when there is more traffic demand in that direction. This is because a lane-direction change at a road segment can affect not only the flow in both directions at the road segment but also the flow at other road segments. Existing solutions for computing lane-direction configurations [4,9,21] do not consider the impact of changes at the network level due the assumption that future traffic dynamics are known beforehand at the beginning of the calculation which is unrealistic for practical applications. More importantly, the computation time can be very high with the existing approaches as they aim to find the optimal configurations based on linear programming, and hence are not suitable for frequent recomputation over large networks.

To address the issues mentioned above: (1) perform in real-time; and (2) having less computational complexity, we propose a multi-agent, scalable, and effective solution, called Coordinated Learning-based Lane Allocation (CLLA), for optimizing lane-directions in dynamic traffic environments. CLLA uses a two-layer architecture. The bottom layer consists of a set of reinforcement learning agents (*RL Agents*) that operate at the intersection level. A RL Agent finds suitable lane-direction changes for the road segments that connect to a specific intersection. The RL Agents use reinforcement learning [17], which helps determine the best changes based on multiple dynamic factors. The RL Agents send the proposed lane-direction changes to the upper layer, which consists of a set of *Coordinating Agents* who evaluate the global impact of the proposed lane-direction changes and decide what changes should be made to the traffic lanes. The decision is sent back to the RL Agents, which will make the changes accordingly. The main contributions of our work are as follows:

- We formalize a lane-direction optimization problem.
- We propose a first-of-its-kind solution, CLLA, for efficient dynamic optimization of lane-directions that uses reinforcement learning to capture dynamic changes in the traffic.
- Our experiments with real-world data shows that CLLA improves travel time by 20% compared to an uncoordinated RL Agent solution.

2   Related Work

2.1   Learning-Based Traffic Optimization

Existing traffic optimization algorithms are commonly based on traffic flow optimization with linear programming [6,7,10]. They are suitable computing optimization solutions if traffic demand and congestion levels are relatively static. When there is a significant change in the network, the solutions normally need to be re-computed from scratch. Due to the high computational complexity of finding an optimal solution, these algorithms are not suitable for highly dynamic traffic environments and not suitable for applications where real-time information are used as an input.

With the rise of reinforcement learning [16], a new generation of traffic optimization algorithms have emerged [13,18,22]. In reinforcement learning, an agent can find the rules to achieve an objective by repeatedly interacting with an environment. The interactive process can be modelled as a finite Markov Decision Process, which requires a set of states S and a set of actions A per state. Given a state s of the environment, the agent takes an action a. As the result of the action, the environment state may change to s' with a reward r. The agent then decides on the next action in order to maximize the reward in the next round. Reinforcement learning-based approaches can suggest the best actions for traffic optimization given a combination of network states, such as the queue size at intersections [1,2]. They have an advantage over linear programming-based approaches, since if trained well, they can optimize traffic in a highly dynamic network. In other words, there is no need to re-train the agent when there is a change in the network. For example, Arel et al. show that a multi-agent system can optimize the timing of adaptive traffic lights based on reinforcement learning [1]. Different to the existing approaches, our solution uses reinforcement learning for optimizing lane-directions which was not considered before.

A common problem with reinforcement learning is that the state space can grow exponentially when the dimensionality of the state space grows linearly. The fast growth of the state space can make reinforcement learning unsuitable for large scale deployments. This problem is known as the *curse of dimensionality* [3]. A common way to mitigate the problem is by using a set of distributed agents that operate at the intersection level. This approach has been used for dynamic traffic signal control [5]. Different to the existing work we use this for dynamic lane-direction configurations.

Coordination of multi-agent reinforcement learning can be achieved through a joint state space or through a coordination graph [8]. Such techniques, however, require agents to be trained on the targeted network. Since our approach uses an implicit mechanism to coordinate (Sect. 4.3), once an agent is trained, it can be used in any road network.

2.2 Lane-Direction Configurations

Research shows that dynamic lane-direction changes can be an effective way to improve traffic efficiency [20]. However, existing approaches for optimizing lane-directions are based on linear programming [4,9,21], which are unsuitable for dynamic traffic environments dues to their high computational complexity. For example, Chu et al. uses linear programming to make lane-allocation plans by considering the schedule of connected autonomous vehicles [4]. Their experiments show that the total travel time can be reduced. However, the computational time grows exponentially when the number of vehicles grows linearly, which can make the approach unsuitable for highly dynamic traffic environments. The high computational costs are also inherent to other approaches [9,21]. Furthermore, all these approaches assume the exact knowledge of traffic demand over the time horizon is known beforehand; this assumption does not hold when traffic demand is stochastic [12]. On the contrary, our proposed approach CLLA is lightweight and can adapt to highly dynamic situations based on reinforcement learning. The reinforcement learning agents can find effective lane-direction changes for individual road intersections even when traffic demand changes dramatically. *To the best of our knowledge, this is the first work for lane-direction allocation by observing real-time traffic information.*

3 Problem Definition

Definition 1. *Road network graph: A road network graph $G_t(V, E)$ is a representation of a road network at time t. Each edge $e \in E$ represents a road segment. Each vertex $v \in V$ represents a start/end point of a road segment.*

Definition 2. *Lane configuration: The lane configuration of an edge e, lc_e, is a tuple with two numbers, each of which is the number of lanes in a specific direction on the edge. The sum of the two numbers is always equal to the total number of lanes on the edge.*

Definition 3. *Dynamic lane configuration: The dynamic lane configuration of an edge e at time t, $lc_e(t)$, is the lane configuration that is used at the time point.*

Definition 4. *Travel cost: The travel cost of a vehicle i that presents at time t, $TC_i(t)$, is the length of the period between t and the time when the vehicle reaches its destination.*

Definition 5. *Total travel cost: The total travel cost of vehicles that present at time t, TTC(t), is the sum of the travel costs of all the vehicles. That is, $TTC(t) = \sum_{(i=1)}^{n} TC_i(t)$, where n is the number of vehicles.*

PROBLEM STATEMENT. Given a set of vehicles at time t and the road network graph $G_{t-1}(V,E)$ from time $t-1$, find the new graph $G_t(V,E)$ by computing dynamic lane configuration ($lc_e(t)$) for all the edges in E such that the total travel cost $TTC(t)$ is minimized.

4 Coordinated Learning-Based Lane Allocation (CLLA)

To solve the optimization problem defined in Sect. 3, we propose Coordinated Learning-based Lane Allocation (CLLA) solution. CLLA uses a two-layer multi-agent architecture, as shown in Fig. 2. The bottom layer consists of a set of RL Agents that are responsible for optimizing the direction of lanes connected to specific intersections. The lane-direction changes that are decided by the RL Agents are aggregated and evaluated by a set of Coordinating Agents at the upper layer, with the aim to resolve conflicts between the RL agents' decisions.

CLLA provides a scalable solution for dynamic lane configuration at the road network level as traffic patterns changes in real-time. CLLA uses reinforcement learning to help optimize lane-direction configurations, which allows optimization in a high variety of real-time traffic conditions. In addition, CLLA achieves coordination between the RL Agents by considering the impact of a potential lane-direction change on different parts of the road network. As detailed later, CLLA only needs to know partial information about vehicle paths in addition to certain real-time traffic conditions, such as intersection queue lengths and lane configuration of road segments, which can be obtained from inductive-loop traffic detectors.

CLLA operates in the following manner. A RL Agent in the bottom layer observes the local traffic condition around a specific intersection. The RL Agents make decisions on lane-direction changes independently. Whenever a RL Agent needs to make a lane-direction change, it sends the proposed change to the Coordinating Agents in the upper layer. The RL Agents also send certain traffic information to the upper layer periodically. The Coordinating Agents evaluate whether a proposed change would be beneficial at the global level based on the received information. The Coordinating Agents may allow or deny a lane-direction change request. It may also decide to make further changes in addition to the proposed changes. After the evaluation, the Coordinating Agents inform the RL Agents of the changes to be made.

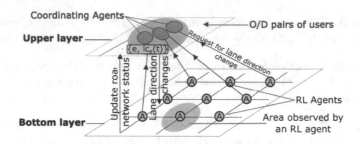

Fig. 2. An overview of the CLLA's architecture

4.1 CLLA Algorithm

Algorithm 1 shows the entire optimization process of CLLA. During one iteration of the algorithm, each RL Agent proposes the lane-direction changes around a specific road intersection using the process detailed in Sect. 4.2. When it is time to evaluate the proposed changes, the system uses the *Global Impact Evaluation* algorithm (Sect. 4.3) to quantify the conflicts between the proposed changes and finds coordinated lane-direction changes (Line 8). The coordinated lane-direction changes are then applied to the road segments (Line 10–11).

4.2 Reinforcement Learning Agent (RL Agent)

In CLLA, the RL Agents use Q-learning technique [19] to find suitable lane-direction changes based on real-time traffic conditions. The Q-learning algorithm aims to find a policy that maps a state to an action. The algorithm relies on an *action value function*, $Q(s, a)$, which computes the quality of a state-action combination. Q-learning tries to find the optimal policy that leads to the maximum action value. Q-learning updates the action-value function using an iterative process as shown in Eq. 1.

$$Q^{new}(s_t, a_t) = (1 - \alpha).Q(s_t, a_t) + \alpha(r_{t+1} + \gamma.max_a Q(s_{t+1}, a)) \tag{1}$$

where s is the current state, a is a specific action, s_{t+1} is the next state as a result of the action, $max_a Q(s_{t+1}, a)$ is the estimated optimal action value in the next state, value r_{t+1} is an observed reward at the next state, α is a learning rate and γ is a discount factor. In CLLA, the states, actions and rewards used by the RL Agents are defined as follows.

States: A RL Agent can work with four types of states as shown below.

- The first state represents the current traffic signal phase at an intersection.
- The second state represents the queue length of incoming vehicles that are going to pass the intersection without turning.
- The third state represents the queue length of incoming vehicles that are going to turn at the intersection.

Algorithm 1: Coordinated Learning Lane Allocation (CLLA)

Input: t_a, time between two coordinating operations
Input: LLC, set of edge-change pairs proposed by the RL Agents
Input: G, Road Network
Input: CLC, set of edge-change pairs given by the Coordinating Agents

1   $t \leftarrow 0$, $t_{step} \leftarrow 0$
2   **while** *True* **do**
3   **foreach** *agent* $\in$ *RL Agents* **do**
4   determine the best lane-direction change for all the edges (road segments) that connect to the vertex $v \in G$ (intersection) controlled by the *agent*
5   **foreach** *edge e that needs a lane-direction change* **do**
6            $LLC.insert(\{e, lc_e(t)\})$
7   **if** $t_a = t_{step}$ **then**
8         $CLC \leftarrow$ **Global Impact Evaluation**(LLC)
9         $LLC \leftarrow \emptyset$, $t_{step} \leftarrow 0$
10   **foreach** $\{e, lc_e(t)\}$ *in* CLC **do**
11   apply the lane-direction change to e
12        $t \leftarrow t + 1$
13     $t_{step} \leftarrow t_{step} + 1$

– The fourth state represents the queue length of outgoing vehicles, i.e., the vehicles that have passed the intersection.

Although it is possible to add other types of states, we find that the combination of the four states can work well because the combination of four states provides; i) information about both incoming and outgoing traffic, ii) from which road to which road vehicles are waiting to move, iii) current traffic signal information.

Actions: We denote the two directions of a road segment as *upstream* and *downstream*. There are three possible actions: increasing the number of upstream lanes by 1, increasing the number of downstream lanes by 1 or keeping the current configuration. When the number of lanes in one direction is increased, the number of lanes in the opposite direction is decreased at the same time. Since a RL Agent controls a specific road intersection, the RL Agent determines the action for each individual road segment connected to the intersection.

We introduced an action restriction mechanism in RL Agents. Changing lane-direction of a road segment takes time as existing vehicles on that road segment should move out before reversing the lane-direction. Therefore, it takes an even longer time to recover from an incorrect lane-direction decision taken by a RL Agent while learning. In order to stabilize the learning, a RL Agent is allowed to take a lane-changing action only when there is a considerable difference between upstream and downstream traffic. The use of this restriction also provides a way to resolve conflicting actions between neighboring RL Agents. When two

RL Agents connected to the same road segment want to increase the number of lanes in different directions, the priority is given to the action, which allocates more lanes to the direction with a higher traffic volume.

Rewards: We define the rewards based on two factors. The first factor is the waiting time of vehicles at an intersection. When the waiting time decreases, there is generally an improvement of traffic efficiency. Hence the rewards should consider the difference between the current waiting time and the updated waiting time of all the vehicles that are approaching the intersection. The second factor is the difference between the length of vehicle queues at different approaches to an intersection. When the queue length of one approaching road is significantly longer than the queue length of another approaching road, there is a higher chance that the traffic becomes congested in the former road. Therefore we need to penalize the actions that increase the difference between the longest queue length and the shortest queue length. The following reward function combines the two factors. A parameter β is used to give weights for the two factors. We normalized the two factors to stabilize the learning process by limiting reward function between 1 to -1. To give equal priority to both factors, we set β to 0.5 in the experiments.

$$
R = (1 - \beta) \times \frac{Current\_wait\_time - Next\_wait\_time}{\max(Next\_wait\_time, Current\_wait\_time)}
$$
$$
- \beta \times \frac{Queue\_length\_difference}{Aggregated\_road\_capacity}
$$

4.3 Coordinating Agent

Given a locally optimized lane-direction change, Coordinating Agents check whether the change can help improve traffic efficiency in surrounding areas based on the predicted traffic demand and the current traffic conditions. If a proposed change is beneficial, it can be actioned. Otherwise, it is not allowed by CLLA.

We first, explain the process of coordinating lane-direction changes using a simple example shown in Fig. 3, where two vehicles are moving from left to right while four other vehicles are moving in the opposite direction. Let us assume that the RL Agent for road segment e_1 proposes to increase the number of lanes from A to B because there is no vehicle in the opposite direction on e1 now. Although such a lane-direction change would help reduce the travel time on e_1, it may conflict with the predicted traffic demand on e_2. The reason is that four vehicles will go through e_2 from right to left (from C to B) but only two vehicles will go through the same road segment from left to right (from B to C). Therefore, the overall traffic demand on e2 will be from right to left (from C to B). However, by increasing the number of lanes from left to right on e_1, the number of lanes in the opposite direction decreases, which is likely to cause a drop of traffic flow speed from B to A. The traffic congestion can eventually propagate to the road segment from C to B. This is not ideal as the overall traffic demand would be from C to B. Consequently, increasing the number of lanes from left to right (from A to B) on e_1 is not beneficial and should not be actioned.

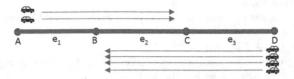

Fig. 3. The vehicles on a road with three road links, e_1, e_2 and e_3. The vehicles will follow the paths shown in arrows.

Due to the dynamic nature of traffic, the Coordinating Agents may not need to consider the full path of vehicles when evaluating the proposed changes based on the predicted traffic demand. This is because the route of vehicles may change dynamically at real time, especially in the era of connected autonomous vehicles when traffic optimization can be performed frequently. Instead of collecting the full path of vehicles, the Coordinating Agents can collect the path within a *lookup distance*. For example, assuming the lookup distance is 200 m, the Coordinating Agents only need to know the road segments that the vehicles will pass within the next 200 m from their current locations.

When there is no conflict between a proposed lane-direction change and the predicted traffic demand, CLLA evaluates the benefit of the proposed change based on the current traffic conditions. Our implementation considers one specific type of traffic condition, the current queue length at road junctions. If a lane-direction change can lead to a lower traffic speed on a road segment, which has a longer queue than the road segment in the opposite direction, the lane-direction change is not allowed. This is because a lower traffic speed can lead to an even longer queue, which can decrease traffic efficiency.

The coordination of lane-direction changes is performed at a certain interval. The time between two coordinating operations is the *assignment interval*, within which the proposed lane-direction changes are actioned, the predicted traffic demand and the current traffic condition are aggregated at the Coordinating Agents.

Global Impact Evaluation Algorithm: The Coordinating Agents use Global Impact Evaluation Algorithm (Algorithm 2) to quantify the conflicts between lane-direction changes. The algorithm takes lane-direction changes that are proposed by the RL Agents as an input (LLC). The input consists of the road and the lane-direction change (lc) proposed by each RL Agent. First, the algorithm finds the neighboring road segments affected by all the changes proposed by the RL Agents (Line 3). For each neighboring road segment, the algorithm finds the predicted traffic flow caused by the proposed lane-direction changes (Line 5). Then the algorithm adds affected neighboring road segments to a queue (Line 7).

Algorithm 2: Global Impact Evaluation (GIE)

Input: LLC, a set of local lane-direction changes (road id, action pair)
           proposed by the RL Agents
Input: t, current time
Output: CLC, a set of (road id, action pair) given by the Coordinating Agents

1   $q \leftarrow \emptyset$; $CLC \leftarrow \emptyset$
2   **foreach** $(r, lc_r(t)) \in LLC$ **do**
3      $roads \leftarrow$ Neighboring road segments affected by the lane-configuration
         $(lc_r(t))$ in r, which are within the lookup distance
4   **foreach** $r_{new} \in roads$ **do**
5   Calculate the predicted traffic flow change in r_{new} due to $lc_r(t)$
6   **if** r_{new} *not in* q **then**
7             $q.add(r_{new})$

8   **foreach** $r_{new} \in q$ **do**
9      $lc_{r_{new}}(t) \leftarrow$ decide the lane-configuration for r_{new} based on predicted traffic
10   **if** $lc_{r_{new}}(t)$ *contains a lane direction change* **then**
11         $CLC.add([r_{new}, lc_{r_{new}}(t)$
12   **if** r_{new} *cannot accommodate predicted traffic flows* **then**
13   mark corresponding change in LLC as a conflict

14 **foreach** $r, lc_r(t) \in LLC$ **do**
15   **if** *no conflicts for* r **then**
16        $CLC.add([r, lc_r(t)])$

In the next step, the algorithm visits each road segment in the queue and determines the appropriate lane-direction configuration ($lc_{r_{new}}(t)$) and the conflicts, where a road segment cannot accommodate the predicted traffic flow (Line 9–13). If a lane-direction change needs to be made, for road segment r_{new}, the road segment is added to coordinated lane changes (CLC) (Line 11). If there is a conflict at road segment r_{new}, corresponding lane-direction change proposed by the RL Agents is marked as a conflict (Line 13).

In the last step, the algorithm adds lane-direction changes proposed by the RL Agents to coordinated lane changes if there is no conflict (Line 14–16).

Complexity of Coordinating Process. Let us use m to denote the number of requests from the RL Agents. The complexity of visiting the relevant road segments is $\mathcal{O}(m \times neb)$ where neb is the number of neighboring road segments that connect to a road segment at a road junction. Since the number of road segments connecting with the same junction is normally a small value, neb can be seen as a constant value with a given lookup distance (l_up). Hence the algorithm complexity can be simplified to $\mathcal{O}(m)$. In the worst case, there is a lane-change request for each road segment of $G(V, E)$, leading to a complexity of $\mathcal{O}(|E|)$.

Distributed Version. Since the execution of Global Impact Evaluation algorithm is independent of the order of requests coming from the RL Agents, requests can be processed in a distributed manner using multiple Coordinating Agents. Every Coordinating Agent traverses first depth neighbors and informs changes to other Coordinating Agents. In such a setting, the complexity of the algorithm is $\mathcal{O}(1)$ with $|E|$ number of Coordinating Agents. In this work, we implemented the centralized version (with one Coordinating Agent); however, when applied to very large road networks, the distributed version can be implemented.

5 Experimental Methodology

We compare the proposed algorithm, CLLA, against three baseline algorithms using traffic simulations. We evaluate the performance of the algorithms using synthetic traffic data and real traffic data. We use SMARTS (Scalable Microscopic Adaptive Road Traffic Simulator) [15], a microscopic simulator capable of changing the travelling directions of lanes, for our experiments.

Datasets. The real traffic data contains the taxi trip records from New York City[1]. The data includes the source, the destination and the start time of the taxi trips in the city. We pick an area of Manhattan for simulation (Fig. 4) because the area contains a larger amount of taxi trip records than other areas. The road network of the simulation areas is loaded from OpenStreetMap[2]. For a specific taxi trip, the source and the destination are mapped to the nearest OpenStreetMap nodes. The shortest path between the source and the destination is calculated. The simulated vehicles follow the shortest paths generated from the taxi trip data.

We also use a synthetic 7×7 grid network to evaluate how our algorithm performs in specific traffic conditions.

We simulate four traffic patterns with the synthetic road network. A traffic pattern refers to generating vehicles to follow a specific path between a source node and a destination node in the road network.

- **Rush hour traffic (RH):** In this setup, traffic is generated so that traffic demand is directionally imbalanced to represent rush hour traffic patterns.
- **Bottleneck traffic (BN):** This setup generates high volume of traffic at the centre of the grid network. This type of traffic patterns create bottleneck links at the center of the network.
- **Mixed traffic (MX):** Mixed traffic contains both **Rush hour traffic** and **Bottleneck traffic** conditions in the same network.
- **Random traffic (RD):** Traffic is generated randomly during regular time intervals. Demand changes over time intervals.

Comparison Baselines. Different to the proposed solution, CLLA, the existing approaches assume future traffic dynamics are known, hence not practical in

---

[1] https://www1.nyc.gov/site/tlc/about/tlc-trip-record-data.page.
[2] https://www.openstreetmap.org.

Fig. 4. The road network of Midtown Manhattan (MM)

real-world applications. Due to the lack of comparable solutions, we define three baseline solutions, which are used to compare against CLLA. In our experiments, the traffic signals use static timing and phasing in all solutions. We conduct comparative tests against the following solutions:

- **No Lane-direction Allocations (no-LA):** This solution does not do any lane-direction change. The traffic is controlled by static traffic signals only.
- **Demand-based Lane Allocations (DLA):** This solution assumes that the full knowledge of estimated traffic demand and associated paths are known at a given time step. DLA computes traffic flow for every edge for both directions by projecting the traffic demand to each associated path. Then it allocates more lanes for a specific direction when the average traffic demand per lane in the direction is higher than the average traffic demand per lane in the opposite direction. Same as CLLA, DLA configures lane-directions at a certain interval, t_a, which is called assignment interval.
- **Local Lane-direction Allocations (LLA):** This solution uses multiple learning agents to decide lane-direction changes. The optimization is performed using the approach described in Sect. 4.2. LLA is similar to CLLA but there is no coordination between the agents.

5.1 Evaluation Metrics

We measure the performance of the solutions based on the following metrics.

Deviation from Free-Flow Travel Time: The free-flow travel time of a vehicle is the shortest possible travel time, achieved when the vehicle travels at the speed limit of the roads without slowing down at traffic lights during its entire trip. Deviation from Free-Flow travel Time ($DFFT$) is defined as in Eq. 2, where t_a is the actual time and t_f is the free-flow travel time. The lowest value of DFFT is 1, which is also the best value that a vehicle can achieve.

$$DFFT = t_a/t_f \tag{2}$$

Average Travel Time: The travel time of a vehicle is the duration that the vehicle spends on travelling from its source to its destination. We compute the

Table 1. Parameter settings

Parameter	Range	Default value
Lookup distance in CLLA	1–7	5
Assignment interval in CLLA/DLA (minutes)	0.5–3	1

average travel time based on all the vehicles that complete their trips during a simulation. A higher average travel time indicates that the traffic is more congested during the simulation. To make the value robust for network size we present results by subtracting free flow travel time from actual travel time. Our proposed solutions aim to reduce the average travel time.

5.2 Parameter Settings

For LLA and CLLA, the learning rate α is 0.001 and the discount factor used by Q-learning is 0.75. The RL agents are pre-trained, based on the traffic at a single intersection before deployed to all the intersections in a road network. For other parameters of the solutions, we use the default values as shown in Table 1.

6 Experimental Results

6.1 Comparative Tests

Average Travel Time: Table 2 shows results with synthetic data. As shown in the results, LLA algorithm performs well in rush hour traffic conditions (**RH**). However, it performs poorly when there are bottleneck traffic links (**BN**). This trend is also observed with DLA. When traffic pattern changes frequently (as in **RD**), DLA is not able to estimate the demand hence perform poorly. In contrast, CLLA algorithm performs well in all traffic conditions.

CLLA algorithm outperforms all other baselines in the Manhattan network, as shown in Table 3. CLLA achieves 5% travel time improvement compared to the next best baseline. In traffic engineering terms, this is a significant improvement. The improvement compared to LLA algorithm is around 20%, which highlights the importance of the coordination between RL Agents.

Table 2. Performance of baselines evaluated using four traffic patterns of the synthetic grid network. **RH, BN, MX, RD** refers to the four synthetic traffic patterns

Baseline	Travel time (s)				% of Vehicles with DFFT>6			
	RH	BN	MX	RD	RH	BN	MX	RD
no-LA	681.08	427.16	506.28	539.89	49.0	4.8	27.7	4.85
LLA	575.59	540.62	561.11	577.6	32.3	24.35	30.5	8.41
DLA	568.02	504.70	493.13	636.51	30.2	16.5	15.5	20.0
CLLA	**568.01**	**428.28**	**449.26**	**523.42**	**32.4**	**5.7**	**14.3**	**3.67**

Table 3. Performance of baselines evaluated using New York taxi data

Baseline	Travel time (s)	% of Vehicles with DFFT > 6
no-LA	604.32	45.9
LLA	585.83	48.6
DLA	496.12	50.7
CLLA	**471.28**	**45.87**

Deviation from Free-Flow Travel Time (DFFT): Table 2 and Table 3 show the percentage of vehicles whose travel time is 6 times or more than their free-flow travel time. The results show that CLLA is able to achieve a lower deviation from the free-flow travel time compared to DLA and LLA.

6.2   Sensitivity Analysis

When the assignment interval t_a of DLA increases, travel time decrease, because it is more likely to get a good estimation of traffic demand when the assignment interval is larger, which can lead to more effective optimizations (Fig. 5a). Different to DLA, the travel time achieved with CLLA grows slowly with the increase of t_a but it is significantly lower than DLA in most cases. The relatively steady performance of CLLA shows that the coordination between lane-direction changes can help mitigate traffic congestion for a certain period of time in the future. If minimizing the average travel time is of priority, one can set t_a to a very low value, e.g.., 0.5 min based on the results.

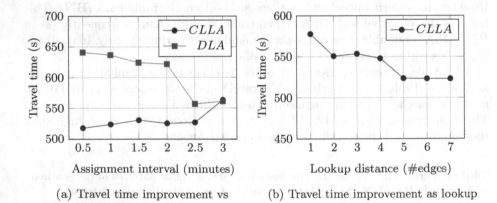

(a) Travel time improvement vs assignment interval

(b) Travel time improvement as lookup distance increases

Fig. 5. Sensitivity analysis with assignment interval and lookup distance

Figure 5b shows that a larger lookup distance can result in a lower average travel time. When the lookup distance increases, CLLA considers more road segments in a vehicle path. This helps identify the conflicting lane-direction changes on the path. Reduction in the average travel time becomes less significant when the lookup distance is higher than 5. This is because the impact of a lane-direction change reduces when the change is further away.

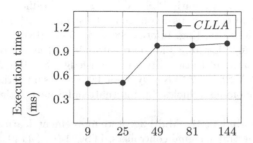

Fig. 6. Execution time for one iteration of GIE algorithm with the road network size. (Lookup distance used − 5)

Figure 6 shows the average execution time of **Global Impact Evaluation** algorithm for one iteration as network size grows. For this test, we build synthetic grid-based road networks. In networks with 9 to 25 nodes, the number of road links on vehicle paths is usually less than the default lookup distance (5). When the number of nodes in a road network is 49, 81 or 144, the number of road links on vehicle paths can be higher than the lookup distance. This is the reason for the increase in execution time when the number of nodes increases from 25 to 49. When the number of nodes is higher than 49, execution is nearly constant, showing that the computation cost does not increase with network size when the lookup distance is fixed.

7   Conclusion

We have shown that effective traffic optimization can be achieved with dynamic lane-direction configurations. Our proposed hierarchical multi-agent solution, CLLA, can help to reduce travel time by combining machine learning and the global coordination of lane-direction changes. The proposed solution adapts to significant changes of traffic demand in a timely manner, making it a viable choice for realizing the potential of connected autonomous vehicles in traffic optimization. Compared to state-of-the-art solutions based on lane-direction configuration, CLLA runs more efficiently, and is scalable to large networks.

An interesting extension would be to incorporate dynamic traffic signals into the optimization process to this work. It would also be interesting to develop solutions that can dynamically change vehicle routes in addition to the lane-direction changes. The dynamic change of speed limit of roads can also be included in an extension to CLLA. Moreover, it is worthwhile to explore how to jointly optimize route allocation and lane directions to improve traffic further.

References

1. Arel, I., Liu, C., Urbanik, T., Kohls, A.: Reinforcement learning-based multi-agent system for network traffic signal control. IET Intel. Transport Syst. **4**(2), 128–135 (2010)
2. Aslani, M., Seipel, S., Mesgari, M.S., Wiering, M.: Traffic signal optimization through discrete and continuous reinforcement learning with robustness analysis in downtown Tehran. Adv. Eng. Inform. **38**, 639–655 (2018)
3. Boutilier, C.: Planning, learning and coordination in multiagent decision processes. In: Theoretical Aspects of Rationality and Knowledge, pp. 195–210 (1996)
4. Chu, K.F., Lam, A.Y.S., Li, V.O.K.: Dynamic lane reversal routing and scheduling for connected autonomous vehicles. In: International Smart Cities Conference, pp. 1–6 (2017)
5. El-Tantawy, S., Abdulhai, B.: Multi-agent reinforcement learning for integrated network of adaptive traffic signal controllers. ITSC **14**(3), 1140–1150 (2012)
6. Fleischer, L., Skutella, M.: Quickest flows over time. SIAM J. Comput. **36**(6), 1600–1630 (2007)
7. Ford, L.R., Fulkerson, D.R.: Constructing maximal dynamic flows from static flows. Oper. Res. **6**(3), 419–433 (1958)
8. Guestrin, C., Lagoudakis, M.G., Parr, R.: Coordinated reinforcement learning. In: International Conference on Machine Learning, pp. 227–234 (2002)
9. Hausknecht, M., Au, T., Stone, P., Fajardo, D., Waller, T.: Dynamic lane reversal in traffic management. In: ITSC, pp. 1929–1934 (2011)
10. Köhler, E., Möhring, R.H., Skutella, M.: Traffic networks and flows over time, pp. 166–196 (2009)
11. Lambert, L., Wolshon, B.: Characterization and comparison of traffic flow on reversible roadways. J. Adv. Transp. **44**(2), 113–122 (2010)
12. Levin, M.W., Boyles, S.D.: A cell transmission model for dynamic lane reversal with autonomous vehicles. Transp. Res. Part C: Emerg. Technol. **68**, 126–143 (2016)
13. Mannion, P., Duggan, J., Howley, E.: An experimental review of reinforcement learning algorithms for adaptive traffic signal control. In: McCluskey, T.L., Kotsialos, A., Müller, J.P., Klügl, F., Rana, O., Schumann, R. (eds.) Autonomic Road Transport Support Systems. AS, pp. 47–66. Springer, Cham (2016). https://doi.org/10.1007/978-3-319-25808-9_4
14. Narla, S.R.: The evolution of connected vehicle technology: from smart drivers to smart cars to... self-driving cars. ITE J. **83**, 22–26 (2013)
15. Ramamohanarao, K., et al.: SMARTS: scalable microscopic adaptive road traffic simulator. ACM Trans. Intell. Syst. Technol. **8**(2), 1–22 (2016)
16. Ravishankar, N.R., Vijayakumar, M.V.: Reinforcement learning algorithms: survey and classification. Indian J. Sci. Technol. **10**(1), 1–8 (2017)
17. Sutton, R.S., Barto, A.G.: Introduction to Reinforcement Learning, vol. 135, 1st edn. MIT Press, Cambridge (1998)
18. Walraven, E., Spaan, M.T., Bakker, B.: Traffic flow optimization: a reinforcement learning approach. Eng. Appl. Artif. Intell. **52**, 203–212 (2016)
19. Watkins, C.J., Dayan, P.: Technical note: Q-learning. Mach. Learn. **8**, 279–292 (1992)

20. Wolshon, B., Lambert, L.: Planning and operational practices for reversible roadways. ITE J. **76**, 38–43 (2006)
21. Wu, J.J., Sun, H.J., Gao, Z.Y., Zhang, H.Z.: Reversible lane-based traffic network optimization with an advanced traveller information system. Eng. Optim. **41**(1), 87–97 (2009)
22. Yau, K.L.A., Qadir, J., Khoo, H.L., Ling, M.H., Komisarczuk, P.: A survey on reinforcement learning models and algorithms for traffic signal control. ACM Comput. Surv. (CSUR) **50**(3), 1–38 (2017)

A Multi-criteria System
for Recommending Taxi Routes
with an Advance Reservation

Jie-Yu Fang[1], Fandel Lin[1], and Hsun-Ping Hsieh[2(✉)]

[1] Institute of Computer and Communication Engineering,
National Cheng Kung University, Tainan, Taiwan
{q36074170,q36084028}@mail.ncku.edu.tw
[2] Department of Electrical Engineering, National Cheng Kung University,
Tainan, Taiwan
hphsieh@mail.ncku.edu.tw

Abstract. As the demand of taxi reservation services has increased, the strategies of how to increase the income of taxi drivers with advanced service have attracted attention. However, the demand is usually unmet due to the imbalance of profit. In this paper, we propose a multi-criteria route recommendation framework that considers real-time spatial-temporal predictions and traffic network information, aiming to optimize a taxi driver's profit when the driver has an advance reservation. Our framework consists of four components. First, we build a grid-based road network graph for modeling traffic network information during the search routes process. Next, we conduct two prediction modules that adopt advanced deep learning techniques to guide a proper search direction in the final planning stage. One module, taxi demand prediction, is used to estimate the pick-up probabilities of passengers in the city. Another one is destination prediction, which can predict the distribution of drop-off probabilities and capture the flow of potential passengers. Finally, we propose our J* (J-star) algorithm, which jointly considers pick-up probabilities, drop-off distribution, road network, distance, and time factors based on the attentive heuristic function. Compared with existing route planning methods, the experimental results on a real-world dataset (NYC taxi datasets) have shown our proposed approach is more effective and robust. Moreover, our designed search scheme in J* can decrease the computing time and make the search process more efficient. To the best of our knowledge, this is the first work that focuses on designing a guiding route, which can increase the income of taxi drivers when they have an advance reservation.

Keywords: Taxi service · Heuristic search · Spatial-temporal predictions · Multi-criteria searching

This work was partially supported by Ministry of Science and Technology (MOST) of Taiwan under grants 108-2221-E-006-142 and 108-2636-E-006-013.

Y. Dong et al. (Eds.): ECML PKDD 2020, LNAI 12460, pp. 308–322, 2021.
https://doi.org/10.1007/978-3-030-67667-4_19

1   Introduction

Taxi service plays an essential role nowadays with the development in modern cities. For example, there are almost ten million requests in Manhattan (New York City) within a month. In such a rapid-paced urban area, the demand of taxi reservation services has increased. However, sometimes taking reservations is less profitable for taxi drivers since they always need to ensure vacancies earlier to avoid missing reservations. This behavior may decrease their time occupying the taxi. It's a common urban policy problem that the supply cannot always meet the demand. To conquer this dilemma, we propose a route recommendation framework to help taxi drivers keep picking up passengers and receive better profit, while also letting drivers successfully arrive at the reservation's location on time.

The key challenge of route recommendation for optimizing taxi drivers' profits with a constraint of advance reservations is multi-criteria consideration. A desirable route should increase occupancy time while preventing taxi drivers from missing the reservation. In previous research on guiding taxi drivers, most of them [7,9,10,14,19,22] only predicted hot spots of taxi demands. Therefore, their solutions are not feasible for the complicated task of this work because there are more spatial and temporal factors that should be considered for a reservation query. On the other hand, in traditional route planning studies [4,6], they are used to be limited as a single-criteria problem (e.g., the shortest path problem). Thus, their works cannot model spatial and temporal correlations effectively. Therefore, under the constraint of reservations, we propose a multi-criteria route recommendation frame-work which takes not only spatial-temporal predictions, but also road network information into consideration. The J* route planning algorithm in the proposed framework is based on a multi-criteria heuristic function and a designed search scheme.

Three primary types of information are considered in J*, including traffic network, the distribution of pick-up probabilities and the distribution of drop-off probabilities. The distribution of pick-up probabilities is predicted to tackle the first factor. Therefore, we exploit a reliable prediction model STDN [20] in our work. STDN is a spatial-temporal neural network for predicting traffic demands in the succeeding time interval, so that it can provide real-time information of predictive taxi demands in the city. The output from this module will be used in J* algorithm for finding passengers precisely. Another prediction model we need is for the distribution of drop-off probabilities. We build a convolutional LSTM model to cope with the second factor, and we adopt it to predict where the passengers may go. Furthermore, our grid-based road network graph can efficiently provide all traffic-related information, including estimated driving distance and estimated travel time between locations. Finally, J* algorithm jointly considers the aforementioned information, and flexibly adopts an attentive heuristic function to generate a desirable route by following the rules of our designed search scheme. Figure 1 is the screenshot of our proposed system interface.

The main contribution of this paper can be summarized as follows:

- We propose a novel framework that intelligently combines two prediction modules, traffic network information and our proposed search algorithm, J* under a reservation constraint. We must emphasize that these two prediction modules can be flexibly replaced by any machine learning models which have high prediction accuracy. The higher the accuracy, the more effective the J* algorithm is.
- We design an attentive heuristic function and a search scheme for J*. These are used to not only find an optimal solution also to decrease searching space by taking advantage of a grid-based road network graph.
- We propose three indicators which evaluate the effectiveness of J*'s generated routes in various aspects. The result shows that our proposed can have better performance than other methods. We also conduct a visualized user study and demo system to show and discuss the effectiveness of our proposed framework comparing others.

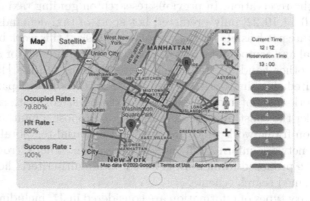

Fig. 1. The screenshot of system interface.

2 Related Work

Taxi Demand Prediction. Several studies have investigated traffic prediction, e.g.. [6,10,19–21]. Among these studies, some of them adopted deep learning approaches to have better performance than traditional machine learning methods. Moreover, some works [6,10,19] incorporate additional static or dynamic features such as meteorological information, time, and textual data with historical traffic data to enhance the predictive ability. In this paper, we claim that our proposed J* algorithm can work well together with the arbitrary methods which can accurately predict taxi demand.

Destination Prediction. Destination prediction recently became popular in location-based services because it is beneficial for urban applications such as recommendation of advertising location, route optimization, and mobile resource arrangement. Two kinds of well-known approaches were used to predict the destination. One is Markov Model based [1,2,6,8,12,16], and the other is machine learning based [5,11,15]. Recurrent Neural Network (RNN) [5,15,23] and Convolutional Neural Network (CNN) [11] are the most popular and effective prediction methods. Most existing approaches predict destination according to existing trip data based on historical trajectories. In this work, we adopt a convolutional LSTM model to predict the destination.

Transportation Route Planning. The work [3] is a bus route planner with multi-criteria, using large-scale taxi GPS traces. For taxi route planning, the work [13] aims to generate the most profitable route for taxi drivers, using the DBSCAN algorithm to gather taxi trajectory data for getting some clusters. However, these works focus on mining trajectory patterns from historical data to generate a general route, it's not practical for the query with reservation. For reservation and dispatch problems, the study [17] proposes a dispatch system which chains reservations as a route with strategies for taxi drivers in Singapore. It focuses on dispatching advance reservations, however, as roadside passengers are the main target for taxi drivers. The aforementioned research is inspiring but seems inadequate to solve our task here. Therefore, we propose a multi-criteria route planning approach, in which details will be discussed in the following sections.

3 Preliminaries

In this paper, we define a gridized map M by splitting a city as n locations ($n = a \times b$) with the resolution of $300\,\text{m} \times 300\,\text{m}$ for each grid, and use $id = \{1, 2, 3, \ldots n\}$ to denote them. In this paper, the term "grid" and "location" are used interchangeably. Furthermore, we map all taxi requests which have their relative pick-up and drop-off coordinates into this gridized map. Hence, all taxi requests can be represented by grids with time labels. We also split the time period (e.g., one day) into m time intervals, where $T = \{1, 2, 3, \ldots m\}$, and each time interval is 30 min.

Definition 1. **Query.** A query in this task consists of start information and reservation information. Start information includes start location g_s and start time t_s; they refer to the current condition of a taxi driver. Reservation information includes a reservation's location g_r and reservation time t_r; they refer a reservation that a taxi driver has. Hence, a query is composed of $<g_s, t_s, g_r, t_r>$.

Definition 2. **Route Information Object (a pick-up or a drop-off object).** A route generated by our method is composed of several predicted pick-up/drop-off pairs O_p, O_d. O_p denotes an object $O_p = \{g_p, t_p\}$ that contains a pick-up location g_p and its predicted pick-up time $t_p, g_p \in M$, and $t_p \in T$, so as O_d denotes an object with drop-off information $O_d = \{g_d, t_d\}$.

Definition 3. **Guiding Route.** A guiding route R is an ordered sequence of route information objects with a start object $O_s(O_s = \{g_s, t_s\})$ and reservation object $O_r(O_r = \{g_r, t_r\})$. Between these start and reservation objects are several recommended pick-up and drop-off objects, i.e., $R = \{O_s, O_{p_1}, O_{d_1}, O_{p_2}, O_{d_2}, \dots, O_r\}$.

Fig. 2. An illustrative example of a recommended route.

Problem Statement. Given the start information O_s of taxi driver and reservation information O_r, our spatial-temporal route planning method aims to generate an effective guiding route $R = \{O_s, O_{p_1}, O_{d_1}, O_{p_2}, O_{d_2}, \dots, O_r\}$ that is determined to pick up more passengers before the reservation time, while arriving at the reservation's location on time. In other words, all (O_p, O_d) pairs indicate that in these locations, we have high probabilities of picking up passengers, while also successfully guiding the entire route to the reserved destination. Thus, the last drop-off location could be close to the reservation's location. Figure 2 is an illustrative example of a recommended route.

4   Methodology

In this section, we provide details of our proposed framework. As shown in Fig. 3, our framework consists of four major parts: (1) Traffic network construction; (2) Pick-up probability prediction; (3) Drop-off probability distribution prediction; (4) Spatial temporal route planner (J* algorithm).

4.1   Traffic Network Construction

Road network plays an important role for route planning, since we can store and associate the information of distance and travel time between grids. Thus, as Fig. 4 shows, we construct a grid-based road network graph by intersecting gridized map with road networks. Those grids which are not associated with the road network will be ignored while searching a route. Therefore, we can estimate the distance and travel time between all grids, as we are able to generate a distance table and a travel time table for generating guiding routes. Furthermore, the grid-based road network graph can save more computing time and space efficiency than the traditional road network, since its numbers of edges and nodes have been reduced. The connected in-formation is remained.

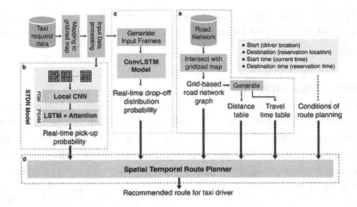

Fig. 3. The architecture of our model.

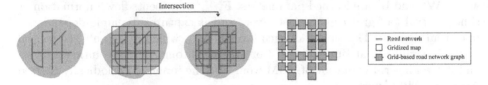

Fig. 4. The process of creating a grid-based network graph. The right-hand side figure shows the connectivity of grids.

Definition 4. **Grid-based Road Network Graph.** Grid-based road network graph is an undirected graph $G = (V, E)$, where V is a vertex set of grids and E is an edge set that denotes connectivity of these grids. A vertex represents a grid. An edge $e_{i,j}$ represents a connected edge between vertex i and vertex j, and the value of the edge is the distance between vertex i and vertex j, which is calculated by a shortest path method (e.g., Dijkstra algorithm).

After construction, we obtain a grid-based graph that reflects the connectivity of locations in the city. Furthermore, we generate distance and travel time for the grid-based road network graph. Distance $d_{i,j}$ represents the shortest length between vertex i and vertex j. $Tr_{i,j}$ represents the estimated travel time between vertex i and vertex j. Among the part of generating travel time table, we adhere to the policy of the target city that the speed limit of vehicles must be under a certain value (e.g., 25 MPH in NYC).

4.2 Pick-Up Probability Prediction

To plan an effective and efficient taxi route before a reservation, an accurate prediction for a passenger pick-up hot spot is necessary. In this work, we utilize the **S**patial-**T**emporal **D**ynamic **N**etwork (**STDN**) model in the work [20], which claims that their STDN model is the state-of-the-art method for predicting real-time traffic. STDN model is a deep learning method for taxi demand prediction, which handles spatial and temporal information via local CNN and LSTM with an attention mechanism. The following sections introduce how we use the STDN model to predict the demand.

4.2.1   Spatial Dynamic Similarity

Flow Gating Mechanism. In the STDN model, Flow Gating Mechanism (FGM) is based on local CNN that aims to capture the dynamic spatial dependency of pick-up volume in each area of the city. Besides tackling the similarity of historical pick-up volume, FGM also takes traffic mobility (including inflow and outflow volume between regions) into consideration. The formulation of each convolutional layer is:

$$\mathbf{Y}_{i,t}^k = ReLU(\mathbf{W}^k * \mathbf{Y}_{i,t}^{k-1} + \mathbf{b}^k) \otimes \sigma(\mathbf{F}_t^{i,k-1}) \tag{1}$$

where $\mathbf{W}^k$ and $\mathbf{b}^k$ are learned parameters, $\mathbf{F}_t^{i,k-1}$ represents flow information at time interval t of grid i in the last layer, and it captures dynamic dependency in spatial domain. The output of each layer is $\mathbf{Y}_{i,t}^k$, which represents the spatial dynamic similarity at time interval t of grid i. Through K convolutional layers and flatten layer, the output of FGM would be the input of Periodically Shifted Attention Mechanism.

4.2.2   Temporal Dynamic Similarity. Periodically Shifted Attention Mechanism

This method considers long-term periodic dependency by modeling relative time interval targets (e.g., the information from the same time of yesterday, and the day before yesterday). Moreover, an attention mechanism was added to address the phenomenon of shifted periodicity. That is, traffic data doesn't follow a consistent pattern. We select $q \in Q$ time intervals from each day to tackle temporal shifting. The following formulation is about temporal information made for each day $p \in P$:

$$\mathbf{h}_{i,t}^{p,q} = LSTM([\mathbf{Y}_{i,t}^{p,q}; \mathbf{e}_{i,t}^{p,q}], \mathbf{h}_{i,t}^{p,q-1}) \tag{2}$$

where $\mathbf{h}_{i,t}^{p,q}$ represents the information of time q in previous day p for the predicted time t in grid i. $\mathbf{e}_{i,t}^{p,q}$ means external features like weather or events. Second, we adopt an attention mechanism in order to make the information of each previous day significantly weighted. In this way, the representation of each previous day $\mathbf{h}_{i,t}^p$ is a weighted sum of each selected time interval q, and the formulation is defined as:

$$\mathbf{h}_{i,t}^p = \Sigma_{q \in Q} \alpha_{i,t}^{p,q} \cdot \mathbf{h}_{i,t}^{p,q} \tag{3}$$

where $\alpha_{i,t}^{p,q}$ represents relative weight for each $\mathbf{h}_{i,t}^{p,q}$, derived by comparing to the short-term memory $\mathbf{h}_{i,t}$. Afterwards, the authors concatenate temporal information as $\mathbf{h}_{i,t}^c$ and feed it to the fully connected layer. The final prediction formulation is:

$$[y_{i,t+1}^s, y_{i,t+1}^e] = tanh(\mathbf{W}_{fa}\mathbf{h}_{i,t}^c + \mathbf{b}_{fa}) \tag{4}$$

where $y_{i,t+1}^s$ and $y_{i,t+1}^e$ represent the predicted value of start and end traffic volume at the next time interval $t+1$ of each grid i respectively. The prediction of $y_{i,t+1}^s$ is the target value that we precisely want to utilize in our work. We normalize the prediction of $y_{i,t+1}^s$ and treat it as the pick-up probability at time slot $t+1$ of each grid i, $i = \{1, 2, 3, \ldots, n\}$, and define it as $A_{i,t+1}$.

4.3 Drop-Off Probability Prediction

In this section, we aim to predict the destinations where those passengers in certain regions (usually referred to as the hot spots we predicted in Sect. 4.2) may go when they were picked up at a certain time slot. The reason is that acquiring the spatial distribution of drop-off probabilities for those demands is helpful for our J* algorithm to avoid having the taxi driver too far from the location of the final reservation.

4.3.1 Urban Drop-Off Prediction Using Convolutional LSTM

To capture the dynamic drop-off distribution of each grid, we use convolutional LSTM (ConvLSTM) [18] to deal with the spatial-temporal sequence problem. We preprocess our original taxi record data as a distribution image the same size as the gridized map. As shown in Fig. 5, one frame $I_{i,t}$ represents a drop-off distribution of grid i at time slot t, and each pixel value represents the drop-off probability transferring from grid i to other grids $j = \{1, 2, 3, \ldots, n\}$. Our approach is to treat this as a spatial-temporal sequence forecasting problem, aiming to predict drop-off probability distribution vector at time $t+1$ for each grid in the city. Figure 6 shows the structure of our ConvLSTM model.

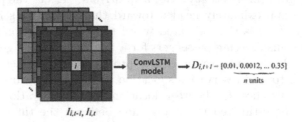

Fig. 5. The schematic flow of predicting drop-off probability distribution.

4.4 Multi-criteria Route Planning (J* Algorithm)

In this section, we introduce our proposed J* algorithm, which intelligently combines the pick-up and the drop-off prediction results together with traffic network information, eventually recommending an effective route efficiently for taxi drivers. We design a search scheme in J* that can reduce unnecessary locations.

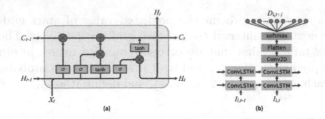

Fig. 6. (a) The structure of a ConvLSTM cell. (b) Our ConvLSTM framework.

Furthermore, similar to the heuristic function of A*, an attentive heuristic function is proposed in our J* algorithm which can enhance the effectiveness and efficiency of route planning. The following sections will introduce the key details of our J* algorithm.

An Attentive Heuristic Function for Guiding Route. To model the relationship between taxi driver's current location and the reservation's location, we need to involve more factors for candidate locations using a multi-criteria heuristic function. The following formulation is our proposed attentive heuristic function:

$$f(i,j) = A_{i,t} \cdot D_{i,t}^{j} \cdot Tr_{i,j}[(1 - \alpha) \cdot rt_{j,r} - \alpha \cdot d_{j,r}] \tag{5}$$

where $A_{i,t}$ represents the pick-up probability of adjacent location i at time interval t, and $D_{i,t}^{j}$ represents the drop-off probability from location i to location j. The benefit of considering $A_{i,t}$ and $D_{i,t}^{j}$ is two-fold. First, drivers can have a high probability to pick up passengers from adjacent locations without unnecessary cruising. Second, as drivers can forecast the destinations where these potential passengers may go, it means that J* can help drivers "select" the potential passengers who are approximately headed toward the reservation's location with high probability. $Tr_{i,j}$ is the estimated travel time from location i to location j. Large $Tr_{i,j}$ means carrying passengers for a long time, which generates more income for taxi drivers in a route. α is a proportional parameter controlling oriented strength toward reservation's location. A higher α means we pay more attention to the distance $d_{j,r}$ between location j and reservation's location r. $rt_{j,r}$ represents the estimated remaining time between the time at location j and the reservation time, in which the estimated arrival time at each location will be updated automatically by travel time table during the search process. The temporal factor, $rt_{j,r}$ and the spatial factor, $d_{j,r}$, are two important values for guiding the search direction properly toward the reservation's location before the searching time is up, to avoid missing the reservation. In this way, $f(i,j)$ represents the guiding value of each candidate location j from adjacent location i connected with current location C. During a search iteration, we select the location with maximum value to extend the route. One of J*'s novel parts is that the heuristic function will dynamically strike a balance between the remaining time and the remaining distance during the search process. If the current time is close to the reservation time, J* will direct the taxi driver toward the reservation's location as soon as possible. On the contrary, if there remains plenty of time, it will mainly focus on seeking locations with high pick-up and drop-off probabilities.

Search Scheme. Our search iteration is divided into two steps, as shown in Fig. 7(a). The first step starts from the current location C, which only considers transferring to its adjacent locations or stay in its current location. The locations without connection in the grid-based graph will be ignored and the searching space pruned. We calculate the pick-up probabilities for the transferable candidate locations. In the second step, J* algorithm considers top u drop-off locations which have high probabilities to transit from these adjacent locations, and then computes their guiding values by the heuristic function $f(i,j)$ respectively (see in Fig. 7(b)). We set drop-off branch factor $u = 10$ in this work. As shown in Fig. 7(b) and (c), we choose the location which has the highest guiding value as the next location to extend branches. J* keeps searching until it reaches the goal (reservation's location) or the search time runs out.

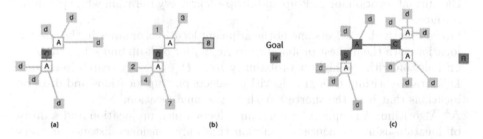

Fig. 7. (a) A search iteration. (b) Calculating values of drop-off locations by heuristic function. (c) After an iteration, extend and update the route.

Route Construction. The route will be initialized as $R = \{O_s, O_r\}$ according to the query. Then, the max-heap iteratively pops out a drop-off location and its relative adjacent location with the highest guiding value $f(i,j)$. The highest pair $\{O_p, O_d\}$ will be selected and added into the guiding route R, where $O_p = \{g_p, t_p\}$, refers to the selected pick-up grid g_p and the timestamp t_p, and $O_d = \{g_d, t_d\}$ refers to the expected drop-off location g_d and the expected drop-off time t_d. After several search iterations, the final recommended route $R = \{O_s, O_{p_1}, O_{d_1}, O_{p_2}, O_{d_2}, \ldots, O_r\}$ is constructed.

5   Experiments

We split the Manhattan area as 13×53 grids, and the size of each grid is 300×300 m. The length of each time slot is 30 min, so we set $m = 48$ in a day.

5.1   Datasets

We evaluate the performance of our framework on real-world taxi trip records from NYC Taxi and Limousine Commission (TLC). The records include pick-up and drop-off dates, times, and GPS locations. In this experiment, we use taxi trip records from 3/1/2015 to 4/30/2015 that contain almost 15 million trips. In this experiment, we evaluate the performance of J* algorithm by our proposed multi-criteria metrics to show its effectiveness and efficiency. Here we choose α value as 0.7 in our attentive heuristic function. According to our experiments, α performs well from 0.6 to 0.9.

5.2   Comparative Methods

To examine the performance and efficiency of our J* algorithm, we consider the following methods for comparison:

– Random Method: It randomly selects one of the adjacent locations or stays in the current location for pick-up and drop-off at every iteration while planning a route.
– Greedy Method: It selects one of the adjacent locations or stays in the current location with the highest probability from $A_{i,t}$ for pick-up and selects a drop-off location with the highest probability from $D_{i,t}$ at every search iteration.
– Dijkstra Algorithm: Dijkstra algorithm selects pick-up locations and drop-off locations that have the shortest path to the final location.
– A* Algorithm: A simple A* algorithm selects a pick-up location and a drop-off location using the heuristic function that only considers distance at every search iteration.
– High-Frequency A* (HF A*): it is a variation from A* that jointly considers the pick-up probabilities of adjacent locations and the distance between current the reservation's location. That is, the heuristic function is a combination of low distance and high pick-up probability.

5.3   Evaluation Metrics

We set up three different evaluation metrics in order to examine the effectiveness of routing methods, including: (1) Occupied rate, (2) Hit rate, and (3) Success rate. First of all, each method will eventually generate a route $R = \{O_s, O_{p_1}, O_{d_1}, O_{p_2}, O_{d_2}, \ldots, O_{p_k}, O_{d_k}, O_r\}$.

Occupied Rate. This evaluation metric aims to evaluate how many passengers that a route can pick up in the same search time for these methods. A higher occupied rate means a route can catch more passengers and lead to a higher profit for a taxi driver. The following is the formulation of the occupied rate:

$$OccupiedRate = \frac{\sum_{i=1}^{k}(t_{d_i} - t_{p_i})}{t_r - t_s} \tag{6}$$

Hit Rate. This evaluation metric verifies whether the recommended route is close enough to real demand. For any pick-up/drop-off pair (O_{p_i}, O_{d_i}) in our recommended route, we then retrieve the real-world requests Q that leave from g_{p_i} at time t_{p_i}. As long as the pair (O_{p_i}, O_{d_i}) exists in Q, it means a hit that the demand we recommend is existed. Assuming that a recommended route contains k pick-up/drop-off pairs, we calculate the ratio of hit counts to k as a hit rate. The formulation is as follows:

$$HitRate = \frac{\Sigma_{i=1}^{k} hit(<O_{p_i}, O_{d_i}> \in Q)}{k} \tag{7}$$

Success Rate. The definition of "success" means a taxi driver picks up the reservation on time. If the ratio of the remaining distance to the remaining time of a route is less than 25 (MPH), it indicates that the taxi driver can arrive at the reservation's location on time under normal speed, meaning it is a "success". A higher success rate means a higher probability that a planning method won't ruin the future order. The formulation is as follows:

$$SuccessRate = \frac{success(\frac{d_{g_{df}, g_r}}{t_r - t_{df}} < 25)}{NumberOfRoutes} \tag{8}$$

Comprehensive Score. Though the aforementioned evaluation metrics are important, each of them cannot fully represent the effectiveness of route planning methods individually. Therefore, we combine these metrics as a score by multiplication; the definition of score is shown below:

$$Score = OccupiedRate \times HitRate \times SuccessRate \tag{9}$$

5.4   Results

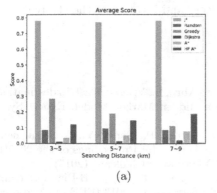
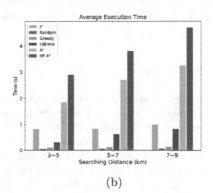

(a)   (b)

Fig. 8. (a) The comprehensive performance. (b) The execution time of all methods.

1) The Comprehensive Performance: As shown in Fig. 8(a), it presents the comprehensive performance of all methods by integrating three evaluation metrics together. The result shows that our proposed method, J*, has the best overall performance. J* takes advantage of taxi demand and destination prediction so that it gains good performance on the hit rate. Furthermore, it always pays attention to both the distance to the reservation's location and remaining time before reservation during searching. Hence, it obtains an excellent occupied rate and meanwhile has a high success rate.

2) Performance on Execution Time: For the route searching problem, the comparison of execution time is particularly important because it represents the practicality of a recommendation system. We set the search time range to 1–3 hours and run 1000 randomly generated testing cases; furthermore, we calculate the average execution time for each method. Figure 8(b) reports the performance of execution time. It shows that the execution time of our J* algorithm is around 1 s. Most significantly, the time doesn't increase a lot with the searching distance. However, the execution time for other shortest-path methods is heavily influenced by the search distance. The result shows that J* algorithm is a stable and practical method for route planning.

6   Conclusions

In this paper, we propose a novel framework including taxi demand prediction, destination prediction, and J* algorithm for taxi drivers to not only earn more profits but also avoid missing the reservation. The proposed multi-criteria route planning method, J*, taking advantage of real-time predictions and traffic network information to generate routes. The evaluation shows that our J* algorithm is a more comprehensive route planner than other methods due to its multi-criteria characteristics; that is, we can intelligently combine multiple factors using the proposed heuristic function. In the future, we plan to integrate more accurate prediction models for inferring pick-up and drop-off probability distributions to improve the performance of J*. Moreover, we plan to extend this work to consider the competition of taxi fleets and investigate the strategies to improve the profit of taxi drivers.

References

1. Alvarez-Garcia, J.A., Ortega, J.A., Gonzalez-Abril, L., Velasco, F.: Trip destination prediction based on past GPS log using a Hidden Markov Model. Expert Syst. Appl. **37**(12), 8166–8171 (2010)
2. Brébisson, A.D., Simon, É., Auvolat, A., Vincent, P., Bengio,Y.: Artificial neural networks applied to taxi destination prediction. In: Proceedings of the 2015th International Conference on ECML PKDD Discovery Challenge (2015)
3. Chen, C., Zhang, D., Zhou, Z.H., Li, N., Atmaca, T., Li, S.: B-Planner: night bus route planning using large-scale taxi GPS traces. In: 2013 IEEE International Conference on Pervasive Computing and Communications (PerCom), pp. 225–233, March 2013

4. Dijkstra, E.W.: A note on two problems in connexion with graphs. Numer. Math. **1**(1), 269–271 (1959)
5. Endo, Y., Nishida, K., Toda, H., Sawada, H.: Predicting destinations from partial trajectories using recurrent neural network. In: Kim, J., Shim, K., Cao, L., Lee, J.-G., Lin, X., Moon, Y.-S. (eds.) PAKDD 2017. LNCS (LNAI), vol. 10234, pp. 160–172. Springer, Cham (2017). https://doi.org/10.1007/978-3-319-57454-7_13
6. Lassoued, Y. Monteil, J., Gu, Y., Russo, G., Shorten, R., Mevissen, M.: A hidden Markov model for route and destination prediction. In: IEEE 20th International Conference on Intelligent Transportation Systems (ITSC) (2017)
7. Luis, M.M., João, G., Michel, F., João, M.-M., Luis, D.: Predicting taxi-passenger demand using streaming data. IEEE Trans. Intell. Transp. Syst. **14**(3), 2013 (2013)
8. Li, X., Li, M., Gong, Y.-J., Zhang, X., Yin, J.: T-DesP: destination prediction based on big trajectory data. IEEE Trans. Intell. Transp. Syst. **17**(8), 2344–2354 (2016)
9. Li, Y., Lu, J., Zhang, L., Zhao, Y.: Taxi booking mobile app order demand prediction based on short-term traffic forecasting. Transp. Res. Rec.: J. Transp. Res. Board **2634**(1), 57–68 (2017)
10. Liao, S., Zhou, L., Di, X., Yuan, B., Xiong, J.: Large-scale short-term urban taxi demand forecasting using deep learning. In: 23rd Asia and South Pacific Design Automation Conference (ASP-DAC) (2018)
11. Lv, J., Li, Q., Sun, Q., Wang, X.: T-CONV: a convolutional neural network for multi-scale taxi trajectory prediction. In: 2018 IEEE International Conference on Big Data and Smart Computing (BigComp) (2018)
12. Manasseh, C., Sengupta, R.: Predicting driver destination using machine learning techniques. In: 16th International IEEE Conference on Intelligent Transportation Systems (2013)
13. Qiu, Y., Xu, X.: RPSBPT: a route planning scheme with best profit for taxi. In: 2018 International Conference on Mobile Ad-Hoc and Sensor Networks, pp. 121–126 (2018)
14. Rodrigues, F., Markou, L., Pereira, F.C.: Combining time-series and textual data for taxi demand prediction in event areas: a deep learning approach. J. Inf. Fusion **49**, 120–129 (2019)
15. Rossi, A., Barlacchi, G., Bianchini, M., Lepri, B.: Modelling taxi drivers' behaviour for the next destination prediction. IEEE Trans. Intell. Transp. Syst. **21**, 2980–2989 (2019)
16. Simmons, R., Browning, B., Zhang, Y., Sadekar, V.: Learning to predict driver route and destination intent. In: 2006 IEEE Intelligent Transportation Systems Conference (2006)
17. Wang, H., Cheu, R.L., Lee, D.H.: Intelligent taxi dispatch system for advance reservations. J. Public Transp. **17**(3), 8 (2014)
18. Xingjian, S.H.I., Chen, Z., Wang, H., Yeung, D.Y., Wong, W.K., Woo, W.C.: Convolutional LSTM network: a machine learning approach for precipitation nowcasting. In: Advances in Neural Information Processing Systems, pp. 802–810 (2015)
19. Xu, J., Rahmatizadeh, R., Bölöni, L., Turgut, D.: Real-time prediction of taxi demand using recurrent neural networks. IEEE Trans. Intell. Transp. Syst. **19**(8), 2018 (2018)
20. Yao, H., Tang, X., Wei, H., Zheng, G., Li, Z.: Revisiting spatial-temporal similarity: A deep learning framework for traffic prediction. In: AAAI Conference on Artificial Intelligence (2019)
21. Yao, H., et al.: Deep multi-view spatial-temporal network for taxi demand prediction. In: AAAI Conference on Artificial Intelligence (2018)

22. Zhang, K., Feng, Z., Chen, S., Huang, K., Wang, G.: A framework for passengers demand prediction and recommendation. In: 2016 IEEE International Conference on Services Computing (SCC) (2016)
23. Zong, F., Tian, Y., He, Y., Tang, J., Lv, J.: Trip destination prediction based on multi-day GPS data. Phys. A: Stat. Mech. Appl. **515**, 258–269 (2019)

Autonomous Driving Validation with Model-Based Dictionary Clustering

Etienne Goffinet[1,2(✉)], Mustapha Lebbah[1], Hanane Azzag[1], and Loic Giraldi[2]

[1] Sorbonne Paris-Nord University, LIPN-UMR 7030,
99 Avenue Jean Baptiste Clément, Villetaneuse, France
etienne.goffinet@lipn.univ-paris13.fr
[2] Groupe Renault SAS, Avenue du Golf, Guyancourt, France

Abstract. Validation of autonomous driving systems remains one of the biggest challenges that car manufacturers must tackle in order to provide safe driverless cars. The complexity of this task stems from several factors: the multiplicity of vehicles, embedded systems, use cases, and the high level of reliability that is required for the driving system to be at least as safe as a human driver. In order to circumvent these issues, large scale simulation that reproduces physical conditions is intensively used to test driverless cars. Therefore, this validation step produces a massive amount of data that needs to be processed. In this paper, we present a new method applied to time-series produced by autonomous driving numerical simulations. It is a dictionary-based method that consists in three steps: automatic segmentation of each time-series, regime dictionary construction, and clustering of produced categorical sequences. We present the time-series specific structure and the proposed method's advantages for processing such data, compared to state-of-the-art reference methods.

Keywords: Autonomous car development · Time series clustering · Mixture models · Dictionary models

1 Introduction

Autonomous car development remains a challenge for car manufacturers. One way to solve this problem is to develop driver assistance systems that are gradually introduced in new car models. This development requires a large amount of data, of good quality, and in large quantities. To provide such data, *Groupe Renault* has made the technical choice to invest in driving simulation technology. This choice led to the development of a dedicated simulation platform that reproduces driving conditions based on car physics, driver behavior, and interaction with a parameterizable environment. This tool allows us to overcome physical simulation limits and to assess an autonomous control law with greater certainty. The simulation process outputs a large amount of information in the form of multivariate time-series. Data size, complexity, and dimensions are considerable: for the validation of the control law, the order of magnitude is $\mathcal{O}(10^6)$

© Springer Nature Switzerland AG 2021
Y. Dong et al. (Eds.): ECML PKDD 2020, LNAI 12460, pp. 323–338, 2021.
https://doi.org/10.1007/978-3-030-67667-4_20

simulations, with $\mathcal{O}(10^3)$ sensors, each recording at $\mathcal{O}(10^4)$ time steps. In total, the validation of a use case requires the production of more than $\mathcal{O}(10^{13})$ data points.

Specific visualization methods are needed to analyze such data. Clustering is a first approach to tackle this problem, which consists in the automatic grouping of "similar" observations into homogeneous groups (clusters). With the help of these tools, the expert has a way to discriminate the time series but also the associated parameters. He can then isolate the effects of the control law parameters and adjust them adequately. Time-series clustering has been widely studied in the past decades. Many dedicated methods have been proposed, each based on specific assumptions on the underlying data structure. These assumptions are crucial as they determine both the clustering results and their interpretability.

In this paper, we present a new method applied to time-series produced by autonomous driving numerical simulations. It is a dictionary-based method that consists of three steps: automatic segmentation of each time-series, regime dictionary construction, and clustering of produced categorical sequences. In this paper's second section, we present the detailed simulation method and the time series structure. In the second part, we discuss the existing approaches and describe our contribution. In the third section, we present the results obtained on public datasets and on an industrial use case: the Autonomous Emergency Braking (AEB) system validation. Finally, we conclude on our method's capabilities and perspectives.

2   Simulating Autonomous Behaviour

Validating an autonomous driving rule is a complicated task, that was for a long time addressed with on-track simulations. The numerical simulation approach allows overcoming the limits of these physical simulations. A large scale simulation reproduces physical conditions is intensively used to test driverless cars. Therefore, this validation step produces a massive amount of time series that needs to be processed.

2.1   Numerical Simulation Assets

Several aspects motivate the use of an autonomous behavior simulation platform. The first motivation is the physical simulation cost, which requires infrastructure, equipment management, and significant human intervention. One digital simulation is estimated 10,000 times cheaper than its physical counterpart. The savings achieved through the use of digital simulation add up to millions of euros. The second motivation comes from the fact that physical simulation is the measurement uncertainty: sensors accuracy, but also initial conditions setting.

Another major disadvantage of physical simulation is the impossibility of producing enough data. A validation objective may be the assessment of vehicle incident odds (e.g. $<10^{-8}$ incidents per hour). With a classical sampling method, estimating such probability would require running prototypes over hundreds of millions of kilometers.

Even if such a large amount of real-life data were available, as is the case in some data science application fields, there would be no guarantees of the data quality or value. In our case, this value lies in the specific driving situation in which to test the control law reaction. These situations are rarely observable in reality, such as the ones of an emergency braking.

2.2 I/O of the Simulation Platform

Assessing a control law reliability requires taking into account every possibility, even the rarest cases. Therefore, validating such system is only feasible with accurate control of each simulation context, operated by a set of parameters divided into five categories:

- Environment parameters: road characteristics, weather conditions, but also driver behavior (cautious or sporty, cooperative or competitive).
- Car physics: weight distribution, engine capacities, etc.
- Sensors to be recorded, including the frequency of observation.
- Control law: triggers reacting to specific conditions (e.g. in the case of emergency braking, the distance to the next car) and with parameterizable effects on the vehicle (e.g. the braking intensity).
- Scenario: a sequence of phases followed by the driver and which puts the car in an experimental context (e.g. reaching a specific speed, then a cruise speed for a specific period).

Several hundreds of parameters, in total, interact to generate simulations and produce time-series. In some use cases, field experts may provide additional labels to help the classification task. However, because of the variety and complexity of the driving situations, drawing up an exhaustive list of the labels is an arduous task. The supervised approach is, therefore, unpracticable.

The scenario is the main factor in time-series construction. Other factors have secondary effects and mainly influence the duration and intensity of the phases (e.g. time to reach cruise speed, braking power, etc.). Therefore, even if several time-series originate from the same scenario, their phases may not be synchronous. Another consequence is that the output time-series differ in length.

2.3 AEB Use Case

In the majority of use cases, the autonomous driving simulation produces a large amount of unlabeled data. To validate our clustering approach, we apply it to the specific AEB use case, in which a ground truth is easily producible. In this situation, the goal is to test the reactions of a car (usually called Ego) equipped with the control law. Ego runs in a straight line towards another vehicle, which moves in the same direction but at a slower pace. We expect the target vehicle detection to trigger the control law, which in turn provokes an emergency braking. The control law objective is to prevent the collision. Three cases can arise:

- The control law is not triggered.
- Target is detected, but braking cannot avoid the collision.
- The target is detected, and braking prevents the collision.

In this illustrating use case, field experts visually assessed the different situations to provide a ground truth. The time series dataset is partitioned in 3 classes according to these labels, depicted in Fig. 1.

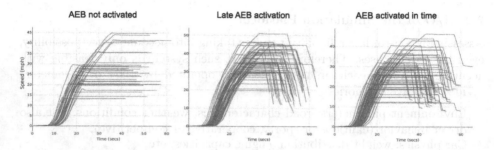

Fig. 1. Time series distribution partitioned by ground truth label.

In order to address this data structure, we developed a clustering workflow independent from the time-series length or regime synchronicity. It relies on the hypothesis of a latent scenario presence.

3   Related Work

Time series Unsupervised classification (or clustering) is a method that aims to partition a dataset into groups of "similar" temporal observations, which is the first step toward understanding its structure. Defining the similarity between times-series is a crucial point as it determines both the clustering results and their interpretations.

3.1   Distance-Based Clustering

The Euclidean distance is one of the most popular for this task. In this case we handle time series as n-size vectors. In practice, this metric is not the most practical as it does not take into account the temporal information and requires aligned series and of equal length.

The Dynamic Time Warping (DTW) [18] measure is another typically relevant metric in the presence of local or uniform temporal scaling (a.k.a. warping). Inspired by the edit distance (used in the context of string comparison), DTW is a measure of the effort required to match two series point-to-point. Although quite resource-intensive originally ($O(n^2)$ complexity), improved version developments over the years have allowed this approach to remain a reference in the domain [11,19].

As presented by [1], choosing a distance is equivalent to determining the invariances to be used for cluster construction. For instance, DTW-based clustering relates to warping invariance. It is, first of all, a hypothesis on the global data structure and a way to cluster shapes. In our case, this approach would not exploit nor conserve the regimes' information. Feature-based clustering makes different assumptions.

3.2 Feature-Based Clustering

The feature-based approach is about designing a way to transform time series into condensed representations. The hypothesis is that this transformation keeps the informative aspects of the data. Two situations can be distinguished: the first when the transformation process is known, the second when it is estimated based on an external criterion (risk, measure, or model assumptions). Either way, this method requires prior knowledge on the time series. The dictionary-based methods family, as proposed by [15] and [21], are based on feature extraction by uniform time step segmentation and are representative of the first case. This approach was first appealing in our application as it allows, to a certain extent, the comparison of similar segments between time series. However, it requires setting arbitrary parameters (including, but not limited to, the dictionary size and the uniform segmentation time step), which is not possible in an unsupervised context. The Time-series Forest method [7] illustrates perfectly the other situation, where feature extraction relies on the supervision of a score based on Entropy and distance. Deep Learning can also be used in this context, as in [13] where the extraction is based on the reconstruction error.

In this application, we make full use of the hidden scenario hypothesis and apply a specific case of feature-based clustering method: a regime-changing time series approach.

3.3 Regime-Changing Time Series Clustering

AEB use case time series are the result of the chaining of distinct phases, also known as regimes. Provided the ability to detect those regimes, it is possible to use their estimated distribution (order, frequency, amplitude...) to characterize the observations and discriminate them. During the last decades, several papers have been proposed to detect optimal regime change points. Those methods sum up to piece-wise polynomial regression models. The common strategy relies on optimizing an approximation error in different ways: sliding windows of increasing size as in [10] and [8], by dynamic programming as in [12], Hidden Markov models in [9] or by regression mixture models in [3]. We selected this last model for two reasons: on the one hand, the benefits of using a mixture model (confidence intervals, model selection strategy...) and on the other hand, the particular performances of this model compared to hidden Markov model approaches and its computational efficiency compared to dynamic programming methods [3, 4].

In *mixRHLP* from [20], the same author combines the piece-wise regression model in a finite mixture to construct a one-step model-based clustering method. The proposed approach aims at regrouping time series with common regimes cut-points. *mixRHLP* also assumes that the number of regimes is known. These assumptions are not the case in our approach.

Our contribution is an attempt to adapt *mixRHLP* to our constraints. It consists of a three-steps workflow with the addition of an original strategy of segmentation model selection. In the first step, we apply Individual time-series segmentation with a polynomial regression mixture. In the second step we build a standard dictionary of regimes by clustering the extracted segments. The clustering of these sequences using Levenshtein distance in categorical sequence space produces the final result. Our method, called *SDLHC* for Segmentation, Dictionary construction, Levenshtein Hierarchical Clustering, has the following advantages:

- Clustering based on regime detection is intuitive and easily interpretable by experts.
- The method can be applied to a dataset of time-series with unequal lengths. Moreover, it is independent of the time-series synchronicity and the regime's moment of appearance synchronicity.
- The segmentation phase can be applied independently on each time-series, which makes the computation an embarrassingly parallel task. This step drastically reduces the data dimension.
- Our segmentation strategy optimizes automatically both the number of segments and polynomial regression on each segment, which allows to get rid of assumptions on the number of regimes and on their optimal order of polynomial regression.

4   A Three-Step Time-Series Clustering Algorithm (SDLHC)

The method *SDLHC* is composed of three steps: segmentation, dictionary construction, and categorical sequence clustering. The first two steps are addressed with the mixture model approach.

4.1   Segmenting Time-Series with a Mixture of Polynomial Regressions

The Regression with Hidden Logistic Process from [3] is based on a polynomial regression model mixture, with time-dependent proportions following a hidden logistic process. Given a time-series $x = (x_t)_T$ and $\phi = (\phi_s(t) = t^s)_{s \in 0, \ldots, S}$ a polynomial basis of size $S \in \mathbb{N}$ (e.g. monomial basis, Legendre basis, Fourier basis, etc.). A Polynomial Regression Model (PRM) of sequence x in the basis ϕ is defined by

$$\tilde{x} = \sum_{s=1}^{S} \beta_s \phi_s(t) + \sigma^2 \epsilon,$$

with $(\beta_s)_{s\in 1,\ldots,S} \in \mathbb{R}^S$, $\sigma \in \mathbb{R}_*^+$ and $\epsilon \sim \mathcal{N}(0,1)$. These PRMs are the segmentation mixture model components.

Given a number of clusters $K \in \mathbb{N}$, let $z = (z_t)_{t\in T}$ be the elements $x = (x_t)_{t\in T}$ cluster membership. At a given time t, z_t follows a Multinomial distribution with parameters $\pi(t) = (\pi_k(t))_{k\in 1,\ldots,K}$. The distribution of x at time t is defined by

$$p(x_t) = \sum_{k=1}^K \pi_k(t) f_{\theta_k}(x_t),$$

and the sequence x log-likelihood,

$$l(x;\theta) = \sum_{t=1}^T log\left(\sum_{k=1}^K \pi_k(t) f_{\theta_k}(x_t)\right), \qquad (1)$$

with $f_{\theta_k}(x_t)$ the density associated to a PRM component. The varying proportions $\pi_k(t)$ can be seen as the parameters of a Multinomial distribution followed by the clusters memberships at a given time t. These proportions vary according to a logistic process. More formally, for $k \in \{1,\ldots,K\}$ and $t \in T$,

$$\pi_k(t) = p(z_t = k) = \frac{\exp(\sum_{s=1}^S w_{k,s}\phi_s(t))}{\sum_{h=1}^K \exp(\sum_{s=1}^S w_{h,s}\phi_s(t))}, \qquad (2)$$

with $w_k = (w_{k,s})_{s\in 1,\ldots,S}$ the associated model parameters. In the following paragraphs, we denote by w the set of parameters $(w_k)_{k\in 1,\ldots,K}$. The complete set of parameters is finally $\theta = (w, \beta, \sigma)$. The log-likelihood (1) optimization requires a specific version of the Expectation Maximization (EM) algorithm described in [6]. The EM algorithm is a standard algorithm for likelihood maximization in the presence of incomplete data. In our case, these missing data are the cluster's membership, denoted by z (the hidden variable). It is an iterative algorithm, each iteration composed of two steps.

Expectation Step (E): Given the parameters θ, the first step of the EM algorithm consists in optimizing the complete log-likelihood defined as below:

$$\mathbb{E}_{x,\theta}\left[l(x,z;\theta)\right] = \mathbb{E}_{x,\theta}\left[\sum_{i=1}^n \sum_{k=1}^K \mathbb{I}_{z_i=k} log\left(p(x_i, z_i = k; \theta)\right)\right]$$

$$= \sum_{i=1}^n \sum_{k=1}^K \tau_{i,k} log\left(\pi_k f_{\theta_k}(x_i)\right). \qquad (3)$$

The development of the Eq. (3) shows that this step is simplified to the estimation of $\tau_{i,k} = p(z_i = k|x_i; \theta)$, the posterior distribution of z conditionally to x. The Bayes theorem gives the following estimation of this quantity:

$$\tau_{i,k} = p(z_i = k|x_i; \theta) = \frac{p(z_i = k, x_i; \theta)}{p(x_i)}$$

$$= \frac{\pi_k f_{\theta_k}(x_i)}{\sum_{h=1}^K \pi_h f_{\theta_h}(x_i)}. \qquad (4)$$

Maximization Step (M): At each iteration, the model parameters are updated during the Maximization step. In this phase, the following decomposition of the complete log-likelihood expectation is maximized:

$$\mathbb{E}_{x,\theta}\left[l(x,z;\theta)\right] = \sum_{t=1}^{T}\sum_{k=1}^{K}\tau_{t,k}log\left(\pi_k f_{\theta_k,t}(x_t)\right)$$

$$= \sum_{t=1}^{T}\sum_{k=1}^{K}\tau_{t,k}log\pi_k + \sum_{t=1}^{T}\sum_{k=1}^{K}\tau_{t,k}log f_{\theta_k,t}(x_t)$$

$$= Q_1(\pi) + Q_2((\theta_k)_{k\in\{1,...,K\}}).$$

with $\tau_{t,k} = p(z_t = k|x_t,\theta)$ the membership posterior distribution estimated in (3) during the expectation step, and $f_{\theta_k,t}$ the density associated to cluster k regression model at time t. This optimization can therefore be achieved by the separate maximization of Q_1 and Q_2. The optimization of Q_2 with respect to the parameters $\theta_k = (\beta_k, \sigma_k)$ provides the following expressions:

$$\tilde{\beta}_k = arg\min_{\beta_k}\sum_{t=1}^{T}\tau_{t,k}(x_t - \sum_{r=1}^{R}\beta_k\phi_r(t))^2, \tag{5}$$

$$\tilde{\sigma}_k^2 = \frac{1}{\sum_{t=1}^{T}\tau_{t,k}}\sum_{t=1}^{T}\tau_{t,k}(x_t - \tilde{\mu}_k(t))^2, \tag{6}$$

with $\tilde{\mu}_k(t) = \sum_{s=1}^{S}\tilde{\beta}_{k,s}\phi_s(t)$ the estimated value of x_t by the regression model of cluster k.

4.2 Adaptive Model Selection Strategy

In the initial model [20], the regression polynomial basis is common to every component, while in our contribution each regression order is specific. Moreover, we do not make a priori assumptions on the segment's number, which is also estimated by our strategy. To estimate both the segment's number and the polynomial regression order on each segment, we combine this model with an innovative top-down strategy. The strategy is iterative and consists, at each step, in identifying the 'worst' component, in terms of the partial likelihood defined as:

$$l_k(x;\theta) = \frac{1}{\sum_{t=1}^{T}\pi_{t,k}}\sum_{t=1}^{T}\pi_{t,k}log\left(f_{\theta_k}(x_t)\right), k \in 1,..,K.$$

This criterion quantifies a component representation quality weighted by the conditional membership probabilities. By improving the component $k_{old} \in \{1,\ldots,K\}$ that minimizes this score, two candidate models are created and compared. Splitting k_{old} in two sub-components, while conserving the other components, produce the first candidate model. We denote by k_1 and k_2 these new clusters. We denote t_m the weighted median of the sequence $\{1, \ldots, T\}$ with weights $\pi_{k_{old}}$, and consider this time as the optimal cut-point for splitting the component $\pi_{k_{old}}$. The observations membership probabilities associated are based on the former component membership probabilities. The component k_1 membership probabilities are defined as follows:

$$\pi_{k_1} = \begin{cases} \pi_{t,k_{old}} & , t \in \{1, \ldots, t_m\} \\ \epsilon & , t \in \{t_m + 1, \ldots, T\} \end{cases}, \qquad (7)$$

with ϵ the threshold precision. The new cluster k_2 membership probabilities are obtained likewise, with inverted time indices. A regularization of the $(\pi_k)_K$ is necessary at this point to enforce the constraint $\sum_{k=1} \pi_{k,t} = 1, \forall t \in \{1, \ldots, T\}$. Increasing the order of k_{old} component regression model by one produces the second candidate. Two runs of EM are then launched, each of them considering one of the candidates as the initial state. After the convergence of both EM, the candidate optimizing the Bayesian Information criterion [22] is selected for the next iteration. This criterion is a score penalizing the likelihood of the model by its complexity. Given a model M with parameters Θ of size C, a set of observation x of size n, the BIC is defined as follows:

$$BIC(X, M, \Theta) = C \ln(n) - 2 \ln(L(X, M, \Theta)). \qquad (8)$$

This strategy is summarized in Algorithm 1.

Algorithm 1: Top down segmentation strategy.

Fix the convergence threshold $c > 0$
Choose an initial state for the first EM run:
$\theta_{old}^{init} := ((w_k, \beta_k, \sigma_k^2)_{k \in \{1, \ldots, K\}})_{old}^{init}$
Compute π_{old}^{init} using Eq. (2)
Estimate θ_{old}^{end} by applying the EM algorithm
while *relative increment in BIC $> c$* **do**
    Construct the first candidate model θ_{addSeg}^{init} with Eq. (7)
    Estimate θ_{addSeg}^{end} by applying the EM algorithm
    Construct the second candidate model θ_{incDeg}^{init} by increasing the least efficient component of the former mixture by one.
    Estimate θ_{incDeg}^{end} by applying the EM algorithm
    $\theta_{old}^{end} = \arg\max_{\theta \in \{\theta_{addSeg}^{end}, \theta_{incDeg}^{end}\}} BIC(\theta)$
end

After convergence of BIC criterion, we estimate the moments of regime change by choosing the maximum of membership probabilities. In Fig. 2, we show the result of segmentation over a few time series from our use case AEB. This segmentation method is applied individually to each time-series and transforms each one in a set of sub-sequences. A segmentation result of an AEB time series is shown in Fig. 2.

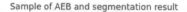
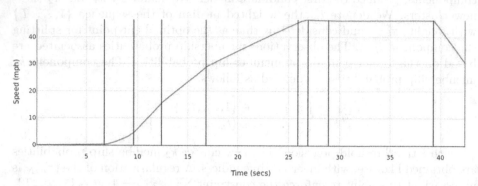

Fig. 2. Segmentation result sample.

4.3   Dictionary Construction

Expressing the extracted segment in a common basis is mandatory to compare and cluster the sequences. This common basis, or dictionary, is constructed with clustering algorithm applied to the dataset composed of all segments from time series. The objective is to encode the original time-series in the new dictionary, as represented in Fig. 3. The sub-segments are first scaled, expressed on a common support, and regressed in a polynomial regression basis. Other informative descriptors can be added depending on the case, as the regime's duration, offset, or variance. These features are then clustered with a Gaussian mixture model (GMM) to produce the dictionary. In Sect. 4.1, we mentioned an implicit assumption based on the segmentation polynomial basis. In this section we make the additional implicit assumption that the GMM is adapted to the regimes density estimation and makes sense from the field expert point of view.

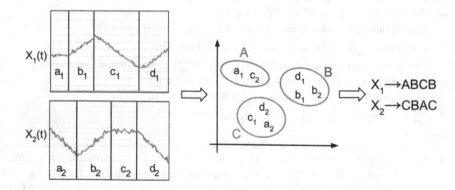

Fig. 3. *SDLHC*: from time-series to categorical sequences.

The EM algorithm is initialized with the K-means++ algorithm, which is a standard approach [2]. At the end of this step, the modes of the Gaussian mixture components are the reference regimes, entitled "patterns" in the following, with which to recode the original time series. The dictionary size is determined by the field experts, assisted by the BIC.

An example of a dictionary with five patterns is shown in Fig. 4. Using this dictionary, Fig. 5 shows the encoded sequence. Two stationary patterns can be recognized (b and c), corresponding to cruise speed phases, as well as two accelerating (d and e) and one decelerating (a).

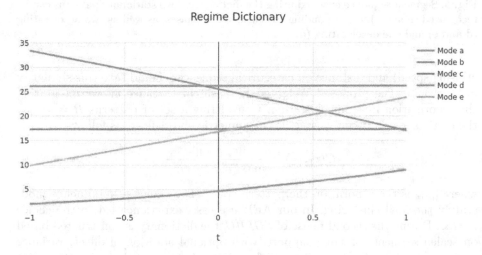

Fig. 4. Dictionary produced in the AEB use case.

After this re-coding phase, data dimension is greatly reduced: for a time-series of size n, the dimension goes from $\mathbb{R}^n$ to D^k, with D the categorical space and k the number of regimes composing the sequences. Clustering these sequences is *SDLHC* third step subject.

4.4 Categorical Sequences Clustering

We use the Levenshtein distance [14] combined with Ward's hierarchical clustering method to obtain the final clusters. Levenshtein distance between two categorical sequences a and b (of size s_a and s_b) is defined as the minimum number of operations (insertion, deletion, substitution) needed to transform a into b. In this categorical space, Levenshtein Distance complexity is $O(s_a \times s_b)$. In the original Levenshtein distance, replacing a symbol with another has a fixed unit cost, independently from the target and replacement symbols. Therefore, it does not account for the fact that the two patterns may be more or less close. Figure 4 shows that some speed patterns are similar (different phases of acceleration of

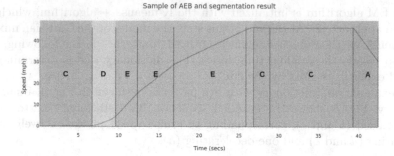

Fig. 5. Segment sequence encoded using the dictionary. Two stationary patterns can be recognized (b and c), corresponding to cruise speed phases, as well as two accelerating (d and e) and one decelerating (a).

cruising speed) and distance on categorical sequences should take this similarity into account. The proposed Weighted Levenshtein Distance allows integrating this information into our clustering. Considering a set of patterns $R = \{r_s\}_S$, the edition cost between $r_a, r_b \in R$ is symmetric and defined as follows:

$$C(r_1, r_2) = \frac{||r_1 - r_2||_p}{\max_{s,t} ||r_s - r_t||_p},$$ (9)

where $||.||_p$ is the p norm on the pattern space. The choice of p influence moderately the final clustering. In our AEB use case, experience led to the choice $p = \infty$. During the second phase of *SDLHC*, the dictionary is constructed based on scaled segments of same support, with optional addition of offset, variance or phase duration. During this part of categorical sequence clustering, the same features can be integrated to the edit operation cost computation.

Once the weighted Levenshtein Distance Matrix computed, Ward's hierarchical clustering method is applied to produce the final clusters. We compare the results with those of other state-of-the-art methods to prove the method capacity to produce a clustering with good performance.

5   Experiments

We present, in this section, the results of several experiments on public datasets and on a real-world use case AEB obtained from Renault's simulation system. The method described in this article was implemented in Scala for the segmentation step and R for the hierarchical step. Code and (public) datasets available at https://github.com/sdlhc-01/SDLHC.

The following baseline methods are selected:

- Three methods based on classical measures (Euclidean distance and DTW) associated with Partitional Around Medoid (PAM) clustering approach. We have also tested the combination of DTW with center construction using the popular Dynamic Barycenter Averaging (DBA) method [17].

- The $K - Shape$ method [16], a partitional clustering using the shape-based distance based on the cross-correlation measure.
- The SAX method, a dictionary-based methods from [15] that builds representations of the time series based on uniform time step segmentation. Based on this representation and associated distance, hierarchical clustering with Ward's criterion produce the clusters.
- In order to compare to the original method we aimed to extend, the results of $mixRHLP$ are also reproduced here.

Whenever baseline methods require it, we interpolate time-series to equal-length sequences. We used the R package $TSclust$'s distance-based and SAX methods implementations and $mixRHLP$ using $flamingos$ R package. Some of these methods depend on parameters, usually estimated by optimizing a risk in a supervised framework. The comparison is based on the Adjusted Rand Index (ARI), a popular score in the clustering validation context. This criterion represents the proportion of correctly grouped and separated observations with respect to the observed classes. The ARIs obtained here are always the maximal ARI obtained when testing the method on a parameter grid, displayed in Table 1, reproducing the results that experts can obtain after fine-tuning.

Table 1. Parameters grid for ARI evaluation.

Method	Parameters	Range
SAX	Number of segments	(5, 10, 20, 30, 40, 50)
	Number of gaussian bins	(2, 3, 5, 7, 10, 20, 30, 40, 50)
SDLHC	Dictionary size	(2, ..., 12)
MIXRHLP	Number of segments	(1, ..., 10)
	Polynomial regression order	(1, ..., 3)

5.1 Public Datasets Results

In order to validate $SDLHC$ adequation to the regime-changing time series clustering problematic, we selected a subset of the UCR archive [5] whose data exhibit regime structure. The ARI score obtained are shown in Fig. 2. Although performant when applied to Renault's dataset (c.f. next subsection), we found that the weighted Levenshtein hierarchical clustering requires fine-tuning to adapt to the considered data characteristics. The test ran in this section therefore use the non-weighted Levenshtein distance. The results confirm that the method perform well when addressing regime-changing time series. In these tests, the considered datasets contain equal-length time-series. However, $SDLHC$ can also be applied, without data preprocessing, to unequal-length time-series, which is the case in our application.

Table 2. Adjusted Rand Index on the UCR archive datasets.

Name	L2.PAM	DTW.PAM	DTW.DBA	K-SHAPE	SAX	MIXRHLP	SDLHC
CBF	0.28	0.66	0.68	0.63	0.46	0.47	0.71
OliveOil	0.46	0.53	0.40	0.50	0.00	0.40	0.55
Trace	0.32	0.40	0.66	0.57	0.32	0.41	0.94

5.2   Real Dataset Results

In the following section, we evaluate the clustering performance of *SDLHC* on an industrial use case: the Autonomous Emergency Braking (AEB) system validation. In this case, a ground truth is available, and it is possible to compare clustering methods based on the similarity between the observed labels and the produced clusters. The clustering methods performances are, as in the previous section, measured by the ARI score. Renault's dataset is composed of 150 time series, with a duration varying from 13 to 52 s and length varying from 415 to 573 data points. The scores are obtained in the same conditions than the previous tests on public datasets, displayed in Fig. 6. Two versions of *SDLHC* are tested: *SDLHC − LEV* and *SDLHC − WLEV* corresponding to the use of the standard and weighted Levenshtein distance in *SDLHC*'s last step. ARI criterion confirms that the *SDLHC − WLEV* method slightly improves the score obtained by *SDLHC − LEV*. Among the distance-based methods, the *K − Shapes* method is the best performer without, however, reaching the ARI threshold of 0.45 regardless of the number of clusters. With high cluster numbers, *SAX* method nearly reaches the performance of *SDLHC − LEV*. This seems logical given the proximity between the proposed workflow and the dictionary-based methods.

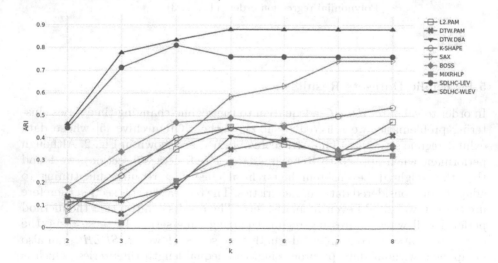

Fig. 6. ARIs scores of various clustering approaches as a function of the number of clusters.

6   Conclusions

In the context of unsupervised classification of regime-changing time-series, we propose a dictionary-based method that consists in three steps: automatic segmentation of each time-series, regime dictionary construction, and clustering of produced categorical sequences. *SDLHC* shows good results when applied to time-series complying to the regime construction assumption, and is competitive with other state-of-the-art methods in this case. The ability to address unequal-length time-series, a-synchronized time-series, and time-series exhibiting asynchronous regimes are its best assets. The current assumptions on the polynomial regression basis for segmentation are adapted to experimental cases, but may not be suited to other physics-oriented use cases. In these circumstances, the Fourier polynomial basis may be another candidate to fit regimes and time-series. In this case, it is possible to re-interpolate the Fourier coefficient to compare regimes on a common basis, and even regimes from different sources, leading to the possibility of multivariate clustering. In this context, our current investigations are focusing on model selection and reduction through co-clustering.

Acknowledgements. The authors want to thank Groupe Renault and the ANRT (french National Association for Research and Technology) for their financial support via the CIFRE convention n° 2018/1114.

References

1. Batista, G.E., Wang, X., Keogh, E.J.: A complexity-invariant distance measure for time series. In: Proceedings of the 2011 SIAM International Conference on Data Mining, pp. 699–710. SIAM (2011)
2. Blömer, J., Bujna, K.: Simple methods for initializing the EM algorithm for Gaussian mixture models. CoRR (2013)
3. Chamroukhi, F., Samé, A., Govaert, G., Aknin, P.: Time series modeling by a regression approach based on a latent process. Neural Netw. **22**(5–6), 593–602 (2009)
4. Chamroukhi, F., Samé, A., Govaert, G., Aknin, P.: A hidden process regression model for functional data description. Application to curve discrimination. Neurocomputing **73**(7–9), 1210–1221 (2010)
5. Dau, H.A., et al.: The UCR time series classification archive, October 2018. www.cs.ucr.edu/~eamonn/time_series_data_2018/
6. Dempster, A.P., Laird, N.M., Rubin, D.B.: Maximum likelihood from incomplete data via the EM algorithm. J. Roy. Stat. Soc.: Ser. B (Methodol.) **39**(1), 1–22 (1977)
7. Deng, H., Runger, G., Tuv, E., Vladimir, M.: A time series forest for classification and feature extraction. Inf. Sci. **239**, 142–153 (2013)
8. Fuchs, E., Gruber, T., Nitschke, J., Sick, B.: Online segmentation of time series based on polynomial least-squares approximations. IEEE Trans. Pattern Anal. Mach. Intell. **32**(12), 2232–2245 (2010)
9. Kehagias, A.: A hidden Markov model segmentation procedure for hydrological and environmental time series. Stoch. Env. Res. Risk Assess. **18**(2), 117–130 (2004)

10. Keogh, E., Chu, S., Hart, D., Pazzani, M.: Segmenting time series: a survey and novel approach. In: Data Mining in Time Series Databases, pp. 1–21. World Scientific (2004)
11. Keogh, E., Ratanamahatana, C.A.: Exact indexing of dynamic time warping. Knowl. Inf. Syst. **7**(3), 358–386 (2004). https://doi.org/10.1007/s10115-004-0154-9
12. Lavielle, M., Moulines, E.: Least-squares estimation of an unknown number of shifts in a time series. J. Time Ser. Anal. **21**(1), 33–59 (2000)
13. Lee, W.H., Ortiz, J., Ko, B., Lee, R.: Time series segmentation through automatic feature learning. arXiv preprint arXiv:1801.05394 (2018)
14. Levenshtein, V.I.: Binary codes capable of correcting deletions, insertions, and reversals. Soviet Phys. Doklady. **10**, 707–710 (1966)
15. Lin, J., Keogh, E., Lonardi, S., Chiu, B.: A symbolic representation of time series, with implications for streaming algorithms. In: Proceedings of the 8th ACM SIGMOD Workshop on Research Issues in Data Mining and Knowledge Discovery, pp. 2–11 (2003)
16. Paparrizos, J., Gravano, L.: k-shape: efficient and accurate clustering of time series. In: Proceedings of the 2015 ACM SIGMOD International Conference on Management of Data, pp. 1855–1870 (2015)
17. Petitjean, F., Ketterlin, A., Gançarski, P.: A global averaging method for dynamic time warping, with applications to clustering. Pattern Recogn. **44**(3), 678–693 (2011)
18. Sakoe, H.: Dynamic-programming approach to continuous speech recognition. In: 1971 Proceedings of the International Congress of Acoustics, Budapest (1971)
19. Salvador, S., Chan, P.: Toward accurate dynamic time warping in linear time and space. Intell. Data Anal. **11**(5), 561–580 (2007)
20. Samé, A., Chamroukhi, F., Govaert, G., Aknin, P.: Model-based clustering and segmentation of time series with changes in regime. Adv. Data Anal. Classif. **5**, 301–321 (2011). https://chamroukhi.com/papers/adac-2011.pdf
21. Schäfer, P.: The boss is concerned with time series classification in the presence of noise. Data Min. Knowl. Disc. **29**(6), 1505–1530 (2015)
22. Schwarz, G., et al.: Estimating the dimension of a model. Ann. Stat. **6**(2), 461–464 (1978)

Applied Data Science: Activity Recognition

Recognizing Complex Activities by a Temporal Causal Network-Based Model

Jun Liao⬤, Junfeng Hu⬤, and Li Liu^(✉)⬤

School of Big Data and Software Engineering,
Chongqing University, Chongqing 401331, China
{liaojun,hjf,dcsliuli}@cqu.edu.cn

Abstract. Complex activity recognition is challenging due to the inherent diversity and causality of performing a complex activity, with each of its instances having its own configuration of primitive events and their temporal causal dependencies. This leads us to define a primitive event-based approach that employs Granger causality to discover temporal causal dependencies. Our approach introduces a temporal causal network generated from an optimized network skeleton to explicitly characterize these unique temporal causal configurations of a particular complex activity as a variable number of nodes and links. It can be analytically shown that the resulting network satisfies causal transitivity property, and as a result, all local cause-effect dependencies can be retained and are globally consistent. Empirical evaluations on benchmark datasets suggest our approach significantly outperforms the state-of-the-art methods. In particular, it is shown that our approach is rather robust against errors caused by the low-level detection from raw signals.

Keywords: Activity recognition · Complex activity · Primitive event · Temporal casual dependence · Network consistency

1 Introduction

A *complex activity* consists of a collection of temporally-composed events of basic actions and movements that can be directly detected from sensors or cameras [11]. For instance, actions like *open fridge* can be observed from an ambient sensor attached on the object; gestures like *shake hand* and *finger stoke* can be inferred by sensors attached to a user's arm or fingers; movements like *walk* or *jump* can be inferred by an accelerometer placed on the user's waist. Techniques so far are mature to recognize these simple actions, gestures and motions together with their durations, which are referred to as *primitive events* that cannot be further decomposed under application semantics.

The main focus of this paper is on complex activity recognition, which faces several key challenges [4]. First of all, understanding complex activities requires not only the detection of primitive events, but also the interpretation of their rich

© Springer Nature Switzerland AG 2021
Y. Dong et al. (Eds.): ECML PKDD 2020, LNAI 12460, pp. 341–357, 2021.
https://doi.org/10.1007/978-3-030-67667-4_21

temporal dependencies. Second, individuals often possess diverse styles of performing the same complex activity, and consequently, a complex activity recognition model should be capable of characterizing the underlying uncertainties over primitive events and their temporal relationships. Third, a complex activity recognition model should be also robust to errors caused by incorrect primitive event detection, due to sensor noise or low-level prediction errors.

Despite being a very challenging problem, in recent years there has been a rapid growth of interest in modeling and recognizing complex activities. They typically fall into two major categories, i.e., knowledge-driven approaches which provide abstract models of common knowledge, while data-driven approaches which exploits the unseen correlations between complex activities and primitive events. The knowledge-driven approaches, such as context-free grammar (CFG) and Markov logic network (MLN), are semantically clear, logically elegant, and easy to interpret. They are capable of representing rich temporal relations among primitive events. Yet formulae and their weights in these models (e.g. CFG grammars and MLN structures) need to be manually encoded, which could be rather difficult to scale up and is almost impossible for many practical scenarios where temporal relations among activities are intricate. On the other hand, the most popular modeling paradigm might be that of the graphical models and neural networks. With the great success being achieved, these data-driven models are capable of handling an astonishing number of correlations between events and are adept at managing uncertainties. However, their results are hard to interpret, and therefore, they are rather limited in further uncovering rich cause-effect relationships among events. For example, *shooting* or *passing a ball* is the cause of the action *foot pushes the floor* in basketball playing. This is because a player must use the appropriate amount of force against the floor when shooting or passing. In the Law of Karma, such pair of cause and effect is also called *action and reaction*. In fact, most of existing data-driven models may find that there is a heavy correlation between *shooting* and *passing* but unfortunately cannot discover the further interpretation that the reaction *foot pushes the floor* is the common effect of these two actions, which leads to their extrinsic association.

Granger causality (or GC) [12] is a way to investigate causality between two primitive events that combines temporal relations with probabilistic description. GC-based model can capture event interactions and their temporal dependencies. Especially, it demonstrates the effectiveness in exploring causal event sets. In the field of human activity recognition existing GC-based models exploit temporal dependencies between time series from raw sensor data and use them to detect primitive events or simple activities. However, since cause and effect are unidirectional, these models have to check triangle relationships to maintain causal consistency, which implies temporal consistency in the meantime. Evaluating all possible causal structures (which relation should be ignored or not) are computationally expensive, and would end up being intractable with the growth of the event sizes. Moreover, it is difficult or even meaningless to understand the causes and effects that are learned from raw time series. It is worth clarifying that Granger causality does not imply "true" causality since the question of "true causality" is deeply philosophical. It can be thought of as a tool of specifying a necessary condition for a temporal causal relation.

To address these issues in complex activity recognition, we present a temporal causal network approach based on Granger causality over primitive events. In particular, our approach considers a principled way of dealing with the inherent temporal causal variability in complex activities. Briefly speaking, to discover causal structures in complex activity such as *basketball playing*, we propose to introduce a temporal causal network (or network for short) generated from Granger causality test among primitive events. Now each resulting network contains its unique set of directed links together with their weights that represent cause-effect relations, characterizing a certain instance of a complex activity that possess similar primitive events and their temporal causal dependencies. Specifically, we optimize the network by leveraging Lasso regression to achieve link sparsity. Note that it enables our model to reduce the network size. In addition, *d-separation* is introduced to ensure causal consistency during the network generation procedure without loss of internal relations. In this way, our network-based approach is more capable of characterizing the inherit causal structural variability in complex activities when compared to existing methods, which is also verified during empirical evaluations to be detailed in later sections.

2 Related Work

Existing approaches for complex activity recognition can be divided into three categories.

2.1 Inferring Complex Activity Directly from Raw Data

Several methods have been proposed for complex activity recognition that operate directly from sensors data or video clips [5,11]. Among them, neural network-based approaches have been at the forefront of this research field. At the beginning, simple deep neural network architectures, such as CNN [23,25], were applied to extract features from raw data. More approaches were introduced to manage temporal dependencies by modifying network structures. Ordonez et al. [22] presented a DeepConvLSTM framework that integrates convolutional and LSTM recurrent layers to characterize activities. Zhao et al. [33] introduced a network architecture of deep residual bidirectional long short-term memory LSTM (Res-Bidir-LSTM) to capture the integrity of human activity. Unfortunately, these approaches are mostly time-point based, often ignoring the inherent structures among events, which hinders them handling intricate temporal relations efficiently. In addition, it is computationally exhaustive to explore representative features directly from raw data, which limits their applications in long-term activity recognition. Furthermore, these approaches are sensitive to sensor noise, and as a result, they suffer from the effects of adding noise during backpropagation training.

Different from the previous work, we design a simple neural network to recognize complex activities by assuming that primitive events and their corresponding intervals have already been recognized from sensors. So far hundreds of approaches have been proposed in the literature to detect simple events from various sensors. We refer the interested readers to the excellent reviews (Aggarwal and Ryoo [1] and Bulling et al. [7]) for recognizing primitive events from sensors.

2.2   Knowledge-Driven Complex Activity Recognition

Semantic-based approaches are one of de-facto knowledge-driven models that can construct semantics and contextual temporal information in activities [28]. Triboan et al. [29] achieved complex activity recognition by segmenting semantics of sensor data streams. Liu et at. [18] proposed a unified framework for semantic query by mining temporal and hierarchical relations over the probability of event occurrence. Safyan et al. [26] focused on semantic segmentation of ontology-based temporal formalisms to identify the concurrency of activities. Many other semantic-based methods [8,27] were proposed to handle temporal relationships through context or semantic information for complex activity recognition. These models are capable of representing rich temporal relations by using ontology or logical representations, but unfortunately they often do not have expressive power to capture uncertainties. Moreover, the semantic rules and their weights are typically hand-coded or based on domain knowledge. In particular, it is not practicable to handcraft the rules whose temporal relations among events are intricate.

2.3   Data-Driven Complex Activity Recognition

Graphical models, such as Dynamic Bayesian Networks (DBN) [10], Hidden Markov Model (HMM) [9] and Conditional Random Fields (CRF) [30], utilize probabilistic network structures to model complex activities. These graphical model-based approaches are capable of managing uncertainties. However, these models are time point-based, resulting in high computationally expense when the number of concurrent activities increases [19]. ITBN [32] and GPA [16] differ significantly from the previous graph model-based methods, as they characterize temporal relations over intervals instead of time points. These graphical-based models are capable of handling intricate temporal relations under uncertainty. However, as aforementioned in the introduction section, they are rather limited in further uncovering rich cause-effect relationships. Therefore, Granger causality [12] are commonly used to explore temporal causal relations in recent years. Avilescruz et al. [3] utilized Granger causality test to detect a user's movement from raw sensor data. Yi et al. [31] introduced a framework that represents human action sequences by Granger causality between joint movements. As aforementioned in the introduction section, a major limitation of these GC-based models concerns that the relationships that are learned from raw data are often hard to understand by human beings. They usually lack the expressive power to capture and propagate rich temporal dependencies in complex activities. Most importantly, these methods often uses GC as a tool to discover temporal dependencies but fail to maintain causal consistency, which are computationally expensive or even intractable in modeling complex activities, where the event size is large. To address the problems in these models, we present our GC-based model to explicitly capture the inherent structural varieties of complex activities by constructing primitive event-based causal networks with temporal dependencies under consistency.

3 Problem Formulation

Given a dataset $\mathcal{D}$ of N samples from a set of M complex activities, a temporal causal network is constructed with respect to the temporal causal relations among primitive events. Each sample is a sequence of T data points measured in time and spaced at uniform time intervals, denoted by $\mathbf{x} =< \mathbf{x}(1), \mathbf{x}(2), \ldots, \mathbf{x}(T) >$. A data point $\mathbf{x}(t)$ is a vector of K event attributes at time point t $(t = 1, \ldots, T)$, with each being associated with a certain primitive event type. We denote it as $\mathbf{x}(t) = [x_{t1}, x_{t2}, \ldots, x_{tK}]$, where x_{ti} is a binary variable that $x_{ti} = 1$ indicates the occurrence of the i-th primitive event e_i at time point t; otherwise, $x_{ti} = 0$ $(t = 1, \ldots, T, i = 1, \ldots, K)$. K denotes the number of primitive event types. In addition, a sequence of t $(t \leq T)$ continuous observations of primitive event e_i in a sample $\mathbf{x}$ is denoted by $\bar{\mathbf{x}}_i(t) =< x_{1i}, \ldots, x_{ti} >$.

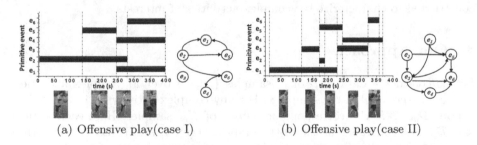

(a) Offensive play(case I)   (b) Offensive play(case II)

Fig. 1. Two example instances of the complex activity *offensive play* and their corresponding networks. $e_1 = walk$, $e_2 = run$, $e_3 = hold\ ball$, $e_4 = jump$, $e_5 = dribble$, $e_6 = shoot$. It is worth emphasizing that the links in the networks do not imply "true" causality, but they can be thought of a kind of "strict" temporal relations among primitive events in complex activity recognition.

A temporal causal network can be used to represent the temporal causal relationships between primitive events, where a node represents a primitive event type and a directed link describes the temporal causal relationship of the two involved events. Denote a temporal causal network $\mathbf{G} = (\mathbf{V}, \mathbf{E})$ the corresponding network of a sample $\mathbf{x}$, a primitive event e_i is a direct cause of e_j if there is a directed link from e_i to e_j in $\mathbf{E}$, denoted by $e_i \rightarrow e_j$, where $e_i, e_j \in \mathbf{V}$. Since causality is transitive, irreflexive and anti-symmetric, it can be verified that the resulting temporal causal network is a directed acyclic graph. Any link $e_i \rightarrow e_j$ in a temporal causal network $\mathbf{G}$ must satisfy *Granger causality test*, which defines e_i as the cause of e_j if the past values of e_i contain helpful information for predicting the future value of e_j. More formally, given the sequences of t observations of e_i and e_j in a sample $\mathbf{x}$ $(t < T)$, i.e. $\bar{\mathbf{x}}_i(t)$ and $\bar{\mathbf{x}}_j(t)$, respectively, e_i is the cause of e_j with respect to time point t if $P(x_{t+1,j} \mid \bar{\mathbf{x}}_i(t), \bar{\mathbf{x}}_j(t)) \neq P(x_{t+1,j} \mid \bar{\mathbf{x}}_j(t))$, and also states that e_i is not the cause of e_j if $P(x_{t+1,j} \mid \bar{\mathbf{x}}_i(t), \bar{\mathbf{x}}_j(t)) = P(x_{t+1,j} \mid \bar{\mathbf{x}}_j(t))$. It is worth mentioning that Granger's definition is not transitive. That is, if any $e_i \rightarrow e_j$ and $e_j \rightarrow e_k \in E$, it does not imply $e_i \rightarrow e_k \in E$. In other words, e_i is an indirect cause of e_k, but e_i may be not a direct cause of e_k under Granger's test, or even inversely e_k is a direct cause of e_i. To address this, a temporal causal network should be *consistent* that the temporal causal relations on every triangle of nodes $\triangle ijk$ in the network satisfy the transitivity property such that if $e_i \rightarrow e_j$ and $e_j \rightarrow e_k$ then $e_k \nrightarrow e_i$. In this paper, the term network refers to the temporal causal network in our definition, and relation refers to the temporal causal relation.

A network however characterizes only a possible style (or an instance) of a complex activity. Figure 1(a) and Fig. 1(b) show two networks that represents the same complex activity *basketball offensive playing* in two different ways of composing primitive events and their temporal causal dependencies. In this way, all the generated networks can form a joint network-based feature space that describes a unique complex activity. This inspires us to present in what follows a GC-based model where these primitive-based networks can be systematically constructed to characterize the complex activities of interests.

4   Our Approach

Let us consider a dataset $\mathcal{D}$ of N samples $\{(\mathbf{x}, c)\}$ over M complex activities, where c is the label of the sample $\mathbf{x}$. For any complex activity $m(1 \leq m \leq M)$, denote $\mathcal{D}_m \subseteq \mathcal{D}$ the corresponding subset of N_m samples. Here each sample $\mathbf{x} \in \mathcal{D}_m$ is an instance of the m-th complex activity, and is associated with a sequence of K primitive events $e = \{e_1, \ldots, e_K\}$ over a length of T time points. Our objective is to build a set of temporal causal networks as being encoded as features for complex activity recognition. The overview of our approach is illustrated in Fig. 2.

4.1   Network Skeleton Measurement

To generate a temporal causal network, we first determine the network skeleton, i.e., which pairs of nodes (primitive events) and their links (temporal causal relations) should be considered as candidates in the network. Our approach begins with measurement of *Granger causality* for any pair of nodes e_i and e_j by leveraging vector autoregressive model to depict the sequences of primitive events.

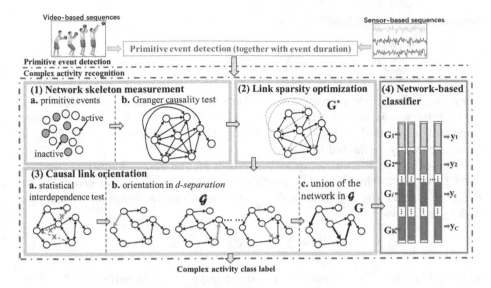

Fig. 2. The framework of our approach. *Active* node refers to the primitive event that occurs at least once in current sample, while *inactive* stands for no occurrence of such primitive event.

Formally, given two sequences of t observations of primitive events $\bar{\mathbf{x}}_i(t)$ and $\bar{\mathbf{x}}_j(t)$ in a sample $\mathbf{x}$, which are individually and jointly stationary, the processes of $\bar{\mathbf{x}}_i(t)$ and $\bar{\mathbf{x}}_j(t)$ can be individually represented by their autoregressive models as follows:

$$x_{ti} = \sum_{\tau=1}^{L} a_{ii}(\tau)x_{t-\tau,i} + \varepsilon_{ti}, \quad \varepsilon_{ti} \sim \mathcal{N}(0, \Gamma_i) \tag{1}$$

$$x_{tj} = \sum_{\tau=1}^{L} a_{jj}(\tau)x_{t-\tau,j} + \varepsilon_{tj}, \quad \varepsilon_{tj} \sim \mathcal{N}(0, \Gamma_j) \tag{2}$$

where $a_{ii}(\tau)$ and $a_{jj}(\tau)$ are regression coefficients, and ε_{ti} and ε_{tj} are their respective regression estimation residuals, which are random Gaussian noises with zero mean and a given standard deviation, where Γ_i is the variance of ε_{ti}, and Γ_j is the variance of ε_{tj}. L is a finite value called lag order, which can generally determined by Akaike Information Criterion (AIC).

According to the definition of *Granger causality*, in prediction for the sequence $\bar{\mathbf{x}}_{ti}$, it can be seen that another sequence $\bar{\mathbf{x}}_j(t)$ causes $\bar{\mathbf{x}}_i(t)$ if adding $\bar{\mathbf{x}}_j(t)$ helps predict $\bar{\mathbf{x}}_i(t)$. Subsequently, the jointly autoregressive model can be expressed as follows:

$$x_{ti} = \sum_{\tau=1}^{L} a_{ii}(\tau)x_{t-\tau,i} + \sum_{\tau=1}^{L} a_{ij}(\tau)x_{t-\tau,j} + \eta_{ti}, \quad \eta_{ti} \sim \mathcal{N}(0, \Sigma_i). \tag{3}$$

Similarly, we can also have

$$x_{tj} = \sum_{\tau=1}^{L} a_{jj}(\tau)x_{t-\tau,j} + \sum_{\tau=1}^{L} a_{ji}(\tau)x_{t-\tau,i} + \eta_{tj}, \quad \eta_{tj} \sim \mathcal{N}(0, \Sigma_j) \quad (4)$$

where η_{ti} and η_{tj} are regression estimation residuals, and $\Sigma_i = var(\eta_{ti})$ and $\Sigma_j = var(\eta_{tj})$. It is said that $e_i \rightarrow e_j$ with respect to t if x_{ti} can be more accurately estimated by Eq. (3) than that by Eq. (1). Inversely, $e_j \rightarrow e_i$ if x_{tj} is more accurately estimated by Eq. (4) than that by Eq. (2).

To measure such accuracy quantitatively from the above definitions, we first need to estimate the regression coefficients and residuals. More generally, we define the vector autoregression model regarding all pairs of primitive events (or nodes) as follows:

$$\mathbf{x}(t) = \sum_{\tau=1}^{L} \mathbf{A}(\tau)\mathbf{x}(t-\tau) + \eta_t, \quad (5)$$

where $\mathbf{A}(\tau)$ is the $K \times K$ coefficient matrix at lag τ where its entry $a_{ji}(\tau) \in \mathbf{A}(\tau)$ is the regression coefficient that indicates the effect of e_i on e_j, and η_t is its corresponding residual vector of size K. There are many ways to estimate these parameters, such as the ordinary least squares procedure or method of moments (through Yule-Walker equations).

4.2 Link Sparsity Optimization

Now it is straightforward to determine $e_i \rightarrow e_j$ by the condition that if any of the coefficients $a_{ij}(\tau) \in \mathbf{A}(\tau)$ at lag τ is non-zero. However, this faces the awkward situation where bidirectional links such as $e_i \leftrightarrow e_j$ largely exist in the network, which need additional estimations on these links to ensure network consistency. As a result, such additional estimation is often exhaustive that does not address the issue of combinatorial explosion, both in the computational and statistical senses. Computationally, it is extremely time-consuming and very sensitive to the number of observations in a sample, that is, $O(K^2T)$ times, and consequently it is not applicable in complex activity recognition where T is often a very large value. On the other hand, the statistical significance tests are conducted sequentially without regard to the possible interactions between them [2].

Here, we adopt a sparse solution of the coefficients by solving the following Lasso problem:

$$\hat{\mathbf{a}}_j = \underset{\mathbf{a}_j}{\arg\min} \sum_{t=L+1}^{T} \|x_{tj} - \sum_{i=1}^{K} \mathbf{a}_{ji}^{\top}\dot{\mathbf{x}}(t,L)\|_2^2 + \lambda\|\mathbf{a}_j\|_1$$

$$= \underset{\mathbf{a}_j}{\arg\min} \sum_{t=L+1}^{T} \|x_{tj} - \sum_{i=1}^{K}\sum_{\tau=1}^{L} a_{ji}(\tau)x_{t-\tau,i}\|_2^2 + \lambda\|\mathbf{a}_j\|_1, \quad (6)$$

where $\mathbf{a}_{ji}$ is the i-th vector of coefficients $\mathbf{a}_j$ modeling the effect of the event e_i on e_j, i.e., $\mathbf{a}_{ji} = [a_{ji}(1), \ldots, a_{ji}(L)]$, and $\dot{\mathbf{x}}(t, L)$ is the concatenated vector of

L lagged observations, i.e. $\dot{\mathbf{x}}(t, L) = [x_{t-L,j}, \ldots, x_{t-1,j}]$. The tuning parameter λ controls the amount of regularization for sparsity. The resulting optimization problem can be solved efficiently by approximate optimization solvers such as sub-gradient method and least angle regression (LARS). In this way, it is capable of eliminating the links that have little impact on cause-effect relations and thereby reduces the density in the network. A network $\mathbf{G}^* = (\mathbf{V}, \mathbf{E}^*)$ is constructed to be an initial skeleton of our network, such that any link $e_i \rightarrow e_j \in \mathbf{E}^*$ if and only if $\hat{\mathbf{a}}_{ji}$ is a nonzero vector.

4.3   Causal Link Orientation

So far we have a network skeleton $\mathbf{G}^*$, which may still contains bidirectional links or inconsistent triangle of nodes on temporal causal relations. We continue to determine the directions of links to ensure network consistency.

First we estimate the statistical interdependence between primitive events e_i and e_j. After obtaining $\hat{\mathbf{a}}_{ij}$ and $\hat{\mathbf{a}}_{ji}$, we can calculate Γ_i, Γ_j, Σ_i and Σ_j according to Eq. (1)–(4) separately. Let Σ_{ij} be the covariance matrix of the residual terms between e_i and e_j, defined as $\Sigma_{ij} = \begin{bmatrix} \Sigma_i & \Upsilon_{ij} \\ \Upsilon_{ij} & \Sigma_j \end{bmatrix}$, where Υ_{ij} is the covariance between η_i and η_j, i.e., $\Upsilon_{ij} = cov(\eta_i, \eta_j)$, where η_i and η_j are their respective residual vectors. We can define the statistical interdependence between e_i and e_j as

$$P(i, j) = \ln(\Gamma_i \Sigma_i / \mid \Sigma_{ij} \mid), \tag{7}$$

where $\mid \Sigma_{ij} \mid$ is the determinant of Σ_{ij}. If e_i and e_j are independent, resulting in $\Upsilon_{ij} = 0$, $\Gamma_i = \Sigma_i$ and $\Gamma_j = \Sigma_j$, then $P(i, j) = 0$; otherwise, $P(i, j) > 0$. We remove all the links $e_i \rightarrow e_j$ with $P(i, j) < \theta$ from the network $\mathbf{G}^*$, where θ is a threshold with small value.

Then, we further orientate the links in $\mathbf{G}^*$ according to the *d-separation* criterion, where if e_i and e_j are *d-separated* by e_k, then e_i and e_j are independent given e_k; otherwise, e_i and e_j are interdependent given e_k. We elaborate four situations of *d-separation* based on the orientation rules [20] in Table 1. In this way, a set of different networks $\mathcal{G}$ can be generated by following the above process. Finally, the resulting network $\mathbf{G} = (\mathbf{V}, \mathbf{E})$ is the union of the networks in $\mathcal{G}$, where $e_i \rightarrow e_j \in \mathbf{E}$ if and only if for every network in $\mathcal{G}$ that link is oriented as $e_i \rightarrow e_j$. Note that every network in $\mathcal{G}$ is a completed partially directed acyclic graph and thus it can be verified that $\mathbf{G}$ satisfies network consistent condition.

Besides, the weight on a link can be estimated in terms of its causal power, as defined by:

$$w_{ij} = \begin{cases} \ln(\Gamma_j / \Sigma_j), & \text{if } e_i \rightarrow e_j \in \mathbf{E} \text{ and } i \neq j \\ 1, & \text{if } \sum_{t=1}^{T}(x_{ti}) > 0 \text{ and } i = j \\ 0, & \text{otherwise.} \end{cases} \tag{8}$$

In fact, Γ_j measures the prediction accuracy of e_j based on its own previous values, whereas Σ_j measures it from the previous values of both e_i and e_j. If Σ_j is less than Γ_j, e_i is said to have a causal influence on e_j. In other words, the

Table 1. Orientation rules in *d-seperation*.

Structure in $\mathbf{E}^*$	$e_i \longleftrightarrow e_j$	$e_i \longleftrightarrow e_j$ $\uparrow$ e_k	$e_i \longleftrightarrow e_j$ $\nwarrow\!\!\nearrow$ $e_k \quad e_l$	$e_i \longleftrightarrow e_j$ $\uparrow\!\searrow$ $e_k \longrightarrow e_l$
Condition	There is a directed path from e_i to e_j in $\mathbf{E}^*$.	$e_k \rightarrow e_j \notin \mathbf{E}^*$	$e_k \rightarrow e_l \notin \mathbf{E}^*$	$e_k \rightarrow e_j \notin \mathbf{E}^*$
Orientation	$e_i \rightarrow e_j$			

causal power is high if adding e_i reduces prediction error of e_j. Theoretically, the larger w_{ij}, the stronger the causal influence. It is worth mentioning that we set $w_{ii} = 1$ if the primitive event e_i occurs at least once in the sample $\mathbf{x}$, which indicates the occurrence of e_i.

4.4 Network-Based Complex Activity Recognition

Now we are ready to build a complex activity recognition classifier by treating these networks as features. In particular, during the training stage, we put all the generated networks together to form a joint network feature space. Within this joint network feature space, each sample $\mathbf{x}$ from the c-th type of complex activity in the training set can be represented by a feature vector of size $K \times K$, and each entry in the feature vector is the weight value of the corresponding link (Eq. (8)). We can feed these feature vectors into any appropriate machine learning models for the recognition task. Here we train a simple neural network model, which consists of two fully-connected layers and one softmax classifier to achieve the tasks of complex activity recognition. In our model, ReLU is used to threshold activation function in each layer. During the testing stage, we encode the feature vector for each testing sample by generating the corresponding network and feed the feature vector into the pre-trained neural network classifier for complex activity recognition.

5 Experiments

5.1 Datasets and Preprocessing

We report the complex activity recognition results from three datasets.

OSUPEL Dataset [6]: This is a video-recorded dataset of real two-on-two basketball games where the players are tracked and labelled with six primitive events, including pass, catch, hold ball, shoot, jump, and dribble, which compose two types of complex offensive play activities with their number of samples being 56 and 16, respectively. We adopted the approach proposed by Zhang et al. [32]

to detect primitive events from videos clips where dynamic Bayesian network models are used to model the six basic basketball play events.

CAD Dataset [15]: It is composed of 693 human body skeleton-based samples captured from RGB-D cameras with 14 actors performing 16 complex activities such as walk while clapping, talk phone and drink, talk phone and answer, and so forth. There are a total number of 26 primitive events, including hand wave, dial phone, among others. Each complex activity contains 3 to 11 primitive events. We employed the hierarchical discriminative model presented by Lillo et al. [15] to recognize those primitive events from RGB-D data.

Opportunity Dataset [24]: It contains a total number of 28,976,744 samples performed by four subjects and recorded in a room with 72 sensors deployed either in objects or on the body. These sensor data samples are grouped into five complex daily living activities (relax, coffee time, early morning, cleanup, and sandwich time), involving a total number of 211 primitive events such as walk, sit, lying, open doors, reach an object, and so on. We utilized the activity recognition chain (ARC) system [7] to implement the primitive event recognition from sensors.

These three datasets contain unique challenges: The OSUPEL dataset comprises of a small number of samples of primitive events with simple temporal relations; The CAD dataset contains a large number of complex activities with diverse forms of temporal relations, while the Opportunity dataset involves a relatively large number of samples of primitive events with intricate temporal relations.

5.2   Baseline Methods

The recognition performance of our approach, named **GC-NN**, is compared against nine established knowledge-driven and data-driven methods:

IHMM [21]: It is a graphical model that identifies interactive events based on time-point.

SCCRF [14]: It is another time point-based model that utilizes the skip chain conditional random field to model concurrent events.

DBN [10]: This model uses DBN structure with event duration for describing interacting events.

ITBN [32]: It is an interval-based model that integrates Bayesian networks with 13 Allen relations to identify complex activities.

GPA Family (GPA-C/-F/-S) [16,17]: They are generative interval-based models that construct Bayesian networks in different ways on link generation (C - chain, F - fully connected, S - learned structure).

CK [13]: It is common knowledge-based that each rule is encoded manually to describe the dependence between the occurrences of an event and a complex activity.

ARF [18]: It is another knowledge-based approach where an event knowledge base is learned from data under uncertainty.

DeepconvLSTM [22]: It infers complex activity directly from raw data.

Res-Bidir-LSTM [33]: It is another raw data-based approach.

The standard evaluation metric of *accuracy* is used, which is computed as the proportion of correct predictions.

5.3   Experimental Results

Comparison Under the Ideal Situation on Primitive Event Detection. We first report the comparison results under idealized situations where the entire sequence of primitive events in a complex activity, including their start-times and end-times, are correctly detected. As shown in Table 2, it is clear that GC-NN outperforms other competing approaches with a large margin on all three datasets. This is mainly due to their abilities to take advantage of the rich temporal causal dependency information between primitive events. In contrast, other models such as HMM and DBN encode simple sequential relations between primitive events only. Although GPA and ITBN can handle the rich temporal relationships among primitive events, they cannot further exploit their intrinsic causal relations. As a result, they are not capable of discovering critical dependencies in the OSUPEL dataset which merely contains a small amount of simple sequential relations. For instance, ITBN discovers a highly temporal correlation between *pass* and *shoot a ball*, but such relation is not real temporal relation due to the fact that a hidden event *foot pushes the floor* is the common effect of the two events, leading to their extrinsic association. GC-NN does not take such disturbing relation into account because it further exploits the cause-effect relation rather than merely temporal relation. CK and ARF can characterize the causal dependence using sematic representation, but it is hard for them to handcraft all the rules thoroughly and efficiently. This might explain why CK gives the worst performance among all comparison methods for the Opportunity dataset where relations among primitive events are intricate.

Table 2. Accuracy comparisons of the competing methods on the three evaluation datasets.

	IHMM	SCCRF	DBN	ITBN	GPA-C	GPA-F	GPA-S	CK	ARF	GC-NN
OSUPEL	0.53	0.67	0.58	0.69	0.79	0.76	0.81	0.69	0.72	**0.95**
Opportunity	0.74	0.94	0.83	0.88	**0.98**	0.96	**0.98**	0.76	0.94	**0.98**
CAD	0.93	0.95	0.95	0.51	0.97	0.98	0.98	0.90	0.97	**0.99**

Robustness Test Under Primitive Event Detection Errors. In practice the accuracy of atomic action recognition will significantly affect complex activity recognition results. We evaluate the performance robustness of competing methods under various atomic activity recognition errors. There are two common errors with primitive event recognition, namely, misdetection errors that the correct primitive event is falsely detected as another event, while duration-detection errors that either the start-time or end-time of a primitive event is falsely detected. To achieve this, we synthetically perturb the low-level predictions, which represent different noise levels for the sensors or cameras. Synthetic misdetection errors are simulated by perturbing the true primitive events with different error rates, while duration-detection errors are simulated by perturbing the start and end times of those primitive events with a varying noise level of 10, 20, and 30% of maximal temporal distances between neighboring events, respectively. We report detailed experimental results on the Opportunity dataset. In addition, we evaluated performances under real detected errors caused by the ARC system (with three classifiers, i.e., kNN, SVM and DT) for low-level recognition. Also, we compared our approach with two models that recognize complex activities directly from raw sensor data without primitive event detection.

Table 3. Accuracies under primitive event errors on the Opportunity dataset. The percentage in the bracket shows the rate of change by taking the corresponding accuracy under the ideal situation (without primitive event error) as a baseline.

Primitive event error rate	Recognition accuracy							
	IHMM	SCCRF	DBN	ITBN	GPA-S	CK	ARF	GC-NN
Under synthetic primitive event misdetection errors								
0.1	0.31(−58%)	0.71(−24%)	0.73(−12%)	0.79(−10%)	0.94(−4%)	0.70(−8%)	0.88(−6%)	**0.96**(−2%)
0.2	0.29(−61%)	0.69(−27%)	0.67(−19%)	0.74(−16%)	0.87(−11%)	0.63(−17%)	0.82(−13%)	**0.94**(−4%)
0.3	0.22(−70%)	0.69(−27%)	0.65(−22%)	0.71(−19%)	0.84(−14%)	0.55(−28%)	0.76(−19%)	**0.86**(−12%)
Under synthetic primitive event duration detection errors								
0.1	0.16(−78%)	0.65(−69%)	0.45(−46%)	0.76(−14%)	0.89(−9%)	0.72(−5%)	0.82(−13%)	**0.92**(−6%)
0.2	0.16(−78%)	0.65(−69%)	0.45(−46%)	0.72(−18%)	0.85(−13%)	0.70(−8%)	0.75(−20%)	**0.92**(−6%)
0.3	0.16(−78%)	0.65(−69%)	0.45(−46%)	0.69(−22%)	0.83(−15%)	0.61(−20%)	0.63(−33%)	**0.92**(−6%)
Under real primitive event detected errors								
0.165 (kNN)	0.66(−11%)	0.69(−27%)	0.62(−25%)	0.54(−39%)	**0.91**(−7%)	0.71(−7%)	0.81(−14%)	0.90(−8%)
0.242 (SVM)	0.58(−22%)	0.10(−89%)	0.54(−35%)	0.46(−48%)	0.85(−13%)	0.72(−5%)	0.79(−16%)	**0.88**(−10%)
0.315 (DT)	0.16(−78%)	0.10(−89%)	0.04(−95%)	0.38(−57%)	0.71(−28%)	0.52(−32%)	0.69(−27%)	**0.75**(−23%)
Recognize complex activities directly from raw sensor data								
Deep Conv LSTM	0.83							
Res-Bidir-LSTM	0.90							

As shown in Table 3, it is obvious that the proposed GC-NN is significantly more robust than other approaches with around 5%–100% performance boost under various low-level detection errors. It indicates that our model can handle noises with various characteristics. It is also clear that CK and ARF perform worse when the noise level lifts up. This is mainly because the sematic rules in these approaches are often obtained from prior knowledge and thus they are not robust to various unknown errors. Notably, ITBN and GPA are more sensitive to duration-detection errors than GC-NN due to the fact that they heavily rely on the interval relations. In fact, it can be seen that the time-point based approaches including GC-NN are more stable than the interval-based approaches against duration-detection errors.

Runtime Comparison. We present three parameters that may affect the runtime, i.e. the number of primitive events types (K), the number of samples per complex activity (N) and the number of primitive events per samples (P). The empirical runtime is tested on different settings by varying one parameter while fixing others. The results of varying these three parameters are respectively shown in Fig. 3. Note that the figure presents the runtime in seconds on a logarithmic scale. It can be seen that our approach outperforms other time point-based methods. Although our approach is not the best on time consumption, overall it is affordable for practical usage in complex activity recognition. Theoretically, the time complexity of our approach is $O(NK^2TL + NMQ_n^2)$, where Q_n is the number of cells set in our neural network classifier.

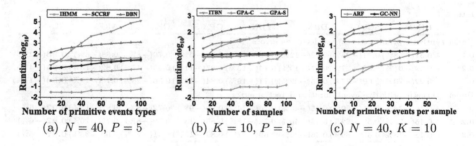

(a) $N = 40, P = 5$ (b) $K = 10, P = 5$ (c) $N = 40, K = 10$

Fig. 3. Runtime comparison at different settings.

5.4   Ablation Study

To sufficiently analyze the effects of two parameters in GC-NN, i.e., lag order L and sparsity regularization λ, we carried out the ablation test in this subsection to answer the intuitive questions: How much will these two parameters affect the performance of GC-NN?

To answer the question, we compared various settings of these two parameters in GC-NN on the three datasets. Here we increase the lag order L from 1 to 10 with a step of 1. Figure 4(a) shows that changing the lag order cannot lead to negative effects on the performance of our model on all the three datasets. Obviously, there is a sharp decline in accuracy when L is set to 2 in the OSUPEL dataset. This is mainly because the duration of primitive events in a complex activity is very short. For instance, *shot* or *pass a ball*, which happens rapidly, are only associated with a handful of frames (time points). Consequently, the vector autoregression model in GC-NN is overfitted when a small lag order is set, leading to over-rejection of the null hypothesis of Granger non-causality. In other words, fake cause-effect relations are remained in the temporal causal network, which is harmful to the GC-NN training. Although the selection of lag order still remains an open issue, we suggest to set the lag order with a value that is slightly larger than the ordinary length of primitive events occurred in a complex activity. Note that a very large value of L may result in computational burden.

The sparsity regularization parameter λ is an important parameter for link sparsity optimization. Its effect on recognition performance on the three datasets is shown in Fig. 4(b) by fixing the lag order to $L = 5$. It is clear that increasing the value of λ strengthens the regularization effect that will shrink more regression coefficients a_{ij} to zero. Normally, when λ grows to a certain large value, the number of links will decrease, which would end up that no links are identified in the network, and as a result, it will lead to accuracy drop. On the other hand, a small value of λ will bring about a great number of noisy links in the network, which may also be unfavorable to the recognition results. It can be viewed that the accuracy drops fast when λ is larger than 10^{-3} in the OSUPEL dataset. This might be due to a small number of relatively simpler temporal causal relations among primitive events contained in this dataset than those in other two datasets. A large λ will lead to insufficient links, which ultimately would result in information loss. As far as these three datasets are concerned, a proper value of λ could be roughly set at the range between $10^{-4.5}$ and $10^{-3.5}$.

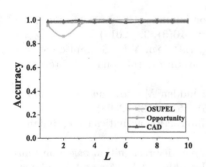
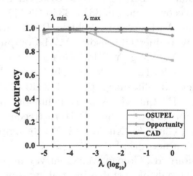

(a) Accuracy changes on different L (b) Accuracy changes on different λ ($L = 5$)

Fig. 4. Investigation of the impact of the parameters lag order and sparsity regularization on GC-NN performance.

6   Conclusion and Discussion

In this paper, we present a Granger causality-based model where primitive event-based networks are constructed to capture the inherit temporal causal varieties of complex activities. It is more efficient and robust than existing methods for complex activity recognition. As for future work, we will further investigate the difference between our model on more datasets, and we will consider relaxing the assumption that a network is causal consistent and will instead learn the network satisfying causal Markov condition over time series data.

Acknowledgement. This work was supported by grants from the National Major Science and Technology Projects of China (grant no. 2018AAA0100703), the National Natural Science Foundation of China (grant no. 61977012), the Central Universities in China (grant no. 2019CDJGFDSJ001).

References

1. Aggarwal, J., Ryoo, M.S.: Human activity analysis: a review. ACM Comput. Surv. (CSUR) **43**(3), 16 (2011)
2. Arnold, A., Liu, Y., Abe, N.: Temporal causal modeling with graphical granger methods, pp. 66–75 (2007)
3. Avilescruz, C., Rodriguezmartinez, E., Villegascortez, J., Ferreyraramirez, A.: Granger-causality: an efficient single user movement recognition using a smartphone accelerometer sensor. Pattern Recogn. Lett. **125**, 576–583 (2019)
4. Bharti, P., De, D., Chellappan, S., Das, S.K.: Human: complex activity recognition with multi-modal multi-positional body sensing. IEEE Trans. Mob. Comput. **18**(4), 857–870 (2019)
5. Bhorge, S.B., Manthalkar, R.: Recognition of vision-based activities of daily living using linear predictive coding of histogram of directional derivative. Ambient Intell. **10**(1), 199–214 (2019)
6. Brendel, W., Fern, A., Todorovic, S.: Probabilistic event logic for interval-based event recognition, pp. 3329–3336 (2011)
7. Bulling, A., Blanke, U., Schiele, B.: A tutorial on human activity recognition using body-worn inertial sensors. ACM Comput. Surv. **46**(3), 33 (2014)
8. Chang, X., Yang, Y., Hauptmann, A.G., Xing, E.P., Yu, Y.L.: Semantic concept discovery for large-scale zero-shot event detection. In: International Conference on Artificial Intelligence, pp. 2234–2240 (2015)
9. Chung, P., Liu, C.: A daily behavior enabled hidden Markov model for human behavior understanding. Pattern Recogn. **41**(5), 1572–1580 (2008)
10. Du, Y., Chen, F., Xu, W., Li, Y.: Recognizing interaction activities using dynamic Bayesian network. In: ICPR (2006)
11. Elbasiony, R., Gomaa, W.: A survey on human activity recognition based on temporal signals of portable inertial sensors. In: Hassanien, A.E., Azar, A.T., Gaber, T., Bhatnagar, R., F. Tolba, M. (eds.) AMLTA 2019. AISC, vol. 921, pp. 734–745. Springer, Cham (2020). https://doi.org/10.1007/978-3-030-14118-9_72
12. Granger, C.W.J.: Investigating causal relations by econometric models and cross-spectral methods. Econometrica **37**(3), 424–438 (1969)
13. Helaoui, R., Niepert, M., Stuckenschmidt, H.: Recognizing interleaved and concurrent activities: a statistical-relational approach. In: IEEE International Conference on Pervasive Computing and Communications (2011)
14. Hu, D., Yang, Q.: CIGAR: concurrent and interleaving goal and activity recognition. In: AAAI (2008)

15. Lillo, I., Soto, A., Niebles, J.: Discriminative hierarchical modeling of spatio-temporally composable human activities. In: CVPR (2014)
16. Liu, L., Cheng, L., Liu, Y., Jia, Y., Rosenblum, D.: Recognizing complex activities by a probabilistic interval-based model. In: AAAI (2016)
17. Liu, L., Wang, S., Hu, B., Qiong, Q., Wen, J., Rosenblum, D.S.: Learning structures of interval-based Bayesian networks in probabilistic generative model for human complex activity recognition. Pattern Recogn. **81**, 545–561 (2018)
18. Liu, L., et al.: A framework of mining semantic-based probabilistic event relations for complex activity recognition. Inf. Sci. **418**, 13–33 (2017)
19. Liu, L., Wang, S., Su, G., Huang, Z., Liu, M.: Towards complex activity recognition using a Bayesian network-based probabilistic generative framework. Pattern Recogn. **68**, 295–309 (2017)
20. Meek, C.: Causal inference and causal explanation with background knowledge. arXiv preprint arXiv:1302.4972 (2013)
21. Modayil, J., Bai, T., Kautz, H.: Improving the recognition of interleaved activities. In: International Conference on Ubiquitous Computing (2008)
22. Ordonez, F.J., Roggen, D.: Deep convolutional and LSTM recurrent neural networks for multimodal wearable activity recognition. Sensors **16**(1), 115 (2016)
23. Pio, G., Ceci, M., Prisciandaro, F., Malerba, D.: Exploiting causality in gene network reconstruction based on graph embedding. Mach. Learn. **109**(6), 1231–1279 (2019). https://doi.org/10.1007/s10994-019-05861-8
24. Roggen, D., Calatroni, A., Rossi, M., Holleczek, T., et al.: Collecting complex activity datasets in highly rich networked sensor environments. In: International Conference on Networked Sensing Systems (2010)
25. Ronao, C.A., Cho, S.: Human activity recognition with smartphone sensors using deep learning neural networks. Exp. Syst. Appl. **59**(59), 235–244 (2016)
26. Safyan, M., Qayyum, Z.U., Sarwar, S., Garciacastro, R., Ahmed, M.: Ontology-driven semantic unified modelling for concurrent activity recognition (OSCAR). Multimedia Tools Appl. **78**(2), 2073–2104 (2019)
27. Saguna, S., Zaslavsky, A., Chakraborty, D.: Complex activity recognition using context-driven activity theory and activity signatures. ACM Trans. Comput. Hum. Interact. (TOCHI) **20**(6), 32 (2013)
28. Tang, Y., Lu, J., Wang, Z., Yang, M., Zhou, J.: Learning semantics-preserving attention and contextual interaction for group activity recognition. IEEE Trans. Image Process. **99**, 1–12 (2019)
29. Triboan, D., Chen, L., Chen, F., Wang, Z.: Semantic segmentation of real-time sensor data stream for complex activity recognition. Pers. Ubiquit. Comput. **21**(3), 411–425 (2017). https://doi.org/10.1007/s00779-017-1005-5
30. Vail, D., Veloso, M., Lafferty, J.: Conditional random fields for activity recognition. In: International Joint Conference on Autonomous Agents and Multiagent Systems (2007)
31. Yi, S., Pavlovic, V.: Sparse Granger causality graphs for human action classification, pp. 3374–3377 (2012)
32. Zhang, Y., Zhang, Y., Swears, E., Larios, N., Wang, Z., Ji, Q.: Modeling temporal interactions with interval temporal Bayesian networks for complex activity recognition. IEEE Trans. Pattern Anal. Mach. Intell. **35**(10), 2468–2483 (2013)
33. Zhao, Y., Yang, R., Chevalier, G., Xu, X., Zhang, Z.: Deep residual Bidir-LSTM for human activity recognition using wearable sensors. Math. Probl. Eng. **2018**, 1–13 (2018)

Unsupervised Human Pose Estimation on Depth Images

Thibault Blanc-Beyne[1,2]($\boxtimes$), Axel Carlier[2], Sandrine Mouysset[2], and Vincent Charvillat[2]

[1] Ebhys, ZA La Cigalière III, 84250 Le Thor, France
[2] Université de Toulouse - IRIT, 2 rue Charles Camichel, 31079 Toulouse, France
{thibault.blanc-beyne,axel.carlier,
sandrine.mouysset,vincent.charvillat}@irit.fr

Abstract. Human pose estimation is a widely studied problem in the field of computer vision that consists in regressing body joints coordinates from an image. Most state-of-the-art techniques rely on RGB or RGB-D data, but driven by an industrial use-case to prevent musculoskeletal disorders, we focus on estimating human pose based on depth images only. In this paper, we propose an approach for predicting 3D human pose in challenging depth images using an image-to-image translation mechanism. As our dataset only consists in unlabelled data, we generate an annotated set of synthetic depth images using a human3D model that provides geometric features of the pose. To fit the challenging nature of our real depth images as closely as possible, we first refine the synthetic depth images with an image-to-image translation approach using a modified CycleGAN. This architecture is trained to render realistic depth images using synthetic depth images while preserving the human pose. We then use labels from our synthetic data paired to the realistic outputs of the CycleGAN to train a convolutional neural network for pose estimation. Our experiments show that the proposed unsupervised framework achieves good results on both usual and challenging datasets.

Keywords: Depth images · Unsupervised learning · Human pose estimation · Image-to-image translation

1 Introduction

Musculoskeletal disorders (MSD) are a leading cause of health issues in the workplace [27]. This condition covers various types of injuries that typically affect tendons, muscles, ligaments, etc. Industrial work is often an important source of MSD, due to the repetitive nature of the tasks that the humans operate as well as the physical constraints that are exerted on their limbs. More generally, monitoring working conditions is becoming an international concern [1]; in this context devising systems that can help preventing MSD is an important challenge.

T. Blanc-Beyne—This work was supported by CIFRE ANRT 2017/0311.

© Springer Nature Switzerland AG 2021
Y. Dong et al. (Eds.): ECML PKDD 2020, LNAI 12460, pp. 358–373, 2021.
https://doi.org/10.1007/978-3-030-67667-4_22

In our work, we study the particular case of the waste industry, in which human operators are especially prone to developing MSD. Both industrial and domestic waste are processed in facilities called waste sorting centers, whose role is to separate different types of waste (for example papers from cardboard and plastics) for later recycling or incinerating. While the sorting process is largely automated, human operators are required to perform negative sorting, i.e. remove unwanted objects to guarantee the purity of the output waste streams. This work involves repetitive movements, with sometimes undesirable limbs angulations, which may eventually lead to MSD.

We aim at designing at system that would help identify risk situations, by monitoring the humans operators posture and revealing harmful angulations. Driven by our industrial context, we should perform these operations under several constraints:

(i)   The monitoring should be non-invasive, to avoid affecting the operators work;
(ii) The lightning conditions can not be controlled: the sensor may face a window or be in a dark corner of the facility;
(iii) The operators may wear reflecting working clothes;
(iv) Anonymity of the workers should be preserved to respect their privacy and avoid abusive uses of the collected data.

For all these reasons, we choose to frame this problem as human pose estimation on depth images.

Human pose estimation, which is the task of localizing anatomical keypoints (joints) or parts, has traditionally been a very challenging task in computer vision, both in 2D and 3D. Most work focus on single-person pose estimation in the 2D setting [2,26] using various architectures of convolutional neural networks (CNN) applied on single RGB images. Another line of work approaches the problem with the goal of detecting multiple persons, still on RGB frames. These more recent methods often use a top-down model, as proposed in [10], where (i) human detection is applied and (ii) single-pose estimation is realized on each detection. However, the most efficient methods rely on bottom-up approaches [7]. These methods make use of ResNet blocks [14] or of a multi-stage CNN to first detect joints, and later match them to perform a complete pose estimation. 3D pose estimation from single images remains a challenge to this day, due to the lack of available labelled data and to the ambiguity of getting 3D information from single images. Using deep learning methods, the 3D pose is often inferred from the 2D estimation [6] as deep neural networks trained on 3D datasets captured in a motion capture context [15] do not generalize well [37]. To tackle this issue, an unsupervised approach based on domain adaptation was recently introduced by [36]. They first train a 2D/3D pose estimator on depth images and a segmentation module on the 2D pose. Using domain adaptation, the authors finally use these two components to train a 2D/3D pose estimator on RGB images.

Even though RGB-D sensors have been used by the general public for many years now (Microsoft released the first version of the Kinect sensor in 2010), only a few research work have focused on how to infer information using depth images only. In particular, most previous work make use of the human segmentation and pose estimation embedded in the Kinect [30], which implicitly assumes that the user is standing in front of the sensor. As previously adopted by [4] in the context of hospitals for action recognition and fall detection, our approach focuses on using only the sensor depth information (disregarding the color information), in order to respect the human operators privacy.

This implies to estimate human pose from unlabelled, noisy depth images. To do so, we propose in this paper to present a framework of unsupervised pose estimation process, described in Fig. 1.

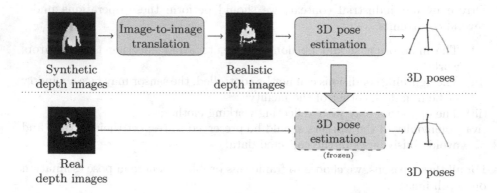

Fig. 1. The framework of our unsupervised pose estimation process.

Our framework relies on the CycleGAN [39] approach, a particular GAN architecture, that is able to preserve the geometric properties of the image while changing its texture. We propose to first generate an annotated set of synthetic depth images using a human 3D model that provides geometric features of the pose (Sect. 3). We then use a modified CycleGAN architecture that alters the generated synthetic depth images to mimic real ones, while preserving geometric content, i.e. the human pose (Sect. 4). In Sect. 5, we present our pose estimation on real images based on a single convolutional neural network that infers the 3D position of body joints. Finally we present our experiments and discuss our results in Sect. 6.

2   Related Work

Human Pose Estimation on Depth Images. Only a few work on human pose estimation discard the RGB images and only analyze the depth values. Among the recent approaches that build on the advances brought by deep learning techniques, Jiu et al. [18] do not output an explicit body pose, but segment body parts using features built by spatially constrained multiscale, CNNs [11]. Wang et al. [33] introduce an inference-embedded multi-task learning framework.

They use a CNN to generate a heatmap of body parts, which is then fed to an inference network seeking the optimal configuration of body parts, using both appearance and geometric compatibility. Similarly, Haque et al. [13] first detect local body parts. A global body pose is then iteratively produced using a leveraged convolutional and recurrent network containing a long short term memory (LSTM) module. Moon et al. [25] propose a voxel-to-voxel network for hand and body pose estimation. They split the 3D space into a grid of voxels, and estimate a per-voxel likelihood for each joint. Marín-Jiménez et al. [23] present a model which uses a CNN to compute weights used to estimate the 3D pose as a linear combination of prototype poses.

Generative Adversarial Networks. Generative adversarial network (GAN) is a new machine learning paradigm introduced by Goodfellow et al. in 2014 [12]. In this framework, two models are simultaneously trained: a generative model G which captures the data distribution, and a discriminative model D that separates the data generated by G from the real training data. The training stability of such system is notoriously difficult to maintain, and improvements were proposed to improve it [29]. GANs were also extended by [24] as conditional GAN to make the generator able to sample data conditioned on class labels. The key principle is the use of an adversarial loss, that ideally makes the generated images indistinguishable from real images at the end of the training. GANs have achieved impressive results in image editing [38], generation [28] and conditional generation [35].

Image-to-Image Translation. GANs are also used for the task of image-to-image translation. Recent approaches rely on a dataset of paired input/output images, such as "pix2pix" [16], which uses conditional GANs to learn a translation function between input and output images, improved to high resolution images by [34] and by the Multimodal Unsupervised Image-to-image Translation (MUNIT) [40] to control the style of translated output. StarGAN [9] performs image-to-image translations for multiple domains using a single network trained on several datasets, while [8] performs image-to-image translation using pose as latent space. A lot of other methods tackle the unsupervised setting and aim to map two different data domains. An approach is to use a weightsharing strategy as in CoGANs [21] or cross-modal scene networks [3]. Another idea is to encourage the input and the output to share specific content while differing in style [31]. Popular methods use transitivity to regularize the data thanks to cycle-consistency [39]. These approaches rely on the fact that, if an image is translated from one domain to the other and back, the resulting image should ideally be the one we started with. This allows to perform image-to-image translation on unpaired datasets, for instance in the task of semantic segmentation [20] or robotics [17].

3   Training Data Generation

In our industrial context in which human activity is monitored to prevent musculoskeletal disorders, we are bounded by several constraints such as sensors positions, uncontrollable lightning conditions, reflective surfaces and clothes, and the overwhelming number of moving objects. To protect workers' privacy and favor our system's acceptability, we choose to acquire data using depth cameras. The challenging environment introduces complexity and noise to the depth images and usual human pose estimation approaches fail to produce exploitable results. We build what we will denote as our real dataset by gathering around 60,000 frames from 8 live-recorded video sessions in a waste sorting center.

Examples of real depth images are depicted in Fig. 2. The operator is segmented from the frame using the approach proposed in [5].

Depth data is directly acquired and stored on 640 × 480, 16 bits images rather than simple 8 bits PNG images, in order to correctly capture the depth value of each pixel. Indeed, precision in depth values is needed to be able to distinguish between the hand of the operator and the object he/she may hold. This kind of storage enables us to perform a better normalization of our data before feeding it to our networks, while PNG images would have provided us with a poorer depth sampling of the operator.

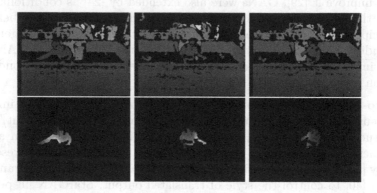

Fig. 2. Examples of images from our dataset (top row) and their associated segmentation (bottom row).

Unfortunately, we can not directly use this data to train our human pose estimation network. Indeed, machine learning often implies a critical need of large amounts of labelled data during the training phase. In our use-case, annotating a 3D human pose just from a depth image is a very difficult and time-consuming task, and in fact almost impossible to realize with a good enough precision as it involves providing a 3D information on a 2D space. In addition, our industrial constraints make it likely that the shooting conditions differ from one sorting center to another. The depth camera may have to be positioned differently or the human operators may wear different reflective clothes, which would result in a new type of data, and would thus require a new annotation campaign to retrain the neural network. We believe this process is not sustainable in our use-case. That is why we decided to rely on a synthetic labelled dataset rather than handcrafted annotations.

The synthetic data is generated using the free and open source 3D creation suite Blender[1]. The rigged 3D human model was created using an free opensource software called MakeHuman[2]. We made use of the Blender scripting tool to generate a highly variable set of poses stored as key frames, while its compositing tool allowed us to render synthetic depth frames. The range of angulations for each joint of our animated model was built thanks to an ergonomic study to closely match those of the sorting activity and thus, those of our real data. These images are encoded the same way we encoded our real depth images, and at the same resolution. Figure 3 shows some examples of depth images from our synthetic dataset, consisting in around $200,000$ images.

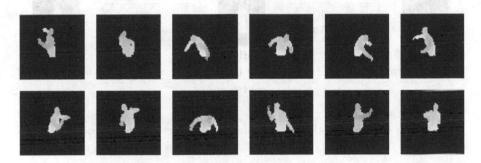

Fig. 3. Examples of synthetic images created using Blender. To match our use case, these images represents only the upper body of the user, in a very variable set of poses.

However, we will see later (see Sect. 6) that training our pose estimator directly with this synthetic data only does not give satisfactory results. We need to transform this data for it to better match our real data. For this task, we chose to use CycleGAN [39].

4   Image-to-image Translation

CycleGAN [39] is an unsupervised approach for image-to-image translation which does not require paired data. The CycleGAN allows to find a mapping between the distributions of the two sets of data. In the following, we present the architecture described by Fig. 4 with training details and provide first results on the synthetic dataset.

4.1   Model Architecture

The CycleGAN is composed of two generative networks and two discriminators. As we need lightweights architectures to achieve real-time performance in our

---

[1] https://www.blender.org.
[2] http://www.makehumancommunity.org.

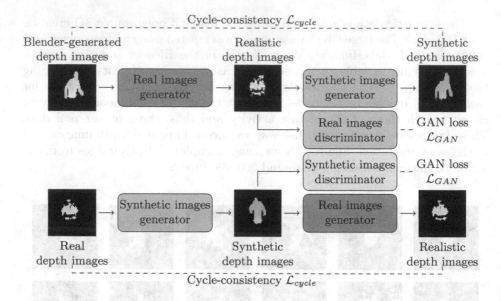

Fig. 4. Overview of the image-to-image translation training process.

industrial use-case, and as the operator is located at the center of the image and only occupies a small part of the depth image, we crop the depth images to only keep the operator and resize it into $64 \times 64 \times 1$ images.

After several tests about the architecture of the generators, the final architecture of both generators is the same and consists of three parts:

- a first descending part composed of three convolutional layers with stride to reduce the spatial dimension of the images, with the number of filter doubling at each convolutional layer from 32 to 128;
- a second part containing three residual blocks [14] of a single 2D convolutional layer of 128 filters;
- a third ascending part consisting of three deconvolutional layers with stride to retrieve the original image size, with the number of filters decreasing from 128 to 32.

All the convolutional and deconvolutional layers have a kernel size of 4×4, with zero-padding, and are followed by a ReLU activation function and an Instance Normalization layer [32]. The final layer is a 2D convolutional layer with a single filter, a kernel size of 4×4 and hyperbolic tangent activation function to obtain our final normalized translated image.

The discriminators are two 64×64 PatchGANs [16,31], which are fully convolutional neural networks whose goal is to classify whether 64×64 overlapping image patches are real or fake by outputting a probability map. More precisely, both discriminators contain a four convolutional layers with stride, an increasing number of filters from 32 to 256, a kernel size of 4×4, with zero-padding, and are followed by a LeakyReLU activation function with $\alpha = 0.2$. They are all followed by an Instance Normalization layer, apart from the first one which is not normalized. A final convolutional layer with one filter outputs the patchs discrimination. These architectures are inspired from the architectures proposed by [39] for both generators and discriminators, and are given in Fig. 5.

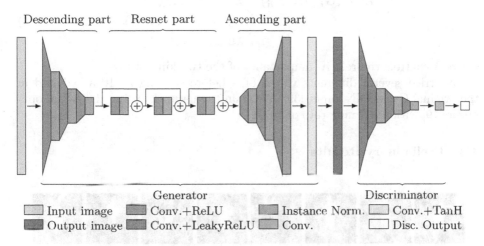

Fig. 5. The architectures of the generators and discriminators of our proposed Cycle-GAN.

4.2 Training Details

To train this model, we use, as described in [39], three losses: two adversarial losses $\mathcal{L}_{GAN}$ and a cycle-consistency loss $\mathcal{L}_{cycle}$.

For the two generators $G_R : S \rightarrow R$, $G_S : R \rightarrow S$ where R and S are the respective Real and Synthetic image datasets and their associated discriminators D_R and D_S, we express the adversarial losses as:

$$\mathcal{L}_{GAN}(G_R, D_R, S, R) = \mathbb{E}_{r \in R}[||D_R(r)||_2] + \mathbb{E}_{s \in S}[||1 - D_R(G_R(s))||_2] \quad (1)$$

where G_R aims to generate images that are similar to synthetic images from S and D_R tries to distinguish real samples r from generated samples, and:

$$\mathcal{L}_{GAN}(G_S, D_S, R, S) = \mathbb{E}_{s \in S}[||D_S(s)||_2] + \mathbb{E}_{r \in R}[||1 - D_S(G_S(r))||_2] \quad (2)$$

where G_S aims to generated images that are similar to real images from R and D_S tries to distinguish synthetic samples s from generated samples. We use an L^2 loss function rather than a negative log-likelihood because this loss is more stable during training and generates higher quality results [22].

In the meantime, the goal of the cycle-consistency loss is to ensure that $G_S(G_R(s)) \approx s$, i.e that a translation cycle should bring s back to the original synthetic image. The same goes with real images, where we want to ensure that $G_R(G_S(r)) \approx r$. This cycle-consistency loss is then expressed as:

$$\mathcal{L}_{cycle}(G_S, G_R) = \mathbb{E}_{s \in S}[||G_S(G_R(s)) - s||_1] + \mathbb{E}_{r \in R}[||G_R(G_S(r)) - r||_1] \quad (3)$$

Finally, the global loss to train our model is:

$$\begin{aligned}
\mathcal{L}(G_S, G_R, D_S, D_R) = & \; \mathcal{L}_{GAN}(G_R, D_R, S, R) \\
& + \mathcal{L}_{GAN}(G_S, D_S, R, S) \\
& + \lambda \mathcal{L}_{cycle}(G_S, G_R)
\end{aligned} \quad (4)$$

where λ controls the relative importance of the two kind of loss.

We tried several differents values of λ before sticking to 10 as advised in the original CycleGAN paper [39]. We train the networks using the Adam optimizer [19] with a learning rate of 10^{-6}.

4.3 Preliminary Results

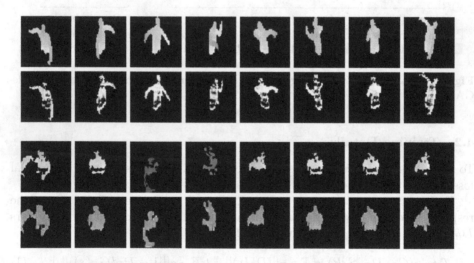

Fig. 6. Examples of image-to-image translation performed by our CycleGAN. We can see that the pose of the user is preserved during both translations. First row: synthese images from our synthetic dataset. Second row: translated realistic images associated to the images from the first row. Third row: real images from our real dataset. Last row: translated synthetic images associated to the images from the third row.

Some results are shown in Fig. 6. We see that the real images generator manages to realistically degrade the synthetic images by adding noise on the human operator's body and removing parts, similarly to a real image. On the other hand, the synthetic images generator seems to fill holes and put together parts from real images to make them look more "synthetic". However, and as expected, both generators neither add nor remove objects around the user.

5   Pose Estimation on Real Images

We perform 3D pose estimation on single depth images using a deep convolutional neural network. Since our real datasets do not contain 3D human pose annotations, we can not directly train our network on this data. However, thanks to our synthetic dataset and the previously trained CycleGAN, we can generate annotated realistic depth images. This process may ideally yield an unlimited set of realistic depth images paired to the 3D human pose annotation corresponding to the synthetic images given as input. We use these couples between realistic depth images and synthetic 3D human pose annotations to train our pose estimation network.

The 3D pose is defined by the following fifteen upper body joints: *BaseSpine*, *MiddleSpine*, *TopSpine*, *Neck*, *Head*, *LeftShoulder*, *LeftElbow*, *LeftWrist*, *Left-Hand*, *RightShoulder*, *RightElbow*, *RightWrist*, *RightHand*, *LeftHip* and *RightHip*. These joints are illustrated in Fig. 7.

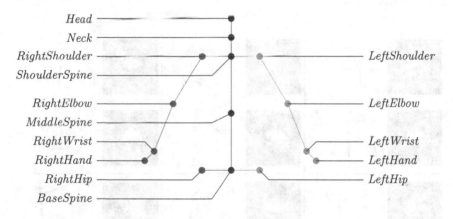

Fig. 7. The fifteen joints we used to define our 3D human pose.

The 3D pose estimation network is a convolutional neural network. Its inputs are 64×64 depth images, and it outputs a 45-dimensional vector representing the 3D position of the fifteen body joints. It contains three blocks of three 2D convolutional layers of 64 4×4 filters, with batch normalization and ReLU activation function. The blocks are separated by max-pooling layers. The final layer is a dense layer of 15×3 neurons, to estimate the 3D pose of the fifteen joints. An illustration of this architecture is given in Fig. 8.

To train this model, we use a mean squared error loss function. We use the Adam optimizer [19], with a learning rate of 10^{-3} and mini-batchs of 32 images.

Examples of our results on pose estimation are depicted in Fig. 9. We can see that our results look consistent with the depth images, however it is highly dependent on the variability of the synthetic dataset and of the quality of the user segmentation in the depth images. Indeed, our synthetic dataset lacks of

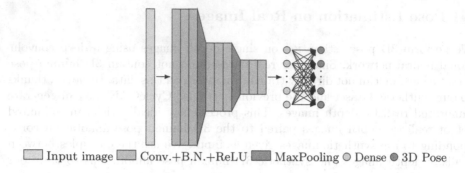

☐ Input image ☐ Conv.+B.N.+ReLU ■ MaxPooling ○ Dense ● 3D Pose

Fig. 8. The architecture of our pose estimation neural network.

synthetic images with the arm up in the air and, as shown in the third row of Fig. 9, the wrongly estimated pose often display this kind of error on flawed segmented frames.

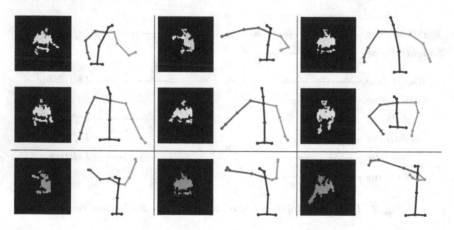

Fig. 9. Examples of real depth images and their associated 3D pose estimation. First two rows: the poses seem consistent with the depth image of the user. Third row: example of failure cases of our approach. After analysis, these errors often occurs when there are flaws in the segmentation, for instance pixels of the background breaking the depth normalization of the image (that we can see here with a different shade of gray in the image of the third row).

6   Experiments

We compare the different results thanks to various commonly used measures [36] such as Percentage of Correct Key-Points (PCK) at 150 mm (PCK@150 mm) and 80 mm (PCK@80 mm), that is the percentage of joints whose distance between predicted and true position lies within a given threshold, the Mean Per Joint

Position Error (MPJPE), which is the mean euclidean distance between ground truths and predictions, and the Procrustes Analysis Mean Per Joint Position Error (PAMPJPE), which is a MPJPE where a similarity transformation is applied before computing the euclidean distance. For the PCK measures, the higher is the better, while for MPJPE and PAMPJPE, given in millimeters, the lower is the better.

In order to be able to measure the quality of our results and since we do not have annotations on real data, we created a new degraded dataset by applying a combination of white noise and image parts removal to build the *noisy and degraded* dataset, consisting of $200,000$ 640×480 synthetic images. We perform an ablation study on this second dataset with three kinds of training: a supervised training where the model is trained on the same kind of data as testing data, a "CycleGAN" training where the model is trained the outputs of the CycleGAN and an unsupervised training where the model is directly trained on the undegraded synthetic data, and tested on the *noisy and degraded* dataset. This ablation study proves that the image-to-image translation block is useful in order to improve the quality of the pose estimation.

As shown in Table 1, using the outputs of our image-to-image translation block clearly improves the quality of the pose estimations compared to directly training the same network on the synthese depth images, even though its performances are still worse than the supervised setting: the PCK@150 mm and PCK@80 mm improves respectively from 55.0% to 93.3% and 0.2% to 46.0%, while the MPJPE and PAMPJPE lower from 148.0 mm to 84.3 mm (43% reduction) and 87.6 mm to 65.9 mm (25% decrease).

Table 1. Results of our pose estimation network under various training conditions. We observe that using a CycleGAN improves the results compared to the unsupervised setting, despite being worse than the supervised learning.

Kind of data	Training type	PCK@150 mm	PCK@80 mm	MPJPE 3D	PAMPJPE 3D
Synthetic	Supervised	100.0	100.0	10.9	9.9
Noisy	Supervised	100.0	100.0	21.4	18.8
&	CycleGAN	93.3	46.0	84.3	65.9
Degraded	Unsupervised	55.0	0.2	148.0	87.6

It shows that our CycleGAN is able to correctly capture some features of degraded depth images and translate them on the synthetic images. It is also important to notice that the real difficulty of our dataset lies in the degradation induced by the industrial context (i.e the holes in the depth image) and not in the imprecision of the sensor (simulated by the white noise).

To be able to perform more measures about the quality of our results, we acquired another real dataset in a controlled setting, simulating a simplified industrial context. These images were acquired with a Kinect V2 sensor in a bird's eye view of the user, allowing us to compare our pose estimation to the pose estimation provided by the sensors. This dataset contains around 13,000 depth frames of several subjects in a wide variety of human poses. However, comparing to this data is difficult for several reasons:

- our model joints are not exactly located at the same place on the human body than those of the Kinect;
- we estimate a 3D human pose, while Kinect V2 only estimates a 2D+Z human pose: the two first coordinates (X,Y) are estimated as pixel coordinates in the depth frame and then translated to the real world space thanks to the camera intrinsic parameters, but the third one is computed by adding a given offset to the depth value of the pixel, which can be far away from the real position of the joint when the sensor is not in front of the user;
- the Kinect V2 pose estimation is calibrated for an entertainment setting where the sensor is in front of the user. In our setting, the sensor has a bird's-eye view on the user, which leads to further deformations of the estimated pose.

An illustration of these issues is given in Fig. 10. In particular, we can see that the Kinect pose estimation is strongly tilted forwards, while ours stays straight, which is closer to the real position of the user.

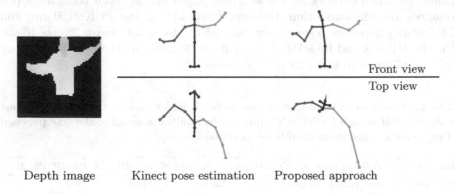

Fig. 10. Comparison of the Kinect pose estimation against the proposed pose estimation. Left: the segmented depth frame. Center: the 2D+Z pose estimation provided by the Kinect sensor. Right: the proposed pose estimation.

However, despite not being able to compare our results to a real ground truth, we confirm the results that we showed on synthetic data (see Table 2). Due to the facts given previously, there is a large gap between the Kinect pose estimation and ours, but the use of the CycleGAN greatly reduces this gap and thus improves the quality of our network's pose estimation. The CycleGAN improves the quality of the pose estimations, by reducing both the MPJPE and PAMPJE, respectively by 39% and 34%.

Table 2. Results of our pose estimation compared to the pose estimation provided by the Kinect V2 sensor.

Training type	MPJPE 3D	PAMPJPE 3D
CycleGAN	268.7	131.6
Unsupervised	442.1	199.6

7 Conclusion

In this paper, we propose an unsupervised framework to provide 3D pose estimation on single depth images to prevent musculoskeletal disorders in an industrial use-case. The approach relies on a synthetic dataset composed of synthetic depth images. Using an unsupervised image-to-image translation CycleGAN preserving geometric features, we refine these images to render realistic depth images. These images, coupled to the body joints annotations from the input synthetic images, allow us to efficiently train a simple 3D pose estimation network. Several experiments performed both on degraded synthetic data and on real data acquired in a simulated industrial context prove the efficiency of this image-to-image translation process to perform unsupervised pose estimation.

In future work, we could exploit temporal information to improve the results of both our CycleGAN and pose estimator. We could also introduce more complicated models.

References

1. Aleksynska, M., Berg, J., Foden, D., Johnston, H.E.S., Parent-Thirion, A., Vanderleyden, J.: Working conditions in a global perspective. Publications Office of the European Union (2019)
2. Alp Güler, R., Neverova, N., Kokkinos, I.: Densepose: dense human pose estimation in the wild. In: Proceedings of the IEEE Conference on Computer Vision and Pattern Recognition, pp. 7297–7306. IEEE (2018)
3. Aytar, Y., Castrejon, L., Vondrick, C., Pirsiavash, H., Torralba, A.: Cross-modal scene networks. IEEE Trans. Pattern Anal. Mach. Intell. **40**(10), 2303–2314 (2017)
4. Banerjee, T., et al.: Monitoring hospital rooms for safety using depth images. AI for Gerontechnology (2012)
5. Blanc-Beyne, T., Carlier, A., Charvillat, V.: Iterative dataset filtering for weakly supervised segmentation of depth images. In: 2019 IEEE International Conference on Image Processing, pp. 1515–1519. IEEE (2019)
6. Bogo, F., Kanazawa, A., Lassner, C., Gehler, P., Romero, J., Black, M.J.: Keep it SMPL: automatic estimation of 3D human pose and shape from a single image. In: Leibe, B., Matas, J., Sebe, N., Welling, M. (eds.) ECCV 2016. LNCS, vol. 9909, pp. 561–578. Springer, Cham (2016). https://doi.org/10.1007/978-3-319-46454-1_34
7. Cao, Z., Simon, T., Wei, S.E., Sheikh, Y.: Realtime multi-person 2D pose estimation using part affinity fields. In: Proceedings of the IEEE Conference on Computer Vision and Pattern Recognition, pp. 7291–7299. IEEE (2017)

8. Chan, C., Ginosar, S., Zhou, T., Efros, A.A.: Everybody dance now. In: Proceedings of the IEEE International Conference on Computer Vision, pp. 5933–5942. IEEE (2019)

9. Choi, Y., Choi, M., Kim, M., Ha, J.W., Kim, S., Choo, J.: StarGAN: unified generative adversarial networks for multi-domain image-to-image translation. In: The IEEE Conference on Computer Vision and Pattern Recognition. IEEE (2018)

10. Fang, H.S., Xie, S., Tai, Y.W., Lu, C.: RMPE: regional multi-person pose estimation. In: Proceedings of the IEEE International Conference on Computer Vision, pp. 2334–2343. IEEE (2017)

11. Farabet, C., Couprie, C., Najman, L., LeCun, Y.: Scene parsing with multiscale feature learning, purity trees, and optimal covers. In: Proceedings of the 29th International Conference on Machine Learning (2012)

12. Goodfellow, I., et al.: Generative adversarial nets. In: Advances in Neural Information Processing Systems, pp. 2672–2680 (2014)

13. Haque, A., Peng, B., Luo, Z., Alahi, A., Yeung, S., Fei-Fei, L.: Towards viewpoint invariant 3D human pose estimation. In: Leibe, B., Matas, J., Sebe, N., Welling, M. (eds.) ECCV 2016. LNCS, vol. 9905, pp. 160–177. Springer, Cham (2016). https://doi.org/10.1007/978-3-319-46448-0_10

14. He, K., Zhang, X., Ren, S., Sun, J.: Deep residual learning for image recognition. In: Proceedings of the IEEE Conference on Computer Vision and Pattern Recognition, pp. 770–778. IEEE (2016)

15. Ionescu, C., Papava, D., Olaru, V., Sminchisescu, C.: Human3. 6m: large scale datasets and predictive methods for 3D human sensing in natural environments. IEEE Trans. Pattern Anal. Mach. Intell. **36**(7), 1325–1339 (2013)

16. Isola, P., Zhu, J.Y., Zhou, T., Efros, A.A.: Image-to-image translation with conditional adversarial networks. In: Proceedings of the IEEE Conference on Computer Vision and Pattern Recognition, pp. 1125–1134. IEEE (2017)

17. James, S., et al.: Sim-to-real via sim-to-sim: data-efficient robotic grasping via randomized-to-canonical adaptation networks. In: Proceedings of the IEEE Conference on Computer Vision and Pattern Recognition, pp. 12627–12637. IEEE (2019)

18. Jiu, M., Wolf, C., Taylor, G., Baskurt, A.: Human body part estimation from depth images via spatially-constrained deep learning. Pattern Recogn. Lett. **50**, 122–129 (2014)

19. Kingma, D.P., Ba, J.: Adam: a method for stochastic optimization. In: International Conference on Learning Representations (2015)

20. Li, Y., Yuan, L., Vasconcelos, N.: Bidirectional learning for domain adaptation of semantic segmentation. In: Proceedings of the IEEE Conference on Computer Vision and Pattern Recognition, pp. 6936–6945. IEEE (2019)

21. Liu, M.Y., Breuel, T., Kautz, J.: Unsupervised image-to-image translation networks. In: Advances in Neural Information Processing Systems, pp. 700–708 (2017)

22. Mao, X., Li, Q., Xie, H., Lau, R.Y., Wang, Z.: Multi-class generative adversarial networks with the L2 loss function. arXiv preprint arXiv:1611.04076 (2016)

23. Marín-Jiménez, M.J., Romero-Ramirez, F.J., Muñoz-Salinas, R., Medina-Carnicer, R.: 3D human pose estimation from depth maps using a deep combination of poses. J. Vis. Commun. Image Represent. **55**, 627–639 (2018)

24. Mirza, M., Osindero, S.: Conditional generative adversarial nets. arXiv preprint arXiv:1411.1784 (2014)

25. Moon, G., Yong Chang, J., Mu Lee, K.: V2v-PoseNet: voxel-to-voxel prediction network for accurate 3d hand and human pose estimation from a single depth map. In: Proceedings of the IEEE Conference on Computer Vision and Pattern Recognition, pp. 5079–5088. IEEE (2018)

26. Newell, A., Yang, K., Deng, J.: Stacked hourglass networks for human pose estimation. In: Leibe, B., Matas, J., Sebe, N., Welling, M. (eds.) ECCV 2016. LNCS, vol. 9912, pp. 483–499. Springer, Cham (2016). https://doi.org/10.1007/978-3-319-46484-8_29

27. Punnett, L., Wegman, D.H.: Work-related musculoskeletal disorders: the epidemiologic evidence and the debate. J. Electromyogr. Kinesiol. **14**(1), 13–23 (2004)

28. Radford, A., Metz, L., Chintala, S.: Unsupervised representation learning with deep convolutional generative adversarial networks. In: International Conference on Learning Representations (2016)

29. Salimans, T., Goodfellow, I., Zaremba, W., Cheung, V., Radford, A., Chen, X.: Improved techniques for training GANs. In: Advances in Neural Information Processing Systems, pp. 2234–2242 (2016)

30. Shotton, J., et al.: Real-time human pose recognition in parts from single depth images. In: 2011 IEEE Conference on Computer Vision and Pattern Recognition, pp. 1297–1304. IEEE (2011)

31. Shrivastava, A., Pfister, T., Tuzel, O., Susskind, J., Wang, W., Webb, R.: Learning from simulated and unsupervised images through adversarial training. In: Proceedings of the IEEE Conference on Computer Vision and Pattern Recognition, pp. 2107–2116. IEEE (2017)

32. Ulyanov, D., Vedaldi, A., Lempitsky, V.: Instance normalization: the missing ingredient for fast stylization. arXiv preprint arXiv:1607.08022 (2016)

33. Wang, K., Zhai, S., Cheng, H., Liang, X., Lin, L.: Human pose estimation from depth images via inference embedded multi-task learning. In: Proceedings of the 24th ACM International Conference on Multimedia, pp. 1227–1236. ACM (2016)

34. Wang, T.C., Liu, M.Y., Zhu, J.Y., Tao, A., Kautz, J., Catanzaro, B.: High-resolution image synthesis and semantic manipulation with conditional GANs. In: Proceedings of the IEEE Conference on Computer Vision and Pattern Recognition, pp. 8798–8807. IEEE (2018)

35. Zhang, H., et al.: StackGAN: text to photo-realistic image synthesis with stacked generative adversarial networks. In: Proceedings of the IEEE International Conference on Computer Vision, pp. 5907–5915. IEEE (2017)

36. Zhang, X., Wong, Y., Kankanhalli, M.S., Geng, W.: Unsupervised domain adaptation for 3D human pose estimation. In: Proceedings of the 27th ACM International Conference on Multimedia, pp. 926–934. ACM (2019)

37. Zhou, X., Sun, X., Zhang, W., Liang, S., Wei, Y.: Deep kinematic pose regression. In: Hua, G., Jégou, H. (eds.) ECCV 2016. LNCS, vol. 9915, pp. 186–201. Springer, Cham (2016). https://doi.org/10.1007/978-3-319-49409-8_17

38. Zhu, J.-Y., Krähenbühl, P., Shechtman, E., Efros, A.A.: Generative Visual manipulation on the natural image manifold. In: Leibe, B., Matas, J., Sebe, N., Welling, M. (eds.) ECCV 2016. LNCS, vol. 9909, pp. 597–613. Springer, Cham (2016). https://doi.org/10.1007/978-3-319-46454-1_36

39. Zhu, J.Y., Park, T., Isola, P., Efros, A.A.: Unpaired image-to-image translation using cycle-consistent adversarial networks. In: The IEEE International Conference on Computer Vision. IEEE (2017)

40. Zhu, J.Y., et al.: Toward multimodal image-to-image translation. In: Advances in Neural Information Processing Systems, pp. 465–476 (2017)

Data Generation Process Modeling
for Activity Recognition

Massinissa Hamidi[✉] and Aomar Osmani

Laboratoire LIPN-UMR CNRS 7030, Univ. Sorbonne Paris Nord,
Villetaneuse, France
{hamidi,ao}@lipn.univ-paris13.fr

Abstract. The dynamics of body movements are often driven by large
and intricate low-level interactions involving various body parts. These
dynamics are part of an underlying data generation process. Incorporating the data generation process into data-driven activity recognition
systems has the potential to enhance their robustness and data-efficiency.
In this paper, we propose to model the underlying data generation process and use it to constrain training of simpler learning models via sample selection. As deriving such models using human expertise is hard,
we propose to frame this task as a large-scale exploration of architectures in charge of relating sensory information coming from the data
sources. We report on experiments conducted on the Sussex-Huawei locomotion dataset featuring a sensor-rich environment in real-life settings.
The derived model is found to be consistent with existing domain knowledge. Compared to the basic setting, our approach achieves up to 17.84%
improvement, by simultaneously reducing the number of required data
sources by one-half. Promising results open perspectives for deploying
more robust and data-efficient learning models.

Keywords: Activity recognition · Domain models · Neural
architectures

1 Introduction

Proliferation of internet of things technologies allows the emergence of sensor-
rich environments where sensing-enabled devices constitute sources of diverse
forms of information describing their surrounding. These sources offer a broad
range of perspectives allowing to perform robust activity recognition [33]. Indeed,
positioned in different places and featuring various sensing modalities, these
sources of information generate a lot of data which, if exploited rightfully, could
provide many advantages like improved signal-to-noise ratio, reduced ambiguity,
and enhanced reliability [22].

Learning tasks that emerge in these sensor-rich environments are profoundly
structured. This is the case of wearable technologies with the considered Sussex-
Huawei locomotion-transportation (SHL) dataset [14] studied in this paper. Our

© Springer Nature Switzerland AG 2021
Y. Dong et al. (Eds.): ECML PKDD 2020, LNAI 12460, pp. 374–390, 2021.
https://doi.org/10.1007/978-3-030-67667-4_23

work focuses on recognizing mobility-related human activities from data sources materialized by on-body sensors placed at different locations of the body following a pre-defined and fixed topology. It has been observed that for a given activity, there is the emergence of dynamics that involve very specific positions of the body parts for which a set of specific modalities can provide complementary information. Primarily, what characterizes these dynamics is the fact that they define largely the activity in question [5,11,25,35].

The dynamics of body movements are part of an underlying data generation process (DGP) and a long line of research, e.g. [8,30,31], proposed to incorporate this kind of prior knowledge into activity recognition models. Specifically, authors in [31] derive 3D body skeleton-based representations while other works encode prior domain knowledge using ontology-based representations [29,36]. These representations are then used to constrain training of activity recognition models. While incorporation of prior knowledge about the dynamics of body movements into learning systems improves performances and is appealing in terms of interpretability, relying solely on human expertise to derive models for these dynamics is hard [39]. Indeed, these dynamics are often driven by large and intricate low-level interactions involving various body parts [21].

In this paper, we propose a novel approach to derive and incorporate DGP into activity recognition models. Our approach enhances the performance of activity recognition models through two major steps. It first constructs a model of the DGP via a large-scale exploration of a neural architecture space. Then, it selects highly confident data sources for inclusion in the final training set using a variance-based importance estimation algorithm.

Our contributions can be summarized as follows. (1) We frame the derivation of the data generation process as an exploration of the neural architecture space; (2) We propose to estimate the relative importance of data sources and their interactions using a variance-based method; (3) Extensive experiments show the effectiveness of combining the data generation process through selection of highly confident data sources. In particular, we achieve improvement of recognition performances of up to 17.84% over the baseline, which is accompanied by a substantial reduction of required data; (4) We perform a comprehensive comparative analysis using different instantiations of the proposed approach (8 exploration strategies) on 4 different representative related datasets.

The rest of the paper is organized as follows. In Sect. 2, we define the problem of data source selection based on the DGP and Sect. 3 presents the details of our approach. In Sect. 4, we detail the empirical evaluation of the proposed approach. We provide a related work in Sect. 5 and finally, Sect. 6 concludes this paper with a summary and future works.

2 Problem Statement

This section defines the problem of modeling the data generation process in the context of activity recognition from sensor-rich environments.

2.1   Preliminaries

We consider settings where a collection S of M sensors (also called data generators or data sources), denoted $\{s_1, \ldots, s_M\}$, are carried by the user during daily activities and capture the body movements. Each sensor s_i generates a stream $\mathbf{x}^i = (x_1^i, x_2^i, \ldots)$ of observations of a certain modality. Furthermore, each observation is composed of channels, e.g. the x, y, and z axes of an accelerometer. In our work, we exploit mainly body-motion modalities that are often used in human activity recognition applications. The goal is to recognize a set $\mathcal{Y}$ of activities, like running or biking, performed in sensor-rich environments.

Definition 1 (Modality). *A modality is a form of perception that conveys a particular perspective about a given phenomenon. E.g. acceleration, gyroscopic and magnetometric observations are different modalities each describing, in a particular way, the motions of the body.*

Definition 2 (Data source). *a given data source (or sensor), denoted s, is characterized by two main attributes: the first is the **modality** being produced by the sensor and the second one is the **position** where the data source is located on the body. A data source is then uniquely defined with these two attributes.*

2.2   Problem Definition

Human activities are largely determined by the dynamics of the gestures. Indeed, each activity is characterized by a different set of gestures which in turn involve specific body parts. In the case of wearable technologies, where these body parts are equipped with data sources, often, focusing on these specific data sources, allows recognizing a given activity precisely. Therefore, our approach attempts to select subsets of data sources that are highly confident and informative with regards to these dynamics, to create a curated training set for model training. In this work, we focus on two different notions that encode these dynamics: importance of a single data source and degree of interaction among a set of data sources.

Definition 3 (Importance). *Given a data source s_i that is attached to a given body part and an activity y, the importance of s_i with regards to activity y, denoted $\mu_i^y \in [0,1)$, is defined as a quantity that represents the relative involvement of that body part in the dynamics of the gestures pertaining to that activity.*

Definition 4 (Interaction). *An interaction involves two or more data sources and is defined as their degree of dependence regarding the relative involvement of the body parts, they are attached to, in the dynamics of the gestures. The greater the degree, the more interacting the data sources. Given a set of interacting data sources, $S \subset \mathcal{S}$, their degree of interaction is denoted by $\mu_S^y \in [0,1)$. Specifically, in the case of two interacting data sources, s_i and s_j, it is denoted μ_{ij}^y.*

Problem 5 (Data source selection based on DGP). Let $DGP : \wp(\mathcal{S}) \times \mathcal{Y} \to [0,1)$ be the data generation process, which gives, for each activity $y \in \mathcal{Y}$ the influence of a set of data sources $S \subset \mathcal{S}$. The goal is to use DGP as an indicator function to select data sources (or samples) that are highly confident and informative to be included in the final training set of activity recognition models. Let $\tau_{imp} \in [0,1)$ and $\tau_{int} \in [0,1)$ be two parameters that determine the thresholds above which a given set of data sources $S \subset \mathcal{S}$ can be selected. It follows that the subsets of interacting data sources pertaining to activity $y \in \mathcal{Y}$, denoted $\mathcal{S}_y$, is defined as
$$\mathcal{S}_y := \{s_i \in \mathcal{S} | DGP(\{s_i\}, y) = \mu_{s_i}^y \geq \tau_{imp}\} \cup \{S \subsetneq \mathcal{S} | DGP(S, y) = \mu_S^y \geq \tau_{int}\}$$

Learning using curated sources of information is widely used in machine learning [17,38]. The DGP in the Problem 5 presents a natural solution for selecting such sources in the context of activity recognition from sensor-rich environments.

3   Approach

Our approach enhances the performance of activity recognition models through two major steps: (1) construct a model of the DGP as described in the Problem 5 using an architecture space as a surrogate model (proxy), and (2) select highly confident and informative data sources for inclusion in the final training set using a variance-based importance estimation algorithm. These two steps are described in the following and Algorithm 1 outlines the complete process.

3.1   Architecture Space as Proxy for the DGP

We use the space defined by multimodal analysis architectures as a proxy for the dynamics of the body movements. The exploration of this architecture space serves, then, to derive the DGP as defined in Problem 5.

An architecture is defined as a set of architectural components responsible for extracting valuable insights, in the form of features, from the observations and efficiently fusing different data sources carrying different modalities and various spatial perspectives. We distinguish four types of architectural components: *feature extraction* (FE), *feature fusion* (FF), *decision fusion* (DF), and *analysis unit* (AU) as defined in [2]. These are illustrated in Fig. 1 (left). An architectural component takes as inputs either raw data, features, or decisions and outputs either a feature or a decision. The way a given component processes each individual input is controlled by a hyperparameter.

It is convenient to represent an architecture as a directed acyclic graph where the architectural components are connected together using valued edges. We associate a value (hyperparameter) h_u^v with every edge in the directed graph that connects two components C_u and C_v. These values control how architectural components process each individual input and by the same occasion their influence on the overall architecture performance. We refer to the set of all hyperparameters of a given architecture by $\mathcal{H}$.

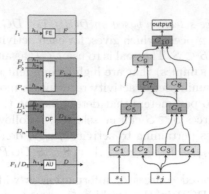

Fig. 1. (Left) feature extraction and multimodal fusion components defined in [2]. Feature extraction (FE), feature fusion (FF), decision fusion (DF), and analysis unit (AU). These building blocks can be combined in order to form feature-level, decision-level, and hybrid multimodal analysis. Additionally, the hyperparameters h_i controlling the effects of each individual input are depicted. (Right) An illustration of an architecture where each node corresponds to a component. An edge from component C_u to component C_v denotes that C_v receives the output of C_u as input.

We focus, particularly, on the insights that stem from tuning and adapting these architectures, through their hyperparameters and specifically those controlling the influence of the data sources. At each layer of a given architecture, setting the right combination of hyperparameters is critical. In particular, choosing the right instantiation for the features learning and sensor fusion components can lead to an architecture capable of building, from the various data sources, an original set of features which is suitable for recognizing a given activity. We take into account the following assumption: let $\mathcal{H}_s \subsetneq \mathcal{H}$ be the set of hyperparameters controlling the impact of a given data source s. The global impact of $\mathcal{H}_s$ on the recognition performances represents the impact of the data source s.

The problem of modeling the DGP becomes, then, an exploration of the architecture (hyperparameter) space. This exploration is determined by three aspects: (1) a search space which defines the architectural components and the type of branching that is allowed for the architectures (e.g. convolutional layers); (2) a search strategy which decides how the exploration of the space should be carried (e.g. Bayesian optimization of the hyperparameters); and (3) a performance estimation strategy (e.g. sequence classification problem) [9].

In the case of convolutional layers, for example, architectures can be constructed by stacking a series of Conv1d/ReLU/MaxPool blocks followed by Fully-Connected/ReLU layers. Denote by ν_k the validation loss of a particular instantiation k of the set of hyperparameters. The exploration strategy tries to find an architecture k^* that minimizes the validation loss $\nu_k^*(w^*)$. The weights w associated with the architecture are obtained by optimizing the weights of the components using, for example, a gradient descent algorithm over a predefined class of functions.

Given an exploration budget B, the exploration strategy yields a series of validation losses $\nu_1, \ldots, \nu_B$ including partial validation losses pertaining to individual activities. The task of modeling the DGP, therefore, reduces to find a link between these validation losses and the impact of each individual data source.

3.2 Variance-Based Importance Estimation

Let $\mathcal{V}$ be a set of validation losses where each validation loss ν_k represents the estimated performance of a particular instantiation of the hyperparameters. To estimate the importance of each individual data source, we decompose the non-linear relation f described by $\mathcal{V}$ as follows

$$f(\mathcal{S}, y) = \mu_0^y + \sum_{i=1}^{M} \mu_i^y(s_i) + \sum_{i \neq j} \mu_{ij}^y(s_i, s_j) + \cdots + \mu_{1...M}^y(s_i, \ldots, s_M) \quad (1)$$

a constant mean μ_0^y plus first-order effects (μ_i^y), plus second-order effects (μ_{ij}^y) and so on. The lower the variance induced by a data source, the higher its influence of on the non-linear relation f. This formulation corresponds to an additive expansion and the variance of each term can be estimated using the functional analysis of variance (fANOVA) [18]. It can be quantified using the efficient implementation proposed in [19] which is based on a linear-time algorithm for computing marginals of random forest predictions.

As we have access to the set of validation losses indexed by the hyperparameters instantiation, in order to estimate the decomposition, we have to determine, first, the correspondence between each individual data source and the set of hyperparameters that controls their influence.

Data Source/Hyperparameters Correspondence. Given an architecture A, we determine a correspondence, $Corr_A : \mathcal{S} \to \wp(\mathcal{H} \times \mathbb{R})$, between each individual data source and the hyperparameters that influence their effects, as follows:

$$Corr_A(s) = \bigcup_{(u,v) \in s \to^* t} < h_u^v, w > \quad (2)$$

where $s \to^* t$ denotes all paths in the architecture that have s as a source and t as sink, h_u^v the hyperparameter associated with edge (u, v), and w corresponds to a weight computed as $w = \frac{\omega_1 \cdot dist(s,v) + \omega_2 \cdot \delta^-(v)}{\omega_1 + \omega_2}$ which ponders the correspondence of a given hyperparameter h_u^v depending on its distance ($dist(s, v)$) to the input s and the number of incoming edges to the component v ($\delta^-(v)$). The weight parameters $\omega_1, \omega_2 \in [0, 1)$ are both set to $\frac{1}{2}$. In the case an edge is shared by many different paths, we sum the weights assigned to the corresponding hyperparameter following each path.

4 Experiments and Results

In this section, we perform empirical evaluation of the proposed approach. We first derive a model of the data generation process from the SHL dataset using

---

Algorithm 1: DGP-based Data Sources Selection

Input : (i) $\{\mathbf{x}^i\}_{i=1}^M$ streams of annotated observations generated by the data sources, (ii) B exploration budget, (iii) τ_{imp}, (iv) τ_{int},
(v) E exploration strategy, (vi) O maximal order of interaction effects

Result: $\mathcal{S}_y \in \wp(\mathcal{S})|_{y \in \mathcal{Y}}$, the sets of most important and interacting data sources for each individual human activity

1 **begin**
2    $\mathcal{V} \leftarrow \varnothing$, $\mathcal{S}_y \leftarrow \varnothing$
3    $(X, Y) \leftarrow$ segmentation($\{\mathbf{x}^i\}_{i=1}^M$) ;   % *preprocess for sequence classif. pblm*
4    $\mathcal{V} \leftarrow E(X, Y, B)$;   % *architecture space exploration*
5   **foreach** $s \in \mathcal{S}$ **do**
6 | $\{(h, w)|h \in \mathcal{H}, w \in \mathbb{R}\}_s \leftarrow Corr_A(s)$;   % *DS/HPs correspondence*
7   **end**
8    $\{\mu_S^y | S \subsetneq \mathcal{S}\} \leftarrow$ QuantifyImportance($\mathcal{V}, \{\{(h, w)\}_s\}, O$) ;   % *Section 3.2*
9   **foreach** *activity* $y' \in \mathcal{Y}$ **do**
10   **foreach** $\mu_S^y \in \{\mu_S^y | S \subsetneq \mathcal{S}, y = y'\}$ **do**
11   **if** $\mu_S^y > \tau_{int}$ **then**
             ; % *use* τ_{imp} *if* $S = s$
12              $\mathcal{S}_y \leftarrow \mathcal{S}_y \cup S$
13   **end**
14   **end**
15   **end**
16   **return** $\{\mathcal{S}_y\}_{y \in \mathcal{Y}}$
17 **end**

---

different space exploration strategies. We, then, demonstrate the effectiveness of incorporating the derived model into four different activity recognition datasets (including SHL). Code to reproduce the experiments is publicly made available[1].

4.1 Datasets

We use the SHL dataset primarily to derive the data generation model. The derived model is then incorporated into the SHL dataset itself and three other datasets including (1) *USC-HAD* [42] containing body-motion modalities of 12 daily activities collected from 14 subjects (7 male, 7 female) using MotionN-ode, a 6-DOF inertial measurement unit, that integrates a 3-axis accelerometer, 3-axis gyroscope, and a 3-axis magnetometer; (2) *HTC-TMD* [41] containing accelerometer, gyroscope, and magnetometer data all sampled 30 Hz from smartphone built-in sensors in the context of energy footprint reduction; and (3) *US-TMD* [6] featuring motion data collected from 13 subjects (9 male, 4 female) using smartphone built-in sensors.

---

[1] Software package and code to reproduce empirical results are publicly available at: https://github.com/sensor-rich/shl-nas.

SHL Dataset. The SHL dataset [15] is a highly versatile and precisely annotated dataset dedicated to mobility-related human activity recognition. In contrast to related representative datasets like [6,41–43], the SHL dataset provides, simultaneously, multimodal and multilocation locomotion data recorded in real-life settings. There are in total 16 modalities including accelerometer, gyroscope, cellular networks, WiFi networks, audio, etc. making it suitable for a wide range of applications. Data collection was performed by each participant using four smartphones simultaneously placed in different body locations: *Hand*, *Torso*, *Hips*, and *Bag*. These four positions define the topology that allows us to model and leverage the dynamics of body movements for activity recognition models. Among the 16 modalities of the original dataset, we select the body-motion modalities to be included in our experiments, namely: accelerometer, gyroscope, magnetometer, linear acceleration, orientation, gravity, and in addition, ambient pressure.

4.2 Training Details

We use Tensorflow [1] for building the neural architectures. In this work, we construct neural architectures by stacking Conv1d/ReLU/MaxPool blocks. These blocks are followed by a Fully Connected/ReLU layers. Architecture performance estimation is based on the validation loss and is framed as a sequence classification problem. As a preprocessing step, annotated input streams from the SHL dataset are segmented into sequences of 6000 samples which correspond to a duration of 1 min. given a sampling rate of 100 Hz. For weight optimization, we use stochastic gradient descent with Nesterov momentum of 0.9 and a learning-rate of 0.1 for a minimum of 12 epochs (we stop training if there is not improvement). Weight decay is set to 0.0001. Furthermore, to make the neural networks more stable, we use batch normalization on top of each convolutional layer [20].

Different exploration strategies will lead to different sets of hyperparameter instantiations. In our experiments, we instantiate our approach with various exploration strategies. We use the Microsoft-NNI toolkit[2] which provides a comprehensive list of exploration strategies, in particular, those based on hyperparameter tuning, including (1) exhaustive search (random search [3], and grid search); (2) heuristic search (naïve evolution [34], anneal [4], and hyperband [23]); and (3) sequential model-based optimization (Bayesian optimization hyperband [10], tree-structured Parzen Estimator [4], and Gaussian process tuner [4]).

We quantify the influence of data sources using the efficient implementation of fANOVA proposed in [19], which is based on a linear-time algorithm for computing marginal predictions in random forests. Interaction structure of the data sources is estimated using fanova-graph [26].

---

[2] https://github.com/microsoft/nni.

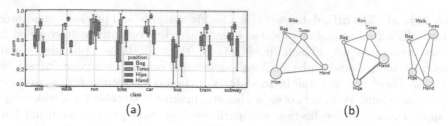

(a)   (b)

Fig. 2. (a) Contribution of the data sources to the overall recognition performances of each human activity. (b) Estimated interaction structure (fANOVA graph [26]) of the data sources for 3 different activities (bike, run, and walk). Data sources are grouped by their respective positions. The circumference of the circles represents main effects (importance), the thickness of the edges represents total interaction effects.

4.3   Performance Evaluation

In our experiments, each architecture is evaluated with a 10-fold meta-segmented cross-validation to avoid the problem of overestimation of the quality of results induced by standard cross-validation procedure [16]. This technique relies on a modified partitioning procedure that alleviates the neighborhood bias, which results from the high probability that adjacent (moreover, overlapping) segments fall into training and test-set at the same time.

We use the f1-score in order to assess performances of the architectures. We compute this metric following the method recommended in [12] to alleviate bias that could stem from unbalanced class distribution. Given the usual definition of precision $Pr^{(i)}$ and recall $Re^{(i)}$ for the ith fold, we compute the f1-score by averaging its different components obtained for each fold as $F_{avg} = \frac{1}{k} \cdot \sum_{i=1}^{k} F^{(i)}$ where $F^{(i)} = 2 \cdot \frac{Pr^{(i)} \cdot Re^{(i)}}{Pr^{(i)} + Re^{(i)}}$, if both $Pr^{(i)}$ and $Re^{(i)}$ are defined. The i-super-scripted measures correspond to measures obtained when the ith fold is used as the test set.

4.4   Evaluation of the Data Generation Model

Here we evaluate the data generation model that is derived using our proposed approach. We specifically assess the plausibility of the derived important data sources and their interactions based on a comprehensive set of studies around activity recognition. These studies are compiled into a data generation model that we refer to as human expertise-based data generation model *HExp*. Furthermore, we instantiate our proposed approach using different space exploration strategies and compare the derived knowledge using each strategy.

Figure 2 shows how data sources grouped by their respective positions contribute to the overall recognition of each human activity. Figure 3 and Table 1 summarize results of the variance-based importance estimation conducted using the fANOVA framework (Sect. 3.2). The estimated first and second order effects of the hyperparameters controlling the importance of each considered modality are illustrated, respectively.

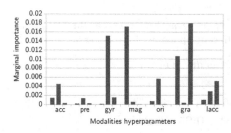

Fig. 3. Individual marginal importance of the kernel size (convolution) hyperparameters controlling the impact of each modality. 3 layers per modality are shown.

Table 1. Most important interactions of the kernel size (ks) hyperparameter. Hyperparameters are grouped by the corresponding modalities that they control.

Hyperparam.	Interaction ($\times 10^{-4}$)
(ks_{gyr}^2, ks_{gra}^2)	9.2778
(ks_{mag}^1, ks_{ori}^2)	7.0166
(ks_{gyr}^2, ks_{ori}^2)	5.5122
(ks_{acc}^1, ks_{mag}^1)	4.0382
(ks_{pre}^1, ks_{gyr}^3)	2.3154
(ks_{gyr}^3, ks_{mag}^1)	2.2472

Data Source Location. Results in Fig. 2 show that the contributions of data sources for recognizing bus, train, and subway related activities are equivalent. More variability appears in the case of the bus activity. Data sources located on the hips, for their part, yield overall the smallest variability. This variability is to some extent more important in the case of bus and run activities but stays in fairly acceptable terms. In the case of car-related activities, relying on the data sources located on the hips seems to be sufficient, this position yielding the best models overall (90%-95% f1 score). The same observation can also be made regarding bike and walk activities where Hips data sources seem to discriminate them accurately. This may be explained by the tight link that exists between these activities and the hips position: biking, walking and conducting a car involve specific repetitive patterns that are their hallmark [6].

Data Source Modality. From modalities perspective, data sources carrying gravity, gyroscope, and magnetometer account for a large part of the variability that is observed on the recognition performances. Surprisingly, another set of modalities emerges from the derived model rather than the accelerometric data which is considered to be one of the most important modalities in representative related work [35,40]. Indeed, the respective individual marginal importance of the accelerometer-related data lies approximately around 0.004 and does not exceed 0.006, while those of gravity, gyroscope, as well as magnetometer, reach 0.01 and almost 0.02 (See Fig. 3). This observation is further confirmed when we analyze the pairwise marginals of the hyperparameters controlling the set of three modalities mentioned above.

Impact of the Space Exploration Strategies. Here, we compare the data generation models obtained using different space exploration strategies. Specifically, we compare the derived subsets of data sources in terms of their level of agreement with those aggregated in the human expertise-based data generation model (HExp). We use for this, Cohen's kappa coefficient [7] which measures the agreement between two raters. We also compare the average recognition performance

Table 2. Degree of agreement with human expertise and average cardinality of the derived sets of data sources obtained using different space exploration strategies.

Exploration strategy	Agreement	ν_k on avg
Exhaustive search		
Random Search	0.156 ± 0.04	67.12%
Grid Search	$\mathbf{0.251 \pm 0.05}$	66.78%
Heuristic search		
Naïve evolution	0.347 ± 0.12	73.35%
Anneal	$\mathbf{0.481 \pm 0.05}$	75.47%
Hyperband	0.395 ± 0.08	74.2%
Sequential Model-Based		
BOHB	0.734 ± 0.03	84.25%
TPE	0.645 ± 0.1	83.87%
GP Tuner	$\mathbf{0.865 \pm 0.02}$	84.95%

ν_k of the explored architectures which can indicate many aspects concerning the exploration strategy, like the concentration of important sets of data sources in regions of the architecture space.

Results in Table 2 show that the sequential model-based exploration strategies are indeed better than heuristic search-based ones. Exhaustive search-based strategies are far behind with an agreement that does not exceed 3. It is worth mentioning that even with a larger exploration budget allowed to exhaustive search, using these kinds of strategies does not allow to derive a valuable data generation process. This could be explained by the fact that important sets of data sources are concentrated in very specific regions the grid search, for example, can not capture. As the GP tuner yields the highest agreement with HExp, in the following, we will, first, use the data generation process derived using this strategy to assess the effectiveness of incorporating such knowledge into activity recognition models.

4.5 Effectiveness of the Data Generation Model

In this second experiment, we incorporate the derived data generation model into activity recognition models via sample selection. We select highly informative data sources to form training sets. During the training phase, activity recognition models are encouraged to concentrate on the provided subsets of data sources to learn the corresponding human activities. We refer to this setting as *w-DGP*, which stands for, with data generation process.

For this, we construct activity recognition models based on neural networks, similar to the architectures used to derive the data generation model, but restricted to 3 Conv1d/ReLU/MaxPool stacked blocks. These blocks are

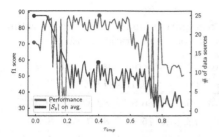

Fig. 4. Recognition performances as a function of the data source importance threshold τ_{imp}. In parallel, the cardinality on average of the subsets $|\mathcal{S}_y|$ used to train the models is shown. The leftmost points correspond to a configuration where all data sources are used, i.e., no DGP.

Table 3. Comparison of different DGP incorporation settings in terms of recognition performances. Scores of column w-DGP correspond to top-performing models selected while varying the data source importance threshold τ_{imp}.

Dataset	Performances		
	wo-DGP	w-HExp	w-DGP
USC-HAD	72.1%	75.38%	89.33%
HTC-TMD	74.4%	77.16%	78.9%
US-TMD	71.32%	80.28%	83.64%
SHL	70.86 %	77.18%	88.7%

followed by a Fully Connected/ReLU layers. The weights of the layers corresponding to all inputs are optimized during training without distinction, the constraining being specified via data augmentation. Indeed, in this setting, for each subset of interacting data sources, we perform data augmentation by assigning values, drawn from a normal distribution, to the unimportant data sources. The goal is to make the neural network insensitive to the remaining inputs. We provide training examples to the neural network according to the given subsets of interacting data sources that we extract from the derived model. Furthermore, we experiment with different values of τ_{int} and τ_{imp} to extract the subsets of data sources.

For comparison, we train the activity recognition models on the whole data sources of each dataset, i.e., without incorporation of the derived data generation model. These models constitute our baselines and we refer to this setting as *wo-DGP*. In addition, we also incorporate the data generation model based on human expertise (HExp). We refer to this setting as *w-HExp*. Table 3 compares recognition performances obtained, on each dataset, using these settings. Overall, we obtain substantial improvements for all datasets when incorporating a data generation process (either w-HExp or w-DGP). It is to note, though, that for HTC-TMD, we get a smaller improvement compared to the other datasets. This could be related to the limited number of modalities and unavailability of the precise location of the data sources.

Figure 4 shows the evolution of the obtained recognition performances depending on the parameters τ_{int} and τ_{imp}. In addition, this figure illustrates the average number of data sources, that are included in the subsets, depending on these two thresholds. In particular, when, for example, τ_{imp} and τ_{int} are set to 0, all data sources are included. We find that the neural networks trained with smaller subsets of data sources perform better than the baseline and most of the settings which rely on a higher number of data sources. Noticeably, we

Table 4. Recognition performances of activity recognition models while incorporating the data generation models derived using different space exploration strategies.

Dataset	Exhaustive search		Heuristic search			Sequential model-based		
	Random	Grid	Naïve	Anneal	HB	BOHB	TPE	GP Tun
USC	79.28%	79.58%	80.76%	83.56%	85.27%	86.66%	82.37%	**89.33%**
HTC-TMD	76.34%	75.17%	74.98%	73.18%	77.45%	75.86%	**80.13%**	78.9%
US-TMD	74.14%	72.21%	79.71%	81.13%	80.80%	79.17%	**84.39%**	83.64%
SHL	72.2%	71.32%	79.46%	84.16%	82.33%	84.22%	86.7%	**88.7%**

get a recognition performance of 88.7% $\pm$ 0.6, measured by the f1-score, using subsets containing on average 12 data sources. Thus, an improvement over the baseline of 17.84% in terms of recognition performances and a reduction of one-half concerning the required quantities of data. Surprisingly, we do not see a lot of bad subsets of interacting data sources for $0.2 \leq \tau_{imp} \leq 0.6$, where the number of data sources per subset is confined between 5 and 12. It is also worthy to note that in some configurations where $|\mathcal{S}_y| = 13$, the trained model performs badly (less than 40%$\pm$0.16 f1-score). In the contrary, for smaller subsets ($|\mathcal{S}_y| \leq 5$), trained models get high recognition performances (more than 80%$\pm$0.05 f1-score). A Deeper inspection of these configurations reveals that the location of selected data sources plays an important role, in particular, the latter subsets are mainly composed of hips data sources.

4.6   Alternative Exploration Strategies

In the previous experiment, we constrain training of activity recognition models using data generation model derived using the Gaussian process tuner as it had the highest degree of agreement with HExp. Since the exploration strategies tend to favor different regions of the architecture space, we hypothesize that the derived models will be characterized by variety in terms of combinations of data sources but will still hold the same property, which is being highly informative with regards to the dynamics of body movements. Here we evaluate the effectiveness of the data generation models derived using the other exploration strategies. Table 4 presents the results obtained for this setting on each individual dataset.

Note that TPE outperforms GP tuner in the case of HTC-TMD and US-TMD datasets. It is also interesting to note that even though exhaustive search strategies have a low degree of agreement with HExp, incorporation of their corresponding data generation models is competitive for both HTC-TMD and USC-HAD, which can be explained by the ability of our approach to derive knowledge that is hardly captured by the sole human expertise.

5 Related Work

In our work, we proposed to derive and incorporate the DGP into activity recognition models. Incorporating domain knowledge into activity recognition models is particularly appealing and attracted lots of research.

A long line of research, e.g. [8,21,30,31,39], proposed to incorporate the 3D body skeleton-based representation into activity recognition models. Specifically, authors in [31] estimate centroids for upper, middle and lower body and use slopes of the segments delimited by these centroids in order to represent the posture in terms of the overall orientation of the upper and lower body. In [30], authors introduced a representation based on the calculation of spherical angles between selected joints and the respective angular velocities. They used their system for real-time tracking of human activities. Other works encode prior domain knowledge using ontology-based representations [29,36] which are then used to constrain training of activity recognition models. While incorporation of domain knowledge into learning processes is beneficial, the way it is done differs substantially from one approach to another. In [21], the obtained ontology serves as a basis for constructing a network of Bayesian inference while in [31], the constructed representations help the neural networks to self-organize.

Beyond activity recognition, many other applications leverage domain models to enforce certain conditions or equations, which are part of prior knowledge, within machine learning models. In [27,37], authors propose to incorporate domain knowledge, like known laws of physics, by constraining neural networks via regularization. Their settings introduce new challenges for encoding knowledge into appropriate loss functions and avoiding trivial solutions in the constraint space. In the same vein, authors in [28] propose to make use of a more experimented model, a proxy, that is responsible for selecting samples in order to train new generations of models in the context of industrial monitoring. New paradigms, like Vapnik's learning using privileged information [38] and Hinton's distilled knowledge [17], propose to incorporate high capacity models, similar to proxy's, called "intelligent teachers" into machine learning models.

A growing volume of work proposes to exploit domain knowledge to improve the performances of machine learning models. Our experiments encourage an even broader range of future applications, where larger and more experienced models like the proposed neural architecture space, form surrogates for prior domain knowledge and provide guidance to simpler models via sample selection.

6 Summary and Future Work

We presented in this paper a novel approach for deriving a model of the data generation process underlying sensor-rich environments. We framed this task as an exploration of the neural architecture space and proposed a variance-based method to estimate the relative importance of data sources and their interactions. Incorporating the derived data generation model into activity recognition models allows us to obtain consistent improvement in recognition performances using a reduced number of data sources. We performed a comprehensive comparative study on various representative datasets using different instantiations of the space exploration strategy. Obtained promising results open perspectives for the development of more robust and data-efficient learning systems pertaining to the internet of things.

In this work, we used exploration strategies based on hyperparameters tuning. An alternative way is to have fine-grained control on the architectural components that make up the neural architectures allowing for more specialized architectures. Recent approaches in neural architecture search, such as ENAS [32] and DARTS [24], enable this kind of granularity. Furthermore, recent advances in weight-agnostic neural architectures [13] and the possibility of building architectures that are completely specialized in a given task and requiring no further weight adjustments open perspectives for these kinds of approaches. As part of our future work, we plan to derive more precise data generation processes using these fine-grained control mechanisms.

References

1. Abadi, M., et al.: Tensorflow: a system for large-scale machine learning. In: OSDI, vol. 16, pp. 265–283 (2016)
2. Atrey, P.K., Hossain, M.A., El Saddik, A., Kankanhalli, M.S.: Multimodal fusion for multimedia analysis: a survey. Multimed. Syst. **16**(6), 345–379 (2010). https://doi.org/10.1007/s00530-010-0182-0
3. Bergstra, J., Bengio, Y.: Random search for hyper-parameter optimization. JMLR **13**, 281–305 (2012)
4. Bergstra, J.S., Bardenet, R., Bengio, Y., Kégl, B.: Algorithms for hyper-parameter optimization. In: NIPS, pp. 2546–2554 (2011)
5. Bevilacqua, A., MacDonald, K., Rangarej, A., Widjaya, V., Caulfield, B., Kechadi, T.: Human activity recognition with convolutional neural networks. In: Brefeld, U., et al. (eds.) ECML PKDD 2018. LNCS (LNAI), vol. 11053, pp. 541–552. Springer, Cham (2019). https://doi.org/10.1007/978-3-030-10997-4_33
6. Carpineti, C., Lomonaco, V., Bedogni, L., Di Felice, M., Bononi, L.: Custom dual transportation mode detection by smartphone devices exploiting sensor diversity. In: International Conference on Pervasive Computing and Communications Workshops, pp. 367–372. IEEE (2018)
7. Cohen, J.: A coefficient of agreement for nominal scales. Educ. Psychol. Measur. **20**(1), 37–46 (1960)
8. Dhiman, C., Vishwakarma, D.K., Aggarwal, P.: Skeleton based activity recognition by fusing part-wise spatio-temporal and attention driven residues. arXiv preprint arXiv:1912.00576 (2019)
9. Elsken, T., Metzen, J.H., Hutter, F.: Neural architecture search: a survey. JMLR **20**(55), 1–21 (2019)
10. Falkner, S., Klein, A., Hutter, F.: Bohb: robust and efficient hyperparameter optimization at scale. In: ICML, pp. 1437–1446 (2018)
11. Foerster, F., Smeja, M., Fahrenberg, J.: Detection of posture and motion by accelerometry: a validation study in ambulatory monitoring. Comput. Hum. Behav. **15**(5), 571–583 (1999)

12. Forman, G., Scholz, M.: Apples-to-apples in cross-validation studies: pitfalls in classifier performance measurement. SIGKDD Explor. Newslet. **12**(1), 49–57 (2010)
13. Gaier, A., Ha, D.: Weight agnostic neural networks. In: NeurIPS (2019)
14. Gjoreski, H., Ciliberto, M., Morales, F.J.O., Roggen, D., Mekki, S., Valentin, S.: A versatile annotated dataset for multimodal locomotion analytics with mobile devices. In: Conference on Embedded Network Sensor Systems, p. 61. ACM (2017)
15. Gjoreski, H., et al.: The university of Sussex-Huawei locomotion and transportation dataset for multimodal analytics with mobile devices. IEEE Access, 6, 42592–42604 (2018)
16. Hammerla, N.Y., Plötz, T.: Let's (not) stick together: pairwise similarity biases cross-validation in activity recognition. In: UbiComp, pp. 1041–1051. ACM (2015)
17. Hinton, G., Vinyals, O., Dean, J.: Distilling the knowledge in a neural network. arXiv preprint arXiv:1503.02531 (2015)
18. Hoeffding, W.: A non-parametric test of independence. Ann. Math. Stat. 546–557 (1948)
19. Hoos, H., Leyton-Brown, K.: An efficient approach for assessing hyperparameter importance. In: International Conference on Machine Learning, pp. 754–762 (2014)
20. Ioffe, S., Szegedy, C.: Batch normalization: accelerating deep network training by reducing internal covariate shift. In: ICML, vol. 37, pp. 448–456. PMLR (2015)
21. Kovalenko, M., Antoshchuk, S., Sieck, J.: Real-time hand tracking and gesture recognition using semantic-probabilistic network. In: International Conference on Computer Modelling and Simulation, pp. 269–274. IEEE (2014)
22. Kurle, R., Günnemann, S., van der Smagt, P.: Multi-source neural variational inference. In: AAAI, vol. 33, pp. 4114–4121 (2019)
23. Li, L., Jamieson, K., DeSalvo, G., Rostamizadeh, A., Talwalkar, A.: Hyperband: a novel bandit-based approach to hyperparameter optimization. JMLR **18**(1), 6765–6816 (2017)
24. Liu, H., Simonyan, K., Yang, Y.: Darts: differentiable architecture search. In: ICLR (2019)
25. Mantyjarvi, J., Himberg, J., Seppanen, T.: Recognizing human motion with multiple acceleration sensors. In: SMC, vol. 2, pp. 747–752. IEEE (2001)
26. Muehlenstaedt, T., Roustant, O., Carraro, L., Kuhnt, S.: Data-driven Kriging models based on FANOVA-decomposition. Stat. Comput. **22**(3), 723–738 (2012). https://doi.org/10.1007/s11222-011-9259-7
27. Nabian, M.A., Meidani, H.: Physics-driven regularization of deep neural networks for enhanced engineering design and analysis. J. Comput. Inf. Sci. in Eng. **20**(1) (2020)
28. Osmani, A., Hamidi, M., Bouhouche, S.: Monitoring of a dynamical system based on autoencoders. In: IJCAI (2019)
29. Ousmer, M., Vanderdonckt, J., Buraga, S.: An ontology for reasoning on body-based gestures. In: SIGCHI EICS, pp. 1–6. ACM (2019)
30. Papadopoulos, G.T., Axenopoulos, A., Daras, P.: Real-time skeleton-tracking-based human action recognition using kinect data. In: Gurrin, C., Hopfgartner, F., Hurst, W., Johansen, H., Lee, H., O'Connor, N. (eds.) MMM 2014. LNCS, vol. 8325, pp. 473–483. Springer, Cham (2014). https://doi.org/10.1007/978-3-319-04114-8_40
31. Parisi, G.I., Tani, J., Weber, C., Wermter, S.: Emergence of multimodal action representations from neural network self-organization. Cogn. Syst. Res. **43**, 208–221 (2017)
32. Pham, H., Guan, M., Zoph, B., Le, Q., Dean, J.: Efficient neural architecture search via parameters sharing. In: ICML, vol. 80, pp. 4095–4104. PMLR (2018)

33. Radu, V., et al.: Multimodal deep learning for activity and context recognition. IMWUT **1**(4), 157 (2018)
34. Real, E., et al.: Large-scale evolution of image classifiers. In: ICML, pp. 2902–2911 (2017)
35. Reddy, S., Mun, M., Burke, J., Estrin, D., Hansen, M., Srivastava, M.: Using mobile phones to determine transportation modes. TOSN **6**(2), 13 (2010)
36. Díaz Rodríguez, N., Wikström, R., Lilius, J., Cuéllar, M.P., Delgado Calvo Flores, M.: Understanding movement and interaction: an ontology for kinect-based 3D depth sensors. In: Urzaiz, G., Ochoa, S.F., Bravo, J., Chen, L.L., Oliveira, J. (eds.) UCAmI 2013. LNCS, vol. 8276, pp. 254–261. Springer, Cham (2013). https://doi.org/10.1007/978-3-319-03176-7_33
37. Stewart, R., Ermon, S.: Label-free supervision of neural networks with physics and domain knowledge. In: AAAI, vol. 1, pp. 1–7 (2017)
38. Vapnik, V., Izmailov, R.: Learning using privileged information: similarity control and knowledge transfer. JMLR **16**(2023–2049), 2 (2015)
39. Vatavu, R.D., Pentiuc, S.G.: Multi-level representation of gesture as command for human computer interaction. Comput. Inform. **27**(6), 837–851 (2012)
40. Wang, S., Chen, C., Ma, J.: Accelerometer based transportation mode recognition on mobile phones. In: APWCS, pp. 44–46. IEEE (2010)
41. Yu, M.C., Yu, T., Wang, S.C., Lin, C.J., Chang, E.Y.: Big data small footprint: the design of a low-power classifier for detecting transportation modes. Proc. VLDB Endow. **7**(13), 1429–1440 (2014)
42. Zhang, M., Sawchuk, A.A.: USC-HAD: a daily activity dataset for ubiquitous activity recognition using wearable sensors. In: UbiComp, pp. 1036–1043 (2012)
43. Zheng, Y., Xie, X., Ma, W.Y.: GeoLife: a collaborative social networking service among user, location and trajectory. IEEE Data Eng. Bull. **33**(2), 32–39 (2010)

Mutual Information Measure for Image Segmentation Using Few Labels

Eduardo H. Sanchez[1,2(✉)], Mathieu Serrurier[1,2], and Mathias Ortner[3]

[1] IRT Saint Exupéry, Toulouse, France
{eduardo.sanchez,mathieu.serrurier}@irt-saintexupery.com
[2] IRIT, Université Toulouse III - Paul Sabatier, Toulouse, France
[3] Airbus, Toulouse, France
mathias.ortner@airbus.com

Abstract. Recently several models have been developed to reduce the annotation effort which is required to perform semantic segmentation. Instead of learning from pixel-level annotations, these models learn from cheaper annotations, e.g. image-level labels, scribbles or bounding boxes. However, most of these models cannot easily be adapted to new annotations e.g. new classes since it requires retraining the model. In this paper, we propose a similarity measure between pixels based on a mutual information objective to determine whether these pixels belong to the same class. The mutual information objective is learned in a fully unsupervised manner while the annotations (e.g. points or scribbles) are only used during test time. For a given image, the unlabeled pixels are classified by computing their nearest-neighbors in terms of mutual information from the set of labeled pixels. Experimental results are reported on the Potsdam dataset and Sentinel-2 data is used to provide a real world use case where a large amount of unlabeled satellite images is available but only a few pixels can be labeled. On the Potsdam dataset, our model achieves 70.22% mIoU and 87.17% accuracy outperforming the state-of-the-art weakly-supervised methods.

Keywords: Mutual information maximization · Weakly supervised learning · Similarity measure · Image segmentation · Satellite datasets

1 Introduction

Most of the successful models for semantic segmentation rely on a supervised learning approach [16]. Even though these models achieve remarkable results, the effort of collecting carefully annotated data to train these models make them impractical to use in many contexts. Generally, these models require a training dataset composed of images with pixel-level annotations, e.g. a class label is assigned to every pixel in the image. Labeling images is very time-consuming, e.g. the reported time to segment a single image from the PASCAL VOC 2012 dataset is around 240 s [2]. Consider the particular case of satellite data, many

© Springer Nature Switzerland AG 2021
Y. Dong et al. (Eds.): ECML PKDD 2020, LNAI 12460, pp. 391–407, 2021.
https://doi.org/10.1007/978-3-030-67667-4_24

missions have been launched to observe the Earth producing massive amounts of satellite images which are absolutely impossible to be annotated by human operators. For example, each of the Sentinel-2 mission satellites [6] provides up to 1.6 TB of images per day. Different methods have been developed to reduce the need of carefully pixel-level annotations for large-scale data analysis. These methods propose a weakly supervised approach for semantic segmentation where the required annotations are less tedious to obtain than pixel-level annotations such as image-level annotations [12], points [2], scribbles [15] or bounding boxes [13]. These annotations are included during the training stage for learning semantic segmentation models. As a consequence, these models are not easily adaptable to new annotations (e.g. to refine the segmentation results or add new class labels) since retraining the models using these new annotations is required. For this purpose, few-shot learning techniques for semantic segmentation have been proposed [24] but it still requires a significant number of labeled samples from seen classes to perform well on unseen classes. Additionally, these methods often produce suboptimal results without providing the user with an interactive way to make corrections without the need to retrain the model.

Recent work has focused on mutual information estimation and maximization for learning representations in an unsupervised manner [3,9,18,19]. The main goal of these unsupervised approaches is to capture the most salient attributes of data to perform downstream tasks using the learned representations. Extensions of the previous models have been proposed using a self-supervised approach to capture the shared attributes from multiple views of a common context [1,21,23]. We think that designing a self-supervised task to learn suitable representations for semantic segmentation is an appealing idea. In particular, our work is inspired by these models [1,9,21] to learn a similarity measure without supervision.

In this work, we take a step forward and propose a model that performs semantic segmentation by computing the similarity between pixels based on a mutual information approach without requiring annotations during training. Using an ideal similarity measure as distance metric, pixels belonging to the same class are close and simultaneously distant from pixels belonging to other classes. A very few pixel-level annotations are only used during test time. Our model computes the mutual information similarity between labeled pixels and unlabeled pixels and then performs a per-pixel nearest-neighbor search from the set of labeled pixels to classify the unlabeled pixels.

Our model offers several advantages. First, there is no need to retrain our model when new annotations are included since the similarity measure is learned using an unsupervised learning approach. Second, our model requires a small amount of annotated data which can be acquired in multiple formats e.g. points, scribbles, bounding boxes. Third, we propose a simple neural network architecture that achieves competitive semantic segmentation results while keeping a reasonable processing time. The following contributions are made in this paper:

- We propose a model that combines a similarity measure based on mutual information between pixels using self-supervised techniques [1,9,21] and a nearest-neighbor search to perform semantic segmentation.

- We show that excellent results can be achieved by labeling less than 0.75% of the total number of pixels in an image.
- We present quantitative results for image segmentation on the Potsdam dataset [10] outperforming the state-of-the-art weakly-supervised methods and qualitative results on Sentinel-2 data [6] to show a real world use case.
- We analyze the impact of using multiple views via data augmentation techniques [1] on the segmentation performance and we perform an ablation study to evaluate the contribution of each element of the model.

2 Related Work

Image Segmentation. Exceptional results have been achieved by fully supervised models on semantic segmentation [16]. To reduce the annotation effort required by supervised learning settings, several methods have been proposed which use cheaper annotations e.g. points [2], scribbles [15], image annotations[12] or bounding boxes [13]. Labels provided by points or scribbles are then propagated to unlabeled pixels during training [2,15]. The main drawback is that these models are not easy to adapt to new annotations for refining the segmentation results or adding new class labels as it requires retraining the whole model. GrabCut [20] performs interactive image segmentation using a bounding box to separate foreground and background. On the other hand, Khoreva et al. [13] propose a semantic segmentation method requiring a costly recursive training where bounding boxes are refined iteratively. Recent work has been presented [24] to segment classes containing few labels in the dataset. However, this method still requires many training examples from the known classes to perform well on the unknown classes.

Self-supervised Learning. In contrast to the prevalent paradigm based on generative or reconstructive models, recent work has been focused on mutual information maximization for representation learning. These models maximize the mutual information between an input and its representation. Mutual information is computed using different estimators based on the Kullback–Leibler [3], Jensen-Shannon [9], Wasserstein [19] divergences or noise-contrastive estimation[18]. Interesting extensions of these mutual information based frameworks have been presented to capture the common attributes from paired images [1,21,23]. Learning representations that capture the most significant attributes of an image from multiple views is useful for semantic segmentation.

Deep Metric Learning. Measuring the similarity between pixels is a useful tool for image segmentation since similar pixels under a given criterion belong to the same class while dissimilar pixels belong to different classes. Generally, raw pixels are mapped to a representation space by a deep neural network and then similarity between pixels is computed in the representation domain [5,8,22]. For instance, Sun et al. [22] propose a neural diffusion distance to perform segmentation. However, it requires labeled data during training to be consistent with a human criterion. For video segmentation, Chen et al. [5] propose a metric based on the triplet loss [4] which is trained in a supervised manner.

In this paper, we propose a model that performs image segmentation in a weakly-supervised setting. The procedure is split into two stages. First, the model learns a mapping from the pixel domain to a representation domain which captures relevant attributes for image segmentation using a mutual information based framework combining the approaches [1, 21]. Using this mapping, our model measures the similarity between pixels in terms of mutual information. In contrast to the models [5, 8, 22], the similarity measure is learned in a fully unsupervised manner. Secondly, our model computes the mutual information similarity between labeled pixels provided by an operator and unlabeled pixels and then performs a per-pixel nearest-neighbor search from the set of labeled pixels to propagate the labels to unlabeled pixels. The labeled pixels are only used during test time instead of training time like the models [2, 12, 13, 15].

3   Background

3.1   Mutual Information

The mutual information between two random variables $X \in \mathcal{X}$ and $Z \in \mathcal{Z}$ is defined in Eq. 1 where $p(x, z)$ is the joint probability density function of X and Z while $p(x)$ and $p(z)$ are the marginal probability density functions of X and Z, respectively.

$$I(X, Z) = \int_{\mathcal{X}} \int_{\mathcal{Z}} p(x, z) \log \left(\frac{p(x, z)}{p(x)p(z)} \right) dx dz \tag{1}$$

It is straightforward to see that $I(X, Z)$ is defined as the Kullback-Leibler divergence between the joint probability distribution $\mathbb{P}_{XZ}$ and the product of the marginal distributions $\mathbb{P}_X \mathbb{P}_Z$, i.e. $I(X, Z) = \mathrm{D}_{KL} (\mathbb{P}_{XZ} \parallel \mathbb{P}_X \mathbb{P}_Z)$. Generally, computing the mutual information between high dimensional variables is a difficult task since the distributions $\mathbb{P}_{XZ}$ and $\mathbb{P}_X \mathbb{P}_Z$ are unknown. Thus, some methods based on deep neural networks have recently been proposed [3, 9, 18, 19].

3.2   Representation Learning

Equation 1 can be used as objective for unsupervised learning where X is a variable corresponding to a given input (image, speech, text, etc.) and Z is the representation of X. The representation Z is extracted by an encoder function defined by a deep neural network of parameters ψ, $E_\psi : \mathcal{X} \rightarrow \mathcal{Z}$, i.e. $Z = E_\psi(X)$.

The Deep InfoMax framework [9] proposes a mutual information estimator $\hat{I}(X, Z)$ based on the Jensen-Shannon divergence instead of the Kullback-Leibler divergence, i.e. $I^{(\mathrm{JSD})}(X, Z) = \mathrm{D}_{JS} (\mathbb{P}_{XZ} \parallel \mathbb{P}_X \mathbb{P}_Z)$.

Intuitively, let X_i and X_j be two observations of X. Let Z_i and Z_j be the representations of X_i and X_j respectively extracted via E_ψ. Therefore, (X_i, Z_i) is an input-representation pair sampled from the joint probability density function $p(x, z)$ while (X_i, Z_j) is an input-representation pair sampled from the product of the marginal probability density functions $p(x)p(z)$.

We define a discriminator function defined by a deep neural network of parameters ρ, $D_\rho : \mathcal{X} \times \mathcal{Z} \to [0,1]$ which represents the probability of a sample (X, Z) coming from $p(x, z)$ instead of $p(x)p(z)$, i.e. the probability that Z is the representation of X. The discriminator D_ρ and the encoder E_ψ are trained to assign a high probability to samples from $p(x, z)$ (close to 1) and a low probability to samples from $p(x)p(z)$ (close to 0) as shown in Eq. 2.

$$\max_{E_\psi, D_\rho} \hat{I}(X, Z) = \mathbb{E}_{p(x,z)} \left[\log D_\rho(X, Z) \right] + \mathbb{E}_{p(x)p(z)} \left[\log \left(1 - D_\rho(X, Z) \right) \right] \qquad (2)$$

By redefining the discriminator function [17] $D_\rho(X, Z) = \frac{e^{-T_\theta(X,Z)}}{1+e^{-T_\theta(X,Z)}}$ where $T_\theta : \mathcal{X} \times \mathcal{Z} \to \mathbb{R}$ is called the statistics network, we obtain the mutual information objective proposed by the Deep InfoMax framework [9] in Eq. 3.

$$\max_{E_\psi, T_\theta} \hat{I}(X, Z) = \mathbb{E}_{p(x,z)} \left[- \log \left(1 + e^{-T_\theta(X,Z)} \right) \right] - \mathbb{E}_{p(x)p(z)} \left[\log \left(1 + e^{T_\theta(X,Z)} \right) \right] \qquad (3)$$

Two mutual information objectives are proposed in the Deep InfoMax framework. Maximizing the mutual information between an input X and a representation Z is called *global mutual information*, i.e. $\mathbf{L}_{\theta,\psi}^{\text{global}}(X, Z) = \hat{I}(X, Z)$. Additionally, maximizing the mutual information between patches of the image X represented by a feature map $C_\psi(X)$ of the encoder E_ψ and a feature representation Z is called *local mutual information* i.e. $\mathbf{L}_{\phi,\psi}^{\text{local}}(X, Z) = \sum_i \hat{I}(C_\psi^{(i)}(X), Z)$.

4   Method

In this paper, we propose a model that combines the mutual information based methods [1,21] to learn a suitable representation domain to measure the similarity between pixels. Our model is trained in a fully unsupervised manner by leveraging large amounts of unlabeled data. Sanchez et al. [21] extends the Deep InfoMax framework to separate the common information and the exclusive information for paired images. Bachman et al. [1] use the Deep InfoMax framework to perform self-supervised representation learning by maximizing the mutual information between representations extracted from multiple views of a shared context, e.g. the context is provided by an image and the multiple views are generated via data augmentation techniques. Learning the common information between images [1,21] provides a way to compute how similar these images are. In Sect. 4.1, we present the mutual information objective to learn the similarity measure and we explain how to use this similarity measure to perform image segmentation in Sect. 4.2.

4.1   Shared Mutual Information

To create a suitable representation domain for image segmentation, we propose to capture the common information between images of the same context (e.g. satellite images from the same forest) into a shared representation. By removing the particular information of each image, we create a representation that distills the class information which is useful for image segmentation. We propose to learn this shared representation by using the principle presented in [1,21]. Let X and Y be two images of the same context and let S_X and S_Y be the respective shared representations extracted by an encoder E_ψ. In order to enforce learning only the common information between images X and Y, the methods [1,21] maximizes the mutual information between the image X and the representation S_Y and similarly, between the image Y and the representation S_X. In order to create pairs of images of the same context, we follow the approach of Bachman et al. [1] and we use data augmentation techniques (rotation, flip, pixel shift, color jitter) to create a second image from a given image, i.e. $X = f(Y)$ where f is a data augmentation function. We use the objective function proposed by Sanchez et al. [21] since it is simpler to optimize. Equations 4 and 5 displays the global and local mutual information maximization objectives.

$$\mathbf{L}_{MI}^{\text{global}}(X,Y) = \mathbf{L}_{\theta,\psi}^{\text{global}}(X,S_Y) + \mathbf{L}_{\theta,\psi}^{\text{global}}(Y,S_X) \tag{4}$$

$$\mathbf{L}_{MI}^{\text{local}}(X,Y) = \mathbf{L}_{\phi,\psi}^{\text{local}}(X,S_Y) + \mathbf{L}_{\phi,\psi}^{\text{local}}(Y,S_X) \tag{5}$$

Sanchez et al. [21] also includes a L_1 constraint to force the shared representations to be identical as shown in Eq. 6. The final objective function is displayed in Eq. 7, where α, β and γ are constant coefficient.

$$\mathbf{L}_1(X,Y) = \mathbb{E}_{p(s_x,s_y)} \left[\|S_X - S_Y\| \right] \tag{6}$$

$$\max_{\psi,\theta,\phi} \mathcal{L}^{\text{shared}} = \alpha \mathbf{L}_{MI}^{\text{global}}(X,Y) + \beta \mathbf{L}_{MI}^{\text{local}}(X,Y) - \gamma \mathbf{L}_1(X,Y) \tag{7}$$

4.2 Mutual Information as Similarity Measure

Similarly to Chen et al. [5], we perform per-pixel retrieval to find the closest pixel from the reference pixel set using the learned representations. A k-nearest-neighbors approach is used to determine the class of unlabeled pixels by propagating the information from labeled pixels. A common way of computing the distance between pixels is to measure the L_1 or L_2 distance between their corresponding representations [5]. Alternatively, we propose to use the global and local mutual information objectives introduced in Sect. 3.2.

During training, the mutual information objective is computed using an image X and a different view of X generated via data augmentation techniques, i.e. $Y = f(X)$. In contrast, during test time the mutual information objective is computed using two different images. Let X_i and X_j be two image patches centered at the pixels i and j respectively and let S_{X_i} and S_{X_j} be the shared representations provided by the encoder function. The similarity between pixels i and j is measured by computing $\mathbf{L}_{\theta,\psi}^{\text{global}}(X,S_X)$ and $\mathbf{L}_{\phi,\psi}^{\text{local}}(X,S_X)$.

After training, our model is capable to predict whether a shared representation S_{X_i} corresponds to the image X_i. Since the shared representation S_{X_i} contains the class information of X_i, it provides a means to identify pixels belonging to the same class of X_i. For example, consider that X_i and X_j are

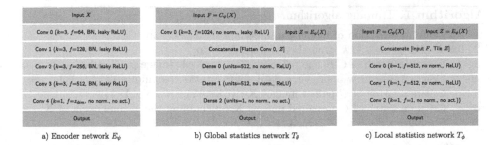

a) Encoder network E_ψ b) Global statistics network T_θ c) Local statistics network T_ϕ

Fig. 1. Network architecture. The encoder and statistics networks are implemented using convolutional and dense layers defined by the number of units, k: kernel size, f: feature maps, BN: batch normalization and activation function. The statistics networks and the encoder share weights: the input F of the statistics network is the output of the Conv 2 layer of the encoder, $C_\psi(X)$.

two different images (e.g. a satellite image from an urban area and another from an agricultural area), the mutual information objective $\mathbf{L}_{\theta,\psi}^{\text{global}}(X, S_X)$ (or $\mathbf{L}_{\phi,\psi}^{\text{local}}(X, S_X)$) achieves a high score since it is easy to distinguish both images. On the other hand, suppose that X_i and X_j are two similar images (e.g. satellite images from the same forest), the mutual information objective $\mathbf{L}_{\theta,\psi}^{\text{global}}(X, S_X)$ (or $\mathbf{L}_{\phi,\psi}^{\text{local}}(X, S_X)$) achieves a low score since it is hard to distinguish the images.

4.3 Implementation Details

Our model is composed of three deep neural networks: the encoder E_ψ, the global statistics network T_θ and the local statistics network T_ϕ. The architecture details are provided in Fig. 1. Every network is trained from scratch by using randomly initialized weights. To optimize the objective $\mathcal{L}^{shared}$ defined in Eq. 7, we use the Adam optimizer with a learning rate of 0.0001, $\beta_1 = 0.9$ and $\beta_2 = 0.999$. We use a batch size of 512 images. Images pairs are created by applying data augmentation techniques (flip, rotation, pixel shift, color jitter). According to Sanchez et al. [21], our baseline model use the following coefficients to weight the terms of the objective function $\mathcal{L}^{shared}$: $\alpha = 0.5$, $\beta = 1.0$ and $\gamma = 0.1$. The size of the shared representation is $z_{dim} = 10$. The training algorithm was executed on a NVIDIA Tesla K80 GPU. The training and image segmentation procedures are summarized in Algorithms 1 and 2. For image segmentation, the number of nearest neighbors is set to $k = 1$. More details are provided in the additional material section.

5 Experiments

5.1 Datasets

Potsdam. The Potsdam dataset [10] contains 8550 aerial images of the city of Potsdam. Each image has a size of $t = 200 \times 200$ pixels and is composed of four

---

Algorithm 1. Training algorithm.

1: Random initialization of model parameters $\psi^{(0)}, \theta^{(0)}, \phi^{(0)}$.
2: **for** $k = 1; k = k + 1; k <$ number of iterations **do**
3:     Sample a batch of C image patches $\{X_1, \ldots, X_C\}$. Image patches have a size s.
4:     Create a new view of X_i via a data augmentation technique $Y_i = f(X_i)$.
5:     Create a batch of C paired images $\mathbf{X} : \{(X_1, Y_1), \ldots, (X_C, Y_C)\}$.
6:     Create a batch of C unpaired images $\tilde{\mathbf{X}}$ by shuffling the Y dimension of $\mathbf{X}$.
7:     Compute $\mathcal{L}^{(k)} = \mathcal{L}^{\text{shared}}(\mathbf{X}, \tilde{\mathbf{X}}, \psi^{(k)}, \theta^{(k)}, \phi^{(k)})$:

$$
\begin{aligned}
\mathcal{L}^{(k)} = \alpha \Big[&- \textstyle\sum_{\mathbf{X}} \text{sp}\left(-T_\theta(C_\psi(X_i), E_\psi(Y_i))\right) - \sum_{\tilde{\mathbf{X}}} \text{sp}\left(T_\theta(C_\psi(X_i), E_\psi(Y_i))\right) \\
&- \textstyle\sum_{\mathbf{X}} \text{sp}\left(-T_\theta(C_\psi(Y_i), E_\psi(X_i))\right) - \sum_{\tilde{\mathbf{X}}} \text{sp}\left(T_\theta(C_\psi(Y_i), E_\psi(X_i))\right) \Big] + \beta \sum_j \Big[\\
&- \textstyle\sum_{\mathbf{X}} \text{sp}\left(-T_\phi^{(j)}(C_\psi(X_i), E_\psi(Y_i))\right) - \sum_{\tilde{\mathbf{X}}} \text{sp}\left(T_\phi^{(j)}(C_\psi(X_i), E_\psi(Y_i))\right) \\
&- \textstyle\sum_{\mathbf{X}} \text{sp}\left(-T_\phi^{(j)}(C_\psi(Y_i), E_\psi(X_i))\right) - \sum_{\tilde{\mathbf{X}}} \text{sp}\left(T_\phi^{(j)}(C_\psi(Y_i), E_\psi(X_i))\right) \Big] \\
&- \gamma \textstyle\sum_{\mathbf{X}} \left(|E_\psi(X_i) - E_\psi(Y_i)|\right)
\end{aligned}
$$

where the softplus function is defined by $\text{sp}(x) = (1 + e^x)$

8:     Update the parameters $\psi^{(k+1)}, \theta^{(k+1)}$ and $\phi^{(k+1)}$ by gradient ascent of $\mathcal{L}^{(k)}$.
9: **end for**

Algorithm 2. Image segmentation algorithm.

1: Select an image X_t from the dataset. Image X_t has a size $t \gg s$.
2: Label a set of L pixels into N classes $P = \{(p_1, c_1), \ldots, (p_L, c_L)\}$ from X_t.
3: where p_i defines the coordinates and c_i is the class of the i-th labeled pixel.
4: **for** unlabeled pixel at $q_j \in X_t$ **do**
5:     **for** labeled pixel at $p_i \in P$ **do**
6:         Select the image patches X_j and X_i of size s centered at q_j and p_i.
7:         Extract the representations $S_{X_i} = E_\psi(X_i)$ and $S_{X_j} = E_\psi(X_j)$.
8:         Extract the feature maps $C_{X_i} = C_\psi(X_i)$ and $C_{X_j} = C_\psi(X_j)$.
9:         Create the image-representation sets $\mathbf{X} = \{(C_{X_i}, S_{X_i}), (C_{X_j}, S_{X_j})\}$
10:        and $\tilde{\mathbf{X}} = \{(C_{X_i}, S_{X_j}), (C_{X_j}, S_{X_i})\}$
11:        Compute the global/local mutual information between X_j and X_i:
12:        $\mathbf{D}_i = \mathbf{L}_{\theta, \psi}^{\text{global}} = -\sum_{\mathbf{X}} \text{sp}\left(-T_\theta(C_{X_k}, S_{X_k})\right) - \sum_{\tilde{\mathbf{X}}} \text{sp}\left(T_\theta(C_{X_k}, S_{X_k})\right)$ or
13:        $\mathbf{D}_i = \mathbf{L}_{\phi, \psi}^{\text{local}} = \sum_j \Big[-\sum_{\mathbf{X}} \text{sp}\left(-T_\phi^{(j)}(C_{X_k}, S_{X_k})\right) - \sum_{\tilde{\mathbf{X}}} \text{sp}\left(T_\phi^{(j)}(C_{X_k}, S_{X_k})\right) \Big]$
14:    **end for**
15:    Assign the pixel q_j the class c_{i^*} of the nearest pixel $i^* = \arg\min_i \{\mathbf{D}_i\}_{i=1}^L$.
16: **end for**

---

channels: red, green, blue and infrared (RGBI). The dataset is split into three parts: 3150 unlabeled images, 4545 training labeled images and 855 test labeled images. Images are labeled into 6 classes (road, car, vegetation, tree, building and clutter). Similarly to [11], we also perform image segmentation using a 3-label version by merging classes (road and car, vegetation and tree and building and clutter). Image patches of size $s = 13 \times 13$ pixels are randomly sampled from the unlabeled images to optimize the model objective (Eq. 7) and the test

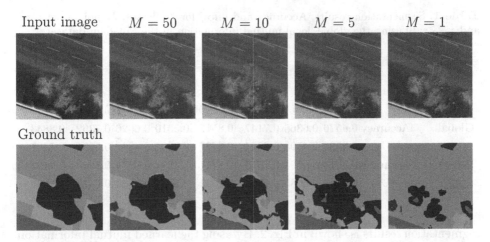

Fig. 2. Image segmentation examples. During test time, only M points per class randomly sampled from the ground truth are used to perform image segmentation. As M increases the accuracy and mIoU are improved. Best viewed in color and zoom-in. (Color figure online)

labeled images to report the experimental results. We use this dataset to provide quantitative results and comparisons to other models.

Sentinel-2. We collected 100 GB of Sentinel-2 time series [6] by selecting several regions of interest on the Earth's surface. Images are acquired at 13 spectral bands using different spatial resolutions. We use the RGBI bands which correspond to bands at 10m spatial resolution. Our dataset is composed of 4200 time series of 12 images acquired at different dates between 2016 and 2018. The size of each image is $t = 512 \times 512$ pixels. Image patches of size $s = 9 \times 9$ pixels are randomly sampled from these images. In addition to data augmentation techniques, the function f creates an image pair by selecting an image patch Y from the same location of X but on a different date. Since there are no labels available, we use this dataset to provide qualitative results in a real world use case where a huge amount of unlabeled data is available and a few annotated pixels are provided by a human operator. Data can be downloaded from the Sentinel Hub [7]. More dataset construction details are provided in the additional material section.

5.2 Image Segmentation on Potsdam

Global and Local Mutual Information. We train our model as described in Sect. 4.3 using the unlabeled images of the Potsdam dataset. Image segmentation is performed on test images where M pixels per class are known. Typically, these annotated pixels are provided by a human operator. To simplify the evaluation, annotated pixels are simulated by randomly sampling M pixels per class from the ground truth. We use several values of $M \in \{1, 5, 10, 50\}$ to evaluate the performance on image segmentation. An example of the impact of M on the

Table 1. Segmentation results. Accuracy and mIoU for N classes, M points per class and $z_{dim} = 10$ using the global/local mutual information in the Potsdam dataset.

Mutual information	Metric	$N = 6$				$N = 3$			
		$M = 1$	$M = 5$	$M = 10$	$M = 50$	$M = 1$	$M = 5$	$M = 10$	$M = 50$
Global	Accuracy	0.4576	0.6366	0.7147	0.8517	0.5310	0.6626	0.7362	0.8843
	mIoU	0.2793	0.4598	0.5354	0.6777	0.3333	0.4888	0.5691	0.7407
Local	Accuracy	0.5013	0.6894	0.7670	0.8717	0.5397	0.7274	0.8045	0.9163
	mIoU	0.3332	0.5085	0.5818	0.7022	0.3632	0.5589	0.6415	0.7866

segmentation results is shown in Fig. 2. By using the learned mutual information based similarity measure, nearest neighbor search is applied to classify pixels into one of $N \in \{3, 6\}$ classes. The performance is reported in terms of mean intersection over union (mIoU) and accuracy. To measure the pixel similarity, we use either the global mutual information objective or the local mutual information objective (see Algorithm 2). Results are reported in Table 1. As expected, the performance is improved as M increases. Our experiments suggest that using the local mutual information objective achieves a better performance than the global mutual information when a few pixels are annotated while the performance is similar when a larger amount of annotated pixels is provided ($M = 50$). Segmentation examples are shown in Fig. 3.

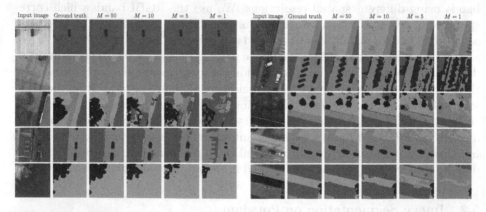

Fig. 3. Image segmentation examples using M points per class to compute the local mutual information. The image segmentation performance is improved as M increases.

Model Comparison. To provide a comparison, we perform image segmentation using different similarity measures to search the nearest neighbor of the unlabeled pixels. First, we compute the nearest neighbor using the L_1 distance between raw pixels. Secondly, we use the L_1 distance between the representations extracted from the VAE model [14], Deep InfoMax model [9] and our model.

Table 2. Model comparison in terms of accuracy and mIoU for N classes, $M = 50$ and $z_{dim} = 10$ using the local mutual information in the Potsdam dataset.

Model	$N = 6$		$N = 3$	
	Accuracy	mIoU	Accuracy	mIoU
Raw pixels (L1)	0.6073	0.3962	0.7267	0.5337
VAE (L1)	0.5844	0.3826	0.7045	0.5230
DIM (L1)	0.4063	0.2103	0.4754	0.2887
Ours (L1)	0.6498	0.4570	0.7391	0.5685
DIM	0.5973	0.4114	0.6497	0.4649
Ours	**0.8717**	**0.7022**	**0.9163**	**0.7866**

Similar images do not necessarily have to be close in the representation domain in terms of the L_1 distance. Therefore, a low performance is expected at image segmentation using the L_1 distance between representations. Finally, we use the mutual information objective of Deep InfoMax [9]. As Deep InfoMax representations keeps all the image information, i.e. more than just class information, we expect this representation to be less appropriate for image segmentation. Table 2 displays the segmentation results. As shown, the local mutual information objective outperforms the other similarity measures for image segmentation. Segmentation examples are shown in Fig. 4.

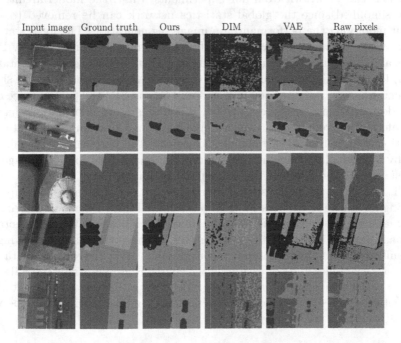

Fig. 4. Image segmentation model comparison. Our model produce the closest predictions to the ground truth using $N = 6$, $M = 50$ and $z_{dim} = 10$ in the Potsdam dataset.

Table 3. Ablation analysis results in terms of accuracy and mIoU for N labels, $M = 50$ and $z_{dim} = 10$ in the Potsdam dataset.

Model	$N = 6$		$N = 3$	
	Accuracy	mIoU	Accuracy	mIoU
Baseline	0.8717	0.7022	0.9163	0.7866
Baseline + $\alpha = 0$	0.8724	0.7068	0.9147	0.7863
Baseline + $\gamma = 0$	0.8636	0.6934	0.9026	0.7655
Baseline + no jitter	0.8767	0.7131	0.9097	0.7800
Baseline + no flip	0.8730	0.7077	0.9123	0.7815
Baseline + no rotation	0.8759	0.7094	0.9114	0.7819
Baseline + no shift	0.7584	0.5710	0.7949	0.6280
Baseline + no SSR	0.7230	0.5405	0.7576	0.5918
Baseline + no SSR + no DA	0.5973	0.4114	0.6497	0.4649
Baseline + random ϕ	0.3994	0.2384	0.5834	0.3986

Ablation Study. We analyze two important factors in our model: the influence of data augmentation techniques to generate multiple views (pixel shift, color jitter, image flip and image rotation) and the importance of some model components, e.g. the statistics networks. Results are displayed in Table 3. Several conclusions can be drawn from our experiments. First, the model architecture can be simplified since the global statistics network can be removed ($\alpha = 0$) without modifying the performance on image segmentation. The local statistics network plays the most important role during training as pointed out by Bachman et al. [1]. Second, removing the L_1 distance between shared representations ($\gamma = 0$) leads to a slightly reduction in the performance. Third, when the shared representations are not swapped in Eqs. 4 and 5 (no SSR) the performance drastically decreases since these representation contains more information than the class information required for image segmentation. Concerning the data augmentation techniques, we surprisingly notice that the performance remains the same by individually removing the color jitter, image rotation and image flip. We believe that the effect of the color jitter is ignored since it is an attribute which is not captured in the shared representation. Additionally, the impact of removing the image rotation or image flip is minimal due to the local information objective where the mutual information is maximized between the representation and image patches instead of the whole image. On the other hand, removing the pixel shift degrades the performance considerably. By removing the data augmentation techniques and not swapping the shared representations (no SSR + no DA) the performance is significantly degraded. We also study the impact of the representation space dimension without noticing significant differences between $z_{dim} = 10$ and $z_{dim} = 32$.

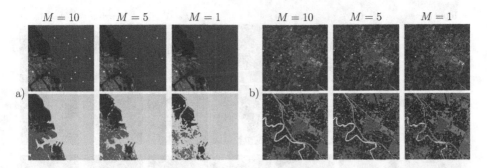

Fig. 5. Image segmentation examples in the Sentinel-2 dataset. A human operator identifies N classes in the satellite image and selects M pixels per label. a) Buenos Aires, Argentina; b) Valencia, Spain. Best viewed in color and zoom-in.

5.3 Image Segmentation on Sentinel-2 Time Series

Since the Sentinel-2 mission does not provide pixel-level annotations for image segmentation, we perform only qualitative experiments. In contrast to the Potsdam case where the annotated pixels are randomly sampled from the available ground truth, now we ask a human operator to label M pixels per class for each image during test time. The reader must note that scribbles, points or bounding boxes can be used for obtaining annotated pixels. As these pixels are annotated under a human criterion, these pixels carry more significant information than pixels randomly sampled from the ground truth and thus the quality of the segmentation results improves significantly using just a few well-selected pixels. As shown in Fig. 5 as the number of pixels per class M increases, the segmentation results considerably improve. Nevertheless, the percentage of annotated pixels remains insignificant. For instance, 60 annotated pixels in a 512×512 pixel image represent less than 0.03% of the total number of pixels. Also the time required for image segmentation is reasonable, an image of 512×512 pixels with 60 annotated pixels takes around 33 s to be segmented.

Segmentation over the Time. Since we maximize the mutual information between images from the same time series, the learned representation ignores the temporal information. As a consequence, by annotating pixels from a single image our model is capable to segment the whole time series the image belongs to. In Fig. 6, it can be seen that the segmentation results are coherent over the time. For instance, agricultural areas are belonging to the same class regardless of whether these areas are grown or harvested.

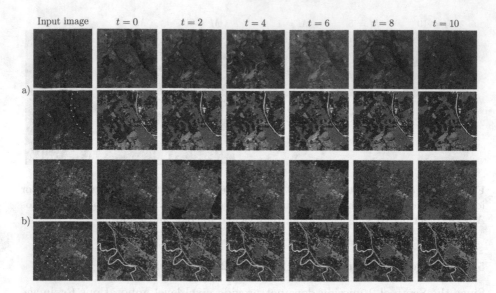

Fig. 6. Image segmentation over the time in the Sentinel-2 dataset. In the first column, the input image and the selected pixels are displayed for $M = 10$. Our method is able to perform image segmentation with few labeled pixels on the entire time series the input image belongs to. The time series and the corresponding predictions are shown in the remaining columns for a) Toulouse, France; b) Valencia, Spain. Best viewed in color and zoom-in.

Segmentation over the Space. In the same manner we perform image segmentation over the time using a single image, our model is able to do it over the space. The annotated pixels provided by a human operator are generally used to perform image segmentation on the image these pixels are extracted from. We also use these annotated pixel to segment other images from the same area achieving satisfactory results as can be seen in Fig. 7. In general, using annotated pixels from a single image we can perform image segmentation on images of the same area independently of the acquisition time.

| Input image | Image 0 | Image 1 | Image 2 | Image 3 | Image 4 |

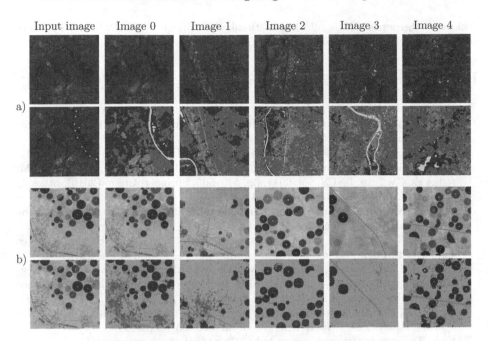

Fig. 7. Image segmentation over the space in the Sentinel-2 dataset. Selected pixels are not only useful for propagating the information from labeled pixels to unlabeled pixels in the same image but also in different images of the same area. a) Toulouse area, France; b) Tubarjal area, Saudi Arabia. Best viewed in color and zoom-in.

6 Conclusion

In this paper, we have proposed to use a mutual information based similarity measure to perform image segmentation. Our approach offers the advantage of learning the proposed similarity measure in an unsupervised manner leveraging large amounts of unlabeled data. Then, per-pixel nearest-neighbor search using the proposed similarity measure is carried out to assign classes to the unlabeled pixels from the labeled pixels provided by a human operator. In particular, we have studied the case of aerial/satellite data where massive amounts of unlabeled images are available while the annotations are scarce. In the Potsdam case, our experiments suggest that the local mutual information objective is useful to measure similarity between pixels. Our approach outperforms other approaches based on state-of-the-art methods demonstrating the usefulness of our learned representation domain. On the other hand, the ablation experiments show that the model can be further simplified as some data augmentation techniques are more relevant and the global mutual information objective can be removed. In the Sentinel-2 case, we have shown that image segmentation can be performed over the time and over the space using a very few amount of annotated pixels, e.g. labeled pixels are less than 0.002% of the total number of pixels of a time series and it can be achieved in a reasonable amount of time.

Acknowledgments. We would like to thank the projects SYNAPSE and DEEL of the IRT Saint Exupéry for funding to conduct our experiments.

References

1. Bachman, P., Hjelm, R.D., Buchwalter, W.: Learning representations by maximizing mutual information across views. In: Advances in Neural Information Processing Systems, vol. 32 (2019)
2. Bearman, A., Russakovsky, O., Ferrari, V., Fei-Fei, L.: What's the point: semantic segmentation with point supervision. In: Leibe, B., Matas, J., Sebe, N., Welling, M. (eds.) ECCV 2016. LNCS, vol. 9911, pp. 549–565. Springer, Cham (2016). https://doi.org/10.1007/978-3-319-46478-7_34
3. Belghazi, M.I., et al.: Mutual information neural estimation. In: Proceedings of the 35th International Conference on Machine Learning (2018)
4. Chechik, G., Sharma, V., Shalit, U., Bengio, S.: Large scale online learning of image similarity through ranking. J. Mach. Learn. Res. **11**, 1109–1135 (2010)
5. Chen, Y., Pont-Tuset, J., Montes, A., Van Gool, L.: Blazingly fast video object segmentation with pixel-wise metric learning. In: Proceedings of the IEEE Conference on Computer Vision and Pattern Recognition (2018)
6. Drusch, M., et al.: Sentinel-2: ESA's optical high-resolution mission for GMES operational services. Remote Sens. Environ. **120**, 25–36 (2012)
7. ESA: The copernicus open access hub. https://scihub.copernicus.eu/
8. Fathi, A., et al.: Semantic instance segmentation via deep metric learning. CoRR (2017). http://arxiv.org/abs/1703.10277
9. Hjelm, R.D., et al.: Learning deep representations by mutual information estimation and maximization. In: International Conference on Learning Representations (2019)
10. International Society for Photogrammetry and Remote Sensing: ISPRS 2D semantic labeling contest. http://www2.isprs.org/commissions/comm3/wg4/semantic-labeling.html
11. Ji, X., Henriques, J.F., Vedaldi, A.: Invariant information clustering for unsupervised image classification and segmentation. In: Proceedings of the IEEE International Conference on Computer Vision (2019)
12. Joon Oh, S., Benenson, R., Khoreva, A., Akata, Z., Fritz, M., Schiele, B.: Exploiting saliency for object segmentation from image level labels. In: Proceedings of the IEEE Conference on Computer Vision and Pattern Recognition (2017)
13. Khoreva, A., Benenson, R., Hosang, J., Hein, M., Schiele, B.: Simple does it: weakly supervised instance and semantic segmentation. In: Proceedings of the IEEE Conference on Computer Vision and Pattern Recognition, pp. 876–885 (2017)
14. Kingma, D.P., Welling, M.: Auto-encoding variational Bayes. In: International Conference on Learning Representations (2014)
15. Lin, D., Dai, J., Jia, J., He, K., Sun, J.: ScribbleSup: scribble-supervised convolutional networks for semantic segmentation. In: Proceedings of the IEEE Conference on Computer Vision and Pattern Recognition (2016)
16. Long, J., Shelhamer, E., Darrell, T.: Fully convolutional networks for semantic segmentation. In: Proceedings of the IEEE Conference on Computer Vision and Pattern Recognition (2015)
17. Nowozin, S., Cseke, B., Tomioka, R.: f-GAN: training generative neural samplers using variational divergence minimization. In: Advances in Neural Information Processing Systems, pp. 271–279 (2016)

18. van den Oord, A., Li, Y., Vinyals, O.: Representation learning with contrastive predictive coding. CoRR (2018). http://arxiv.org/abs/1807.03748
19. Ozair, S., Lynch, C., Bengio, Y., van den Oord, A., Levine, S., Sermanet, P.: Wasserstein dependency measure for representation learning. In: Advances in Neural Information Processing Systems, vol. 32 (2019)
20. Rother, C., Kolmogorov, V., Blake, A.: "GrabCut" interactive foreground extraction using iterated graph cuts. ACM Trans. Graph. (TOG) **23**, 309–314 (2004)
21. Sanchez, E.H., Serrurier, M., Ortner, M.: Learning disentangled representations via mutual information estimation. CoRR (2019). http://arxiv.org/abs/1912.03915
22. Sun, J., Xu, Z.: Neural diffusion distance for image segmentation. In: Advances in Neural Information Processing Systems, vol. 32 (2019)
23. Tian, Y., Krishnan, D., Isola, P.: Contrastive multiview coding. CoRR (2019). http://arxiv.org/abs/1906.05849
24. Xian, Y., Choudhury, S., He, Y., Schiele, B., Akata, Z.: Semantic projection network for zero-and few-label semantic segmentation. In: Proceedings of the IEEE Conference on Computer Vision and Pattern Recognition (2019)

Applied Data Science: Hardware and Manufacturing

FlowFrontNet: Improving Carbon Composite Manufacturing with CNNs

Simon Stieber$^{(\boxtimes)}$ ⓘ, Niklas Schröter ⓘ, Alexander Schiendorfer ⓘ,
Alwin Hoffmann ⓘ, and Wolfgang Reif ⓘ

Institute for Software and Systems Engineering, University of Augsburg,
Universitätsstr. 6a, 86159 Augsburg, Germany
{stieber,schiendorfer,hoffmann,reif}@isse.de,
niklas.schroeter@student.uni-augsburg.de

Abstract. Carbon fiber reinforced polymers (CFRP) are light yet strong composite materials designed to reduce the weight of aerospace or automotive components – contributing to reduced emissions. Resin transfer molding (RTM) is a manufacturing process for CFRP that can be scaled up to industrial-sized production. It is prone to errors such as voids or dry spots, resulting in high rejection rates and costs. At runtime, only limited in-process information can be made available for diagnostic insight via a grid of pressure sensors. We propose FlowFrontNet, a deep learning approach to enhance the in situ process perspective by learning a mapping from sensors to flow front "images" (using upscaling layers), to capture spatial irregularities in the flow front to predict dry spots (using convolutional layers). On simulated data of 6 million single time steps resulting from 36k injection processes, we achieve a time step accuracy of 91.7% when using a 38×30 sensor grid 1 cm sensor distance in x- and y-direction. On a sensor grid of 10×8, with a sensor distance of 4 cm, we achieve 83.7% accuracy. In both settings, FlowFrontNet provides a significant advantage over direct end-to-end learning models.

Keywords: Process monitoring · Convolutional neural networks · Digital twin · Manufacturing · Industrial automation · Resin transfer molding · Carbon composites

1 Introduction to Composite Manufacturing via RTM

Carbon fiber reinforced polymers (CFRP) are extremely strong composite materials despite their low weight. That makes them attractive for the construction of lighter aerospace and automotive parts (conventionally made from steel or aluminum) to reduce fuel consumption and CO_2 emissions [5]. In essence, these composites are made from a so-called polymer matrix that is reinforced with textiles containing carbon fibers. To produce CFRP parts industrially, resin transfer molding (RTM, [1]) is a commonly applied manufacturing process for medium volumes (1,000s to 10,000s of parts) and is depicted in Fig. 1a: A liquid thermoset polymer (called a resin) is injected under pressure into a mold cavity that

© Springer Nature Switzerland AG 2021
Y. Dong et al. (Eds.): ECML PKDD 2020, LNAI 12460, pp. 411–426, 2021.
https://doi.org/10.1007/978-3-030-67667-4_25

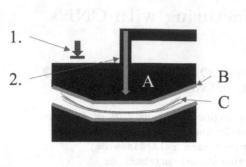 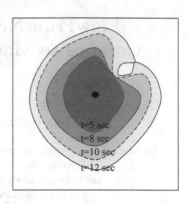

(a) Front view of the hardware involved in RTM.   (b) Top view of the evolving flow front inside (B), with a dry spot.

Fig. 1. Overview of resin transfer molding (RTM): The resin (2) is injected into a mold cavity (B) filled with textiles containing carbon fibers (C). A press (A) applies the necessary pressure (1).

contains reinforcement material such as textiles with carbon fibers. This results in a "flow front" that separates impregnated material from dry material, shown in Fig. 1b. Thermoset resins are converted from a liquid to a solid through heat – they are "cured".

During the RTM process, several errors can render the result useless and, thus, make the overall production expensive [4,9,17]. They occur, inter alia, due to high input variances of the fiber contents in the textile (the "preform"). *Dry spots* refer to areas of the preform that are not impregnated by the liquid, as shown in the top right of Fig. 1b. In some cases, these dry spots irreparably invalidate the stability and stiffness properties required for the manufactured part. In others, they can be repaired manually. Either way, automated process monitoring based on sensors applied to the mold would significantly improve the quality assurance – called in-situ monitoring. These sensors (e.g., temperature, dielectric, ultrasound, or pressure) are able to track the flow front of the fluid and, consequently, predict *spatial* deviations from proper RTM runs. This may indicate problems such as dry spots or voids (i.e., enclosures where air is trapped), providing diagnostic insight, or even control actions to avoid rejects.

In this paper, we propose to use machine learning, in particular convolutional neural networks (CNN), to get a binary classifier $f : \mathbb{R}^n \to \{0, 1\}$ from n mold-integrated sensors to labels describing whether, at a given point in time, there is a dry spot present or not. As an intermediate step, we train the network to generate higher-resolution images of the flow front (extracted from an RTM simulation in PAM-RTM[1]) from sensor data. These images capture spatial irregularities in the flow. Since such full images are only available during simulation or specialized

---

[1] https://www.esi-group.com/software-solutions/virtual-manufacturing/composites/pam-composites-pam-rtm-composites-molding-simulation-software.

permeability studies [3] and not in real-world closed molds, a trained model could substantially enhance the spatial information transmitted by actual flow front sensors in productive settings – in the sense of a digital twin [8]. In this paper, we focus on detecting dry spots from sensor input obtained from simulated data as a first step towards a transfer to real data. These are major quality concerns during the injection process [7].

Our model, *FlowFrontNet*, first uses several deconvolutional and convolutional layers to create the image representation from pressure sensor data and proceeds to perform the binary classification using convolutional and dense layers. The overall classification is performed individually for every frame of a simulated injection sequence such that a warning of dry spots can be issued at any time. Our central scientific question is if we can detect dry spots from simulated sensor data and whether the spatial flow-front information captured in convolutional layers improves the classification. We compare the accuracy the simulation-enhanced FlowFrontNet achieves to that of a standard feed-forward network (both based on sensor data) and find major improvements in Sect. 4. Our goal is to offer FlowFrontNet as a starting point for future research in composite manufacturing: code[2], checkpoints [19] and data [20] are available.

Following a discussion of related work, we present the data regime including input variation for simulation and automated label acquisition in Sect. 2, present the neural network models and training methodology in Sect. 3, and conclude with experimental results in Sect. 4.

1.1 Related Work

To cover the relevant related work, two aspects have to be addressed: The technical process and the machine learning models.

There have been several publications on in-situ monitoring the RTM process. Pantelelis et al. [12] present a system on how to detect the curing state but only mention ML-based analyses as future work. Zhang et al. [22] use simulated and actual sensor data to detect the curing rate of the process. They use aggregated shallow neural networks to achieve this based on two sensors. They further use the model to adapt the heating of the process. Our work, by contrast, focuses on the resin injection and detect dry spots in the flow front from multiple sensors. Leveraging the spatial capabilities inherent to CNNs has never been applied to dry spot detection in the RTM process before.

Unrelated to composite production, deconvolutional layers that increase the spatial dimensions have been used for image enhancement tasks. Xu et al. [21] used them to deblur images. Shi et al. [15] approach superresolution, the task of upscaling pictures from low to high resolution, with deconvolutional networks. For semantic segmentation, Noh et al. [11] used a combination of convolutional and deconvolutional networks. They compress the input image with a VGG-16 to afterward decompress the encoding with deconvolutional layers.

---

[2] https://github.com/isse-augsburg/rtm-predictions.

Out of these approaches, [15] comes closest to our task of extracting information from a grid of sensors since the compressed data is known before and not a learned encoding as in [11]. However, our approach works with more heterogeneous data in that it generates images from sensors rather than improving the resolution of conventional photos. Furthermore, the dimensions of the input are enlarged and not kept the same as in [21].

2   Creating Training Data from Simulation

To train a detector for dry spots from RTM sensor data, we need to observe sufficiently many training instances. Generally speaking, industrial ML use cases based on sensor data from actual production cycles tend to suffer from limited data quality and quantity. Moreover, setting up a new process (including design of the mold, choice of the resin, etc.) takes time until sufficiently many training runs have been processed, not to mention the material costs. Simulation is a proven remedy for the lack of data [13] which is our focus in this paper. Basic RTM processes are well supported by existing engineering tools [3].

The lack of real high-quality data is only one reason to opt for simulation. Another more pressing reason is that we can create a spatial representation of the flow front, i.e., a "flow front image" (see Fig. 3a), that is not observable in real closed molds. Those images will serve as the target for the generative Deconv/Conv part of FlowFrontNet, described in Sect. 3.

Being able to simulate the process might raise the question of why one has to apply machine learning at all – instead of just running the simulation online. First, the trained models will encapsulate only those aspects of the simulation that are needed to make good dry spot predictions for a given RTM setup. An online simulation would take much longer and be infeasible for real-time monitoring. Second, we can reasonably anticipate that real runs will contain aspects that are not properly captured by simulation. For instance, variability in the process parameters such as the textile might not be observable. Using an ML-model will eventually enable us to add real data to the training.

In the following, we describe our process of obtaining enough (6 million) domain-specific training instances from simulations executed with randomized initial conditions (variances in the input textile) and automatically deriving the corresponding dry spot labels that are images with the filling level as intensities.

2.1   Simulated RTM Runs in PAM-RTM

PAM-RTM is a software package designed for laying out RTM processes, including fluid dynamics simulation. Modeling flow transport in porous media mathematically is most commonly carried out based on Darcy's law [2] for one-dimensional flow:

$$v_x = -\frac{k_x}{\eta} \frac{\Delta P}{\Delta x} \qquad (1)$$

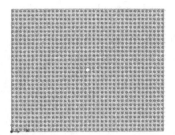 

Fig. 2. Simulated composite plate with a full sensor grid and central injection point - 1140 sensors located at 1 cm in x and y. On the right, a fiber volume content (FVC) map is given with local perturbations (rectangular and circular).

where v_x is the 1D flow velocity, k_x represents the permeability value of the textile (corresponding to how "easily" fluids can permeate it), η denotes the fluid viscosity, and $\frac{\Delta P}{\Delta x}$ expresses the pressure drop along a specific flow length [3].

PAM-RTM includes other features as well, including draping simulation, the meshing of CAD parts, and distortion during curing which makes it a standard tool in composite manufacturing. For this paper, we only needed the fluid simulation part modeling resin injection. Getting the flow front prediction and, subsequently, dry spot classification right is a high-impact quality goal.

Setting up a particular RTM simulation consists of defining the geometrical 3D-model of the part to be constructed, the viscosity and permeability parameters of the resin and textile, respectively, as well as temperature and pressure of the injection. A single simulation *run* then represents a whole injection that continues to pump resin until the mold cavity is filled entirely.

As our running example, we choose a simple rectangular composite plate with dimensions of 38×30 cm. Adding a simulated pressure sensor for every centimeter in x and y direction yields a total of 1140 sensors, as Fig. 2 shows. The sensors can be placed at any location in the simulated composite plate, independent of the mesh grid of the plate. To ensure that the chosen parameters are realistic, this configuration follows the setup for experimental permeability characterization available at the Processing of Composites Group of Montanuniversität Leoben as presented by Grössing et al. [3]. For the textile, the setup assumes a natural fiber fabric with a fiber volume content (FVC) of 0.268 and a 3.5 mm thickness. Finally, the resin used in the model experiment in Leoben was a plant oil with a viscosity of $0.065 \frac{\text{N} \cdot \text{s}}{\text{m}^2}$.

In reality, dry spots emerge from local variations in the textiles (thicker or thinner areas) that lead to regions of low or high fluid permeability, respectively. To recreate that effect in our simulation and obtain varied training data, we altered the fiber volume content in certain areas, as shown in Fig. 2. These perturbations are responsible for the distribution our training and test RTM runs are drawn from. For simplicity, every perturbation corresponds to one circle with a random diameter and one rectangle with random side lengths – with varying strengths of FVC perturbation. All of these perturbations were automatically

Table 1. Data set sizes – train/val/test-split

Data set	# Runs	# Samples	% (Samples)
Training	29,663	5,067,352	81.11
Validation	756	131,072	2.09
Test	6,244	1,048,576	16.78
All	36,663	6,247,000	100

created for 40k simulation setups that were run in parallel on 10 hosts with 32 cores each. We repeated the simulations with different random strengths of FVC perturbations: setting them to 0.1–0.8 covers a wide range of cases (including mild cases that will eventually not produce dry spots), increasing them to 0.2–0.8 or even 0.3–0.8 provokes more severe perturbations leading to dry spots.

For training our models, Table 1 shows how we split the data obtained from those FVC-perturbed simulation runs. Although we consider individual frames (corresponding to time steps of runs), the splits do not break up runs, i.e., a run is fully contained in either train, validation, or test set. That allows for reusing the data when considering sequence models in the future.

When working with simulations, we have to double-check the time series resolution. We noted that PAM-RTM distinguishes between multi-state results for the simulated pressure sensors and single-state results for the simulation results at every node of the mesh of the plate such as filling status and pressure. While multi-state results are present for every simulated time step (including the respective simulation times), single state results are saved every k-th second in simulation time or every i-th simulation step. To keep the simulation efforts tractable and produce enough runs, we selected an approximate time resolution of 0.5 s and dispose of all pressure sensor results that do not correspond to time steps with available filling status. Finally, before feeding the pressure sensors' values to the ML-models, we divide them by 10^5 to obtain numerically well-behaved training dynamics in the early layers of the neural networks.

2.2   Dry Spot Label Creation

Varying the FVC locally in the input textiles and recording the simulated runs provides us with pressure value time series and flow front developments as "images" where every pixel corresponds to the filling level at a given point in time. To classify dry spots in a supervised manner, we also need labels indicating whether a flow front image contains a dry spot (e.g., Fig. 3a) or not (e.g., Fig. 3c). And we need those labels for all 6 million frames from 36k runs.

An area counts as a dry spot if it is a non-filled area that is enclosed by resin. Comparing Fig. 3a and Fig. 3b, we might be tempted to directly derive binary labels from the modified FVC maps that served as simulation input. However, the matter is more complicated. First, an observed dry spot tends to be smaller than the original perturbation since the outskirt still gets permeated

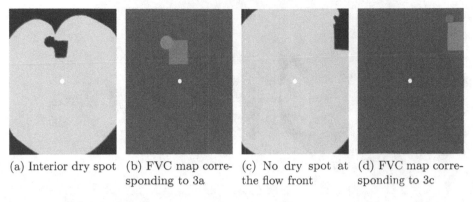

(a) Interior dry spot   (b) FVC map corre-   (c) No dry spot at   (d) FVC map corre-
                        sponding to 3a   the flow front   sponding to 3c

Fig. 3. Exemplary flow fronts resulting from locally perturbed FVC-maps extracted from specific time steps (Color figure online)

by resin. Second, an FVC perturbation would be the same for all frames of a run even though the dry spot only becomes apparent after the flow front reached it geometrically. Third, a regular flow front might be jagged such as the one in Fig. 3c. Such areas should not be counted (yet) as dry spots since they tend to eventually be filled. Thus, we devised a heuristic that combines the perturbed FVC-maps with computer vision techniques on the flow front images to output label maps indicating if a pixel belongs to a dry spot or not.[3]

First, we extract the already filled pixels from each frame, e.g., the light (yellow) pixels in Fig. 3a. The negative of that image only contains the dry parts. Since dry spots cannot lie within the filled regions, we focus on the latter for finding contours. We use OpenCV[4] to find contours in those negative images. The contours are constrained to be smaller than the whole flow front but at least larger than the central injection point. However, not all of the contours identified as dry spot candidates must be dry spots. Instead, they could emerge from jagged flow fronts as apparent in Fig. 3c) that are not enclosed by resin. To distinguish between these two cases, we enhance the pure computer vision operations (such as contour or hole detection) with the perturbed FVC maps. Since a high FVC corresponds to low permeability and vice versa, we overlap the dry spot candidate contours with the FVC map. For every contour in the dry area, its probability of being a dry spot is estimated proportionally to the percentage of the overlap with an FVC perturbation. To reduce the number of incorrectly labeled dry spots, we look at runs, i.e., sequences of frames, and discard candidates that only occur in a single frame. This is justified since the flow front is expanding continuously and, thus, the probability of a dry spot has to be similar for multiple frames in a run. Some edge cases (e.g., simulation errors or dry spots in areas other than

---

[3] Note that we did not use the "air entrapment" feature in PAM-RTM since that would prematurely end simulation runs, produces lagging information, and cause the experimental setup to diverge from the model setup in Leoben.

[4] https://opencv.org/.

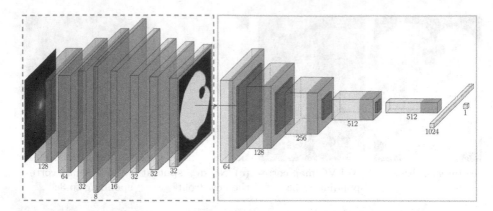

Fig. 4. FlowFrontNet: The first part (dashed, green) is a Deconv/Conv network that maps from sensors to images. The second part (solid, orange) is the DrySpotNet consisting of five convolutional/max-pooling layers and two fully-connected layers to perform the final classification task. Numbers denote the resulting feature maps. (Color figure online)

the FVC perturbations) lead to unexpected jumps in dry spot probabilities over consecutive frames. To maintain high-quality data, we excluded whole runs with such phenomena – approximately 9.16% of all runs which leaves us with 36,663 valid runs (cf. Table 1).

3   Approach - Model and Training

After generating data for different FVC contents in a sufficient amount and labeling them in an automated manner, the next step is to present *FlowFrontNet*, the main model of our approach, which is shown in Fig. 4. Before going into detail, we present the key points of this network.

The overall FlowFrontNet maps pressure sensor inputs to classification decisions. It consists of the generative part that upsamples from sensor grids to flow front images (called Deconv/Conv) and the binary classification part (called DrySpotNet). By learning to produce a flow front from sensor input, the network learns a representation of fluid dynamics. The Deconv/Conv part itself is useful for other use cases down the road, say, exact dry spot localization or pixel-wise flow front detection. The later DrySpotNet performs a binary classification on the generated images (see Figs. 5b and 5c for example inputs to that part). By adding spatial fluid dynamics knowledge, we aim to surpass the performance of a conventional feed-forward classifier achieves based on the same sensor input.

We present the version of FlowFrontNet that is used for 1140 sensors. The other sensor resolutions (see Table 2) only require slight changes in kernel size and layer count. Those are necessary since the smaller input sizes lead to smaller outputs with the hyperparameters used for the 1140-sensors Deconv/Conv network. Due to the resulting poor image resolution, the results would not be adequate for detecting dry spots. In the following, we describe the intricacies of each network and their training processes.

Deconv/Conv Network: Sensor Data to Flow Front Images. The first part of FlowFrontNet, *Deconv/Conv*, is a fully convolutional neural network [14] (see Fig. 4, left part) that receives the low-resolution sensor grid values and outputs a high-resolution flow front image. The first four layers are deconvolutional to extract features from the sensor array. As opposed to convolutional layers, deconvolutional layers increase the spatial dimensions of an image, known from image segmentation or superresolution [11,15].

Our approach is similar in this regard: the pressure sensor grid has a low resolution and constitutes a compressed representation of the flow front. We apply deconvolutional layers to increase the resolution of the sensor array while simultaneously filling the spaces between the sensors, i.e., interpolating missing values. Afterward, we utilized five standard convolutional layers (providing the required amount of non-linearity) to create and shape the final flow front image (e.g., with spatial dimensions 117×149).

The first step is to pre-train the Deconv/Conv network to produce images of the flow front from sensors, as shown in the left part of Fig. 4. As mentioned before, this pre-training needs simulated data since flow fronts are hidden in real-world closed molds. All hidden layers are followed by the rectified linear unit (ReLU) activation, while the output layer is activated by the sigmoid function to make sure the generated flow front images lie within the range $[0, 1]$.

DrySpotNet: Flow Front Image to Binary Dryspot Classification. The second part of FlowFrontNet, DrySpotNet, receives the generated images and classifies them concerning dry spots. The dry spot labels are obtained as described in Sect. 2.2. The architecture for this classification follows a standard convolutional classification network [6] (see the latter part of Fig. 4). The final output yields soft classification scores using the sigmoid function (commonly interpreted as class probabilities) that are eventually thresholded to achieve a hard classification.

FlowFrontNet: Training End-to-End Sensor Data to Dryspot. After pretraining the Deconv/Conv net using flow front images, we append the untrained DrySpotNet for the final classification. To avoid changing the already trained weights of the generative Deconv/Conv layers, the pre-trained layers are "frozen", meaning that their weights are not updated during training. This is identical to the practice of freezing convolution layers in fine-tuning for special purpose image classification [16]. After training the newly appended output layer, it can be useful to also "unfreeze" the pre-trained weights of early hidden layers during backpropagation. The data is exchanged and parts of the original network are used as the backbone for a new image processing task. Here, our approach is different: we change the objective and targets from generating images to that of doing dry spot binary classification, leaving the early layers fixed.

Table 2. Sensor layouts

Sensors	Layout	Sensor distance - x and y
1140	38×30	1 cm
80	10×8	4 cm
20	5×4	8 cm

When combining the generative part of the Deconv/Conv net with the DrySpotNet, low sensor input numbers may lead to comparably "blurry" flow front images (e.g., Figs. 5b to 5d) before handing them to the classification part. To obtain higher-contrast flow front images and, consequently, better classification results, we add a non-differentiable hard *pixel-threshold* to the forward pass of the network after the Deconv/Conv net. This operation sets all values below the threshold to 0 and all above to 1. Since the flow-front-generating layers of the networks that lie before the pixel-threshold are frozen and need no gradients, we can easily incorporate this operation.

Feed-Forward Network: Baseline Classifier. To judge the merits of FlowFront-Net, we design a basic end-to-end feed-forward network as a baseline classifier. It consists of two ReLU-activated, fully-connected hidden layers and a sigmoid output for the dry-spot probabilities. As input, it receives the sensor values described in Sect. 2 without performing upsampling to the flow front image.

For training all models, we utilized an Nvidia DGX-1 with 8 Tesla V100 GPUs. This machine is able to train with a batch size of 2048 for both steps of the training process. Especially the Deconv/Conv network is very resource-intensive in terms of its parameters, with the highest consumption of computation power when using 1140 sensor values as input. For this training, the DGX-1 reached its maximum load with batch size 2048, for training runs with smaller sensor grids, the same batch size yielded the best performance compared to larger batch sizes.

4   Experimental Evaluation

To put FlowFrontNet to the test, we devised three central evaluation hypotheses. The first is if we can predict a dry spot from simulated pressure sensors at all. The second and central hypothesis is, whether the intermediate step of flow front image generation can improve dry spot classification. This also involves testing the pixel thresholding introduced to obtain sharper internal flow front representations. As a third point that is interesting to evaluate, we investigate the number of sensors that are necessary to classify dry spots sufficiently well, with and without the intermediate flow front generation. We investigate models for 1140, 80, and 20 sensors, to estimate the prediction quality achievable by a reduced number of sensors, see also Table 2.

These particular numbers emerge from taking every 4-th or every 8-th sensor of the full 1140 grid. Taking 1140 sensors corresponds to a sensor distance of

Table 3. Accuracy values on three different sensor resolutions for feed-forward baselines and FlowFrontNet

# Sensors	Feed-forward		FlowFrontNet	
	Threshold	Accuracy	Threshold	Accuracy
1140	0.54	82.74%	0.49	**91.68%**
80	0.52	79.57%	0.57	**83.69%**
20	0.49	74.68%	0.51	**75.22%**

1 cm, which is an unrealistically high number. To come closer to reality, we focused on a sensor distance 4 cm for the 80 sensors because that is within range of physical feasibility and set a baseline with even fewer sensors 8 cm distance.

4.1   Results

Can Machine Learning Predict Dry Spots Based on Sensor Inputs? To start with the first question, we obtained a baseline from the feed-forward network performing the classification task directly from sensor inputs. When optimizing this architecture, we found that larger batch sizes positively affected the evolution of the validation loss, with a sweet spot found at $32,768 = 2^{15}$ training instances per batch. The results from experiments suggest the following baselines: the best feed-forward network with two hidden layers achieves an accuracy of 82.73% with 1140 sensors, an accuracy of 79.51% with 80 sensors, and an accuracy of 74.6% with 20 sensors, see Table 3. Unsurprisingly, the more sensors we use, the better the accuracy gets for a network unaware of the underlying fluid dynamics.

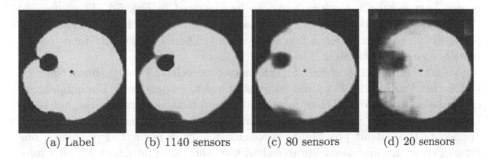

(a) Label   (b) 1140 sensors   (c) 80 sensors   (d) 20 sensors

Fig. 5. Exemplary flow front predictions in the Deconv/Conv part of FlowFrontNet based on the different sensor counts.

Does the Deconv/Conv Net Improve the Classification Accuracy? For the second question – the improvement of the classification by generating flow front

Table 4. Confusion matrices for 1140 and 80 sensors: Feed-forward vs. FlowFrontNet

For 1140 sensors

		Actual				Actual	
		¬ Dry spot	Dry spot			¬ Dry spot	Dry spot
Pred.	¬ Dry spot	40.44%	8.59%	Pred.	¬ Dry spot	44.60%	4.43%
	Dry spot	8.66%	42.31%		Dry spot	3.89%	46.98%

For 80 sensors

		Actual				Actual	
		¬ Dry spot	Dry spot			¬ Dry spot	Dry spot
Pred.	¬ Dry spot	38.97%	10.06%	Pred.	¬ Dry spot	41.60%	7.44%
	Dry spot	10.37%	40.06%		Dry spot	8.87%	42.10%

images using deconvolutional and convolutional layers – the answer is a clear yes, with only a slight limitation. Figure 5 allows us to visually inspect exemplary results from the learned sensor-to-flow-front mapping on the test set. Even for 20 sensors, a rough idea of the underlying flow front is obtained, for 80 and more sensors, a fairly accurate image can be reconstructed. Based on that internal representation of the flow front, the accuracy of FlowFrontNet with 1140 sensors as input increases to 91.68%, which is a 9% advantage over the pure Feed-forward Network. For 80 sensors (with pixel-thresholding), the accuracy can be enhanced from 79.51% to 83.69%, a margin of 4%. Furthermore, the 80 sensor FlowFrontNet performs better than the feed-forward network with 1140 sensors which shows that sensor investments could be reduced in favor of encoded simulation knowledge.

Alas, the improvement over the feed-forward network decreases even more with 20 sensors: 75.22%, which is less than 1% of accuracy boost. The spatial information density in a 5×4 input sensor grid turned out too low to still get a useful representation, which can also be observed in Fig. 5d. The image of the flow front is not clear at all and it appears as though there is no dry spot enclosed by resin but rather a jagged flow front. Therefore, we only focus on the models equipped with 80 or 1140 sensor inputs.

To get a more comprehensive performance overview, Table 4 shows the confusion matrices for the 1140 and 80 sensors input, respectively. The models are rather balanced regarding their false positive and false negative shares, but these values are – as always – object to modification by the classification threshold applied to the output probability score. Especially in industrial processes, one type of error can be more favorable than the other, depending on whether false positives (e.g., causing disruptions in the process) or false negatives (e.g., leading to undetected errors in the produced parts) are more acceptable. For the behavior of the classifiers under various classification thresholds, consider the ROC curve in Fig. 6. The confusion matrices here are given for the classification threshold value that gives the highest validation accuracy, as per Table 3.

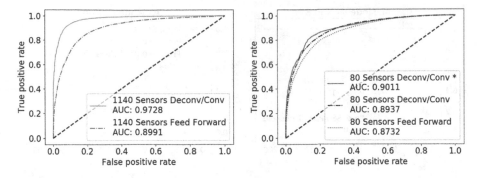

Fig. 6. ROC Curves for different models and sensor inputs. * pixel-threshold at .8

Is the Pixel-Thresholding Step in FlowFrontNet Useful? By experimenting with possible pixel thresholds (see Sect. 3) of 0.2, 0.5, and 0.8, we found that 0.8 yielded the best results and is used for the 80-sensor results (also in the already presented results). The accuracy increased from 81.48 to 83.69%. We observed that the training loss decreases steeper and farther without pixel-thresholds but the validation loss increases to a greater extent. By contrast, the validation loss for the pixel-threshold model declined steadily – indicating that the thresholding of flow front images regularizes.

Moreover, training more than one epoch without pixel-thresholds produced heavy overfits. Even with an exponential learning rate scheduler and a very low initial learning rate, the training dynamics did not improve. While the validation loss was decreasing, it was at a higher base level than before. With pixel-thresholds in place, training got easier, even without scheduling the learning rate but using a fixed value of 10^{-4} with AdamW [10]. Alas, for the smallest 20 sensor grid, pixel-thresholding did not give better results.

The ROC curves and the corresponding area under the curve (AUC) values are shown in Fig. 6. The AUC for the .8-pixel-thresholded model is the best of all 80 sensor models, but only by a small margin. The other curves do not hold any surprises, with the 1140 sensor model outperforming all and all other models with similar AUCs. Only the 80 sensor feed-forward net is underperforming.

4.2 Discussion - Metrics on Run Level

Our previous evaluations exclusively considered metrics per frame that are each drawn from many independent injection runs. Such a dry spot classification for every frame and, thus, point in time is desirable to closely monitor the RTM process (in a "digital twin"-manner) to intervene at process execution time by adjusting process parameters. Additionally, practitioners care about a judgment concerning the process quality of a single run, e.g., a single produced composite plate. This is similar to lifting single-frame classification to video classifications, e.g., how many dry spot frames make a dry spot run? The first difficulty is to

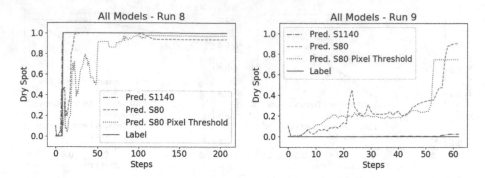

Fig. 7. Different classifications of two runs by three different models

decide when a run counts as a failure and doing so automatically for both the label maps and predictions for all 36,663 runs.

The naive approach would take the last few frames to determine if a run counts as having produced a dry-spot. However, we would miss dry spots that only occur in the middle of a run. These might hint at problems in reality and only "close" due to simulation artifacts (e.g., unrealistically increasing the pressure). Another possible criterion counts all dry spot frames and prescribes a minimal amount to mark a run as failed. Alternatively, we could require the dry spot sequences to be contiguous, to avoid listing too many runs as failed if there are single dry spots in the label maps or predictions which could also be artifacts from the label generation process described in Sect. 2.2. Figure 7 offer some insight into how different run classifiers would behave for two exemplary runs, given the frame-wise ground truth and predictions.

Here, it becomes apparent that the models perform very differently and also confirms that the models taking 80 sensors as input cannot be compared to the 1140 sensor input models. The 1140-sensor-models adhere closely to the label most of the time whereas both 80-sensor-models produce more noise, in different ways. We anticipate further experiments as future work, especially combining the single-frame predictions with another sequence model to classify whole runs – provided that we can obtain or define meaningful run labels.

5   Conclusion and Future Work

We presented FlowFrontNet, a deconvolutional/convolutional neural network suitable for detecting dry spots in simulated RTM processes, i.e., improved process monitoring. In doing so, we showed that it is possible to learn the intermediate representation of flow fronts from sensor data by upsampling via a deconvolutional network. This enabled us to reliably classify dry spots on individual frames substantially better than with a feed-forward network using the same sensor input. The classifier makes it possible to intervene during the manufacturing of a single composite plate. We also investigated that the prediction quality decreases with the number of sensors in use, and found that acceptable accuracy requires at least 80 sensors, a sensor grid 4 cm distance. That is a realistic order of magnitude for real mold sensor layouts whereas 1140 sensors cannot be placed as closely due to cost and wiring issues.

Future work can be divided into short and long term goals that focus on use cases with simulated and actual data, respectively. In the short run, we plan to use the flow front image generated from sensors for other classification or object detection tasks and use sequence models for predicting full runs, with their temporal information. Our long term goal is to use models pre-trained on simulated data for real RTM process data [18], much like sim-to-real applications are being used in reinforcement learning and robotics [13]. In addition to the costly process of actually producing composite plates in a sufficient number for training, reality confronts us with heterogeneous and noisy sensors other than pressure alone. Eventually, we want to generate feedback for process control to avoid rejects. The first step towards this goal, learning the flow front from sensor data and applying it to dry-spot classification, has been successfully achieved.

Acknowledgments. This research is funded by the Bavarian Ministry of Economic Affairs, Regional Development and Energy in the project CosiMo. We thank Ewald Fauster from Montanuniversität Leoben for his expert advice on the RTM process and Frederic Masseria from ESI for supporting our RTM-simulations.

References

1. Babb, D.A., et al.: Resin transfer molding process for composites (1998)
2. Darcy, H.P.G.: Les Fontaines publiques de la ville de Dijon. V. Dalamont (1856)
3. Grössing, H., Stadlmajer, N., Fauster, E., Fleischmann, M., Schledjewski, R.: Flow front advancement during composite processing: predictions from numerical filling simulation tools in comparison with real-world experiments. Polym. Compos. **37**(9), 2782–2793 (2016). https://doi.org/10.1002/pc.23474
4. Heuer, H., et al.: Review on quality assurance along the CFRP value chain - nondestructive testing of fabrics, preforms and CFRP by HF radio wave techniques. Compos. Part B Eng. **77**, 494–501 (2015). https://doi.org/10.1016/j.compositesb.2015.03.022
5. Heywood, J., et al.: On the road toward 2050: report Massachusetts Institute of Technology potential for substantial reductions in light-duty vehicle energy use and greenhouse gas emissions. Massachusetts Institute of Technology (2015). http://mitei.mit.edu/publications/
6. LeCun, Y., Bottou, L., Bengio, Y., Haffner, P.: Gradient-based learning applied to document recognition. Proc. IEEE **86**(11), 2278–2324 (1998)
7. Lee, D.H., Lee, W.I., Kang, M.K.: Analysis and minimization of void formation during resin transfer molding process. Compos. Sci. Technol. **66**(16), 3281–3289 (2006)
8. Lee, J., Bagheri, B., Kao, H.A.: A cyber-physical systems architecture for industry 4.0-based manufacturing systems. Manuf. Lett. **3**, 18–23 (2015)
9. Liu, B., Bickerton, S., Advani, S.G.: Modelling and simulation of resin transfer moulding (RTM) - gate control, venting and dry spot prediction. Compos. Part A Appl. Sci. Manuf. **27**(2), 135–141 (1996). https://doi.org/10.1016/1359-835X(95)00012-Q

10. Loshchilov, I., Hutter, F.: Decoupled weight decay regularization. In: 7th International Conference on Learning Representations, ICLR 2019, November 2019
11. Noh, H., Hong, S., Han, B.: Learning deconvolution network for semantic segmentation. In: Proceedings of the IEEE International Conference on Computer Vision, vol. 2015, pp. 1520–1528, May 2015. https://doi.org/10.1109/ICCV.2015.178
12. Pantelesis, N., Bistekos, E.: Process monitoring and control for the production of CFRP components. In: SAMPE Conference, pp. 5–9 (2012)
13. Peng, X.B., Andrychowicz, M., Zaremba, W., Abbeel, P.: Sim-to-real transfer of robotic control with dynamics randomization. In: 2018 IEEE International Conference on Robotics and Automation (ICRA), pp. 1–8. IEEE (2018)
14. Shelhamer, E., Long, J., Darrell, T.: Fully convolutional networks for semantic segmentation. IEEE Trans. Pattern Anal. Mach. Intell. **39**, 640–651 (2017). https://doi.org/10.1109/TPAMI.2016.2572683
15. Shi, W., et al.: Real-time single image and video super-resolution using an efficient sub-pixel convolutional neural network. In: Proceedings of the IEEE Computer Society Conference on Computer Vision and Pattern Recognition, vol. December 2016, pp. 1874–1883, September 2016. https://doi.org/10.1109/CVPR.2016.207
16. Shin, H.C., et al.: Deep convolutional neural networks for computer-aided detection: CNN architectures, dataset characteristics and transfer learning. IEEE Trans. Med. Imaging **35**(5), 1285–1298 (2016). https://doi.org/10.1109/TMI.2016.2528162
17. Sorg, C.: Data Mining als Methode zur Industrialisierung und Qualifizierung neuer Fertigungsprozesse für CFK-Bauteile in automobiler Großserienproduktion. Ph.D. thesis, Technische Universität München (2014)
18. Stieber, S.: Transfer learning for optimization of carbon fiber reinforced polymer production. Organic Computing: Doctoral Dissertation Colloquium 2018, pp. 1–12 (2018)
19. Stieber, S.: FlowFrontNet Checkpoints (2020). https://doi.org/10.6084/m9.figshare.12102714.v1. https://figshare.com/articles/FlowFrontNet_Checkpoints/12102714
20. Stieber, S.: FlowFrontNet Data: Sensor to Flowfront/Dryspot (2020). https://doi.org/10.6084/m9.figshare.12063480.v4. https://figshare.com/articles/FlowFrontNet_Data_Sensor_to_Flowfront_Dryspot/12063480
21. Xu, L., Ren, J.S., Liu, C., Jia, J.: Deep convolutional neural network for image deconvolution. Adv. Neural Inf. Process. Syst. **2**(January), 1790–1798 (2014)
22. Zhang, J., Pantelelis, N.: Modelling and optimisation control of polymer composite moulding processes using bootstrap aggregated neural network models. In: Proceedings of 2011 International Conference on Electric Information and Control Engineering, ICEICE 2011, vol. 1, no. 2, p. 3 (2011). https://doi.org/10.1109/ICEICE.2011.5777841

Learning I/O Access Patterns to Improve Prefetching in SSDs

Chandranil Chakraborttii[✉] and Heiner Litz

Department of Computer Science, University of California Santa Cruz,
11156 High Street, Santa Cruz, CA 95064, USA
{cchakrab,hlitz}@ucsc.edu

Abstract. Flash based solid state drives (SSDs) have established themselves as a higher-performance alternative to hard disk drives in cloud and mobile environments. Nevertheless, SSDs remain a performance bottleneck of computer systems due to their high I/O access latency. A common approach for improving the access latency is prefetching. Prefetching predicts future block accesses and preloads them into main memory ahead of time. In this paper, we discuss the challenges of prefetching in SSDs, explain why prior approaches fail to achieve high accuracy, and present a neural network based prefetching approach that significantly outperforms the state-of-the-art. To achieve high performance, we address the challenges of prefetching in very large sparse address spaces, as well as prefetching in a timely manner by predicting ahead of time. We collect I/O trace files from several real-world applications running on cloud servers and show that our proposed approach consistently outperforms the existing stride prefetchers by up to 800× and prior prefetching approaches based on Markov chains by up to 8×. Furthermore, we propose an address mapping learning technique to demonstrate the applicability of our approach to previously unseen SSD workloads and perform a hyperparameter sensitivity study.

Keywords: Prefetching · Neural network · Flash

1 Introduction

Solid state drives (SSDs) have become the primary storage device technology for mobile devices and high-performance servers. SSDs have replaced the spinning disks (HDDs) for many applications in cloud services due to their higher I/O performance [47], lower failure rate [37], and better endurance [34]. Nevertheless, although SSDs deliver significantly higher speeds than HDDs, SSDs still remain a performance bottleneck of computing systems [24], as processors and DRAM technologies support three orders of magnitude lower access latency. Two common approaches to hide the high access latency of storage devices are caching and prefetching. Caching utilizes less dense but faster types of memory to store frequently used data items, filtering out many accesses of the slow SSDs. Examples include Linux's page cache [12] and filesystem caches [43]. Prefetching [9]

© Springer Nature Switzerland AG 2021
Y. Dong et al. (Eds.): ECML PKDD 2020, LNAI 12460, pp. 427–443, 2021.
https://doi.org/10.1007/978-3-030-67667-4_26

approaches read data from SSDs in advance, in order to serve the later demand accesses from the cache with low latency. Prefetching can be implemented either in software, e.g., within operating system [29,35] or within the SSD itself [48].

Existing prefetching mechanisms [38,39] are limited by the computational complexity and difficulty of correctly predicting future I/O accesses. For instance, the read-ahead prefetcher [15,26] is limited to prefetching the next data item within a file to accelerate sequential accesses. More advanced prefetchers [5,17] that can learn complex I/O access patterns have been dismissed because of their computational cost. Recently, storage vendors, including Samsung, have proposed SmartSSDs [14,33], adding computational capabilities to SSDs. These devices offer new opportunities as they enable offloading of prefetching to hardware, removing the burden from the host CPU. While this approach addresses the compute overhead of prefetching, predicting future I/O accesses accurately remains a challenge. Real-world applications not only perform sequential accesses, but also exhibit complex workload patterns [7]. Applications are frequently used by multiple users simultaneously, performing independent tasks, resulting in a mix of sequential and random I/O requests which are difficult to model and challenging to predict. Furthermore, in existing systems, I/O accesses need to traverse a deeply layered software stack, transforming the easy to predict accesses on the application side into seemingly random accesses on the SSD level. Predicting future memory accesses from multiple interleaved I/O access streams on the SSD device layer hence represents a challenging problem.

Modern SSDs and operating systems offer a wide range of telemetry data for analysis. Utilizing I/O access tracing in hardware and software enables the collection of large, clean, and automatically labeled datasets that can fuel powerful machine learning models. In this work, we leverage Long Short-Term Memory (LSTM) [19] based sequence-to-sequence neural networks to learn spatial I/O access patterns of applications from block level I/O traces collected from a diverse set of data center applications. LSTMs are capable of capturing long-term dependencies in data and can address sequences of different lengths. LSTMs integrate model training and representation learning together, without requiring additional domain knowledge, enabling the discovery of unseen patterns in the data to improve generalization capability of a model. In this work, we leverage LSTMs to deliver the following contributions. First, our model provides high accuracy even in the presence of complex interleaved I/O streams. Second, it addresses the challenge of timeliness by predicting multiple I/O accesses ahead of time. Third, to cope with the dynamic behavior of applications and to improve the reusability of our model, we propose an address mapping learning (AML) technique enabling our model to predict different types of workloads. To demonstrate the practicality of our approach, we build a simulator enabling us to measure timeliness in addition to prediction accuracy. We utilize I/O traces to train the neural network models offline and predict future logical block addresses (LBAs) at runtime using the simulator. To reduce address space, we take the l_1 norm between a pair of consecutive memory accesses as input to the model in addition to the requested I/O size. This enables the model to also predict the *size* of the incoming I/O request, representing the amount of data blocks to be prefetched ahead of time. We show that our approach enables predicting LBAs sufficiently far ahead to compensate for the read latency of accessing flash as well as for the inference latency of our model. We present an analysis of the impact of predicting N steps ahead into the future and evaluate the impact of cache size on the performance of our prefetcher. We compare our work with three baselines, a naive approach that only prefetches the most frequently accessed LBAs, a stride prefetcher [23], and the Markov chain based prefetcher [11,26,48], showing an improvement of up to 800× over the stride prefetcher and up to 8× over the Markov chain prefetcher.

2   Background

2.1   Flash Device Architecture

NAND flash drives or SSDs are non-volatile memory devices storing individual bits on floating gate transistors. Floating gate transistors are arranged in large bit cell arrays increasing not only the storage capacity, but also the access latency. Furthermore, flash cells suffer from limited endurance and frequent bit errors, which are exacerbated by transistor scaling and the introduction of techniques such as multi-level cells [31]. To ensure data integrity, multiple reads using different reference voltages need to be performed, and the controller needs to perform error detection and correction as part of each read, further increasing the read latency. As a result, the I/O access latency of SSDs (~100 μs) is three orders of magnitude higher than the latency of reading DRAM (~100ns). Hence, a mechanism that prefetches the data into DRAM provides significant performance gains.

2.2   Prefetching

Prefetching in storage systems is the process of preloading data from a slow storage device into faster memory, generally DRAM, to decrease the overall read latency. Accurate and timely prefetching can effectively reduce the performance gap between different levels of memory [30]. There are three important metrics used to compare prefetchers including coverage, accuracy, and timeliness of prefetchers [23]. Coverage is the ratio of the number of SSD reads that can be prefetched to the total number of SSD reads. Accuracy is the ratio of number of data blocks being prefetched to the number of prefetched data blocks that were actually requested by the application. Timeliness requires data blocks to be prefetched sufficiently ahead of time so that the data is present in DRAM whenever the read request is performed by the application. If the prefetched data blocks are not available when they are needed, the application is required to stall, rendering prefetching ineffective. Furthermore, if the data is prefetched too early, it may not be available anymore when it is actually needed, due to the eviction from the capacity-limited cache. Inaccurate prefetches that read in unneeded

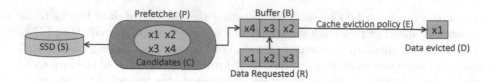

Fig. 1. Demonstration of prefetching at time, t

data are harmful as they waste I/O bandwidth and DRAM capacity. If prefetching is performed too conservatively, coverage is low and the overall performance gains are limited. Hence, the ideal prefetcher has high coverage, high accuracy, and executes prefetching timely so that the data is fetched exactly when it is needed. A basic prefetching mechanism is shown in Fig. 1. The SSD prefetcher (P) is responsible for predicting candidate data blocks (C) to prefetch from the SSD (S) into a fast cache (DRAM) buffer (B) of size s. The cache eviction policy (E) is responsible for evicting the data blocks from B in order to make space for new incoming data. In this example, at time t, P determines candidates x1, x2, x3, and x4 for prefetching, but the actual data requested at time t is x1, x2, and x3. Here, x1 was prefetched too early while x4 was inaccurately prefetched, resulting in cache miss in both the cases. Candidates x2 and x3, however, were present in the cache when requested, and hence, result in cache hit.

2.3 Neural Network Based Prefetching

While most work on I/O prefetching has focused on conventional techniques, some prior works have explored using machine learning techniques. Hashemi [18] used neural network based sequence models for prefetching DRAM accesses. The models proposed in this work, however, cannot be applied to our problem as prefetching I/O accesses differs significantly from prefetching DRAM accesses. First, I/O accesses do not contain instruction information to enable stream disambiguation, second, I/O accesses do not have a fixed size like DRAM accesses, third, I/O accesses and DRAM accesses interact differently with the OS, and fourth, I/O prefetching models need to account for timeliness. A second line of work utilized Markov chains [11] for prefetching data from SSDs [26,48]. We compare our approach with these prior works in Sect. 6, confirming prior observations that Markov chain based prefetchers perform poorly on real world applications where the I/O streams are more complex [44].

3 Problem Statement

We assume a digital system that consists of the following components. A flash based digital storage device (SSD) that provides high capacity but low performance, and a high access latency. A central processing unit (CPU) that can process data at orders of magnitude faster than the SSD. In addition, the system is comprised of a cache (usually DRAM) that is placed in between the CPU and the SSD. The CPU can access data with low latency from the cache, however, the cache capacity is orders of magnitude smaller than the SSD capacity. Reads access a specific logical block address (LBA) and are generally more performance critical than writes, as future operations depend on the data supplied by the reads, which is why this work focuses on reads. The goal we aim to achieve is to accurately predict future LBAs so that they can be prefetched into the cache, enabling low latency accesses by the CPU. In addition to the LBA, we also need to predict the size of the I/O, as prefetching only parts of an I/O access is useless. Thus, an efficient prefetching mechanism requires optimizing three metrics, particularly, the *coverage, accuracy,* and *timeliness.*

Coverage or recall refers to the ratio of future memory accesses that are attempted to be prefetched. Prefetching of an LBA is accurate if the same LBA is subsequently accessed by a demand read. Accuracy is hence defined as the ratio of accurate prefetches to executed prefetches. A prefetch is timely if it is executed sufficiently ahead of time of the demand read. In particular, $T_{cand} + T_{read} < PA * T_{arrival}$ must hold, where T_{cand} represents the time to compute a prefetch candidate, T_{read} represents the time to perform a read from the SSD, $T_{arrival}$ represents the inter arrival time between demand reads, and PA represents prefetch-ahead, which is the number of accesses we need to predict into the future. Executing prefetches too early is generally of a lesser concern as prefetches can be stored for a finite time in the cache. As a result, the time that a prefetch can be executed too early is bounded only by the cache capacity.

Storage accesses to an LBA are generally handled by the operating system. User applications, however, generally communicate with the storage devices by reading and writing files. Consequently, the filesystem layer within the OS needs to map file accesses to LBA accesses before they can be submitted to the storage device. Furthermore, to improve performance, the OS maintains several caching layers in the filesystem and logical block layer, aiming to filter out a significant fraction of all application accesses. The result of this architecture is that even a seemingly easy to predict operation on the application layer, such as reading a file sequentially, may result in a very hard to predict access patterns on the LBA level, as perceived by the SSD. Finally, the storage device is generally accessed by different application threads simultaneously, resulting in multiple interleaved I/O streams that are indistinguishable by the SSD. In summary, the existing storage stack architecture renders predicting future I/O accesses a challenging problem. Predictive models need to be able to separate multiplexed I/O streams and then predict future LBAs from within the hard to predict sequences. In addition, they need to provide information on the number of data blocks to prefetch, starting from the initial predicted LBA.

4   Proposed Prefetching Technique

Learning SSD storage accesses for prefetching is a challenging task for the following reasons. As SSDs are increasing their storage capacity to 16TB and beyond, drives are now supporting billions of logical block addresses. As prefetching is only successful if every bit of the logical block address is predicted accurately, models are required to predict which LBA to prefetch with perfect accuracy within a very large LBA space. This space is often sparse, as the operating system allocates blocks within the filesystem layer, and hence, even sequential data within files may be mapped to arbitrary LBAs within the SSD. Furthermore, as prefetches need to be timely, predicting only the next LBA and the requested I/O size is not sufficient, and it is required to predict several accesses into the future. Finally, to support dynamically changing workloads, we evaluate our proposed address mapping learning technique to determine whether prefetching models can learn generalized patterns within complex I/O access patterns.

4.1   Data Preparation for Reducing the Output Label Space

We preprocess the input dataset to address the problem of large logical block address space. The number of unique memory addresses within an SSD is typically of the order of billions, rendering a separate class for each memory address impractical. To reduce the address space, we take the l_1 norm of each pair of consecutive LBAs (LBA delta). For example, if consecutive I/O accesses starting from LBA 10000 are requested as 10001, 10003 and 100006, the corresponding LBA deltas were recorded as 1, 2, and 3, respectively. This significantly reduces the number of classes that our model needs to predict. We identify the top 1000 frequently occurring LBA deltas and assign each one of them to a class in decreasing order of frequency. All remaining LBA deltas are assigned to a separate class representing a "no prefetch" operation, thus limiting the number of classes for model to predict to 1001. The reason for choosing LBA deltas over actual addresses is to increase the coverage of LBA deltas in the data. For example, for Microsoft Research Cambridge traces [27] (MSR_1), the top 1000 most frequently occurring LBAs covered only 2.77% of all the LBA accesses, whereas the top 1000 most frequently occurring LBA deltas covered 91.66% of all LBA accesses. The coverage of top 1000 frequent LBA deltas for the datasets used in this study ranged between 54% and 92%, as seen in Table 1. Expanding the number of classes to beyond 1000 is possible with more computational power, however, for our datasets, we chose 1000, as it provides a considerable coverage for LBAs and is a sufficiently large size to prove the practicality of our approach.

The requested I/O sizes for the analyzed real world applications ranged from 4KB to several MBs with up to 10,000 different I/O sizes for an individual

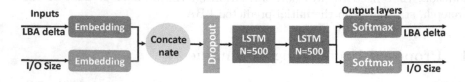

Fig. 2. Model architecture

application. In order to reduce the number of possible target I/O size values, we round off each observed I/O size to the nearest number that is a power of 2, 2^n, and use n as an I/O size class. This reduces the number of possible target I/O sizes for most applications to 16 while still supporting requests of size up to 64MB. A limitation of this approach is that, in the worst case, roughly twice as many as required 4KB blocks may be prefetched from the SSD.

4.2   Model Architecture

We designed our proposed neural network model to predict both the I/O size and LBA deltas at the same time. The model has two separate input layers, one for I/O size and one for LBA delta, where each input layer is an embedding layer [49] consisting of 500 neurons. The inputs to the model are categorical, one-hot, representation of the two features, LBA deltas and I/O size, each being fed to a separate embedding layer. The model has two hidden LSTM layers, where each LSTM layer has 500 hidden nodes. The outputs of the two embedding layers are first concatenated and then fed to the shared LSTM layers. The final output layer is split into two branches, where each branch is a dense layer consisting of softmax [32] nodes. The number of neurons in the LBA delta output layer is 1001, representing top 1000 LBA deltas and a "no prefetch" LBA delta, and the number of neurons in the I/O size output layer ranged between 12 and 20, depending on the I/O sizes present in each dataset. The model architecture is shown in Fig. 2. The number of neurons in each of the first three layers of the model was set to 500 to ensure a good representation of input features, and we used a dropout [16] of 0.2 to prevent overfitting of the model. Having an initial embedding layer facilitates better representation of the input features and helps the subsequent LSTM layers to learn effectively from sequential data.

4.3   Timeliness

As discussed in Sect. 3, a prediction from the prefetcher is timely only if the following equation holds: $T_{cand} + T_{read} < PA * T_{demand}$. We empirically determined T_{cand} to be 734 μs by measuring the inference latency of our model. We measured the latency of accessing an Intel P3600 NVMe based SSD using the flexible I/O tester (FIO) [6] to be 300 μs on average under 80% workload. For the traces that we examined, the average time between two successive I/O requests ranged between 800 μs and 1200 μs, and the minimum time was 10 μs. As a result, a good PA value is in the range of $5 > PA > 100$. We evaluate a range of PA values and its impact on prediction accuracy in Sect. 6. Predicting further ahead in the future typically reduces the accuracy due to the increased uncertainty. We find that, in order to increase the accuracy in case of a high PA value, training the model with longer history of sequences can improve performance.

4.4   Address Mapping Learning

Different workloads show similar I/O access patterns due to shared design patterns and commonly used data structures. For instance, array-based data

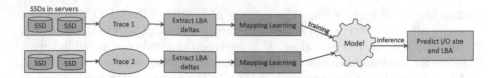

Fig. 3. Block diagram of the Address Mapping Learning process

structures used by applications generally entail sequential I/O access patterns. Furthermore, as most applications leverage the same underlying filesystem, it is likely that I/O accesses show common patterns. An ideal prefetcher would be trained once, on a varied set of applications, providing high performance even for previously unseen applications. Such a prefetcher is also likely to be more robust with respect to dynamically changing data inputs or code changes to the original application. To test the idea that applications share common patterns that can be learned, we train the model on traces from one dataset (source) and evaluate the performance of the prefetcher on another dataset (recipient). The mapping of addresses to labels is done by sorting the frequency distribution of LBA deltas from *both* the source and recipient traces and assigning them labels in decreasing order of frequency of occurrences. We call the process of extracting the LBA deltas, training the model on source dataset, and using the model to predict LBA deltas and I/O sizes for the recipient dataset as Address Mapping Learning (AML) and present the block diagram of this process in Fig. 3.

5   Methodology and Experimental Setup

5.1   Model Training

For our experiments, we used a total of 10 block-level I/O traces from three different sources running applications in live production servers. The datasets included traces describing enterprise storage traffic in commercial office virtual desktop infrastructure (VDI) [27], as well as traces from live production servers at Microsoft SNIA [22] and Microsoft Research Cambridge [36]. We did not use any synthetic benchmarks, as used in previous work [26,48], as those traces do not accurately represent the complexity and interleaved patterns exhibited in real applications. The utilized trace files are open-source and can be obtained online [1,2,22]. Table 1 provides information about the datasets used in this study. From the table, we see that the coverage of top 1000 LBA deltas is consistently higher than direct memory addresses (offset), and hence it was selected as one of the features for training the model. The datasets also contained other information such as the I/O size, response time, filename, file location, etc. In this work, we only used the timestamp, offset (LBA), and I/O size as features. We trained our model using Google's Tensorflow [3] library on a Intel Xeon server with 8 CPU cores running 1.7 GHz containing 96GB of DRAM. The server also had 4 NVIDIA Tesla 2080TI GPUs for training the model. We split the dataset

Table 1. Dataset description

Trace source	Dataset name	Represented name	Num obs	Coverage offset (%)	Coverage LBA delta (%)
VDI	2016022315.csv	VDI_1	5226120	58.76	66.96
VDI	2016030817.csv	VDI_2	4443487	63.94	70.08
VDI	2016030819.csv	VDI_3	2902328	68.94	69.8
VDI	2016031115.csv	VDI_4	2408227	68.65	72.35
MSR	proj_3.csv	MSR_1	2244642	2.77	91.66
MSR	mds_0.csv	MSR_2	1211034	63.46	76.94
MSR	src1_1.csv	MSR_3	45746222	28.6	77.7
MSR	usr_1.csv	MSR_4	45283980	2.64	82.12
Microsoft	buildserver-2.csv	MS_1	1600430	2.77	28.84
Microsoft	buildserver-7.csv	MS_2	1714151	8.97	55.49

into training and test set, where the training set contained the first 70% of the I/O accesses, and the test set contained the last 30% of the I/O accesses. The sequence of LBA deltas, ordered by timestamps, is fed to the model for training. For all the experiments, we trained our model using Adam optimizer [46] with a cross-entropy loss function, and a learning rate of 10^{-3} for up to 1000 epochs, and stopped model training if there was no improvement in validation loss, with validation loss not decreasing by at least 10^{-5} for five consecutive epochs.

5.2 Prefetcher Simulation Environment

To enable the comparison of our prefetcher against prior baselines, evaluating only recall and precision is not sufficient. As motivated before, analyzing the prefetcher's timeliness is required to evaluate the end-to-end performance gains of prefetching, as even the most accurate prefetcher will not improve the performance if it lacks timeliness. As shown in Sect. 4.3, in order to compensate for the model's prediction latency and the latency to perform a read from the SSD, it is required to generate predictions ahead of time (PA). We evaluate the end-to-end performance as follows. As we iterate through the test dataset, the evaluation models continuously generate prefetch candidate predictions that are inserted into the cache.

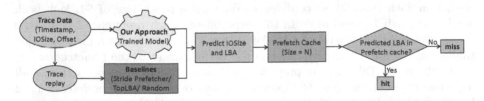

Fig. 4. Block diagram of the evaluation process using our simulator

Every I/O access is checked against the cache to see if the LBA is present, where the access is recorded as a hit, otherwise it is recorded as a miss. We utilize the Least Recently Used (LRU) [42] eviction policy for our experiments. The architecture of the simulator is presented in Fig. 4. We choose variable cache sizes of LBAs for the stride, Markov-based, and our proposed prefetcher, and run experiments to provide a comparative study in Sect. 6.

5.3 Baselines

We compare our proposed prefetcher to three baselines. The first, naive prefetcher, baseline always predicts the most common delta of a trace. The second baseline implements a Markov chain predictor [26,48]. This method treats each LBA access as a state and predicts the next LBA based on the previous state by computing a probability distribution over the probabilities of transition from one state to another. The third baseline is a stride prefetcher which is commonly used in software and hardware systems. The stride prefetcher concurrently observes 128 I/O access streams. Each access is mapped to a stream based on hashing the most significant bits of the LBA. For each stream, the stride prefetcher tracks the last three I/O accesses. If the difference between the three I/O accesses match, the prefetcher detects a stride and prefetches the next access. Note that the stride prefetcher's results are optimistic, as it only prefetches one access ahead of time and does not compensate for timeliness. In the next section, we evaluate our proposed prefetcher in terms of prediction accuracy, timeliness, and capability to generalize to different workloads.

6   Results

6.1   Prefetcher Accuracy, Precision and Recall

Table 2 shows the comparative performance of our neural network based prefetcher against the three chosen baselines. The table lists the dataset name, number of samples in the dataset, and the accuracy for the three chosen baselines, Naive prefetcher, Stride prefetcher, and Markov chain based prefetcher. For our approach, we provide the accuracy, precision, and recall results. For each sample, our prefetcher predicts both LBA and I/O size in increments of 4KB blocks, as the minimum block size for a drive operation in SSD is typically of 4KB size [33]. We only count the actual blocks that are correctly prefetched. For each data sample, we prefetch only the top predicted LBA and I/O size using the prediction with the highest confidence. We used a batch size of 64, look back of 64, and predict-ahead of 64 in this experiment. Each prefetcher has a cache size of 1000 for this experiment. In the next section, we present a more detailed analysis of the impact of cache size on the performances of the prefetchers.

As shown in Table 2, our proposed prefetcher consistently outperforms all three baselines delivering up to 11× improvement over the stride prefetcher using Microsoft SNIA traces with the same cache size. For VDI traces, our proposed

Table 2. Performance comparison of our proposed prefetcher against baselines

Dataset name	No. samples	Naive prefetcher	Stride prefetcher	Markov prefetcher	Our (accuracy)	Our (precision)	Our (recall)
VDI_1	5226120	0.17	0.01	0.09	0.73	0.76	0.71
VDI_2	4443487	0.21	0.01	0.07	0.59	0.75	0.49
VDI_3	2902328	0.19	0.02	0.12	0.66	0.73	0.57
VDI_4	2408227	0.21	0.05	0.09	0.73	0.77	0.69
MSR_1	2244642	0.14	0.01	0.21	0.41	0.66	0.31
MSR_2	1211034	0.09	0.21	0.17	0.49	0.65	0.33
MSR_3	45746222	0.12	0.001	0.16	0.79	0.89	0.46
MSR_4	45283980	0.33	0.007	0.15	0.53	0.66	0.38
MS_1	1600430	0.27	0.02	0.25	0.63	0.79	0.53
MS_2	1714151	0.41	0.003	0.07	0.77	0.83	0.61

prefetcher achieves the highest accuracy, providing 800× improvement over the stride prefetcher. Our prefetcher also achieved the highest precision and recall compared to the baselines. The Markov chain based prefetcher performed considerably worse compared to our prefetcher, with the accuracy ranging between 7% and 25%, performing even worse than the Naive prefetcher in several cases.

6.2   Impact of Cache Size, Look-Back, and Predict-Ahead

In this section, we present an analysis of the impact of look back, predict-ahead, and cache size on our proposed prefetcher's performance. In order to ensure the availability of data in the cache when the data block is requested, we trained the model to predict N steps ahead for varying values of N, and evaluated the performance of the prefetcher. Higher values of N typically resulted in lower accuracy due to the increased uncertainty in predicting further ahead in the future, while improving timeliness. To improve our prefetcher's predict-ahead performance, we found that it is necessary to increase the look back size for increasing values of PA, where, as described in Sect. 4.3, good values for PA are in the range of $5 < PA < 100$. Low values (< 5) of PA result in cache misses as the data cannot be fetched soon enough, whereas higher values of PA (> 100) result in untimely predictions as the data gets evicted before requested. Table 3 shows the performance of our prefetcher for different values of PA showing the accuracy of predicting the LBA and I/O size, as well as the cache hit ratio (Net Hit ratio). We measured accuracy as the actual number of 4KB data blocks that were correctly prefetched for three different values of PA, 32, 64, and 128.

In general, the accuracy of predictions decreases as we predict further ahead, producing the worst performance when predicting 128 samples ahead. For MS SNIA traces, the performance was comparable for PA equal to 32 and 64, and the accuracy degraded significantly for $PA = 128$, whereas for VDI and MSR Cambridge traces, the performance degradation was gradual. These results show

Table 3. Impact of different predict values on our prefetcher performance

Dataset	Predict ahead = 32			Predict ahead = 64			Predict ahead = 128		
	Accuracy (LBA)	Accuracy (Size)	Net hit ratio	Accuracy (LBA)	Accuracy (Size)	Net hit ratio	Accuracy (LBA)	Accuracy (Size)	Net hit ratio
VDI_1	0.72	0.65	0.71	0.69	0.65	0.73	0.42	0.6	0.33
VDI_2	0.76	0.51	0.58	0.64	0.51	0.59	0.41	0.42	0.29
VDI_3	0.73	0.88	0.69	0.48	0.88	0.66	0.42	0.67	0.37
VDI_4	0.71	0.66	0.71	0.71	0.66	0.73	0.32	0.34	0.31
MSR_1	0.65	0.49	0.41	0.65	0.49	0.41	0.34	0.19	0.29
MSR_2	0.59	0.69	0.49	0.59	0.69	0.49	0.19	0.61	0.33
MSR_3	0.95	0.67	0.66	0.91	0.61	0.79	0.13	0.61	0.19
MSR_4	0.59	0.77	0.51	0.49	0.77	0.53	0.49	0.47	0.28
MS_1	0.93	0.67	0.61	0.93	0.52	0.63	0.62	0.52	0.49
MS_1	0.89	0.71	0.73	0.88	0.69	0.77	0.57	0.69	0.47

Table 4. Impact of cache size on the accuracy of our and two baseline prefetchers

Dataset name	Cache size = 10			Cache size = 100			Cache size = 1000		
	Markov prefetcher	Stride prefetcher	Our prefetcher	Markov prefetcher	Stride prefetcher	Our prefetcher	Markov Prefetcher	Stride prefetcher	Our prefetcher
VDI_1	0.05	0.001	0.68	0.05	0.001	0.69	0.09	0.011	0.73
VDI_2	0.05	0.0001	0.55	0.05	0.0001	0.55	0.07	0.0015	0.59
VDI_3	0.04	0.0001	0.64	0.04	0.0001	0.64	0.12	0.0014	0.66
VDI_4	0.01	0.006	0.7	0.01	0.006	0.71	0.09	0.005	0.73
MSR_1	0.12	0.00005	0.39	0.12	0.00005	0.39	0.21	0.0011	0.41
MSR_2	0.09	0.1	0.41	0.09	0.1	0.41	0.17	0.21	0.49
MSR_3	0.07	0.0002	0.75	0.07	0.0002	0.76	0.16	0.001	0.79
MSR_4	0.06	0.0005	0.51	0.06	0.0005	0.51	0.15	0.007	0.53
MS_1	0.16	0.004	0.57	0.16	0.004	0.57	0.25	0.02	0.63
MS_1	0.02	0.0003	0.71	0.02	0.0003	0.71	0.07	0.003	0.77

that our approach is successful in prefetching SSD accesses, as PA equal to 32 or 64 is generally sufficient to ensure timeliness in real-world settings. Nevertheless, to support upcoming storage devices that support even higher request ratios, reducing the inference latency and predicting even further ahead will be required.

Table 4 presents the impact of varying cache size on our prefetcher's performance. The table shows the accuracy of our approach compared to the Markov and Stride prefetchers for cache sizes of 10, 100, and 1000 LBAs, respectively. From the table, we can see that our prefetcher consistently outperforms the baselines for each cache size, and the performance improvement using VDI traces is as high as 800× over the Stride prefetcher, and 8× over the Markov prefetcher. While the baselines show marginal improvements using larger cache sizes, our prefetcher benefits significantly from a larger cache size. This suggests that while our prefetcher provides high accuracy and coverage, its timeliness can still be improved. For a large cache, prefetched blocks remain in the cache for a longer

Table 5. Performance of Address Mapping Learning (AML)

	Similar source					Dissimilar source		
Source Trace	MSR_3	MSR_1	MS_1	VDI_1	VDI_3	MSR_3	MS_1	VDI_4
Recipient Trace	MSR_2	MSR_4	MS_2	VDI_2	VDI_4	VDI_3	VDI_1	MSR_2
Accuracy on Source Trace	0.95	0.63	0.93	0.75	0.87	0.92	0.92	0.82
Acuracy on Recipient Trace	0.59	0.59	0.87	0.72	0.75	0.75	0.75	0.72
AML Accuracy	0.37	0.39	0.84	0.52	0.47	0.31	0.22	0.35

time and hence, prefetching exactly at the time when the LBA is requested is less important. Achieving perfect timeliness would require adjusting PA dynamically, as the inter-arrival time between requests varies at runtime.

6.3 Evaluation of Address Mapping Learning

In this section, we evaluate whether our prefetcher can learn common patterns among workloads to predict accesses for previously unseen workloads. In the previous sections, we obtained the training and test datasets from different portions of the same workload and the trace file. In this section, we define two types of dataset sources. *Similar* sources are those where the training and test data are from the same application, however, with different data inputs, different execution times, and only small run time modifications in applications. *Dissimilar* sources are those where the training and test data are from completely different applications. Table 5 shows the prediction accuracy for different types of sources. We show the accuracy of the model when it is trained and tested on *similar* source traces, and also when it is trained and tested on the *dissimilar* source traces. In Table 5, for our proposed AML technique, the model is trained on the source trace and tested on the recipient trace. For instance, when training on MS_1 and evaluating on MS_2 trace files, the accuracy of our address mapping approach is 84% which is only 3% less than training and evaluating both on MS_2 (fourth column). The overall effectiveness of AML depends on the frequency distribution of LBA deltas in the two datasets. The results in Table 5 show that our approach can be applied to diverse workloads, as long as they share some similar characteristics. This increases the practicality of our approach, as we can train specific models for various workloads, and expect at least a moderate increase in performance for other workloads.

7 Related Work

Machine learning techniques have been applied to the prefetching problem in multiple domains such as web caching [4] and memory prefetching [18,50]. While previous work also utilized neural networks for determining prefetch candidates, they operate on very different datasets, as DRAM accesses differ significantly from I/O accesses. For instance, I/O accesses are not tagged with the source instruction for stream disambiguation, I/O accesses do not have a fixed size [41] and, in contrast to I/O, memory accesses are not intercepted by the OS. Prior work on SSD prefetching utilized algorithmic approaches, typically using a data-range-table to detect usable strides and memory access streams [23]. Several variations of stride prefetchers have been proposed [20,25] taking into account the spatial locality [20], feedback [40], and context [8]. However, as we showed in this work, algorithm based prefetchers do not perform well on real world applications due to their limited ability to learn complex patterns. The only prior research we are aware of that applies machine learning for prefetching in SSDs is based on Markov chains [26,48], which we used as a baseline in this work. Finally, machine learning techniques have been applied to improve SSDs in other ways, for instance, by optimizing garbage collection [45], for predicting device failures [21,37], for improving SSD virtualization [13], for managing SSDs in large clusters [28], and for improving the quality of service of SSDs [10]. These prior works are orthogonal to our work.

8   Conclusion

In this paper, we showed how to leverage neural network models to predict future storage I/O accesses to improve SSD performance via prefetching. We addressed several challenges such as the large and sparse logical block address space, ensuring timeliness of prefetching, predicting both the address and size of I/O accesses, as well as the challenge of training predictive models that can generalize across different workloads. We achieved generalization across workloads by leveraging a large set of real world cloud application traces. We compared the performance of our prefetcher to existing techniques and used an in-house simulator developed to test the accuracy, coverage, and timeliness of our proposed prefetcher. Our proposed model outperforms prior approaches such as the stride prefetcher by up to 800× and Markov chain based prefetcher by up to 8×.

Acknowledgements. This work was supported in part by Samsung Semiconductor, Inc. and in part by NSF grants CCF-1823559 and CCF-1942754.

References

1. Microsoft snia: Traces. http://iotta.snia.org/traces/4928
2. Msr cambridge traces. http://iotta.snia.org/traces/388
3. Abadi, M., et al.: Tensorflow: large-scale machine learning on heterogeneous distributed systems. arXiv preprint arXiv:1603.04467 (2016)
4. Ali, W., Shamsuddin, S.M., Ismail, A.S., et al.: A survey of web caching and prefetching. Int. J. Adv. Soft Comput. Appl **3**(1), 18–44 (2011)
5. Averbouch, I., Birnbaum, A.J., Hsieh, J.T., Shum, C.L.K.: Automatic pattern-based operand prefetching, 10 Feb 2015. uS Patent 8,954,678
6. Axboe, J.: Fio-flexible i/o tester synthetic benchmark (2005). https://github.com/axboe/fio. Accessed 13 June 2015

7. Boboila, S., Desnoyers, P.: Performance models of flash-based solid-state drives for real workloads. In: 2011 IEEE 27th Symposium on Mass Storage Systems and Technologies (MSST), pp. 1–6. IEEE (2011)
8. Bradford, J.P., Kossman, H.F., Mullins, T.J.: Context switch instruction prefetching in multithreaded computer, 10 Nov 2009. uS Patent 7,617,499
9. Callahan, D., Kennedy, K., Porterfield, A.: Software prefetching. ACM SIGARCH Comput. Architect. News **19**(2), 40–52 (1991)
10. Chakraborttii, C., Sinha, V., Litz, H.: SSD QOS improvements through machine learning. In: Proceedings of the ACM Symposium on Cloud Computing, p. 511 (2018)
11. Chung, K.L.: Markov Chains. Springer, New York (1967). https://doi.org/10.1007/978-3-642-62015-7
12. Da Zheng, R.B., Szalay, A.S.: A parallel page cache: IOPS and caching for multicore systems. In: Proceedings of the 4th USENIX conference on Hot Topics in Storage and File Systems, p. 5 (2012)
13. Dartois, J.E., Boukhobza, J., Knefati, A., Barais, O.: Investigating machine learning algorithms for modeling SSD I/O performance for container-based virtualization. IEEE Trans. Cloud Comput (2019)
14. Do, J., Kee, Y.S., Patel, J.M., Park, C., Park, K., DeWitt, D.J.: Query processing on smart SSDS: opportunities and challenges. In: Proceedings of the 2013 ACM SIGMOD International Conference on Management of Data, pp. 1221–1230 (2013)
15. Fengguang, W., Hongsheng, X., Chenfeng, X.: On the design of a new linux readahead framework. ACM SIGOPS Oper. Syst. Rev. **42**(5), 75–84 (2008)
16. Gal, Y., Ghahramani, Z.: Dropout as a Bayesian approximation: representing model uncertainty in deep learning. In: International Conference on Machine Learning, pp. 1050–1059 (2016)
17. Han, W.S., Whang, K.Y., Moon, Y.S.: A formal framework for prefetching based on the type-level access pattern in object-relational DBMSS. IEEE Trans. Knowl. Data Eng. **17**(10), 1436–1448 (2005)
18. Hashemi, M., et al.: Learning memory access patterns. arXiv preprint arXiv:1803.02329 (2018)
19. Hochreiter, S., Schmidhuber, J.: Long short-term memory. Neural Comput. **9**(8), 1735–1780 (1997)
20. Iacobovici, S., Kadambi, S., Chou, Y.C.: Multi-stride prefetcher with a recurring prefetch table, 3 Feb 2009. uS Patent 7,487,296
21. Iwasaki, T.O., Ning, S., Yamazawa, H., Sun, C., Tanakamaru, S., Takeuchi, K.: Machine learning prediction for 13x endurance enhancement in reram ssd system. In: 2015 IEEE International Memory Workshop (IMW), pp. 1–4. IEEE (2015)
22. Kavalanekar, S., Worthington, B., Zhang, Q., Sharda, V.: Characterization of storage workload traces from production windows servers. In: 2008 IEEE International Symposium on Workload Characterization, pp. 119–128. IEEE (2008)
23. Ki, A., Knowles, A.E.: Stride prefetching for the secondary data cache. J. Syst. Architect. **46**(12), 1093–1102 (2000)
24. Kim, H., Ramachandran, U.: Flashfire: overcoming the performance bottleneck of flash storage technology. Technical report, Georgia Institute of Technology (2010)
25. Kondguli, S., Huang, M.: T2: a highly accurate and energy efficient stride prefetcher. In: 2017 IEEE International Conference on Computer Design (ICCD), pp. 373–376. IEEE (2017)
26. Laga, A., Boukhobza, J., Koskas, M., Singhoff, F.: Lynx: a learning linux prefetching mechanism for SSD performance model. In: 2016 5th Non-Volatile Memory Systems and Applications Symposium (NVMSA), pp. 1–6. IEEE (2016)

27. Lee, C., Kumano, T., Matsuki, T., Endo, H., Fukumoto, N., Sugawara, M.: Understanding storage traffic characteristics on enterprise virtual desktop infrastructure. In: Proceedings of the 10th ACM International Systems and Storage Conference, pp. 1–11 (2017)

28. Li, B., Deng, C., Yang, J., Lilja, D., Yuan, B., Du, D.: HAML-SSD: a hardware accelerated hotness-aware machine learning based SSD management. In: 38th IEEE/ACM International Conference on Computer-Aided Design, ICCAD 2019, p. 8942140. Institute of Electrical and Electronics Engineers Inc. (2019)

29. Li, M., Varki, E., Bhatia, S., Merchant, A.: Tap: table-based prefetching for storage caches. In: FAST, vol. 8, pp. 1–16 (2008)

30. Liu, C.C., Ganusov, I., Burtscher, M., Tiwari, S.: Bridging the processor-memory performance gap with 3D IC technology. IEEE Design Test Comput. **22**(6), 556–564 (2005)

31. Liu, R.S., Yang, C.L., Li, C.H., Chen, G.Y.: DuraCache: a durable SSD cache using MLC NAND flash. In: Proceedings of the 50th Annual Design Automation Conference, pp. 1–6 (2013)

32. Liu, W., Wen, Y., Yu, Z., Yang, M.: Large-margin softmax loss for convolutional neural networks. In: ICML, vol. 2, p. 7 (2016)

33. Mehra, P.: Samsung smartSSD: accelerating data-rich applications. Flash Memory Summit

34. Mohan, V., Siddiqua, T., Gurumurthi, S., Stan, M.R.: How i learned to stop worrying and love flash endurance. HotStorage **10**, 3 (2010)

35. Mowry, T.C., Demke, A.K., Krieger, O., et al.: Automatic compiler-inserted I/O prefetching for out-of-core applications. In: OSDI, vol. 96, pp. 3–17 (1996)

36. Narayanan, D., Donnelly, A., Rowstron, A.: Write off-loading: practical power management for enterprise storage. ACM Trans. Storage (TOS) **4**(3), 1–23 (2008)

37. Narayanan, I., et al.: SSD failures in datacenters: What? When? and Why? In: Proceedings of the 9th ACM International on Systems and Storage Conference, pp. 1–11 (2016)

38. Nijim, M.: Modelling speculative prefetching for hybrid storage systems. In: 2010 IEEE Fifth International Conference on Networking, Architecture, and Storage, pp. 143–151. IEEE (2010)

39. Nijim, M., Zong, Z., Qin, X., Nijim, Y.: Multi-layer prefetching for hybrid storage systems: algorithms, models, and evaluations. In: 2010 39th International Conference on Parallel Processing Workshops, pp. 44–49. IEEE (2010)

40. Srinath, S., Mutlu, O., Kim, H., Patt, Y.N.: Feedback directed prefetching: improving the performance and bandwidth-efficiency of hardware prefetchers (2006)

41. Pike, R.: Storage mechanism with variable block size, 13 Mar 2014. uS Patent App. 13/612,968

42. Puzak, T.R.: Analysis of cache replacement-algorithms (1986)

43. Rodeh, O., Bacik, J., Mason, C.: BTRFS: The linux b-tree filesystem. ACM Trans. Storage (TOS) **9**(3), 1–32 (2013)

44. Santos, J.R., Muntz, R.R., Ribeiro-Neto, B.: Comparing random data allocation and data striping in multimedia servers. ACM SIGMETRICS Perform. Eval. Rev. **28**(1), 44–55 (2000)

45. Smith, K.: Garbage collection. SandForce, Flash Memory Summit, Santa Clara, CA, pp. 1–9 (2011)

46. Tato, A., Nkambou, R.: Improving Adam optimizer (2018)

47. Wu, G., He, X.: Reducing SSD read latency via NAND flash program and erase suspension. In: FAST, vol. 12, p. 10 (2012)

48. Xu, R., Jin, X., Tao, L., Guo, S., Xiang, Z., Tian, T.: An efficient resource-optimized learning prefetcher for solid state drives. In: 2018 Design, Automation & Test in Europe Conference & Exhibition (DATE), pp. 273–276. IEEE (2018)
49. Xue, B., Fu, C., Shaobin, Z.: A study on sentiment computing and classification of Sina Weibo with Word2vec. In: 2014 IEEE International Congress on Big Data, pp. 358–363. IEEE (2014)
50. Zeng, Y.: Long short term based memory hardware prefetcher (2017)

Interpretable Dimensionally-Consistent Feature Extraction from Electrical Network Sensors

Laure Crochepierre[1,2]([✉]), Lydia Boudjeloud-Assala[1], and Vincent Barbesant[2]

[1] Université de Lorraine, CNRS, LORIA, 57000 Metz, France
{laure.crochepierre,lydia.boudjeloud-assala}@univ-lorraine.fr
[2] Réseau de Transport d'Electricité (Rte) R&D, Paris, France
{laure.crochepierre,vincent.barbesant}@rte-france.com

Abstract. Electrical power networks are heavily monitored systems, requiring operators to perform intricate information synthesis before understanding the underlying network state. Our study aims at helping this synthesis step by automatically creating features from the sensor data. We propose a supervised feature extraction approach using a grammar-guided evolution, which outputs interpretable and dimensionally consistent features. Operations restrictions on dimensions are introduced in the learning process through context-free grammars. They ensure coherence with physical laws, dimensional-consistency, and also introduce technical expertise in the created features. We compare our approach to other state-of-the-art feature extraction methods on a real dataset taken from the French electrical network sensors.

Keywords: Grammar-Guided Genetic Programming (GGGP) · Supervised learning · Feature extraction · Interpretability · Electrical power system

1 Introduction

Electric transmission power grids are large complex systems monitored and operated in real-time, 24/7, by highly trained control room operators (also called dispatchers). Their task is mainly to ensure that the overall system, critical in modern societies, remains in a secure state at all times to conduct electricity from producers to consumers. In particular, they watch over the *electrical flow* on each line to keep it under its *thermal limit*; a physical threshold above which a short-circuit could happen, risking for the safety of property and people nearby. To accomplish this monitoring, a large number of sensors placed throughout the electrical network provide them measurements relayed by a large number of screens in the control room. From these measurements, operators continuously perform information synthesis to prepare their strategy upstream and plan preventive actions (mainly changing the network topology) to redirect the power

© Springer Nature Switzerland AG 2021
Y. Dong et al. (Eds.): ECML PKDD 2020, LNAI 12460, pp. 444–460, 2021.
https://doi.org/10.1007/978-3-030-67667-4_27

flow before it reaches its limit. However, even if their ability to run the power system is well established, highlighted by the absolute absence of any significant blackout recently, Transmission System Operators (TSO) have noticed a steep rise in the complexity of real-time operations [8]. This trend is mainly linked to market dynamics, increased renewable energy sources connected to the grid, and the development of electrical interconnections with other European countries. As a consequence, power lines are operated closer to their thermal limit, and dispatchers have to go through their decision-making process faster to keep time to handle more critical situations. Today, the information synthesis step is computer-assisted by some hand-crafted aggregation indicators and computationally massive simulations calculated from the network measurements. Historically created by operators with the use of their expert knowledge, these few indicators aren't exhaustive and can not confirm the safety of all situations regarding electrical flows. Also, these indicators might need to be revised more frequently than they are now, given the system's current dynamics (e.g., newly installed renewable power plants or cross-border flow thresholds adjusted by the markets). Besides, the simulations are quite long to compute and cannot cover all possible forecasts of the future. Consequently, operators still perform some parts of this information synthesis by themselves using their knowledge of the system and the outputs of the simulations to synthesize measurements, results, and information about the connection of the lines in the grid.

A recent study on the French electrical transmission power system [31] proposed an exploratory dimensionality-reduction method, to identify interactively some factors influencing atypical consumption behaviors. In this experiment, knowledge was introduced by conditioning autoencoder with input features experts thought to be causing the output behavior. As in our application data is given with different physical dimensions such as voltage and active power and needs to respect physical properties, we were interested in finding ways to inject a different kind of knowledge coming from the field of physics (Ohm's law for example). Therefore, the aim of our work is to propose an automatic feature extraction method to *explain* electrical flow with physically consistent and intelligible indicators created from sensor data. From this point on, dispatchers could directly then use these indicators as a surrogate of the status of the power network zone they look after.

In this context, we investigate how to perform *feature extraction* with expert constraints for power line flow explanation, by creating relevant and *potentially non-linear combinations* of features from the initial dataset. The proposed approach relies on Grammar-Guided Genetic Programming [34], often abbreviated as GGGP (or G3P), to extract human-readable combinations of features. Our contribution is twofold. First, we propose a feature creation method which integrates domain-knowledge from power system experts using a context-free grammar build interactively with them. More specifically, this grammar includes some physical properties of electrical systems and prevents from using worthless combination operators, which helps reduce the search space. Finally, the created features are analyzed by a human expert who provides insights on what is correct and what would be expected from operators. The second contribution presented in the experiments is the interpretability evaluation of the outputs. Following the terminology used by Doshi-Velez and Kim in [9], we performed both "Functionally-Grounded" and "Human-Grounded" evaluations of the proposed approach by comparing it to other interpretable state-of-the-art methods and give created feature for expert analysis. In this paper, we also use a correlation-based optimization objective as a metric to evaluate individuals. We compare this metric to distance-based metric Mean Squared Error (MSE) to detail why both metrics can't be used equivalently. Throughout this article, we use the real-world dataset created from measurements of the French power grid, from January 2014 to December 2018 at 5-min intervals.

This paper is organized as follows. Section 2 summarizes related works on the topics of feature extraction, interpretability, and Grammar-Guided Genetic Programming. Then, Sect. 3 details the data used to develop our method. In Sect. 4, we described the proposed approach. The experimental evaluations and their results are presented in Sect. 5, and finally, Sect. 6 concludes this paper and introduces some future works.

2   Related Works

2.1   Feature Extraction

Real-world applications often produce data in a very high-dimensional space [25], but they are very sparse, redundant, and their underlying structure is often representable in a much lower dimension. In this context, dimensionality reduction (DR) techniques can be of great support to visualize data or improve a classifier performance [49] and are even used for data compression [6]. Thus, DR is an important preprocessing step in many machine learning pipelines. More formally, DR can be defined as the set of techniques taking inputs X with a high number of features D, and mapping it to a reduced set of features X' with size d, such as $d << D$ while retaining as much information from the original structure as possible. Exhaustive reviews on this topic can be found in [17,27,47].

DR is mainly done using two types of approaches: feature selection or feature extraction (also called sometimes feature transformation, augmentation, or creation). While feature selection only selects the most informative features from the dataset, feature extraction tries to effectively *combine features* from the original dataset to produce more expressive ones. Among the feature extraction methods, another distinction can be made between linear and non-linear methods (also called manifold-learning methods). Linear methods have been used for a long time. They include methods such as Principal Component Analysis (PCA) [16] which finds axis by variance maximization, Laplacian Eigenmaps [3], Non-Negative Matrix Factorisation [24] or Locally Linear Embedding (LLE) [36]. PCA has the advantage of providing quite interpretable results using the selected principal components [48]. However, in many cases, the data structures are too complex, and linear mappings cannot retain enough information from the initial feature space. In this context, non-linear mappings are considered to represent the original data as closely as possible. Among non-linear algorithms, Isomap [46] is one of the widely used methods. Other methods include Kernel PCA [43] (non-linear extension of PCA), t-SNE [30] or UMAP [33] two dimensionality-reduction methods for data visualization, or deep learning methods such as Variational Autoencoders [18] or Conditional Autoencoders [45].

However, these DR methods were initially presented in an unsupervised setting and did not use any supervision scheme. This element is an issue in our application, as they can't take into account the valuable knowledge of available target values. More recent works focus either on how to extend classical methods to supervised configurations (supervised-LE) [38] or on how to take advantage of some target features to structure the new feature space: for example by integrating an additional optimization objective (i.e., as second loss term in the neural network [28]) or producing multiple transformations one for each class.

2.2 Interpretability

One of the recurring concerns about DR is the lack of interpretability of the axes in the feature space. As we discussed above, linear methods are often considered as interpretable (even when they add up different dimensions), but it is not the case for non-linear ones. It has been observed that many DR methods construct the new feature space "upon arbitrary combinations of many uncorrelated physical dimensions" [15], leading to the non-usability in many industrial processes. Some promising works propose interpretable DR methods based on kernel dimensionality reduction [15,50], which are able to project the embedding dimensions on the label-space to make interpretations.

In the supervised machine learning community, the interpretability of the results is a key challenge to improve the user *trust and acceptation* of the created model, and the proposed results. The existing interpretability methods are roughly divided into two categories: interpretable models and post-hoc interpretability. Interpretable models include Linear Regression, Decision Trees [5], Generalized Additive Models [13], or Rule Fit [12]. They produce interpretable outputs, but they are often considered as sub-optimal regarding complex classification or regression tasks. On the other side, post-hoc interpretability is often used for more complex models which produce more accurate results such as Deep Neural Networks. This problem is called the *accuracy-interpretability trade-off* [4]. With the omnipresence of deep learning models, several works focus today on the post-hoc interpretation of models using model-agnostic methods such as LIME (Local Interpretable Model agnostic Explanations) [40] or model-specific ones SHAP (SHapley Additive exPlanations) [29].

However, recent works by Laugel et al. [23] warns about the "risk of having explanations that are a result of some artifacts learned by the model instead of actual knowledge from the data". They also suggest that further research needs to be done to provide satisfying post-hoc explanations, both faithful to the predictor and to ground-truth data. Rudin et al. [41] rather suggests to start with interpretable models and only shift to black-box models if no sufficient solution has been found. They also suggest asking for *strong* explanations of the created models. Interpretability is therefore increasingly required, whether for safety, fairness, specification issues, or scientific understanding. However, there is no complete consensus for now about how to define and how to evaluate interpretability. To answer this problem, Doshi-Velez and Kim [9] proposed a three-level evaluation: the first one is an "Application-grounded" evaluation where humans evaluate interpretability on an exact application task; the second, called "Human-grounded", uses a similarly applied evaluation but on a simpler task; finally, the "Functionaly-grounded" category uses an even simpler evaluation on a simple task and involves no human. This final category assesses, for example, multiple interpretable algorithms on the same metric to identify which one performs best. We'll detail interactivity in our experiment using this taxonomy.

2.3   Grammar-Guided Genetic Programming

Recently there has been new application perspectives for Genetic Programming (GP) regarding the increasing need for interpretable results. GP was for example used to provide interpretable policies in reinforcement learning [14], to learn manifolds [25], to create visualizations [26] or to explain complex deep learning models [10]. For dimensionality reduction tasks, GP has also been used a lot as a feature construction method [35]. It presents some advantages in comparison with other methods presented in Sect. 2.1. As identified in [25]:

- they try building a global learner unlike local methods such as t-SNE [30],
- they do not require a differentiable fitness function (unlike Autoencoders) and thus can be used with a great variety of objective functions,
- they intrinsically produce an interpretable mapping.

Genetic programming was initially introduced by Koza [20,21], who identified that many problems could be reformulated as program induction. Unlike Genetic Algorithms, which evolves a population of fixed-length binary vector, GP evolves a population of programs represented as *trees*. Each tree consists of a combination of initial features using several operations taken in a list of allowed functions (e.g., $+$, $-$, $\times$, %). Initial features are represented in leaves and functions in nodes.

Nowadays, there are many different variants and implementations of GP for program creation. The three major ones are the following : the first and most classical approach, tree-like GP; the second Linear GP [2] represents programs as linear sequences to perform imperative program evolution; the third is grammatical evolution [42], where the representation language uses a Backus Naur Form grammar [19] and programs-trees are derived from this grammar.

As in some cases, the search space in GP may be too large, thus preventing the algorithm from converging, some alternatives have been proposed among grammatical evolution strategies. For example, in Grammar-Guided Genetic Programming (GGGP) [22], the search space is constrained by a set of *rules* to create features. These rules are defined using grammar written in Backus-Naur Form (BNF) [19] as the one provided in Fig. 1. A comprehensive review of Grammar-Guided Genetic Programming can be found in [34]. GGGP has been identified as a way to enforce expert domain knowledge into the learning procedure [39] using *ontologies*. For instance, it has been used as a way to impose constraints on the *dimensions* of variables in a classification problem [7]. This is what led us to consider this method to find explanatory variables.

3 Data Description

The electrical power network can be represented as a graph $G = (N, L)$ with a set of nodes N and lines L. Lines represent here electrical transmission power lines, and nodes are the locations where lines can be physically connected. Measurements are acquired in nodes and line extremities. In each node $n \in N$ the observed quantities are active power p_n and reactive power q_n while for a line l measures contain information about the line connection $connected_l^{or}$, $connected_l^{ex}$ and also an $neighbor\_id_l$ key corresponding to the list of neighboring lines at this timestep.

Using these measurements as inputs, simulations can be done to estimate voltage magnitude v_n and angle θ_n in nodes $n \in N$ using Netwon-Raphson based power-flow analysis [44] and eventually compute flow values at each line extremities (with origin *or* and extremity *ex*), i_l^{or} and i_l^{ex}, $\forall l \in L$. In our case, a solver computes these quantities for each timestep. In the rest of this article, flow values i_l^{or} and i_l^{ex} will constitute the output of our different methods and experiments. We can thus consider the studied system as a *closed system* without time dependency as the target features $i_l^{or,ex}$ provided by our simulator only depends on measures and expert hyperparameters setting used to calibrate the power-flow calculus.

First, we exhaustively describe the graph with variables reflecting the different electrical links ($connected_l^{or,ex}$ a boolean representing the connection of line l at its origin or its extremity, and $neighbor\_id_l$ an id used as a key to represent all lines electrically connected to line l). We move the measured variables from the nodes of the graph to each origin/end of the line connected to this node. Although this representation implies redundancy in the data, we can now reason only in terms of power lines and forget about nodes objects. From now on, we'll refer to measured and simulated variables by $X = (X_l)_{l \in L}$ and target flow variables $y = (y_l)_{l \in L}$ where:

$$\forall l \in L, X_l = ((p, q, v, \theta)_{n\_or(l), n\_ex(l)}, connected_l^{or,ex}, neighbor\_id_l)$$
$$y_l = (i_l^{or}, i_l^{ex})$$

In our case, we focus on the network operated by the French TSO called Rte (Réseau de Transport d'Electricité). The French power grid as a whole is a very complex system, with up to 6500 nodes, 12000 power lines, and many interactions both with other European networks and within it. A common approach, to control how these interactions influence studies of the grid, is to divide into sub-zones within which the elements have a high-mutual influence on each other [32]. This approach, historically done by TSOs, allows several operators to work simultaneously on separated zones of an acceptable size they can control.

Thus, we focus on a specific mountainous valley where the escarpment intrinsically constrains interactions with the rest of the network. This selection restricts the study perimeter to 69 nodes and 92 lines, where 9 lines connect the zone with other parts of the network. By collecting measurements from January 2014 to December 2018, we obtained 365 165 timestep observations and target flow variables.

4   Proposed Approach

As explained in the introduction, we are interested in finding *"explanations"* about flow variable y_l, $\forall l \in L$. More formally, given a set of observed features $X \in \mathbb{R}^D$, we want to extract relevant and *potentially non-linear combinations* of these features $\forall l \in L, X_l^{prime} \in \mathbb{R}^d$ with $d << D$ and so that X_l is relevant to explain y_l. We chose to focus on Grammar-Guided Genetic Programming (GGGP) methods to propose a custom grammar for electrical data and a new correlation-based metric.

4.1   Grammar Description

Grammar construction plays a crucial role in GGGP methods as it defines the search rules in the feature space: a too-loose grammar would have a too-wide space to search, while a too-constraint grammar would be limited to sub-relevant zones. For the sake of comprehension, we provide a simplified version of our grammar in Fig. 1. The complete version of the grammar uses all features described in Sect. 3 including topological variables, and a wider variety of functions on each dimension. The grammar is iteratively constructed with experts-in-the-loop withdrawing or adding constraints such as new variables or new operations.

Dimensions Definition. Firstly, we need to define the dimensions the grammar can handle. The physical dimensions taken into account in this grammar can either be active power p (with dimension : watt W), reactive power q (volt-ampere reactive VAR), apparent power s (volt-ampere VA), voltage magnitude v (voltage V) or intensity i (ampere A). From there, we can define the square of all dimensions: $p2$, for example, is the squared value of p with dimension W^2.

Grammar Structure. The first step to build a grammar is to define the output structure, here `<expr>`. The output can be one of the elements separated by the character "|". This formulation allows to enforce expert knowledge on the output dimension. In this example, `<expr>` can be of dimension q, p^2, q^2 or v^2.

The second step is to *impose* the dimensional consistency of the output variable, by defining which operations can be performed on each dimension and how to combine two dimensions. To jump from one dimension to the next, laws of physics such as the power triangle $s = \sqrt{p^2 + q^2}$ or some variation of Ohm's law $i = \frac{s}{v}$ have been expressed in the grammar. Input variables are defined here using

```
# 1) Create unitary expressions (allowed returned dimensions)
<expr> ::= <p> | <s> | <i> | <f> * <expr> | <p>/<v>
# 2) Define legal operations on each dimension
<p>   ::= <p>-<p> | <pop>(<p>, <p>) | <sop>(<p>) | <p_var>
<q>   ::= <q>-<q> | <pop>(<q>, <q>) | <sop>(<q>) | <q_var>
<v>   ::= <v>-<v> | <pop>(<v>, <v>) | <sop>(<v>) | <v_var>
<p2>  ::= <p>*<p> | square(<p>)
<q2>  ::= <q>*<q> | square(<q>)
<s>   ::= sqrt(<p2> + <q2>) |<v> * <i>
<i>   ::= <s>/<v> | <pop>(<i>,<i>) | <sop>(<i>) | <i_frontier_var>
<f>   ::= <f>*<f> | <p>/<p> | <q>/<q> | <v>/<v>
# 3) Define operations returning variable with the same dimension
<pop>::= sum | minimum | maximum # Functions with two arguments
<sop>::= abs | neg | pos # Functions with only one argument
```

Fig. 1. Grammar example. "|" represents separation between each possibility of replacement of the element located at the beginning of the line, before "::=". Input variables are <p_var> <q_var>, <v_var> and <i_frontier_var>.

observations <p_var>, <q_var>, <v_var>, and <i_frontier_var> ($i\_frontier$ corresponding to the flow for 9 cross-border lines to model interactions with outside the zone) with corresponding dimension p, q, v, i. They can be combined to produce a new variable with either the same dimension (for example <p> - <p> produces a new variable with dimension p), or a different dimension (<p>/<p> has no dimension while square(<p>) has dimension p^2).

Finally, we define licit operations to perform on a single or a pair of variables, which are repeated for almost all dimensions.

4.2 Methodology Description

We base our method on GGGP algorithms and introduce human-experts in the grammar construction process. First, as in GGGP methods, we use a population-based search on the space by performing multiple times operations (selection, crossover, mutation) on a group of individuals, as described in Algorithm 1. The particularity of GGGP methods is to ensure individuals are still consistent with grammar rules after each crossover and mutation operation. In order to select the parents of the next generation, the performance of each individual in the current generation is assessed using an optimization objective (called fitness function). The algorithm stops either when reaching the maximum number of iterations or when a tree-program has a higher score than a satisfaction threshold.

To constrain the search space, we asked operators to look at the variables created by the algorithm, to extract relevant characteristics from them and to propose new grammatical rules. These iterations allowed us to create the grammar described above.

4.3 Objective Function

We use the absolute value of the correlation coefficient r as a measure of the efficiency of each individual. This measure, based on the linear Pearson

---

Algorithm 1: Interactive Evolutionary Search Algorithm

---

input : observations X, target y
grammar ← initialize_grammar()
while operator_not_satisfied do
 population ← create_population(grammar)
 evolutionary_search_condition_not_met ← True
 while evolutionary_search_condition_not_met do
  parents, best_individual ← parents_selection(population, X, y)
  offspring ← crossover(parents)
  offspring ← mutation(offspring)
  population ← replacement(population, offspring)
  evolutionary_search_condition_not_met ← test_conditions(best_individual)
 grammar ← operator_grammar_update(grammar)
return grammar, best_individual

---

correlation, lies between 0 and 1, where a r value of 1 indicates that the predictions $yhat$ perfectly matches the behavior of the target y. While the linear correlation coefficient mainly measures the strength of the linear relationship between two variables, it also has a clear advantage to be able to compare the behavior of two variables which range on different scales.

This measure seems particularly well suited in this particular case because we are not interested here in predicting the exact flow value but rather to understand the *global underlying relationship* between input variables X and flow output y. Moreover, from an operator viewpoint, it is as interesting to look at one feature F as its rescaled value $10 \times F$. To understand the advantages of using this metric, we also compare it to the distance-based metric Mean Squared Error (MSE).

5   Experiments

5.1   Target Selection

As we are only interested in analyzing flows y on *sensitive* lines, we first selected a subset of all 92 lines on which we'll perform feature extraction. This preprocessing step ensures that we won't look at residual information or negligible effects on the target. To do so, we identified which are the *most frequently loaded lines* by selecting the ones above a designated percentage $percent\_i\_threshold$ (100, 90, 80, and 70%) of the line thermal limit $i\_threshold$ during a percentage of total time $percent\_time$ (0.05 or 0.1%). The final target selection is defined as the union of lines identified as loaded by one of each combination of hyperparameters ($percent\_i\_threshold$, $percent\_time$). Using this method, we detected 24 lines.

5.2   Settings

Experimental Protocol. For each line identified as sensitive, with a corresponding target flow y, we now search for a tree-like variable to represent it. We evolve a large population of 2 000 individuals over 200 generations by following the grammatical rules defined above. A large population is proved necessary due to the high number of constraints defined in the grammar. The initial dataset containing 365 165 observations is split following a 80/20% ratio between the train and test sets. Each line feature search is launched 30 times with random population initialization, and only the top features are kept for manual inspection. All hyperparameters are provided in Table 1 for reproducibility, and were selected during preliminary cross-validation experiments.

Table 1. Evolutionary parameters chosen for the experiment.

Parameter	Setting
Generation	200
Population size	2000
Initialization	PI Grow with max initial depth 10
Selection	Tournament with a size of 2
Crossover	Type variable one point (0.9 probability)
Mutation	Type int flip per codon (0.1 probability)
Elitisme	Top 10
Replacement	Generational
Fitness	Absolute Pearson correlation or MSE

For the two fitness metrics (correlation and MSE) in both experiments, we insert an additive regularization term, relative to the depth of the feature-tree: `individual_depth`. The depth is the maximum number of nodes in the feature from the root to any leaf. This constraint aims at preventing the tree size explosion (called bloating phenomenon) [37] and is slightly weighted to only remove redundant nodes without constraining the search-space too much. Eventually our fitness function is: `fitness = selected_metric + 10^-8 * individual_depth`.

Implementation. To take advantage of their parallelized implementation, we used the open-source implementation of GE in Python PonyGE2 [11] as backbone code. We inserted correlation-based error-metrics, a specialized data processing, custom evolutionary step, and the full grammar tailored to our problem.

5.3   Results

Experiment 1 : Metrics Comparison. In this first experiment, we conduct two parallel trials, where we only vary the fitness metric used to evolve and assess the top-individuals performance line per line. Our objective is to identify the adequate fitness to our problem, choosing between distance-based methods such as Mean Squared Error (MSE) and correlation-based methods such as absolute Pearson correlation. Obtained results on the test set are summarized in Figs. 2 and 3.

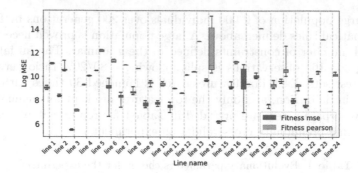

Fig. 2. Comparison of two evolution strategies, either using MSE (in blue) or Pearson Correlation (in orange) as fitness on 30 runs. Each boxplot summarizes the log-MSE scores for each line's best individuals (i.e., for each run, the one with the highest fitness score at the end of the evolution). Figures are best seen in color. (Color figure online)

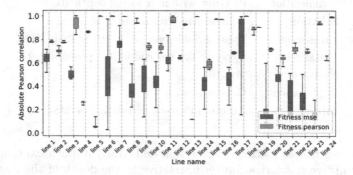

Fig. 3. Comparison of two evolution strategies on the same 30 runs presented in Fig. 2. Each boxplot now represents the absolute Pearson Correlation score for the best individuals of each line. Individuals compared in Figs. 2 and 3 are identical.

Figure 2 details the results of the two trials (MSE or Pearson as fitness) for 30 runs and compares them using MSE metric on a log-scale. In this Figure, two boxplots are associated to each line. They are placed on each side of a vertical dotted line: on the left the blue boxplot corresponds to the top-individuals evolved with MSE fitness ; on the right an orange boxplot contains scores of correlation-based evolutions. Similarly, Fig. 3 compares the same two trials regarding the absolute value of the Pearson correlation of the top-individual.

By comparing power line per power line the correlation-based and distance-based evolutions using correlation (Fig. 3) and MSE (Fig. 2) metrics, we identify that the two evolutions produce contrastive individuals, with top MSE-fitness individuals usually performing far worse than correlation-fitness individuals on the Pearson correlation scale (and conversely on the log-MSE scale). Except for few lines, where both fitness metrics performed well (ex. lines 3 and 15 on a log-MSE scale, 15 and 17 on correlation scale which have close scores whatever the evolution fitness metric), the 2 metrics seem to explore the feature space in opposite directions and can't be used equivalently. Moreover, when analyzing the features produced using MSE fitness, we identify that they tend to select combinations of features preferentially from the initial dataset with value ranges similar to the target one (although their behavior is different). Unlike individuals created with the correlation fitness, features obtained with MSE fitness wouldn't have a physical/technical interest and would mainly use the feature with intensity variables i. This point is critical to use distance-based fitness metrics because we can't normalize our data to have similar ranges as it would result in the impossibility to respect physical laws. For example, after normalizing each input features independantly, Kirchhoff's circuit laws don't apply anymore. Based on these observations, the only acceptable strategy is then to use fitness-based metrics such as Pearson correlation.

Regarding the outputted features, this first experiment also identified that we couldn't have constructed fewer features than the number of lines without loosing in performance, because extracted variables are very different from one line to another.

Experiment 2: Comparison to Other Methods. In the second experiment, we compare individuals from experiment 1 (obtained through evolution with a correlation-based fitness) with the output of algorithms such as LASSO Lars with Bayes Information criterion (Lasso Lars-BIC) [51], depth-3 Decision Tree [6]. These algorithms were selected because of their capacity to give outputs with a comparable level of interpretability, thus falling under the "Functionally-grounded" evaluation [9]. We also show the most correlated feature from the original dataset as a baseline. The obtained results are presented in Fig. 4. In this figure, the results associated with each line are displayed along a dotted vertical line and labeled at the bottom by their corresponding line name. Thus, for line 24 on the right, we have from the bottom to the top: the most correlated feature from the initial dataset (marked by a red star); the correlation to Lasso Lars-AIC output (pink rectangle); the correlation to a Decision Tree output (blue diamond) and the boxplot of our GGGP method (orange box).

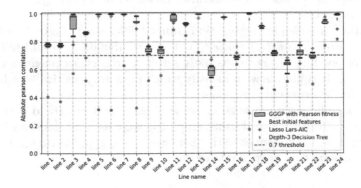

Fig. 4. Comparison between GGGP and state-of-the-art-algorithms. (Color figure online)

As presented in Fig. 4, all GGGP outputs have a higher correlation to the target than features from the initial dataset. Furthermore, by looking at only at boxplots, we identify that only 3 out of 24 highest values in boxplots were under the 0.7 threshold, under which correlation is usually not considered strong enough for the feature to be significant. We finally compare our approach in terms of correlation with partially interpretable ones such as Lasso Lars or a depth-constrained Decision Tree. From these experiments, we can highlight that GGGP outputs with significant correlation are at least as highly and sometimes even more correlated to the target than outputs from other methods. In the next paragraph, we'll look in detail at the produced combination categories.

Human-Expert Output Analysis and Pieces of Advice. Eventually, the relevance of top features obtained with GGGP-method is technically assessed by power system experts to identify whether the obtained formulas are conclusive from a technical and physical perspective, could be useful to operators, and would make sense to them. This experiment aims at confirming outputs interpretability from a human perspective by performing a "Human-Grounded" evaluation [9]. The first conclusions are that all features above a 0.8 correlation are relevant even if some of them could be improved. Thus, 0.8 could then be used as an acceptation threshold below which features created would be rejected. Indeed, extracted features with a very high score (above 0.9) show a small discrepancy between runs and could be useful as-is. However, for features under 0.8, experts would have found it interesting to intervene during the learning process by removing, replacing, or adding nodes or leaves in the evolved trees to increase their score. Uncovering the literal expression of features, we identify groups of features with intriguing expert interpretations:

- some features are variations around the expression $\sqrt{p^2 + q^2}$ (such as $\frac{\sqrt{p^2+q^2}}{v}$ or $\sqrt{(p1 + p2)^2 + (q1 + q2)^2}$)
- some others are the sum, minimum or maximum of a list of active powers, tweaked using absolute value, positive or negative parts. In some cases, these non-linearities are also found useful to tackle outliers coming from sensor or simulation errors.
- few features are also an aggregation (sum or difference) of cross-border flows (ex. $i\_frontier_1 - abs(i\_frontier_2)$). These combinations would tell us that the corresponding target line is more sensitive to a *global* phenomenon than other lines. It could allow to identify which lines are sensitive to a high flow coming inside or leaving the zone.
- using a specific grammar version, some of those features could even include multiple features combined using graph topology as conditions, such as :
    ```
    if{line_1 is connected} then{monitor gggp_feature_1}
    else{monitor gggp_feature_2}
    ```

6 Conclusion and Future Works

We have shown that the proposed interactive GGGP method achieved promising results on *interpretable feature extraction*. We marked the first milestone with first the production and then both qualitative and quantitative validation of a custom context-free grammar, which could interactively include some power system knowledge. Our experiments also provide some insights on the interpretability of our method from "Human-Grounded" and "Functionally-Grounded" [9] perspectives. By introducing expertise and physical properties in the grammar rules, we obtained explainable features. Some of them were also found relevant enough by power system operators to be included in hyper-vision tools. However, a few target features were still tricky to handle with only one dimension. Indeed, for these few lines, a 1D-manifold is surely too restrictive, and we envision that building a multi-dimensional space could highly increase the representativity of the reduced space. We also identified the use of Probabilistic Grammars to enhance more precise space exploration.

Moreover, as these new features are made to be used by operators, the very next step will be to introduce interactivity with non-machine-learning experts, directly inside evolutionary runs [1]. We would allow them to provide insights and technical information that could significantly help to create more insightful representation: either selecting/removing individuals, inforcing constraints on the search space by iteratively changing the grammar over the generations.

Undergoing works also focus now on applying the same method to a wider geographical zone only on high voltage power lines (low voltage power lines modeled as aggregated consumptions). This perspective brings us closer to performing an "Application-Grounded" evaluation of our method with humans on a more complex task. Eventually, to release reproducible results on power systems test cases and open-source our code, we plan to use the open-source framework Grid2Op[1] developed to test machine learning strategies for power grid operations.

References

1. Amershi, S., Cakmak, M., Knox, W.B., Kulesza, T.: Power to the people: the role of humans in interactive machine learning. Ai Mag. **35**(4), 105–120 (2014)
2. Banzhaf, W., Francone, F.D., Keller, R.E., Nordin, P.: Genetic Programming: An Introduction: On the Automatic Evolution of Computer Programs and Its Applications. Morgan Kaufmann Publishers Inc., Boston (1998)
3. Belkin, M., Niyogi, P.: Laplacian eigenmaps for dimensionality reduction and data representation. Neural Comput. **15**(6), 1373–1396 (2003)
4. Bohanec, M., Bratko, I.: Trading accuracy for simplicity in decision trees. Mach. Learn. **15**(3), 223–250 (1994). https://doi.org/10.1023/A:1022685808937
5. Breiman, L., Friedman, J.H., Olshen, R.A., et al.: Classification and Regression Trees. Wadsworth, Belmont (1984)

---

[1] Grid2Op github repository : https://github.com/rte-france/Grid2Op.

6. Charrier, C., Lézoray, O.: Color VQ-based image compression by manifold learning. In: Elmoataz, A., Lezoray, O., Nouboud, F., Mammass, D., Meunier, J. (eds.) ICISP 2010. LNCS, vol. 6134, pp. 79–85. Springer, Heidelberg (2010). https://doi.org/10.1007/978-3-642-13681-8_10

7. Cherrier, N., Poli, J., Defurne, M., Sabatié, F.: Consistent feature construction with constrained genetic programming for experimental physics. In: IEEE Congress on Evolutionary Computation, CEC, pp. 1650–1658 (2019)

8. Donnot, B., Guyon, I., Schoenauer, M., Panciatici, P., Marot, A.: Introducing machine learning for power system operation support. In: IREP Symposium (2017)

9. Doshi-Velez, F., Kim, B.: Towards a rigorous science of interpretable machine learning. Stat **1050**, 2 (2017)

10. Evans, B.P., Xue, B., Zhang, M.: What's inside the black-box?: A genetic programming method for interpreting complex machine learning models. In: Proceedings of the Genetic and Evolutionary Computation Conference, pp. 1012–1020 (2019)

11. Fenton, M., McDermott, J., Fagan, D., et al.: Ponyge2: grammatical evolution in python. In: GECCO (Companion), pp. 1194–1201 (2017)

12. Friedman, J.H., Popescu, B.E., et al.: Predictive learning via rule ensembles. Ann. Appl. Stat. **2**(3), 916–954 (2008)

13. Hastie, T.J.: Generalized Additive Models. Wiley, Hoboken (2017)

14. Hein, D., Udluft, S., Runkler, T.A.: Interpretable policies for reinforcement learning by genetic programming. Eng. Appl. Artif. Intell. **76**, 158–169 (2018)

15. Hosseini, B., Hammer, B.: Interpretable discriminative dimensionality reduction and feature selection on the manifold. In: Brefeld, U., Fromont, E., Hotho, A., Knobbe, A., Maathuis, M., Robardet, C. (eds.) ECML PKDD 2019. LNCS (LNAI), vol. 11906, pp. 310–326. Springer, Cham (2020). https://doi.org/10.1007/978-3-030-46150-8_19

16. Jolliffe, I.T.: Principal Component Analysis. Springer, Heidelberg (1986). https://doi.org/10.1007/978-1-4757-1904-8

17. Khalid, S., Khalil, T., Nasreen, S.: A survey of feature selection and feature extraction techniques in machine learning. In: 2014 Science and Information Conference, pp. 372–378. IEEE (2014)

18. Kingma, D.P., Welling, M.: Auto-encoding variational Bayes. In: ICLR (2014)

19. Knuth, D.E.: Backus normal form vs. backus naur form. Commun. ACM **7**(12), 735–736 (1964)

20. Koza, J.R.: Concept formation and decision tree induction using the genetic programming paradigm. In: Schwefel, H.-P., Männer, R. (eds.) PPSN 1990. LNCS, vol. 496, pp. 124–128. Springer, Heidelberg (1991). https://doi.org/10.1007/BFb0029742

21. Koza, J.R.: Hierarchical automatic function definition in genetic programming. In: Proceedings of the Second Workshop on Foundations of Genetic Algorithms, pp. 297–318. Morgan Kaufmann (1992)

22. Koza, J.R., Keane, M.A., Streeter, M.J., Mydlowec, W., Yu, J., Lanza, G.: Genetic programming IV: Routine Human-Competitive Machine Intelligence, vol. 5. Springer, Heidelberg (2006)

23. Laugel, T., Lesot, M., Marsala, C., et al.: The dangers of post-hoc interpretability: unjustified counterfactual explanations. In: IJCAI, pp. 2801–2807 (2019)

24. Lee, D.D., Seung, H.S.: Algorithms for non-negative matrix factorization. In: Advances in Neural Information Processing Systems, pp. 556–562 (2001)

25. Lensen, A., Xue, B., Zhang, M.: Can genetic programming do manifold learning too? In: Sekanina, L., Hu, T., Lourenço, N., Richter, H., García-Sánchez, P. (eds.)

EuroGP 2019. LNCS, vol. 11451, pp. 114–130. Springer, Cham (2019). https://doi.org/10.1007/978-3-030-16670-0_8

26. Lensen, A., Xue, B., Zhang, M.: Genetic programming for evolving a front of interpretable models for data visualization. IEEE Trans. Cybern, 1–15 (2020). https://doi.org/10.1109/TCYB.2020.2970198

27. Li, J., Cheng, K., Wang, S., et al.: Feature selection: a data perspective. ACM Comput. Surv. (CSUR) **50**(6), 1–45 (2017)

28. Li, Y., Pan, Q., Wang, S., et al.: Disentangled variational auto-encoder for semi-supervised learning. Inf. Sci. **482**, 73–85 (2019)

29. Lundberg, S.M., Lee, S.I.: A unified approach to interpreting model predictions. In: Advances in Neural Information Processing Systems, vol. 30, pp. 4765–4774. Curran Associates, Inc. (2017)

30. Maaten, L.v.d., Hinton, G.: Visualizing data using t-SNE. J. Mach. Learn. Res. **9**(Nov), 2579–2605 (2008)

31. Marot, A., Rosin, A., Crochepierre, L., Donnot, B., Pinson, P., Boudjeloud-Assala, L.: Interpreting atypical conditions in systems with deep conditional autoencoders: the case of electrical consumption. In: Brefeld, U., Fromont, E., Hotho, A., Knobbe, A., Maathuis, M., Robardet, C. (eds.) ECML PKDD 2019. LNCS (LNAI), vol. 11908, pp. 638–654. Springer, Cham (2020). https://doi.org/10.1007/978-3-030-46133-1_38

32. Marot, A., Tazi, S., Donnot, B., Panciatici, P.: Guided machine learning for power grid segmentation. In: 2018 IEEE PES Innovative Smart Grid Technologies Conference Europe (ISGT-Europe), pp. 1–6. IEEE (2018)

33. McInnes, L., Healy, J., Melville, J.: Umap: uniform manifold approximation and projection for dimension reduction. arXiv preprint arXiv:1802.03426 (2018)

34. McKay, R.I., Hoai, N.X., Whigham, P.A., Shan, Y., O'Neill, M.: Grammar-based genetic programming: a survey. Genet. Program Evol. Mach. **11**(3–4), 365–396 (2010). https://doi.org/10.1007/s10710-010-9109-y

35. Neshatian, K., Zhang, M., Andreae, P.: A filter approach to multiple feature construction for symbolic learning classifiers using genetic programming. IEEE Trans. Evol. Comput. **16**(5), 645–661 (2012)

36. Polito, M., Perona, P.: Grouping and dimensionality reduction by locally linear embedding. In: NIPS, pp. 1255–1262. MIT Press (2001)

37. Purohit, A., Bhardwaj, A., Tiwari, A., Chaudhari, N.S.: Handling the problem of code bloating to enhance the performance of classifier designed using genetic programming. In: IICAI, pp. 333–342 (2011)

38. Raducanu, B., Dornaika, F.: A supervised non-linear dimensionality reduction approach for manifold learning. Pattern Recogn. **45**(6), 2432–2444 (2012)

39. Ratle, A., Sebag, M.: Genetic programming and domain knowledge: beyond the limitations of grammar-guided machine discovery. In: Schoenauer, M., et al. (eds.) PPSN 2000. LNCS, vol. 1917, pp. 211–220. Springer, Heidelberg (2000). https://doi.org/10.1007/3-540-45356-3_21

40. Ribeiro, M.T., Singh, S., Guestrin, C.: "why should I trust you?": explaining the predictions of any classifier. In: Proceedings of the 22nd ACM SIGKDD International Conference on Knowledge Discovery and Data Mining, pp. 1135–1144 (2016)

41. Rudin, C.: Stop explaining black box machine learning models for high stakes decisions and use interpretable models instead. Nat. Mach. Intell. **1**(5), 206–215 (2019)

42. Ryan, C., Collins, J.J., Neill, M.O.: Grammatical evolution: evolving programs for an arbitrary language. In: Banzhaf, W., Poli, R., Schoenauer, M., Fogarty, T.C. (eds.) EuroGP 1998. LNCS, vol. 1391, pp. 83–96. Springer, Heidelberg (1998). https://doi.org/10.1007/BFb0055930

43. Schölkopf, B., Smola, A., Müller, K.R.: Nonlinear component analysis as a kernel eigenvalue problem. Neural Comput. **10**(5), 1299–1319 (1998)

44. Sereeter, B., Vuik, C., Witteveen, C.: On a comparison of Newton-Raphson solvers for power flow problems. J. Comput. Appl. Math. **360**, 157–169 (2019)

45. Sohn, K., Lee, H., Yan, X.: Learning structured output representation using deep conditional generative models. In: NIPS, pp. 3483–3491 (2015)

46. Tenenbaum, J.B., De Silva, V., Langford, J.C.: A global geometric framework for nonlinear dimensionality reduction. Science **290**(5500), 2319–2323 (2000)

47. Van Der Maaten, L., Postma, E., Van den Herik, J.: Dimensionality reduction: a comparative. J. Mach. Learn. Res. **10**(66–71), 13 (2009)

48. Vilenchik, D., Yichye, B., Abutbul, M.: To interpret or not to interpret PCA? This is our question. In: ICWSM, pp. 655–658. AAAI Press (2019)

49. Vlachos, M., Domeniconi, C., Gunopulos, D., et al.: Non-linear dimensionality reduction techniques for classification and visualization. In: KDD, pp. 645–651 (2002)

50. Wu, C., Ioannidis, S., Mario, S., et al.: Iterative spectral method for alternative clustering. In: Artificial Intelligence and Statistics (2018)

51. Zou, H., Hastie, T., Tibshirani, R., et al.: On the "degrees of freedom" of the lasso. Ann. Stat. **35**(5), 2173–2192 (2007)

Automatic Remaining Useful Life Estimation Framework with Embedded Convolutional LSTM as the Backbone

Yexu Zhou$^{(\boxtimes)}$, Michael Hefenbrock, Yiran Huang, Till Riedel,
and Michael Beigl

Telecooperation Office, Karlsruhe Institute of Technology, Karlsruhe, Germany
{zhou,hefenbrock,yhuang,riedel,michael}@teco.edu

Abstract. An essential task in predictive maintenance is the prediction of the Remaining Useful Life (RUL) through the analysis of multivariate time series. Using the sliding window method, Convolutional Neural Network (CNN) and conventional Recurrent Neural Network (RNN) approaches have produced impressive results on this matter, due to their ability to learn optimized features. However, sequence information is only partially modeled by CNN approaches. Due to the flatten mechanism in conventional RNNs, like Long Short Term Memories (LSTM), the temporal information within the window is not fully preserved. To exploit the multi-level temporal information, many approaches are proposed which combine CNN and RNN models. In this work, we propose a new LSTM variant called embedded convolutional LSTM (ECLSTM). In ECLSTM a group of different 1D convolutions is embedded into the LSTM structure. Through this, the temporal information is preserved between and within windows. Since the hyper-parameters of models require careful tuning, we also propose an automated prediction framework based on the Bayesian optimization with hyperband optimizer, which allows for efficient optimization of the network architecture. Finally, we show the superiority of our proposed ECLSTM approach over the state-of-the-art approaches on several widely used benchmark data sets for RUL Estimation.

Keywords: Multivariate time series prediction · Remaining useful life · Predictive maintenance · Embedded convolutional LSTM

1 Introduction

As system complexity and efficiency requirements continue to increase, the strategy of machine maintenance has changed. Where in the past, breakdown corrective maintenance or scheduled preventive maintenance was the standard, now, more intelligent approaches, like predictive maintenance (PM), are strived for. Unlike previous maintenance strategies, PM uses the machine's historical time series sensor data to evaluate the condition. The goal is to proactively maintain the machines before failures occur and therefore minimize down-times. One

© Springer Nature Switzerland AG 2021
Y. Dong et al. (Eds.): ECML PKDD 2020, LNAI 12460, pp. 461–477, 2021.
https://doi.org/10.1007/978-3-030-67667-4_28

critical part of PM is the estimation of the remaining useful life (RUL). By quantifying the remaining time until a component loses functionality, downtimes and costs of premature maintenance can be avoided by replacing only components that will fail soon.

Over the past decade, deep learning (DL) has achieved remarkable results in this task. By reviewing the related works, we found that there are still some challenges.

- Temporal information preservation and utilization in state-of-the-art algorithms: Recently, researchers focused mainly on constructing various types of neural networks for RUL. However, we found that in many works, the flatten mechanism (layer) is applied. This obfuscates temporal information and potentially leads to under-utilization (This will be further explained in the Sect. 2). While recent works, such as [18,27], build complex structures to fully preserve and utilize this information, they are often not usable for RUL estimation because of the high model complexity.
- Flexibility: As RUL estimation is used for various components with different degradation patterns, such as lithium batteries [17], Rolling Bearing [1] and complex power generation systems [9], experts have designed specialized deep learning model structures and often employed task-specific feature engineering. However, the design of the model architecture and the setting of hyper-parameters may be challenging for non-experts. Thus a universal automatic prediction framework is of huge benefit for practical applications.

To preserve and exploit the multi-level temporal information, we design a novel LSTM variant called embedded convolutional LSTM (ECLSTM). Instead of prepending a convolutional layer, it allows inputs of each time step to be 2-dimensional or 3-dimensional. The convolution can make full use of local temporal information and the complexity of the model does not depend on the size of the window but only the width of the kernel. In order to address the difficulty of setting the hyper-parameters of the model, we propose an automatic deep learning framework for RUL estimation. In the framework, we apply stacked ECLSTM as the backbone to extract features. By using Bayesian Optimization and Hyperband (BOHB) [7], this framework can automatically adapt all hyper-parameters involving in the whole data analysis pipeline without expert knowledge. We validate the performance of our approach on a number of real-world public RUL data sets. The results show, that the proposed ECLSTM has a superior prediction ability over competing state-of-the-art approaches.

2 Related Work

The methods of RUL estimation can be roughly categorized into model-based methods and data-driven methods. Deriving a model-based method is difficult, as it requires an accurate understanding of the underlying physical phenomena. Data-driven methods, on the other hand, model the degradation characteristic based on historical sensor data. Among the data-driven methods, DL-based

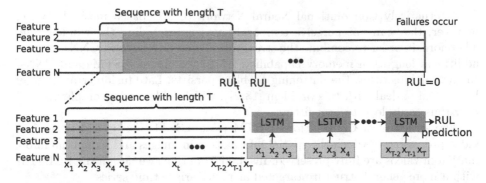

Fig. 1. The sliding window method to turn a multivariate time series data into a supervised learning problem (top). Features in each labeled window are sequentially fed into the (unrolled) LSTM (bottom right).

approaches attract the most attention due to their ability to learn task-specific features from time series.

In order to train DL models, the time series problem is reframed as a supervised regression problem using the sliding window method (see Fig. 1). Time series data is segmented into overlapping, fixed-length sequences where each sequence is assigned a label (the RUL). In RUL estimation tasks, the aim is to predict the corresponding RUL at a given point in time.

In order to predict the RUL, early works applied simple RNNs [11]. However, to address the problem of learning long-term time dependencies in RNN, LSTM models [23,25] consume the values at each point in time sequentially. In order to provide more context and enhance the feature extraction, the sliding window method (see Fig. 1) can be used [26]. Here, all values in the window are fully connected to an LSTM layer. Due to the fully connected layer in the LSTM (FCLSTM), the input of each time step must be one-dimensional, so that temporal and feature structure need to be projected (see Fig. 2). Furthermore, the model complexity (number of weights) increases linearly with the window size, which is unfavorable.

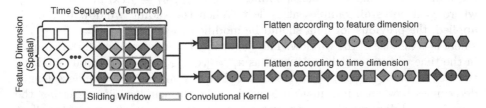

Fig. 2. Multiple features (shapes) at multiple sampling times (colors) within one window must be flattened into a 1D vector for LSTM input destroying the natural sequence. Alternatively, a convolutional kernel can be applied to aggregate local information. (Color figure online)

Alternatively Convolutional Neural Network (CNN) based methods [3,20], however, they can not perceive temporal patterns exceeding the window size. Therefore, in order to combine the strengths of shared weights in CNNs with an additional long term memory capability, [15,21] combined LSTMs and CNNs. In such architectures, the flattening problem persists. Late fusion strategy may be applied to deal with this problem [18,27], which adds model complexity and may thus easily lead to over-fitting.

The structure of Convolutional LSTM (ConvLSTM) [24] is widely used in video processing in which the spatial information in images and sequential information in videos are fully preserved. Inspired from that, we design the ECLSTM with a more general structure targeted at multivariate time series.

Automated machine learning is now a very popular research direction. There are many open-source tools such as auto-sklearn [8] and auto-weka [13]. These tools help users automatically select the best model and the best hyper-parameters. Based on the existing automated machine learning framework, we build a tool that can automatically perform RUL estimation. The implementation of the framework is available online[1].

3   Problem Definition

By reviewing various open-source data sets, we consider the RUL estimation problem in a more general form that takes round-robin sampling strategy into account. In actual production or run-to-failure experiments, the sensor information is not always recorded as storing large amounts of high-precision floating-point data is expensive and requires extremely high bandwidth to transmit. Therefore, the data is typically collected using a round-robin sampling strategy, which records values only once per predefined cycle. For example, in the C-MAPSS data set [9], only one sample is recorded in each cycle, while in the FEMTO-ST bearing data set [19], values are recorded at a frequency of 25.6 KHZ 0.1 s per each cycle (10 s).

Formally, RUL estimation can be described as a sequence to target problem. Given a T length sequence time series $\mathbf{X} = (\mathbf{x}_t \mid t = 1, \cdots, T)$ with $\mathbf{x}_t \in R^{n \times m}$, where n is the number of sensors and m the number of samples per cycle. Now, the aim is to predict the corresponding output y_T, $y_T = f(\mathbf{x}_t \mid t = 1, \cdots, T)$, where $\mathbf{x}_t$ denotes all samples at cycle t. When the sliding window method is applied, the previous formula should be modified to $y_T = f(\mathbf{x}_t^w \mid t = w, \cdots, T)$, where w is the size of the window. The vector $\mathbf{x}_t^w$ thereby contains all the samples in the time window which is denoted as $\mathbf{x}_t^w = (\mathbf{x}_{t-w+1}, \cdots, \mathbf{x}_t)$. In our settings, the size of sliding step is always set to 1. For any model, the sequence length determines how past information is used, while the window size describes the complexity of dynamic features over time. As both parameters can greatly affect the model performance, both should be considered when optimizing the model.

---

[1] https://github.com/YexuZhou/AutoRUL.

4   Embedded Convolutional LSTM

Inspired by the work ConvLSTM [24], we propose an extension of FCLSTM, in which a group of different 1D convolutions is embedded into the LSTM structure, which we call Embedded Convolutional LSTM (ECLSTM). We assume that such ECLSTM architecture is more powerful than FCLSTM in handling multivariate time series tasks.

In order to preserve the temporal information within the window, the input should be kept as a 2-dimensional tensor. This can be achieved by replacing the full connection in the FCLSTM with the convolutional operation. The equations of ECLSTM are then given by

$$
\begin{aligned}
i_t &= \sigma(W_i * [x_t, h_{t-1}] + b_i) \\
f_t &= \sigma(W_f * [x_t, h_{t-1}] + b_f) \\
o_t &= \sigma(W_o * [x_t, h_{t-1}] + b_o) \\
C_t &= f_t \circ C_{t-1} + i_t \circ \tanh(W_C * [x_t, h_{t-1}] + b_C) \\
h_t &= o_t \circ \tanh(C_t),
\end{aligned}
\tag{1}
$$

where $*$ indicates the convolution operator and $\circ$ the element-wise product. There are three benefits to using the convolution operator in LSTM. Firstly, the convolution parameters are only related to the defined kernel size and the number of filters and not to the size of the window. When the window size is large, the complexity of the model does not increase with it. Secondly, the hidden state H and the memory C also become 2D tensors. This means that they implicitly inherit and preserve the temporal relationship. Thirdly, the input, hidden state, and memory can even maintain a 3-dimensional shape, as this will not affect the operation of the convolution. Keeping the three-dimensional shape allows for more different convolutions.

The stacking of convolutional layers allows a hierarchical decomposition of the raw data and combinations of lower-level features. In order to get more complex features, the convolutions in (1) can be stacked as convolutional cells in a chain structure. If three convolutional layers are stacked in the cell, we call the ECLSTM as 3-depth-ECLSTM. Taking the input gate in ECLSTM as an example, the activation can be calculated as

$$
i_t = \sigma(W_i^3 * \sigma(W_i^2 * \sigma(W_i^1 * [x_t, h_{t-1}] + b_i^1) + b_i^2) + b_i^3).
\tag{2}
$$

Other gates have the same structure, but do not share the weights.

Moreover, the results of many multivariate time series analysis works like [18] indicate that different fusions strategies affect the performance. Inspired by that, the convolution cell can be composed of the following three different 1-dimensional convolutions, which are shown in the Fig. 3. The first is the early fusion convolution, which is same as conventional 1D convolution. Here, the features are extracted from all sensory information jointly. The second is the late fusion convolution. In late fusion, the features are extracted separately from each sensor. The third is hybrid fusion convolution, where features are separately extracted from each sensor but weights are shared.

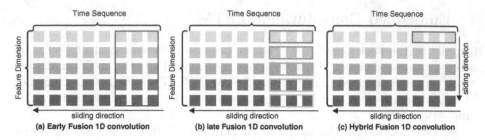

Fig. 3. In the early fusion convolution, the kernel height is fixed, that is, the same as the number of features. The sliding direction of the convolution kernel is along the time axis. In the late fusion convolution, the kernel height is 1. Each feature has its own convolution kernel. The convolution kernel also has only one sliding direction, namely the time axis. In hybrid fusion convolution, the kernel height is also 1. But it has two sliding directions, one is the time axis and the other is the feature axis. Because of weight sharing, it can save many parameters. It should be noted that when the number of filters is greater than 1, the output of the early fusion convolution is 2-dimensional. The outputs of the two remaining convolutions are 3-dimensional.

5   Automatic Prediction Framework

Because the above architecture is sensitive to hyper-parameters that are difficult to choose even by domain experts, we introduce the automatic prediction framework. Our assumption is that such a framework will outperform typical state of the art time series prediction algorithms when applied to RUL prediction.

The framework's structure is designed based on a summary of the structures from other related works that use neural networks to handle multivariate time series tasks. This framework consists of three parts, namely the pre-processing, feature extraction and RUL prediction parts. The structure and configuration space of each part will be introduced separately. Then the optimization process of the entire framework will be shown at the end of this section.

Pre-processing. When the sampling frequency in one cycle is too high, it is impractical to directly use the raw data as the input to the recurrent neural network. Processing so many values requires a larger kernel width or a relatively deep convolutional network. Both will lead to increased model complexity i.e. parameters and lead to over-fitting. Traditional methods to reduce the mode complexity are feature extraction or down-sampling. Since down-sampling leads to a loss of information, and manual feature extraction needs to be done on a per task basis, we avoid both for our framework. In order to control the model complexity while still being able to extract sophisticated features, we resort to convolutions with a kernel height of 1. Through this, the dimensionality can be reduced while useful features can be extracted automatically (illustration see Fig. 4). By adjusting the kernel width, the stride, and the dilation rate, the intensity of the dimensional reduction can be adjusted. Note that in this step the conventional 1D convolution is applied and the weights used by the convolution in each cycle and each feature are shared.

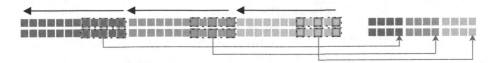

Fig. 4. Convolutions applied to a window size of 3 over 3 cycles. Each color represents a cycle of 12 samples. All convolutions are performed simultaneously in each cycle and feature with shared weights. A kernel width of 3, the dilation rate of 2, and the stride of 2 reduce the dimensionality to 4 samples per cycle each feature.

The configuration space for this section is shown in Table 1. There are dependencies between parameters. If the number of layers is defined as 0, it means that no pre-processing is performed. Only when the number of levels is greater than 0, other parameters are used. These parameters are not defined separately for each layer. As a rule of thumb, we define simple rules here. All layers are initialized with the same number of filters, stride and dilation rate. The kernel width is halved in each subsequent layer. Additionally, whether to perform 1d max-pooling between layers can be selected.

Feature Extraction. The main part of a neural network model is the backbone network, which is responsible for extracting features. Our framework uses stacked ECLSTM as the backbone. By default, batch normalization is applied between layers. Table 1 also shows the configuration space for this part. We define the hyper-parameters on a per-layer basis. Although this is a relatively simple setup, it guarantees the diversity of the backbone network where subsequent layers can have different structures. Due to the requirement for the optimized time budget, the training time for each configuration is not expected to be too long. Therefore, we limit the maximum number of stacked ECLSTM layers to 4. After the number of stacked layers is determined, the depth of the convolution needs to be defined for each layer. Finally, the type of convolution, the number of filters and the kernel width are set for each convolution.

RUL Prediction. The RUL prediction part is composed of stacked fully connected layers. The input is the feature extracted by the previous backbone. The last layer outputs the final prediction. As shown in Table 1, each layer needs to define the number of nodes, the activation function, and the size of the dropout.

Hyper-parameter Optimization. In addition to the structure parameters introduced earlier, also the sequence length, window size, and training batch size need to be determined. One common solution is the use of random search, but that is very inefficient. For such a highly conditional configuration space, the

Table 1. Configuration space of the proposed framework.

Part	Conditioned on	Name	Range	Type
Pre-processing	–	No. of layers	[0, 5]	int
	No. of layers	Kernel width	[2, 1024]	int
		Strides	[1, 20]	int
		Dilation rate	[1, 10]	int
		Activation	{"Sigmoid", "ReLU", "Leaky", "Linear", "hard sigmoid"}	cat
		No. of filters	[1, 20]	int
		1D max pooling	{"True", "False"}	cat
Feature Extraction	–	No. of layers	[1, 3]	int
	No. of layers	Dropout in layer	[0.0, 0.99]	float
		Depth of cell	[1, 4]	int
	Depth of cell	Activation	{"Sigmoid", "ReLU", "Leaky", "Linear", "hard sigmoid"}	cat
		Convolutional type	{"early", "hybrid", "late"}	cat
		No. of filters	[4, 64]	int
		Kernel width	[2, 32]	int
RUL Prediction	–	No. of layers	[1, 4]	int
	No. of layers	No. of units	[8, 1024]	int
		Activation	{"Sigmoid", "ReLU", "Leaky", "Linear", "hard sigmoid"}	cat
		Dropout in layer	[0.0, 0.99]	float

proposed framework uses the BOHB optimizer [7], which consists of the Bayesian optimization (BO) and the Hyperband (HB). By repeatedly calling Successive-Halving (SH), HB can efficiently identify the best of n randomly-sampled configurations. More time budget will be invested in promising configurations and the configurations with poor performance will be stopped early. BO uses a kernel density estimator (KDE) to model the objective function of configurations. It describes the regions of high performance in the configuration space. In our framework, the objective function is the performance of 3-folds cross validation.

6 Experiments

In this paper we have stated two hypotheses that we want to support with our experiments: (H1) Our ECLSTM architecture is more powerful than FCLSTM when applied to common multivariate time series prediction tasks and (H2) that when included into an automated prediction framework it can outperform state-of-the-art multivariate time series prediction algorithms on RUL prediction benchmark data sets. Further we want to explore if such an approach might generalize to similar time series prediction problems.

In this section, we evaluate the proposed framework on four benchmark data sets, three from the domain of predictive maintenance and one human activity recognition task. We compare our results to state-of-the-art approaches and perform an ablation study to provide insights into the effectiveness of the different elements of the framework. All experiments are run on a single GPU(RTX 2080 8 G RAM). Models are trained with the Adam optimizer [12].

6.1   C-MAPSS Data Set

C-MAPSS data set [9] which contains turbofan engine degradation data is a widely used prognostic benchmark data for predicting the RUL. This data set is simulated by the tool Commercial Modular Aero Propulsion System Simulation (C-MAPSS) developed by NASA. Run-to-failure simulations were performed for engines with varying degrees of initial wear but in a healthy state. During each cycle in the simulation, one sample of all 21 sensors such as the physical core speed, the temperature at fan inlet and the pressure at fan inlet etc. will be recorded once. As the simulation progresses, the performance of the turbofan engine degrades until it loses functionality.

Table 2. Description of four sub-data sets from C-MAPSS.

Data set	FD001	FD002	FD003	FD004
Training set	100	260	100	249
Test set	100	259	100	248
Operational conditions	1	6	1	6
Fault conditions	1	1	2	2

Data Description. C-MAPSS data consists of four sub-data sets with different operational conditions and fault patterns. As shown in Table 2, each sub-data set has been split into a training set and a test set. The training sets contain sensor records for all cycles in the run-to-failure simulation. Unlike the training sets, the test sets only contain partial temporal sensor records which stopped at a time prior to the failure. The task is to predict the RUL of each engine in the test sets by using the training sets with the given sensor records. The corresponding RUL to test sets has been provided. With this, the performance of the model can be verified. It should be noted that the four sub-data sets are of varying complexity.

Evaluation Metrics. There are two performance metrics utilized to measure the quality of our proposed model and other benchmark models. One is the commonly used root mean squared error (RMSE) and the other is the scoring function which was used in the PHM08 prognostics challenge competition. RMSE is a symmetric loss that assigns the same penalties for over- and under-prediction. However in practice, over- and under-prediction will lead to different value consequences. The scoring function assigns more penalty when the predicted RUL is larger than the true RUL. Due to being based on such a prediction, the maintenance plan will be delayed. The following equations show these two evaluation metrics.

$$RMSE = \sqrt{\frac{1}{N} \sum_{i=1}^{N} (\hat{r}_i - r_i)^2} \tag{3}$$

$$score = \begin{cases} \sum_{i=1}^{N} (e^{-\frac{\hat{r}_i - r_i}{13}} - 1), & \text{if } \hat{r}_i < r_i \\ \sum_{i=1}^{N} (e^{-\frac{\hat{r}_i - r_i}{10}} - 1), & \text{if } \hat{r}_i \geq r_i \end{cases} \tag{4}$$

where N denotes the total number of engines in the test sets. $\hat{r}$ and r_i represent predicted RUL and true RUL respectively.

Data Preparation. The data preparation for the C-MAPSS data set mainly consists of two aspects, one is normalization and the other is the definition of the RUL objective function. In order to do the following ablation study and comparison with other published work, the data preparation work is consistent with the works as [15,26]. Due to the different distribution of the values from each sensor, these values are normalized through the z-score normalization method which is described in the following equation $x'_i = (x_i - \mu_i)/\sigma_i$. μ_i represents the mean of i-th sensor and σ_i the corresponding standard deviation. The RUL objective function assigns each cycle an RUL label. Here we adopt the piece-wise linear RUL target function, where the maximum RUL is defined as 130.

Ablation Study. To validate whether the ECLSTM can improve performance with the sliding window method, in this study, we investigated two fundamentally similar neural network architectures evaluated on the four sub-data sets. One neural network model, called FCLSTM, was taken from [26], which is often cited as a baseline by other related works. The network has a five-layer architecture. The first layer is a LSTM layer with 32 hidden nodes. The second layer is the same, but contains 64 hidden nodes. The third and fourth layers are forward fully connected layers, each with 8 hidden nodes. The last layer is a 1-dimensional output layer which predicts the RUL. This architecture has been optimized on the four sub-data sets in [26]. The other neural network model is composed of the proposed ECLSTM. The difference from the former network is that the first and second layers are replaced by 2-depth-ECLSTM layer. To determine the hyper-parameters (convolution kernel size and filters number) in ECLSTM layers we apply simple rules. The number of filters in these two layers is simply defined as 10. The size (width) of the convolution kernel varies with the window size, that is, the rounded value of the window size divided by 4. The window sizes can be 1, 5, 10, 15 and 20. According to the work [26], the sequence length is defined as 30. These two architectures were trained 10 times for each window size. Their performance on the test set was recorded.

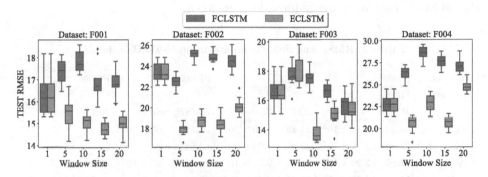

Fig. 5. Ablation study comparing ECLSTM against FCLSTM models on the four sub data sets of C-MAPSS using different window sizes on RMSE metric (lower is better).

In most cases, the performance of FCLSTM deteriorates with the increase of the window size. This can be seen in Fig. 5. Only on the FD002 and FD003 sub-data, when the window sizes are equal to 5 and 20 respectively, the performance of FCLSTM improves, but it is not significant. The reason is that with flatten, the temporal relationship within the window is ignored. In contrast when the window size becomes larger, the performance of ECLSTM, in most cases, is improved. We also observe that the performance of ECLSTM keeps increasing from window size equals 1 to 15 but drops at 20. This is likely due to less temporal information between windows. Because the sequence length is fixed, if the window size is too large, the time-steps (number of windows) becomes smaller. Smaller time-steps means less temporal information between windows, which leads to worse model performance.

Here we conducted a significance test and the results show that the perfor-mance of ECLSTM compared to FCLSTM with window size 1 is significant. Also, on sub-data FD002, ECLSTM has already achieved the best result com-pared to other state-of-the-art methods. It is worth noting that the input sensor is not being selected and the structure of the network can be further optimized. Therefore, the ECLSTM's performance still has room for improvement. Because the difference between the two structures lies only in the first and second layers, and ECLSTM achieves better performance on all four sub-data sets. So through this experiment, it proves that when the window size is not equal to 1, ECLSTM can more efficiently extract temporal information.

Results of Proposed Framework. To get better results, the proposed framework is used to optimize the architecture of the model. The search space consists of the hyper-parameters listed in Table 1. The optimization was run on GPU RTX 2080 for 24 h. The final optimized model is trained 10 times with different seeds. The average results on the test set are listed in Table 3. The approaches listed in the table are all recently published. Among them [16] is currently the best performer on C-MAPSS data set. It can be seen that the optimized model achieved the best results on all four sub-data sets, whose RMSE values of the 4 sub-data sets are the lowest. Especially on the FD002 and FD004 sub-data set, the estimation accuracy has improved significantly.

Table 3. RMSE and Score comparison on C-MAPSS data.

Datasets	FD001		FD002		FD003		FD004	
Methods	Metric							
	$RMSE$	$Score$	$RMSE$	$Score$	$RMSE$	$Score$	$RMSE$	$Score$
LSTM [26]	16.14	3.38×10^2	24.49	4.45×10^3	16.18	8.52×10^2	28.17	5.55×10^3
DCNN [20]	12.61	2.74×10^2	22.36	1.04×10^5	12.64	2.84×10^2	23.31	1.25×10^5
DAG [15]	11.96	2.29×10^2	20.34	2.73×10^3	12.46	5.35×10^2	22.43	3.37×10^3
AdaBN [16]	11.94	2.20×10^2	19.29	2.25×10^3	12.31	2.60×10^2	22.14	3.63×10^3
Proposed	**11.03**	**2.16×10^2**	**15.95**	**1.44×10^3**	**11.23**	**1.93×10^2**	**16.21**	**1.4×10^3**

6.2 PHM 2008 Data Set

This data set is similar to C-MAPSS data set. It was used for the prognostics challenge competition at the International Conference on Prognostics and Health Management (PHM) in the year 2008. It contains only one failure mode and therefore shows lower complexity. Associated true RUL values to test trajectories are not revealed. After the model is trained on the training trajectories, the results on the test trajectories need to be uploaded to the website, which will return a final score value as Eq. (4). Data preparation is the same as it on C-MAPSS data set. We use piece-wise linear RUL as the target function, where the maximal RUL is 130. The values of each sensor are normalized through the z-score normalization. We let our framework optimize for 12 h and select the model with the best validation performance as the final model. The score results are shown in Table 4.

6.3 FEMTO-ST Bearing Data Set

The FEMTO-ST data set [19] was used in the PHM Challenge in the year 2012 for the RUL estimation of bearings. As shown in Table 5, run-to-failure experiments were performed on 17 bearings. There were three different conditions for the experiment. In each condition, the data of 2 bearings were used as the

Table 4. *Score* of prediction results compared to PHM '08 rognostics challenge

Methods	Year	Score
Competition rank 1	2008	436.841
Competition rank 2	2008	512.426
Competition rank 3	2008	737.769
Competition rank 4	2008	809.757
Proposed framework	2020	823.341
Competition rank 5	2008	908.588
Competition rank 15	2008	1557.61
Deep LSTM [26]	2017	1862
Deep CNN [3]	2016	2056

Table 5. Data sets of IEEE 2012 PHM Prognostic Challenge.

Operation Conditions	Training set	Test set
1800 rpm and 4000 N	Bearing1_1, Bearing1_2	Bearing1_3, Bearing1_4, Bearing1_5, Bearing1_6, Bearing1_7
1650 rpm and 4200 N	Bearing2_1, Bearing2_2	Bearing2_3, Bearing2_4, Bearing2_5, Bearing2_6, Bearing2_7
1500 rpm and 5000 N	Bearing3_1, Bearing3_2	Bearing3_3

training set. The model is trained on 6 bearings and then predicts the RUL of the remaining 11 bearings. Two accelerators are mounted on the outer ring of the bearing and vertical and horizontal vibrations are recorded. The sampling frequency is 25.6 KHZ. It Records 0.1 s every 10 s. That is, 2560 samples are recorded per cycle. To evaluate the performance of methods, the absolute percent error of predicted results are used, which was also applied in the challenge. It is defined as $Er_i = abs(\frac{ActRUL_i - RUL_i}{ActRUL_i} \times 100\%)$, where $ActRUL_i$ is the actual RUL and RUL_i the predicted RUL of the i-th testing bearing.

After the values of the two sensors are normalized by z-score normalization, we let the optimizer BOHB of the framework run for 24 h. The model with the best validation performance is trained on all training data. its performance on the test set is shown in Table 6. The work in [22] has the best performance among existing studies, in which the original data is first converted into time-frequency images by continuous wavelet transform. Lei Y. et al. [14] proposed to apply the mutual information from multiple time series to construct the health indicator. Guo L. et al. [10] applied a set of selected features based on expert knowledge as input to the RNN model. For our framework, no expert knowledge is used for pre-processing and we reach comparable results (see Table 6).

Table 6. FEMTO-ST RUL prediction mean absolute percentage error for each model.

Testing	Actual	Predicted	RNN-HI [10]	WMQE[14]	CCG-HI [22]	Proposed
Bearing1_3	5730	5883	43.28%	0.35%	25.09%	2.67%
Bearing1_4	339	495	67.55%	5.6%	16.22%	46.02%
Bearing1_5	1610	1357	22.98%	100.00%	15.34%	15.71%
Bearing1_6	1460	1715	21.23%	28.08%	26.30%	17.47%
Bearing1_7	7570	7227	17.83%	19.55%	6.68%	4.53%
Bearing2_3	7530	5810	37.84%	20.19%	31.23%	22.84%
Bearing2_4	1390	1804	19.42%	8.63%	25.39%	29.78%
Bearing2_5	3090	3855	54.37%	23.3%	41.65%	24.76%
Bearing2_6	1290	1510	13.95%	58.91%	11.24%	17.05%
Bearing2_7	580	670	55.17%	5.17%	12.41%	15.52%
Bearing3_1	795	950	3.66%	40.24%	3.05%	19.50%
Mean of Error			34.28%	28.18%	18.51%	19.62%

6.4 UCI Human Activity Recognition (HAR)

In order to further verify if the proposed approach may also be applicable to general multivariate time series prediction task aside from RUL estimation, we conduct an experiment on the UCI-HAR data set [2]. HAR has the same problem definition as RUL estimation. Given a fixed length sequence multivariate time series, the aim is to predict the corresponding activity. This data set consists of sensor signals (accelerometer and gyroscope) gathered from a smartphone by 30 volunteer subjects. The volunteers performed six activities (walking, walking upstairs, walking downstairs, sitting, standing, laying). A total of 10298 sequences were collected. The data has been pre-processed and each sequence has 128 samples. Moreover, UCI-HAR data set provides a train-test partition. 21 volunteers were selected for generating the training data and 30% the test data. To evaluate the performance, balanced accuracy is used as the metric. After the values are normalized through z-score normalization, we let the optimization run for 12 h. The model with the best cross validation performance is trained 10 times with different seeds. The balanced accuracy on the test data set is listed in Table 7.

Table 7. Comparison of different models (publication year) using mean balanced accuracy on UCI-HAR data set

CNN-LSTM [5] (2019)	HCF+CNN [4] (2020)	Proposed (2020)
93.40%	93.80%	94.69% ± 0.42%

7   Conclusion and Future Work

In this work, an automatic RUL estimation framework is proposed. The backbone of the framework is the ECSLTM, which effectively combines the strengths of CNNs and LSTMs. In order to show the effectiveness of the framework, we evaluated its performance on different, real-world benchmark data sets. From our experiments, the following conclusions can be drawn:

1. The window size can affect the performance of the model. Compared to LSTMs, the proposed ECLSTM has better performance with an increase in window size.
2. Our framework has achieved state-of-the-art results on the three benchmark RUL estimation data sets. Especially on the C-MAPSS data set, a significant improvement compared to other recently published methods was achieved. Furthermore, the entire process does not require expert knowledge for architecture and parameter tuning and is therefore user-friendly for non-experts.
3. To check if the proposed framework works for general multivariate time series tasks, We additionally validate the framework's capabilities on an activity recognition data set. The result shows that our proposed framework achieves state-of-the-art performance. This framework can be applied to other similar multivariate time series tasks.

For future work, the diversity of the framework should be increased through e.g. adding attention mechanisms and including properties from bidirectional ECLSTM. At the same time, we will try to design a more reasonable configuration space. The configuration space defines how the architectures can be represented, which also determines the difficulty of the hyper-parameter optimization problem. A well-designed configuration space can simplify the optimization and may discover better models faster [6].

References

1. Ali, J.B., Chebel-Morello, B., Saidi, L., Malinowski, S., Fnaiech, F.: Accurate bearing remaining useful life prediction based on Weibull distribution and artificial neural network. Mech. Syst. Sig. Process. **56**, 150–172 (2015)
2. Anguita, D., Ghio, A., Oneto, L., Parra, X., Reyes-Ortiz, J.L.: A public domain dataset for human activity recognition using smartphones. In: ESANN (2013)
3. Sateesh Babu, G., Zhao, P., Li, X.-L.: Deep convolutional neural network based regression approach for estimation of remaining useful life. In: Navathe, S.B., Wu, W., Shekhar, S., Du, X., Wang, X.S., Xiong, H. (eds.) DASFAA 2016. LNCS, vol. 9642, pp. 214–228. Springer, Cham (2016). https://doi.org/10.1007/978-3-319-32025-0_14
4. Cruciani, F., et al.: Feature learning for human activity recognition using convolutional neural networks. CCF Trans. Pervasive Comput. Interact. **2**(1), 18–32 (2020)
5. Deep, S., Zheng, X.: Hybrid model featuring CNN and LSTM architecture for human activity recognition on smartphone sensor data. In: 2019 20th International Conference on Parallel and Distributed Computing, Applications and Technologies (PDCAT), pp. 259–264. IEEE (2019)

6. Elsken, T., Metzen, J.H., Hutter, F.: Neural architecture search: a survey. arXiv preprint arXiv:1808.05377 (2018)
7. Falkner, S., Klein, A., Hutter, F.: BOHB: robust and efficient hyperparameter optimization at scale. arXiv preprint arXiv:1807.01774 (2018)
8. Feurer, M., Klein, A., Eggensperger, K., Springenberg, J., Blum, M., Hutter, F.: Efficient and robust automated machine learning. In: Advances in Neural Information Processing Systems, pp. 2962–2970 (2015)
9. Frederick, D.K., DeCastro, J.A., Litt, J.S.: User's guide for the commercial modular aero-propulsion system simulation (C-MAPSS) (2007)
10. Guo, L., Li, N., Jia, F., Lei, Y., Lin, J.: A recurrent neural network based health indicator for remaining useful life prediction of bearings. Neurocomputing **240**, 98–109 (2017)
11. Heimes, F.O.: Recurrent neural networks for remaining useful life estimation. In: 2008 International Conference on Prognostics and Health Management, pp. 1–6. IEEE (2008)
12. Kingma, D.P., Ba, J.: Adam: a method for stochastic optimization. arXiv preprint arXiv:1412.6980 (2014)
13. Kotthoff, L., Thornton, C., Hoos, H.H., Hutter, F., Leyton-Brown, K.: Auto-WEKA 2.0: automatic model selection and hyperparameter optimization in WEKA. J. Mach. Learn. Res. **18**(1), 826–830 (2017)
14. Lei, Y., Li, N., Gontarz, S., Lin, J., Radkowski, S., Dybala, J.: A model-based method for remaining useful life prediction of machinery. IEEE Trans. Reliab. **65**(3), 1314–1326 (2016)
15. Li, J., Li, X., He, D.: A directed acyclic graph network combined with CNN and LSTM for remaining useful life prediction. IEEE Access **7**, 75464–75475 (2019)
16. Li, J., Li, X., He, D.: Domain adaptation remaining useful life prediction method based on AdaBN-DCNN. In: 2019 Prognostics and System Health Management Conference (PHM-Qingdao), pp. 1–6. IEEE (2019)
17. Miao, Q., Xie, L., Cui, H., Liang, W., Pecht, M.: Remaining useful life prediction of lithium-ion battery with unscented particle filter technique. Microelectron. Reliab. **53**(6), 805–810 (2013)
18. Münzner, S., Schmidt, P., Reiss, A., Hanselmann, M., Stiefelhagen, R., Dürichen, R.: CNN-based sensor fusion techniques for multimodal human activity recognition. In: Proceedings of the 2017 ACM International Symposium on Wearable Computers, pp. 158–165 (2017)
19. Nectoux, P., et al.: PRONOSTIA: an experimental platform for bearings accelerated degradation tests (2012)
20. Ren, L., Sun, Y., Wang, H., Zhang, L.: Prediction of bearing remaining useful life with deep convolution neural network. IEEE Access **6**, 13041–13049 (2018)
21. Wang, B., Lei, Y., Yan, T., Li, N., Guo, L.: Recurrent convolutional neural network: a new framework for remaining useful life prediction of machinery. Neurocomputing **379**, 117–129 (2020)
22. Wang, Z., Fan, W., Zhang, H., Zhou, Y.: Remaining useful life estimation of bearings based on nonlinear dimensional reduction combined with timing signals. Int. J. Electron. Commun. Eng. **13**(7), 484–491 (2019)
23. Wu, Y., Yuan, M., Dong, S., Lin, L., Liu, Y.: Remaining useful life estimation of engineered systems using vanilla LSTM neural networks. Neurocomputing **275**, 167–179 (2018)
24. Xingjian, S., Chen, Z., Wang, H., Yeung, D.Y., Wong, W.K., Woo, W.C.: Convolutional LSTM network: a machine learning approach for precipitation nowcasting. In: Advances in Neural Information Processing Systems, pp. 802–810 (2015)

25. Zhao, R., Wang, J., Yan, R., Mao, K.: Machine health monitoring with LSTM networks. In: 2016 10th International Conference on Sensing Technology (ICST), pp. 1–6. IEEE (2016)
26. Zheng, S., Ristovski, K., Farahat, A., Gupta, C.: Long short-term memory network for remaining useful life estimation. In: 2017 IEEE International Conference on Prognostics and Health Management (ICPHM), pp. 88–95. IEEE (2017)
27. Zhu, W., et al.: Co-occurrence feature learning for skeleton based action recognition using regularized deep LSTM networks. In: Thirtieth AAAI Conference on Artificial Intelligence (2016)

On-Site Gamma-Hadron Separation
with Deep Learning on FPGAs

Sebastian Buschjäger[1]([✉]), Lukas Pfahler[1], Jens Buss[1], Katharina Morik[1],
and Wolfgang Rhode[2]

[1] Artificial Intelligence Group, TU Dortmund, Dortmund, Germany
{sebastian.buschjager,lukas.pfahler,jens.buss,
Katharina.Morik}@tu-dortmund.de
[2] Astroparticle Physics Department, TU Dortmund, Dortmund, Germany
wolfgang.rhode@tu-dortmund.de

Abstract. Modern high-energy astroparticle experiments produce large
amounts of data everyday in continuous high-volume streams. The First
G-APD Cherenkov Telescope (FACT) aims at detecting particle showers
of gamma rays, because cosmic events can be derived from the energy and
angle of gamma rays. The separation of gamma rays from background
noise, which is inevitably recorded, is called the Gamma-Hadron separation problem. Current solutions heavily rely on hand-crafted features.
The current approach computes these features in a long data processing pipeline and trains a random forest classifier for the Gamma-Hadron
separation. The overall machine learning pipeline is executed on commodity computer hardware after an event has occurred. In this paper,
we propose an alternative approach which applies (Binary) Convolutional Neural Networks (B-CNN) directly to the raw feature stream of
the telescope's camera. We investigate if these models can be executed
on commodity hardware available at the telescope to handle its datas-
tream in real time. For fully Binary Neural Networks we also study the
use of FPGAs for inference. Our experiments show that this approach
outperforms hand-crafted features and random forests by a large margin, while still being applicable in real-time for moderate sized models.
Furthermore, we show that our approach does not only work well on
simulated data, but also on real cosmic events originating in the Crab
Nebula, a supernova remnant.

Keywords: Deep learning · Binary Neural Networks · Astroparticle
physics · FPGA

1 Introduction

To study our universe, modern astronomy observes high energy beams emitted from celestial objects in order to categorize the sources together with their

Electronic supplementary material The online version of this chapter (https://
doi.org/10.1007/978-3-030-67667-4_29) contains supplementary material, which is
available to authorized users.

© Springer Nature Switzerland AG 2021
Y. Dong et al. (Eds.): ECML PKDD 2020, LNAI 12460, pp. 478–493, 2021.
https://doi.org/10.1007/978-3-030-67667-4_29

key characteristics. For example, different types of supernovae can be found by observing high energy beams [5]. Large international collaborations deploy a wide variety of detector hardware including telescopes [2,10,19] to observe different ranges of electromagnetic beams. A central problem in all these detectors is the distinction between gamma rays which indicate a celestial object and background noise which is mostly produced by cosmic rays from hadrons that do not allow to conclude on a particular source – the *gamma-hadron separation problem*. Data analysis has been established as an effective tool for analyzing modern high energy particle experiments and solving the gamma-hadron separation problem [4]. Current approaches use a basic trigger, i.e., they begin the recording of an event when sufficient energy hits the detector. Then, a complex analysis pipeline calibrates the data and extracts pre-defined features which are used for classifying the stored event as hadron or gamma rays.

We wonder, whether we could replace the long pipeline by a deep learning process. Deep learning is supposed to decrease the burden of feature engineering, as the network already learns a suitable feature representation. Is this true for gamma-hadron-separation? Deep Learning is widely known to be resource hungry requiring not only vast amounts of training data, but also GPU hours to train and apply models. Future monitoring facilities will be installed around the world for a round-about view of the sky [21]. They need to detect an interesting gamma event fast, so that they can notify the other telescopes, which then turn in order record the event from their angle. In modern multi-messenger astrophysics, even different types of detectors inform each other so that the same event can be verified by different measurements. This includes places where running and maintaining a server with GPUs is not so easy. Hence, we investigate whether deep learning models can be executed on a small device that needs little maintenance. We tackle these questions in the context of the First G-APD Cherenkov Telescope (FACT) telescope [2]. The contributions of our interdisciplinary research are the following:

- A Deep Learning model is trained on the raw data which accurately predicts gamma events.
- A Binary Neural Network (BNN) is constructed as a resource-efficient version of this classifier.
- An efficient implementation of BNNs on commodity hardware is presented by the means of code generation which outperforms current solutions and forms the basis of our FPGA implementation.
- The BNN model is implemented on Field Programmable Gate Arrays (FPGA) allowing the direct incorporation into the telescope as a hardware trigger.

This paper is organized as follows. Section 2 introduces the data driven astronomical research using the FACT telescope and surveys related work. We explain the data gathering, the pre-processing, the overall workflow, the simulations used for labeling the data. Section 3 shortly characterizes Deep Learning for Cherenkov astronomy, particularly Convolutional Neural Networks (CNN) and BNNs. Section 4 explains the implementation of a BNN model by the means of code generation. Section 5 shows the experiments of learning from raw data in comparison to learning from the long pre-processing pipeline. First, simulation data are used. Second, we apply learning to real-world data, where the true event is known, namely the Crab Nebular. Third, we investigate the FPGA implementation regarding runtime.

2   Data Analysis in Cherenkov Astronomy

Celestial objects of several of hundred million light-years away from the earth are recognized by observing the energy beams emitted by these sources. The energy beams have an effect on a detector medium. For example, particles interact with the earth's atmosphere, producing cascading air showers. These showers emit Cherenkov light which, in turn, can be measured by telescopes such as FACT. Figure 1 shows an air shower triggered by some cosmic ray beam, emitting Cherenkov light that is captured by the FACT telescope (left side). The telescope can be viewed as a camera with 1440 pixels arranged in hexagonal form. Each pixel consists of a small light sensor which samples light pulses at 2 GHz frequency. The right side of Fig. 1 shows the resulting images taken by the telescope. Green indicates the telescopes surface, whereas blue indicates the amount of light hitting the sensors (the shower). Red indicates padding pixels which are used to form quadratic images from the telescope (discussed in more detail later). The camera continuously samples all the pixels into a ring-buffer and a hardware trigger initiates a write-out to disk storage, if some pixels exceed a specified threshold indicating that a shower is hitting the telescope. Upon trigger activation, a series of camera samples which amount for a time period of 150 ns, called the region of interest (ROI), are written to disk. This time-series of sensor voltages represents an event and corresponds to the light cone induced by the airshower. The FACT telescope records roughly 60 events per second, where each event amounts up to 3 MB of raw data, resulting in a rate of about 180 MB/s.

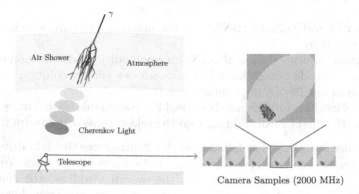

Fig. 1. An air shower produced by a particle beam hitting the atmosphere (left) and the corresponding measurements (right side). Picture was taken and modified from [4]. (Color figure online)

The processing pipeline that is used to analyze the data performs multiple steps as depicted in Fig. 2. First, the data are calibrated in order to account for environmental changes such as, e.g., the day-night cycle. Second, the resulting raw data are cleaned, i.e., values from broken sensors are corrected and artifacts are removed. Third, the pipeline extracts high level features based on hand-crafted rules from domain experts. Finally, the ML model for the gamma-hadron separation is applied. For a detailed explanation of the over 80 individual steps involving this pipeline and its very fast execution on commodity hardware we refer interested readers to [4]. Unfortunately, this approach does not scale well with higher data rates and larger telescopes. For example, the Cherenkov Telescope Ring (CTR) aims to connect multiple telescopes around the globe thereby increasing the data-rate orders of magnitudes [21]. Moreover, we note that even a Mac mini (as used in [4]) requires up to 85W as well as space and cooling. In this paper, we want to move beyond the given solution and try novel ways which might later on be applied on CTR and remotely deployed telescopes in resource-constraint environments.

The current approach heavily relies on hand-crafted features. A more philosophical question is whether this introduces some bias into the data. Ultimately, we want to enhance our understanding of the universe by measuring the energy beams of celestial objects. However, we often have some hypothesis about our data which lead to these higher-level features. In turn, the ML classifier which has only access to these biased features will indirectly confirm our hypothesis, whereas the raw data might not really support them.

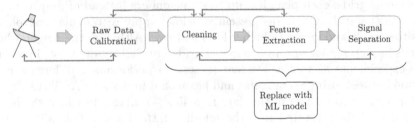

Fig. 2. Data processing steps from raw data acquisition to signal separation. The classic workflow of using a simple trigger with data calibration/cleaning as well as feature extraction now to be replaced by a model learned from the raw observations. Picture was taken and modified from [4].

Labeled Data by Probabilistic Simulation. When applying machine learning in astrophysics, it is difficult to obtain labeled data since particles from outer space can come from any source. A common approach to solve this problem is to combine Monte Carlo simulations with a careful training of the classifier. Astrophysics has a profound understanding of particle interactions in the atmosphere: Given the energy and direction of some parent particle (gamma, proton, etc.), its interaction can be described by a probabilistic model which gives a probability for particle collisions, possibly resulting in secondary particles, which again may interact with each other. This results in a cascade of levels of interactions that form the air shower, which can be simulated by particle simulation software like CORSIKA [8]. The output is a simulated air shower, which needs to be run through a simulation of the telescope and camera device to produce realistic raw data mimicking a shower that would have been recorded using the telescope. We can simulate interesting particles (e.g. gamma) and uninteresting particles (e.g. proton) and label the resulting raw-data accordingly.

Pre-processing: As mentioned before, we want to move the machine learning model closer to the telescope. Nevertheless, a minimal pre-processing consisting of three steps is necessary. *Sensor calibration:* The detectors' sensors behave differently in different environmental situations, e.g., the temperature has an effect on the sensor which should be corrected. This calibration involves the correcting of sensor values by multiplicative constants and biases and can therefore easily be performed either by an FPGA or a digital signal processor. *Extracting photon counts:* The FACT telescope produces 1440 time series each with a length of 150 ns. We remove noise from the time series and focus on a time window of 50 ns which contains most of the relevant Cherenkov photons. Calibration measurements for the sensors depict a typical voltage-curve when a single photon hits the sensor. This baseline measurement is subtracted as often as possible from the actual measurement until there is no signal left [14]. The number of subtractions can be considered the number of photons which arrived during the time series. The resulting image then shows the photon counts for each sensor in each pixel. *Image mapping:* The FACT sensors are arranged in a hexagonal form. In hexagonal grids, each pixel has up to six neighbors instead of four as in regular Euclidean grids. CNNs (as presented below) apply rectangular convolution filters to extract and generate higher level features. Although the neighborhood of pixels in rectangular and in hexagonal grids are slightly different, in a series of pre-experiments as well as student theses, no performance difference could be found between using rectangular and hexagonal filters for FACT [13,22]. We therefore choose to transform the data into 45×45 images in which the hexagonal grid is slightly rotated into the middle of the image. This allows us to use regular CNNs architectures and filters together with common frameworks. Figure 1 (right side) depicts the sensor mapping. Here, the red color denotes unused pixels which are always '0', whereas green (and blue) pixels are mapped to the corresponding sensors.

3   Deep Learning for Cherenkov Images

The observations from the telescope are 2-dimensional images where each pixel represents the number of photons arrived at each sensor. We apply CNNs to classify images as produced by hadron or gamma rays. CNNs are feed-forward networks which repeatedly apply convolutional filters to extract high level features. The goal of learning is to find a suitable weight configuration for the network. Let $\mathcal{D} = \{(x_1, y_1), \ldots, (x_N, y_N)\}$ denote the training data obtained by simulation where $x_i \in \mathbb{N}^{45 \times 45}$ is a photon count image and $y_i \in \{0, 1\}$ is its corresponding label.

3.1   Binary Neural Networks

CNNs are often over-parameterized and thus require not only a lot of memory but also a lot of computational power for the inference. While GPUs are the natural choice for model inference in this setting they are not ideal to be deployed in the field. GPUs not only require a lot of energy, but also require space and cooling. We propose to use BNNs as an alternative which can be executed on small devices using less energy, space and cooling so that they can be deployed in remote areas. BNNs are a subclass of feed-forward nets which use binary weights $\mathbb{F}_2 = \{-1, +1\}$. Therefore, BNNs require 32 times less memory as do single-precision floating point nets. Moreover, the inference of BNNs can be implemented using simple bit-level and integer operations making them ideal candidates to be implemented on an FPGA. Before discussing the inference and FPGAs implementation of these networks we will revisit the training of BNNs. Recall that regular floating-point CNNs are trained with stochastic gradient methods and backpropagation: For a batch of examples the forward-pass of the network is computed to obtain its current prediction. Then we compute the gradient with respect to the models weights via backpropagation, which is then used to update the weights accordingly.

For training BNNs the same general methodology can be used. Recall that CNNs are usually trained with gradient-based approaches which apply small changes to the current weight tensors at each iteration to better fit the training data. In the case of BNNs, we cannot perform gradient-based optimization directly for two reasons: First, the space of weights is discrete and thus the parameter-vector obtained by taking a small step in the opposite direction of the gradient is almost certainly not binary. Second, the sub-gradient of the sign-function is zero almost everywhere. Thus, arguably the most direct method to train BNNs is to store weights as floating point numbers during training, but round both - activation as well as weights - to $\mathbb{F}_2$ during forward computations [9]. More formally, Hubara et al. [9] propose a scheme that during training stores weights as floating point numbers constrained to values between -1 and 1 and then *binarizes* the network during the forward pass. Let $b: \mathbb{R} \to \mathbb{F}_2$ be a binarization function with

$$b(x) = \begin{cases} 1 & x > 0 \\ -1 & \text{else} \end{cases}$$

and let $B(W)$ denote the element-wise application of b to a tensor W, then we simply apply B during the forward pass to each weight tensor. During the backward pass, the authors propose to use full floating point precision. To mitigate the second problem – b is not differentiable – they replace the gradient of b with the so-called straight-through estimator. Consider the forward computation $Y = B(X)$. Let $\nabla_Y \ell$ denote the gradient with respect to Y. The straight-through estimator approximates

$$\nabla_X \ell := \nabla_Y \ell,$$

essentially pretending that B is the identity function. Algorithm 1 summarizes this approach.

Algorithm 1. Binarized forward pass for a network with L layer each with weight tensors W^l performing a generic operation $\circ^l$ (e.g. a convolution in CNNs).

```
1: function FORWARD(model, x)
2:     for l ∈ {1, ..., L} do
3:         x ← B(B(W^l) ∘^l x)
4:     return x
```

4   On-Site BNN Execution with FPGAs

Combining the computation power of FPGAs with Deep Learning models has a long history (see, e.g., [11, 25] for overviews). Focusing on the implementation of BNNs on FPGAs, there are two trends. Some approaches map singular building-blocks such as Matrix-Vector multiplication to FPGAs which are applicable for training and testing. For example, Nurvitadhi et al. compare the implementation of Binarized Matrix-Vector operations to their float-point siblings in [15] on an Intel Startix 10 FPGA. They highlight that GPUs excel in dense floating-point matrix multiplication, whereas BNNs naturally involve sparse operations which is better suited for FPGAs. They indicate some possible speed gains by combining BNNs and FPGAs.

The second trend focuses on fast model application of pre-trained models. Zhao et al. exemplify this in [27] by presenting a hardware accelerator design for a fixed CNN architecture which outperforms existing approaches by up to a factor of 7. For each layer type used in their model architecture, they provide a highly optimized implementation which is then fine-tuned for the specific parameters used in the model. The specialized implementations are then pipelined by leveraging high level synthesis (HLS) programming. Similarly, Fraser, Umuroglu and others present the FINN accelerator in [7, 24] which also leverages High Level Synthesis tools to generate optimized FPGA code, but is not limited to a single network architecture.

These approaches all show the potential advantages of implementing BNNs on FPGAs, but are unfortunately not yet ready for production. For example, FINN aims to support all major types of layers, but currently[1] only supports fully connected layers, completely.

In order to still study the application of BNNs on FPGAs for our application, we use the possibly simplest solution and expect performance to improve in the near future. We generate C-Code for each layer with minimal optimization and include the learned weights into the generated code. Then, we rely on HLS tools for optimizations, so that we gain acceptable performance.

Let us now describe our solution in more detail. We assume that we are given an ONNX[2] model definition file which defines the computation graph of the (binarized) neural network including its weights. For each layer, the code generator uses a template which is then instantiated with the specific layer parameters

---

[1] Date March 17, 2020. https://github.com/Xilinx/finn.
[2] https://onnx.ai/.

and correct data types. Once the code for each layer is created, we connect each layer's output to the input of subsequent layers, thus describing the complete computation graph. We generate C-style code without dynamic data structures or pointers. We automatically check for type-correctness and use the smallest appropriate data type in each processing step. All weights of each layer are stored in static arrays at the beginning of the code. Buffers for intermediate outputs of each layer are also statically allocated. The resulting code is then synthesized either by a HLS tool in case of an FPGA or compiled using a compiler in case of a regular CPU. Template substitutions are managed by jinja2[3].

We provide optimized template code and perform a post-code-generation design space exploration using the HLS tool, e.g., to unroll or pipeline certain loops. The only code optimization we perform across layers during the generation phase is the shift of activation thresholds by batch normalization as explained below. The implementation follows the established best-practices in literature for implementing BNN using bit-level and integer operations whenever possible [7,24]:

- **Fully Connected/Convolutional Layers:** Recall that for BNNs, we need to compute a binary dot-product. The multiplication of two binary values can be implemented with XOR operations, whereas their sum represents the number of 1 in the resulting bitstring (after the XOR operation). This operation is also known as popcount and CPUs usually ship specific instructions for this operation. On FPGAs, we can perform the XOR operation in parallel for each value inside the bitstring and then simply count the number of set-bits, e.g. by combining multiple look-up tables.
- **Activation:** For BNNs, we use the step function as activation, which can be directly implemented by comparing its input against 0.
- **Batch Normalization:** Batch Normalization (BN) re-scales and shifts the input to follow a normal distribution. Note, that if a BN layer precedes the activation (as done in our experiments) the scaling does not change the activation, but only shifting of the input does. Thus, we simply need to move the comparison threshold according to the BNs running mean value to implement the combination of BN+Activation. Also note that even though the running mean of BN is not binary, this is no problem, because float comparison can be implemented by comparing their binary representation [1]. Thus, from an FPGAs perspective, it does not matter if we compare against a float or an integer threshold.

5   Experiments

Experiments are designed to answer three key question: First, can (binarized) CNNs replace hand-crafted features with reasonable accuracy on simulated data? Second, will these models generalize well enough to be used for real data? Third, can we execute these models on-site and in real-time using commodity hardware

---

[3] https://jinja.palletsprojects.com/en/2.11.x/.

and/or FPGAs? We now tackle each question individually. For the first two question we want to emphasize, that Random Forest are still the state of the art for the FACT data due to their resilience against overfitting and thereby overcoming the gap between simulation and real-world data.

5.1 Models

For our study we investigate one small and one large VGG-style neural network architecture [23]. These networks are composed of blocks, where each block consists of two 3×3 convolutional layers each followed by a batch normalization layer and a ReLU activation as well as a single max-pooling layer of stride 2. The small network is composed of two of these blocks, while the large network is composed of four. The number of channels in the convolution layers starts with 128 in the first block and then increases by 128 with every following block. Finally we compute a linear layer of size 128 or 512, respectively. We apply batch normalization, before we compute the class probabilities using a final softmax layer. We have also experimented with residual neural networks, as suggested by Zhang et al. for training BNNs [28], but found that they did not outperform our purely convolutional models (see supplementary material).

We train our neural network models using the AMSGrad optimizer [20] with a batch size of 128 examples and minimize the cross entropy loss. We train our models for 100 epochs and use an initial learning rate of 0.001 which we reduce by a factor of 0.1 every 25 epochs. The neural networks learn from raw photon counts (denoted by PhC).

Since the state of the art is the Random Forest (RF) learner, it is applied for comparison. It learns from the hand-crafted high-level features (denoted by DL2) as well as from PhC data. Our RFs consist of 128 decision tree estimators of unlimited depth. Its decision trees are built using bootstrap samples of the training data and each of its splits is selected by maximizing the Gini score on a random subset of features of size $\sqrt{d}$, where $d = 22$ (DL2) or $d = 45^2$ (PhC). For deep learning we use PyTorch [17] and for fitting RF we use scikit-learn [18].

5.2 Experiments on Simulation Data

The Monte Carlo simulator CORSIKA has produced 200k training data and 100k test data, each set with perfectly balanced class frequency of hadron and gamma events. The simulation comes in two variants, with or without quality cuts. The simulation designers have identified regions, where the simulation is inaccurate and does not resemble real events sufficiently well. The so-called quality-cuts eliminate such unrealistic simulated events. We train our models on both variants of the simulation data and summarize the findings in Table 1. Neural networks beat the RF baseline trained on high-level features by a large margin. Possibly due to the neatness of the simulation data, a RF trained on photon counts also beats the baseline. As the results show, the dataset with quality cuts poses an easier classification problem where higher accuracies are achieved. Float models show overfitting after 100 epochs. Hence, we also tried training

Table 1. Accuracy on simulation data. We distinguish models trained on simulations with and without quality cuts (QC). For the neural networks, we also report accuracies for models trained with early-stopping after 10 epochs.

Model	Data	Accuracy, no QC		Accuracy, QC	
		Epochs:100	Epochs:10	Epochs:100	Epochs:10
RF	DL2	0.70959		0.78483	
RF	PhC	0.74711		0.78839	
CNN (small)	PhC	0.90825	0.88867	0.93441	0.93846
BNN (small)	PhC	0.90861	0.88644	0.90440	0.88866
CNN (large)	PhC	**0.91094**	0.90251	0.93735	**0.94228**
BNN (large)	PhC	0.90011	0.89925	0.93112	0.91369

with early-stopping after 10 epochs, which results in a small positive effect for floating points trained with quality cuts. BNNs, however, need more training epochs to achieve good test accuracy. Overall, our BNNs perform slightly worse than their floating point counterparts. However, for the large models this difference is small. In the next section, we investigate if our findings carry over from simulation data to data recorded by a real telescope.

5.3 Experiments on Real-World Crab Nebula Observations

Now we evaluate our trained models from the last section on real-world data collected by the FACT telescope at the Observatorio del Roque de los Muchachos (La Palma, Canary Islands, Spain). The telescope has been directed once towards a known gamma source, the Crab Nebula which emits large amounts of gamma rays. On these recorded data, we run the full source detection pipeline of the FACT experiment and investigate the influence of the gamma-hadron separation models on the overall quality of the source detection.

The evaluation proceeds in the following steps: We take the publicly available[4] Crab Nebula observation data [2,3], which consist of 17.7 h or 3,972,043 recorded events. Our gamma-hadron classification models are applied to classify which events are gamma rays. Then, an established model estimates the direction of the incoming gamma rays [16]. From the direction, we compute the angle between the trajectory of any incoming ray and the known direction of the Crab Nebula. The number of gamma rays can be regarded as a distribution that depends on the angle. High counts are to be expected for small angles. The distribution with respect to the direction of the Crab Nebula is called an *on-distribution* [6]. Contrasting the distribution of counts with distributions for five different positions with no known gamma sources yields the *off-distributions*. A uniform distribution of counts over angles is expected, where the majority of counted rays can be attributed to misclassified hadronic rays. We state the null-hypothesis that on-distribution and off-distribution follow the same distribution,

---

[4] https://fact-project.org/data/.

or, intuitively, that there is no gamma source at the direction of Crab Nebula. The margin by which a significance test rejects this null-hypothesis gives us a significance of detection $S_{Li\&Ma}$, reported by the number of standard deviations σ [12]. This $S_{Li\&Ma}$ is the performance metric for gamma-hadron classifiers, where larger numbers are better.

The trained classifiers output probabilities that an event is a gamma ray and we can control the classification behavior by varying the threshold for actually predicting gamma. A large threshold yields less events and also less misclassified events, because the classifier is more certain. If we set the threshold too large, we get too few total events which results in a small statistical significance. In contrast, if we decrease the threshold, we obtain more events, but also more misclassifications. If we set the value too small, we count too many noise events, the difference between on- and off-distribution shrinks, which also yields a low significance of detection.

To ensure that the output probabilities of the models are meaningful estimates of the classifiers' confidence, we apply isotonic probability calibration [26] using the simulated test-examples as calibration data. If not explicitly mentioned otherwise, the RFs use a threshold of 0.85, CNNs use a threshold of 0.6 and BNNs a threshold of 0.75. A careful analysis of the influence of these thresholds on the significance of detection is performed in the supplementary material.

In Table 2 we summarize the results for all models. Using the established RF classifier on high-level features, we obtain a significance of detection of 23.8σ. Small and large float CNNs outperform the baseline with a significance of 24.12σ and 24.20σ, respectively. BNNs achieve a slightly smaller significance of 22.96σ and 22.35σ. We hypothesise that this is again due to overfitting. To investigate this, during training of our models, we compute the validation loss on the simulated test data after each epoch and use the best epoch for classification. When we inspect the $S_{li\&ma}$ scores for the epoch with best loss, we see that these more carefully selected models indeed perform better: Both the small BNN and CNN now achieve significances over 25.8σ. For large models, however, we do not see the same benefits, further analysis is needed to better understand the connection between loss on simulation data and detection significance on real data. Last,

Table 2. Significance of detection

Model	Data	$S_{li\&ma}$, no QC		$S_{li\&ma}$, QC	
		Epoch: 100	Epoch: best loss	Epoch: 100	Epochs: best loss
RF	DL2	22.86σ		23.82σ	
RF	PhC	2.09σ		3.35σ	
CNN (small)	PhC	24.09σ	25.83σ	24.12σ	24.89σ
BNN (small)	PhC	19.55σ	$\mathbf{25.87\sigma}$	22.96σ	21.67σ
CNN (large)	PhC	23.68σ	24.64σ	24.20σ	23.17σ
BNN (large)	PhC	22.70σ	22.92σ	22.35σ	22.26σ

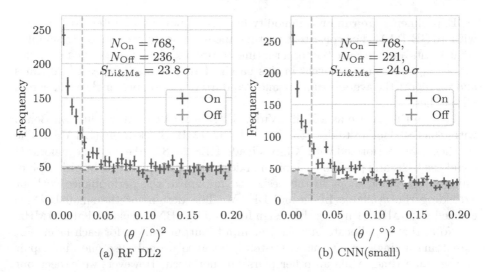

Fig. 3. Histogram of the frequencies as a function of the squared angular distance between the trajectory of any incoming ray and a position in the sky. On-events show the frequency with respect to the position of the Crab Nebula, while Off-events are w.r.t. positions with no known sources. The significance of detection test only considers angles smaller than 0.025 (left of the dashed vertical line).

we see that the random forests trained on photon counts are not useful at all for real-world data.

Visual inspection of the on- and off-distribution in Fig. 3 reveals the different classification behavior of the random forest baseline and our float CNN: We see that the random forest has a uniform off-distribution, while the CNN has a decaying distribution with smaller counts for larger angles. This suggests that our classifier is biased to predict gamma at the camera positions used for the off-distributions. Indeed, the gamma rays in the simulation data are not generated uniformly, which can explain this bias.

5.4 Proof-of-Concept with FPGAs

We want to measure the impact of using BNNs compared to floating-point nets running on different hardware. Since FACT produces data at a rate of roughly 60 events per second, on average we cannot spend more than 16ms to classify a single event.

As explained above, we generate c-code for each model and compile this for the target architecture, either by High Level Synthesis or by a regular compiler for CPUs. For CPUs, we enabled the most aggressive optimizations -Ofast -march=native -mtune=native using gcc version 8.3. We compare our results with the deep learning inference engine ONNX Runtime[5], which is optimized towards real-time model inference.

---

[5] https://microsoft.github.io/onnxruntime/.

Experiments are run on commodity hardware, namely an Intel i7-6700 CPU with 16 GB RAM. For consistent runtime measurements, we randomly sampled 1000 events and measure the total runtime to process these events, then compute the average runtime per event in this batch. This process is repeated 20 times and we report the average runtime and standard deviation per single event across all batches.

For the FPGA, we us a Xilinx Virtex UltraScale VCU110 Evaluation Board with 805,680 lookup-tables (LUTs) and 132.9 MB Block-Ram (BRAM). The synthesis was performed with Xilinx Vivado HLS 2018.3. We used the generated c-code as basis and performed a minimal design space exploration by either pipe-lining or unrolling loops in the design to maximize the performance without exceeding the available LUTs and BRAM. The design for the small BNN is clocked at 25 MHz[6] whereas the design for the large BNN is clocked at 100 MHz.

Note, that we allocate independent input/output buffers for each layer. For classifying a single event this is wasteful, because we only use one buffer pair at the same time, while all other pairs are not used. However, we expect our design to run continuously so that a stream of events is available. This enables efficient pipe-lining of the entire design: For each event we process one layer, so that the classification of the first event is delayed by the number of layers L in the entire network. Processing a single layer is much quicker than processing all layers, which means that despite the initial delay we can classify events at a faster overall rate.

Table 3 shows the latency of the different neural network configurations using different inference engines. We see that ONNX Runtime offers the fastest clas-sification rate for small and large float networks, whereas our code generator outperforms Onnx Runtime in case of BNNs. It was not possible to synthesize a working FPGA design for float networks, because they utilize too much BRAM. The FPGA offers the fastest (small BNN, pipelined) and slowest execution time (large binary, pipeline) for BNNs depending on the specific configuration. In sum-mary, for floating point networks Onnx Runtime is the fastest method, whereas for smaller BNNs the FPGA is the fastest and for larger BNNs our generated code seems to be the best method. The reasons for this are three-fold: Onnx Runtime is highly optimized for floating-point operations utilizing vectorization instructions to their fullest. In contrast, the code generator relies on the compiler to vectorize loops. Looking at BNNs, the situation reverses. Where our imple-mentation exploits the specific structure of BNNs to gain performance, Onnx Runtime does not support this. Finally, small models have the lowest latency on FPGAs, because large parts of the network can be unrolled so that they fit entirely on the FPGA. In contrast, larger models which do not fit well on the FPGA suffer tremendously. If most loops cannot be unrolled, the result is a very slow design. For the application at the Cherenkov telescope, large CNNs are not an option with either inference system, since none of them meet the required 16 ms latency. It is interesting to note that small float nets can be executed slightly quicker than BNNs on commodity hardware. We attribute this fact to the

---

[6] We found that using fewer clocks improved latency because loops can be unrolled.

Table 3. Latency of different neural net configurations using different inference engines. The best inference engine for float and binary is marked in bold. Smaller is better.

System	Type	Runtime [ms/event]	
		Float	Binary
ONNX Runtime	Large	**21.083 ± 0.078**	26.642 ± 0.100
	Small	**0.957 ± 0.020**	1.861 ± 0.037
Generated Code	Large	78.583 ± 1.704	**11.250 ± 0.077**
	Small	2.757 ± 0.026	1.574 ± 0.014
FPGA	Large	-	561.588 ± 0.000
	Small	-	4.221 ± 0.000
FPGA pipelined	Large	-	72.657 ± 0.000
	Small	-	**0.662 ± 0.000**

floating point vectorization instructions available on current Intel CPUs. If these are not available, BNNs are a very attractive alternative especially for large networks. All in all, the generated code satisfies the most scenarios enabling small CNNs, small BNNs and large BNNs making it the best overall choice.

6   Conclusion

Machine Learning is one of the basic building blocks of modern high-energy astroparticle experiments. One important problem in this application domain is the gamma-hardron separation problem which aims at separating interesting gamma events from hadronic background noise. The FACT telescope measures Cherenkov light emitted in earth' atmosphere when hit by gamma beams. The telescope measures at a rate of 60 events per second demanding fast processing pipelines. This is even more crucial for large telescope arrays in one location or a global distribution of sites that are possibly being deployed in resource constraint and remote locations. Current approaches solve the gamma-hadron problem for FACT by using long processing pipelines that extract hand-crafted features which are then used by RFs. In this paper, we successfully replaced the long pipeline by a CNN. For a resource-efficient alternative, BNNs were applied. Extensive experiments with larger and smaller VGG networks on state of the art astrophysical simulation data with and without quality cuts showed that our large CNNs is the by far best performing model with BNNs second. Both methods outperform RF with hand-crafted feature set. Hence, learning from the raw telescope data is now possible.

Morcover, the learned models were applied to data of a known source of gamma rays, the supernova remnant Crab Nebula. Again we see that CNNs outperform RF on the real-world data, whereas BNNs are on-par with RF only using raw measurements.

A third contribution of this paper is the careful implementation on different hardware. This is important for future astrophysical applications, when monitoring explorations in a ring of telescopes demand fast notification of events from places around the world. We presented a novel way to implement neural network inference by the means of code generation. This approach utilizes the fact, that neither the weights nor the structure of a network changes during inference and therefore statically creates the entire network code. Then the same code can either be compiled for execution on CPUs or synthesized for FPGAs using HLS tools. Our experiments show, that this approach outperforms current inference for BNN execution and to the best of our knowledge, our implementation of BNN inference is the fastest implementation available for CPUs. With respect to runtime we showed that large CNNs cannot be executed on CPUs or FPGAs in a timely manner as is dictated by FACT. Here, smaller CNNs and BNNs in general show superior performance. We achieved the fastest execution with small BNNs on FPGAs, whereas larger BNNs are best executed on regular CPUs. However, we expect more performance gains in the future when larger FPGAs become available fitting the entire BNN network on a single chip.

Acknowledgement. This work has been supported by Deutsche Forschungsgemeinschaft (DFG) within the Collaborative Research Center SFB 876 "Providing Information byResource-Constrained Analysis", project A1 and C3 (http://sfb876.tu-dortmund.de/).

References

1. IEEE standard for floating-point arithmetic. IEEE Std 754-2008, pp. 1–70 (2008)
2. Anderhub, H., et al.: Design and operation of FACT-the first G-APD Cherenkov telescope. J. Instrum. **8**(6), P06008–P06008 (2013)
3. Biland, A., et al.: Calibration and performance of the photon sensor response of FACT - the first G-APD Cherenkov telescope. J. Instrum. **9**(10), P10012–P10012 (2014)
4. Bockermann, C., et al.: Online analysis of high-volume data streams in astroparticle physics. In: Bifet, A., et al. (eds.) ECML PKDD 2015. LNCS (LNAI), vol. 9286, pp. 100–115. Springer, Cham (2015). https://doi.org/10.1007/978-3-319-23461-8_7
5. Carroll, B.W., Ostlie, D.A.: An Introduction to Modern Astrophysics. Cambridge University Press, Cambridge (2017)
6. Fomin, V.P., Stepanian, A.A., Lamb, R.C., Lewis, D.A., Punch, M., Weekes, T.C.: New methods of atmospheric Cherenkov imaging for gamma-ray astronomy. I. The false source method. Astropart. Phys. **2**(2), 137–150 (1994)
7. Fraser, N.J., et al.: Scaling binarized neural networks on reconfigurable logic. In: PARMA-DITAM 2017, pp. 25–30 (2017)
8. Heck, D., Knapp, J., Capdevielle, J.N., Schatz, G., Thouw, T.: CORSIKA: a Monte Carlo code to simulate extensive air showers. Forschungszentrum Karlsruhe GmbH, Karlsruhe (Germany) (1998)
9. Hubara, I., Courbariaux, M., Soudry, D., El-Yaniv, R., Bengio, Y.: Binarized neural networks. In: Lee, D.D., Sugiyama, M., Luxburg, U.V., Guyon, I., Garnett, R. (eds.) NIPS, vol. 29, pp. 4107–4115. Curran Associates, Inc. (2016)

10. Kieda, D., Collaboration, V., et al.: Status of the veritas ground based GeV/TeV gamma-ray observatory. Bull. Am. Astron. Soc. **36**, 910 (2004)
11. Lacey, G., Taylor, G.W., Areibi, S.: Deep learning on FPGAs: past, present, and future. arXiv preprint arXiv:1602.04283 (2016)
12. Li, T.P., Ma, Y.Q.: Analysis methods for results in gamma-ray astronomy. Astrophys. J. **272**, 317–324 (1983)
13. May, M.: Effizente Bildverarbeitung hexagonaler Strukturen mittels Deep Convolutional Neural Networks. Master's thesis (2018)
14. Mueller, S., et al.: Single photon extraction for FACT's SiPMs allows for novel IACT event representation. In: Proceedings of Science (2017)
15. Nurvitadhi, E., et al.: Can FPGAs beat GPUs in accelerating next-generation deep neural networks? In: ACM FPGA 2017, pp. 5–14 (2017)
16. Nöthe, M.: Improving the Angular Resolution of FACT Using Machine Learning. Tu Dortmund University, Dortmund, Technical report (2017)
17. Paszke, A., et al.: PyTorch: an imperative style, high-performance deep learning library. NeurIP **32**, 8024–8035 (2019)
18. Pedregosa, F., et al.: Scikit-learn: machine learning in Python. J. Mach. Learn. Res. **12**, 2825–2830 (2011)
19. Petry, D.: The magic telescope-prospects for GRB research. Astron. Astrophys. Suppl. Ser. **138**(3), 601–602 (1999)
20. Reddi, S.J., Kale, S., Kumar, S.: On the convergence of adam and beyond. In: International Conference on Learning Representations (2018)
21. Ruhe, T., Elsässer, D., Rhode, W., Nöthe, M., Brügge, K.: Cherenkov telescope ring - an idea for world wide monitoring of the VHE Sky (2019)
22. Rötner, S.: Deep Learning on Raw Telescope Data. Master's thesis (2017)
23. Simonyan, K., Zisserman, A.: Very deep convolutional networks for large-scale image recognition. arXiv preprint arXiv:1409.1556 (2014)
24. Umuroglu, Y., et al.: FINN: a framework for fast, scalable binarized neural network inference. In: ACM FPGA 2017, pp. 65–74 (2017)
25. Venieris, S.I., Kouris, A., Bouganis, C.S.: Toolflows for mapping convolutional neural networks on FPGAs: a survey and future directions. ACM Comput. Surv. **51**(3), 39 (2018)
26. Zadrozny, B., Elkan, C.: Obtaining calibrated probability estimates from decision trees and naive Bayesian classifiers. In: In Proceedings of the 18th International Conference on Machine Learning, pp. 609–616. Morgan Kaufmann (2001)
27. Zhao, R., et al.: Accelerating binarized convolutional neural networks with software-programmable FPGAs. In: ACM FPGA 2017, pp. 15–24 (2017)
28. Zhuang, B., Shen, C., Tan, M., Liu, L., Reid, I.: Structured binary neural networks for accurate image classification and semantic segmentation. In: Proceedings of the IEEE Conference on Computer Vision and Pattern Recognition, pp. 413–422 (2019)

Applied Data Science: Spatiotemporal Data

Benchmarking Tropical Cyclone Rapid Intensification with Satellite Images and Attention-Based Deep Models

Ching-Yuan Bai[1(✉)], Buo-Fu Chen[2], and Hsuan-Tien Lin[1]

[1] Department of Computer Science and Information Engineering,
National Taiwan University, Taipei, Taiwan
b05502055@ntu.edu.tw, htlin@csie.ntu.edu.tw
[2] Center for Weather Climate and Disaster Research,
National Taiwan University, Taipei, Taiwan
bfchen777@gmail.com

Abstract. Rapid intensification (RI) of tropical cyclones often causes major destruction to human civilization due to short response time. It is an important yet challenging task to accurately predict this kind of extreme weather event in advance. Traditionally, meteorologists tackle the task with human-driven feature extraction and predictor correction procedures. Nevertheless, these procedures do not leverage the power of modern machine learning models and abundant sensor data, such as satellite images. In addition, the human-driven nature of such an approach makes it difficult to reproduce and benchmark prediction models. In this study, we build a benchmark for RI prediction using only satellite images, which are underutilized in traditional techniques. The benchmark follows conventional data science practices, making it easier for data scientists to contribute to RI prediction. We demonstrate the usefulness of the benchmark by designing a domain-inspired spatiotemporal deep learning model. The results showcase the promising performance of deep learning in solving complex meteorological problems such as RI prediction.

Keywords: Atmospheric science · Tropical cyclone · Rapid intensification · Spatiotemporal data · Deep learning · Attention

1 Introduction

The tropical cyclone (TC) is one of the most devastating weather systems on Earth, characterized by intense and rapidly rotating winds around a low-pressure center and associated with eyewall clouds and spiral rainbands producing heavy rainfall. In order to reduce and respond to damage caused by TCs, the past half-century has seen much effort devoted to improving the forecast of TC track, intensity, and the associated rainfall and flooding. Although TC track forecasts has improved significantly during the past few decades, prediction of TC rapid

© Springer Nature Switzerland AG 2021
Y. Dong et al. (Eds.): ECML PKDD 2020, LNAI 12460, pp. 497–512, 2021.
https://doi.org/10.1007/978-3-030-67667-4_30

intensification (RI) remains challenging, which affects the subsequent production of TC structure and rainfall forecast [5]. TC intensity is defined as the maximum sustained wind in the TC inner-core region, and rapid intensification (RI) is defined as a TC experiencing an intensity increase surpassing a threshold (25–35 knots) within a 24-h period. The range of thresholds represents the 90 to 95 percentiles of 24-h TC intensity changes in different basins.

Accurately predicting the onset of RI is particularly crucial because reacting to an off-shore RI event before TC landfall requires sufficient time; delayed reaction has caused some of the most catastrophic TC disasters. For instance, hurricane Harvey in August 2017 caused 107 casualties and cost approximately $125 billion USD as it rapidly intensified to a category 4 hurricane only hours before landfall. However, improvement in RI prediction has been slow partly because RI events are rare. Additionally, favorable environmental conditions are generally necessary but do not guarantee RI onset [7]. Thermo-dynamical processes within the TC in response to these environmental forces are believed to be even more critical. For example, the development of up-shear convective burst [11,14] and asymmetric surface fluxes and boundary-layer inflow associated with the background flow [2,13] are factors that control the rainfall distribution and, in turn, TC intensification. Thus, a successful RI prediction scheme must accurately depict both environmental conditions (in which a TC is embedded) and vortex-scale features such as the distribution of precipitation or inner-core TC structure.

The goal of this paper is to tackle TC RI prediction from a data-science perspective. We propose a new benchmarking procedure with rigorous practices common to data science which includes a satellite image based dataset to be publicly released. We experiment with deep learning methods for this task. In Sect. 2, we briefly review related work on TC intensity and RI prediction. In Sect. 3, we introduce our newly devised benchmark and the major improvements over previous benchmarks. In Sect. 4, we propose an attention-based deep learning approach for this task and highlight the connection between model design and meteorological domain knowledge. In Sect. 5, we present the experimental results and attempt to interpret what the model has learned based on attention mechanisms. We conclude our findings in Sect. 6.

2   Related Work

Previous studies on predicting TC RI focus on utilizing predictors as features (specifically predictors published by the SHIPS project, to be introduced in Sect. 2.1). Predictors are high level statistics (e.g., mean or standard deviation) that summarize collected or simulated atmospheric data in a span of time and space. The high level physical meaning carried by predictors allow human to easily comprehend and design prediction models accordingly. However, the major downside of studies relying on predictors is the lack of exploration for new techniques that better exploit the raw data. The loss of detail in predictors has bottlenecked improvements in prediction performance.

In this work, instead of relying on predictors, we directly utilize raw data collected from satellite sensors with deep learning models. Our model architecture is inspired by domain insights from the Advanced Dvorak Technique (ADT) [12], an operational technique for TC intensity forecasting based on raw data. Details of ADT will be discussed in Sect. 2.2.

2.1 Statistical Hurricane Intensification Predictive Scheme

The SHIPS project has developed a series of statistical models for probabilistic prediction of RI based on atmospheric predictors [8,15,16]. The SHIPS RI index (SHIPS-RII [8]) predicts the probability of a TC intensifying by at least 25, 30, and 35 kt within 24 h. This scheme uses simple linear discriminant analysis to determine the RI probability based on a relatively small number (<10) of predictors describing mainly environmental factors and limited aspects of internal TC structure observed by meteorological satellites. Candidate predictors (~20) for SHIPS-RII [8] are subjectively determined by human intelligence, and the final predictors used for linear discriminant analysis are basin dependent. The final model is not publicly available, but the details of model design are well documented in publication and thus technically reproducible.

A subsequent work [15] uses Bayesian inference (SHIPS-Bayesian) and logistic regression (SHIPS-logistic) to predict RI probability. The authors show that both SHIPS-Bayesian and SHIPS-logistic exhibit forecast performance that generally exceeds the skill of SHIPS-RII; blending the three models further improves performance. Another study [16] integrates an additional 4 to 6 predictors derived from satellite passive microwave (PMW) observations into the SHIPS-logistic model and demonstrates a relative performance improvement from 53.5% to 103.0% in the Atlantic compared to the original model. More details will be discussed in Sect. 3.1. Note that these SHIPS-RII techniques are only applicable in the Atlantic and eastern Pacific basins. Thus, a new technique applicable for all global TCs is demanded.

2.2 Advanced Dvorak Technique

ADT [12] is an automated technique for real-time TC intensity estimation based on satellite image analysis. The technique replaces several human steps in the Dvorak Technique (DT) with automated procedures. The ADT has two stages: scene type classification and intensity post-processing. A satellite image instance is first classified into one of many pre-determined scene types according to the cloud distribution with respect to the TC center. Scene scores are derived from the characteristic matching test for each scene type, and the image is classified as the scene type with the highest corresponding scene score. For instance, the curved band cloud scene characteristic matching test measures how well the curve of the cloud bands matches the 10° log spiral. For each scene type, there is a unique method of deriving the intensity of the satellite image. In the second stage, the predicted intensity is post-processed by applying heuristic rules to clip the intensity into a reasonable range, after which the value is smoothed by applying a weighted average over the intensity of previous time steps. Although ADT is currently used only to estimate TC intensity, it inspires our proposed model in (a) focusing on individual frames to relate cloud features and TC intensity in the first stage; (b) averaging to take the time information into account in the second stage. More details are presented in Sect. 4.

3   Benchmark

A fair and reproducible benchmark is the key to validating model performance for the continuous improvement of RI prediction. This includes adequately chosen metrics to reflect model performance and well-defined, disjoint training and testing datasets to evaluate how well the model generalizes on holdout data. In this section, we first review the limitations of the current benchmark within the development of SHIPS-RII. Then, we propose a revised benchmark that solves the limitations and is better aligned with data science practices.

3.1   Existing Benchmark Within SHIPS-RII Development

The metric adopted in the meteorology domain for benchmarking RI prediction is the Brier score (BS), which is the mean square error (MSE) between the $\{0, 1\}$-valued ground-truth RI labels and the probabilistic ($[0, 1]$-valued) predictions. Leave-one-year-out cross validation of TC data from 1995 to 2013 is applied for model training and hyperparameter tuning, with the Brier score serving as both the optimization objective and evaluation metric. Cross-validated performance is reported with no special mention of holdout data for testing.

The dataset for building SHIPS-RII is constructed as follows. First, numerous features are collected from TCs, including climatology features (statistical properties calculated from historical data), real-time measured atmospheric/oceanic features, and features extracted from large-scale numerical model simulations of current atmosphere conditions. Then, a cherry-picked subset of TCs is removed for feature extraction to improve model performance. Finally, a subset of summarized features is selected by linear discriminant analysis and heuristics to serve as the final set of features (i.e. predictors in Sect. 2.1).

Given the metric and dataset within the current benchmark above, we observe four main issues:

1. The original data for deriving the features is not easily accessible to the public.
2. The cherry-picking data cleaning is subjective and not easily reproducible.
3. There is no holdout test set to gauge the true generalization performance of models.
4. The BS metric is knowingly sensitive to class imbalance, but the RI prediction problem is class-imbalanced in nature.

3.2   Proposed Benchmark

Next, we proposed a new benchmark to conquer the four issues. The new benchmark includes a more adequate evaluation metric and a publicly accessible dataset with a reproducible construction procedure and training/test splitting.

Metric. Different metrics are appropriate for evaluating different problems. For instance, for problems that involve outliers in the data that one would like to ignore, mean absolute error (MAE) is more suitable than mean square error (MSE), which puts greater emphasis on outlying data points. For the TC RI task, one major property of the data is its highly imbalanced classes. Due to the nature of the TC life cycle, RI mostly occurs only during the early to mid stages of a TC when it grows in strength in a short period of time. Thus, the number of RI to non-RI timeframes is approximately 1:20 (the positive class comprises 5% of the total data). Naturally, the performance metric should account for such data imbalance and should avoid rewarding highly biased classifiers that only output the majority class.

Unfortunately, the Brier score does not take class imbalance into consideration. For instance, a naive constant classifier that predicts only the majority class (in this case, no-RI) with class imbalance of 1:20 yields a very low Brier score of 0.05. As we are more interested in discerning the minority class, it is difficult for the Brier score to differentiate skillful models from unskillful models.

The Heidke Skill Score (HSS) is a metric commonly used in meteorology that accounts for class imbalance. It is defined as how much better the model prediction accuracy is than the standard forecast:

$$\text{HSS} = \frac{\text{ACC} - \text{SF}}{1 - \text{SF}}, \quad \text{ACC} = \frac{\text{TP} + \text{TN}}{N}, \quad \text{and}$$

$$\text{SF} = \frac{\text{TP} + \text{FN}}{N} \times \frac{\text{TP} + \text{FP}}{N} + \frac{\text{TN} + \text{FN}}{N} \times \frac{\text{TN} + \text{FP}}{N},$$

where N is the number of total instances, ACC is the accuracy, SF is the standard forecast, defined as correct by chance with class proportion, TP is the number of true positives, TN is the number of true negatives, FP is the number of false positives, and FN is the number of false negatives. It is more difficult for the standard forecast to correctly predict under class imbalance as a skewed class proportion increases the odds of being correct by chance, Thus, using standard forecast as the baseline makes HSS class-imbalance aware.

Other common metrics for handling imbalanced data include the F1 score (the harmonic mean of precision and recall), the Matthews correlation coefficient, the precision-recall area under the curve (PR-AUC), and the receiver operating characteristic area under the curve (ROC-AUC) (the latter two are only for binary classes). We select PR-AUC as the main evaluation metric, since precision and recall are both important for this task, coupled with HSS as a side metric to make it easier for meteorology experts to gauge performance.

Dataset. Since the raw form of most TC features is too complicated for people to analyze and comprehend, only summarized statistical properties are released for public usage and used by SHIPS-RII. For example, one common feature is the standard deviation of 100–300 km infrared brightness temperatures whereas the entire distribution of infrared brightness temperatures over the whole geographical location is available but not fully utilized. Another common feature is the time-averaged potential TC intensity estimated by simulation, even when data is available for the entire duration of the TC. Information loss due to statistical properties taken over time or space is particularly harmful for RI prediction because location- and timing-specific properties of the tropical cyclone are known to cause rapid intensification.

Thus, we present a new dataset derived from [3,4], which consists of satellite images of tropical cyclones (four channels) with data available every three hours.[1] Each TC sequence is split into 24-h segments. Each segment is associated with a label indicating the occurrence of RI during the period. Preprocessing of satellite images follows typical meteorology practices with emphasis on the "objectiveness", i.e., the entire procedure involves no human intervention. The data consists of TCs from 2003 to 2017 and is split into two parts for training (2003 to 2014) and testing (2015 to 2017). A disjoint testing set allows for the evaluation of model generalization on holdout data; cross-validation is avoided because it is particularly infeasible in terms of efficiency for deep learning models.

Satellite images, unlike features generated by predictive simulations, are considered lower-level data, i.e., they involve less processing. On the other hand, predictive features are generated by large-scale dynamic atmospheric simulations. Depending on predictive features poses a consistency issue because the simulation models are updated yearly; thus they may change over a short period of time. The fact that previous studies all depend on mixed data underlines the importance of making better use of satellite images.

To highlight the contribution of our proposed benchmark, we inspect whether the issues with the existing benchmark are resolved. First of all, the satellite image data including the scripts for preprocessing are released for public use. The preprocessing procedure is fully automated and does not require any human intervention. A disjoint set of data is reserved for testing to avoid presenting overfitting benchmark results. Finally, PR-AUC and HSS are adopted as evaluation metrics, both better reflect model performance under class imbalance.

4   Proposed Method

The new benchmark allows us to design and evaluate the potential of a domain-inspired model that takes in satellite images as inputs. The model belongs to the family of spatiotemporal deep learning methods, using deep convolutional neural networks (CNNs) to extract essential features from image data in an autonomous manner. As models generally suffer from the curse of dimensionality, complex spatiotemporal data requires manual feature extraction (such as hand-carved kernels) to compress it into smaller chunks. Such manual processing discards a large amount of information and is inefficient in finding optimal features because all feedback comes from trial and error. Currently the meteorology community is attempting to break through the performance bottleneck by experimenting with modalities such as lightning strike occurrence.

---

[1] Tropical Cyclone Rapid Intensification with Satellite Images: https://www.csie.ntu.edu.tw/~htlin/program/TCRISI/.

We opt for an orthogonal path by attempting to extract more from the data that we already have, specifically TC satellite images with infrared (IR1), water vapour (WV), visible light (VL), and passive microwave (PMW) channels. Table 1 shows that satellite images contain sufficient if not significantly more information in comparison to the features adopted in SHIPS-RII. One thing to note is the overlapping dependence on the infrared and water vapor channels, with infrared strictly dominating water vapor. Another is the lack of connection with the visible light channel as the signal is useful for only half of the time (the other half is nighttime). This is aligned with the feature selection in [3], which conclude that using only the IR1 and PMW channels yields the best performance. We follow the same practice to use only IR1 and PMW channels in our study.

Table 1. Potential satellite channels implicitly related to SHIPS-RII environmental conditions

SHIPS-RII feature	Satellite channel
200-hPa divergence from 0–1000-km radius	IR1
850–700-hPa relative humidity from 200–800-km radius	IR1/WV
Total precipitable water averaged from 0–100-km radius	IR1/WV
Std dev of 50–200-km IR cloud-top brightness temperatures	IR1

4.1 Model Architecture

To recapitulate, the data for this task is spatiotemporal and highly class-imbalanced while relatively lacking in quantity, with only approximately 1900 tropical cyclone instances in total (from 2003 to 2017). The constraint posed by data properties guides our model design. As the tropical cyclone is a well-studied phenomenon in meteorology and many techniques have been proposed and withstood the test of time, translating the accumulated domain knowledge into data science is a difficult but rewarding task. Specifically, the Advanced Dvorak Method (ADT) is the inspiration for the attention modules in our model, which also allows us to observe what the model is focusing on. More details regarding individual components of the model are discussed later in this section.

4.2 Base Structure

Convolutional neural networks (CNNs) are suitable for modeling data with local spatial dependency, as the convolution operator operates on neighboring pixels, making it the base structure of our model. We further extend the model to take into account the sequential nature of data. Currently there are three main neural network families that deal with time series: 3D-CNN, recurrent neural networks (RNNs), and Transformer (from low to high model complexity). 3D-CNN applies convolution on the time dimension in addition to the two spatial dimensions. RNN reuses the same recurrent block to process each single time frame in a series and summarizes past time frames to feed into the next time frame. The long-short term memory network (LSTM) is one of the most popular implementations of RNN with an additional memory state for storing long-term dependencies in sequences. It can be coupled with CNNs to form a convolutional LSTM (ConvLSTM) [18], replacing all matrix multiplications with convolution operations to simultaneously model spatial and temporal aspects of data. Transformer relies on an attention mechanism to determine dependency between each input and output time frame, calculating a similarity score between each input time frame (a query) and a set of context (keys) to serve as importance weighting for a linear combination of the values that correspond to the individual keys. The query approach decouples the similarity and information aspect as the keys and values—in contrary to past attention mechanisms—can now be different. It also allows the output at each time frame to pull information from any (masked) input time frame, thus excelling in modeling sequential dependency. However, Transformers are huge architectures requiring large amounts of data to train, which exceeded the resources of the authors at the time this research was conducted.

Most pretrained computer vision neural networks are pretrained on large-scale image classification datasets such as CIFAR-10 or ImageNet. As the domain for satellite images is fundamentally different from those represented in publicly-available pretrained models, it is difficult to apply common image models (for instance ResNet50 [6]) to this task. Due to the lack of data, training from scratch was not feasible for larger models (as verified empirically). In the end, we opted for the minimalist model architecture presented in Fig. 1. The base structure is a ConvLSTM model with convolutional layers in the front to enrich features and a dense block in the end to compress the three-dimensional features to one single probabilistic scalar output.

4.3   Incorporating Meteorological Knowledge

Neural network models in deep learning are notorious for their lack of explainability and interpretability. This may work without major drawbacks in common tasks such as image classification or language translation, as the purpose of these tasks is to autonomize previously human-labor-intensive jobs. The relatively lower cost of making mistakes in these tasks allows for the application of black-box models, as deep neural networks outperform past approaches. However, for other tasks with more at stake (for instance in the medical domain), being able to peek into the black box becomes significantly more important than pure performance. Understanding when and why a model makes mistakes helps people build trust the model output and evaluate the risk of incorporating it into the workflow. For instance, saliency maps [17] are utilized to trace which part of the image contributes most to the classification result.

One of the most intuitive methods of understanding models is to directly design the model according to previously acquired domain knowledge. By exploiting such know-how, the learning is guided by expert insight which allows analysis of whether the models learn as intended. As mentioned in earlier sections, the Advanced Dvorak Technique used for TC intensity estimation inspired the attention components of our model. The major purpose of these additional components is to help diagnose and understand the model, as opposed to improving its performance.

Cross Channel Attention (CCA). The first of the two key points of ADT is to locate the TC eye and perform scene-type analysis. The extreme convection in the core area of TCs is known to be critical for RI. We integrate this into our model by applying cross channel attention between the CNN encoder block and recurrent block to reinforce the key locations for the model to focus on. Specifically, cross channel attention takes the multiple channel feature and apply a 2-dimensional importance weighting mask that is shared across every channel. In our model, the importance mask is implemented as a two-layer CNN with sigmoid function in the end for normalization. The global view across all different channels allows incorporation of information from different sources to facilitate an overall importance evaluation of where in the hidden map the model should focus its attention. Given our domain knowledge, we expect a successful model to focus mainly on the TC core.

Sequence Self-attention (SSA). The second key point of ADT is to apply time averaging on the predicted intensity, showcasing the importance of taking earlier information into account. In ADT the averaging scheme calculates the non-weighted mean intensity over a 3-hr window (empirically determined). Directly translating the concept to RI prediction is problematic, as (1) our model predicts the occurrence of a future event, so the ground truth for the recent past is still not available; and (2) averaging the RI probability may not be as reasonable as averaging TC intensity. Also, we are not satisfied with the heuristic of applying the non-weighted average over a fixed window as, intuitively, the optimal contribution of different time steps is certainly different and should be determined dynamically. As a result, we incorporate the widely adopted sequence attention mechanism [10] into our recurrent block. Of the different styles of sequence attention, we base our design on Bahdanau style self-attention [1]. For each time step, features with information for the current time step is concatenated with a weighted average of features from all time steps prior to form the recurrent block input. The weighting is calculated by evaluating the similarity between features for the current time step and prior features. In our model, we use the projected cosine distance as the similarity metric, which is implemented as first passing the two given features through a two-layer CNN for feature extraction, then applying normalized inner product. The CNN for feature extraction before calculating the cosine distance is essential due to the self-rotation motion of tropical cyclones. Pixel misalignment caused by rotation is remedied by adding the additional feature extractor, which implicitly learns to transform the features into higher level ones with rotation-invariance.

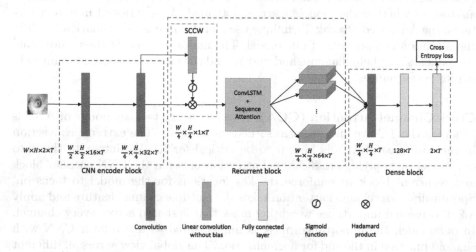

Fig. 1. Model architecture. The proposed model is split into three main components: CNN encoder block, recurrent block, and dense block.

5   Experiment and Analysis

For the experiments, we trained the model on the training set for 500 epochs, validating on the validation set every 10 epochs. Hyper-parameter tuning and epoch selection were achieved by selecting the model with best validation PR-AUC. We then evaluated the trained model on the test set and report the test performance in this paper. Hyper-parameters were minimally tuned only for the base ConvLSTM model such that the model converges and does not collapse to predicting the majority class only; for all other variants, they were kept identical. Tuned parameters include

- Batch size: 256
- Learning rate: 5e−4
- L2 regularization factor: 1e−5
- Dense layer dropout rate: 0.2.

All weights are initialized identically according to convention except for the final logits layer bias, which was carefully set to $\log(\frac{\#\text{of positive class instances}}{\#\text{of negative class instances}})$ to output a closer guess initially before any training given class imbalance. We chose the Adam optimizer [9] to optimize the weighted binary cross entropy loss with class weights corresponding to the reciprocal of the number of class instances (1:20). We experimented with both class weighting and minor class bootstrapping to prevent model collapse and found that the latter caused the model to overfit. The batch size was set as large as possible to mitigate large variances in gradient magnitude when the number of sampled instances was different for each mini-batch. The input satellite images were augmented by random rotation to simulate the natural rotation of tropical cyclones. This is the key to making deep learning possible, as without rotation augmentation, even the most shallow neural network model easily overfits the training set. All the code and trained models are publicly released.[2]

5.1 Input Time Length Experiment

The input time length is how long prior to the time step should be included as the input to the model. It may seem that the longer the better, as the model can actively choose what features to depend on. However, tropical cyclones are fundamentally chaotic systems where small changes in the initial value lead to largely bifurcated results. In SHIPS-RII, 24 h of data are fed into the model. The trade-off for the input time length is that with a shorter length, the model can start predicting in the very early stages of TC development but may use more information from earlier stages, whereas with a longer length, the situation is the exact opposite, and the training time is extended linearly with longer gradient backpropagation. As ConvLSTM can take arbitrary-length sequence input, we experimented with dynamic- and fixed-length input and found that dynamic-length resulted in inefficient training (each sequence cannot be stacked uniformly) and poor performance, suggesting that fixed-length input is better. We experimented with input lengths of 12, 24, 36, and 48 h; the results are presented in Fig. 2. For PR-AUC, an input length of 24 h yields the best performance, identical to the empirically determined length adopted in past studies. An input length of 36 h performs satisfactorily as well, indicating the possibility of the model benefiting from more data. For HSS, an input length of 36 h yields the best performance, in contrast to the empirically determined length in past studies, with 24 and 48 h not far behind. Thus, input lengths of 24 or 36 are in the range of optimal input time lengths. Interestingly, this alignment with meteorology-domain knowledge suggests that approaches from past studies and our model, as different as they might be, find important features from a similar range in history. For the sake of faster training and less data required for prediction, we conducted the following experiments with 24 h as input.

5.2 Ablation Study

We conducted an ablation study on the two ADT-inspired components attached to the base ConvLSTM model. Recall that the purpose of CCA and SSA is to assist in the understanding of the model prediction with gains in model performance as a secondary goal. As a result, the hyperparameters are not specifically tuned for each model, nor is the architecture. Performance gains via optimizing

---

[2] https://github.com/jybai/TCRI-Benchmark.

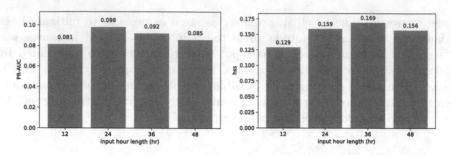

Fig. 2. Performance given different input time lengths (left: PR-AUC, right: HSS). These results show that input lengths of 24 or 36 are in the range of optimal input time lengths.

these details is left as future work. The PR-AUC and HSS results of the different variants are shown in Table 2.

ConvLSTM, ConvLSTM + CCA, and ConvLSTM + SSA perform similarly in both metrics, which shows that the additional domain-inspired components do not harm prediction quality. Note however that the combination of CCA and SSA results in clearly worse performance as the model size becomes too large, causing the model to overfit on the training set. This is an example of the most difficult issue we encountered when designing the model: balancing having enough capacity to model spatiotemporal relations and being compact enough to not overfit on the small dataset. We will discuss the takeaways from CCA and SSA in the following sections.

Table 2. Results of ablation study

	PR-AUC	Heidke Skill Score
ConvLSTM	0.098	0.159
ConvLSTM + CCA	0.095	0.164
ConvLSTM + CCA	0.099	0.161
ConvLSTM + CCA + SSA	0.089	0.152

5.3   Cross Channel Attention Analysis

The purpose of CCA is to combine information across all channels to produce a shared importance mapping. In Fig. 3 we randomly sampled four instances for visualization and compare these with existing domain knowledge in meteorology regarding factors relevant to rapid intensification. We observe that CCA helps the neural network to focus on coherent convective features within the TC core region. High importance is given to areas with extreme convection characterized by high PMW rain rates (warm color in the figures) and extremely low IR1 brightness temperatures (dark blue in the figures), indicating that cross channel information is jointly considered. The strongly subsidence and dry area in the TC outer region characterized by zero PMW rain rate and very high IR1 brightness temperatures are also highlighted by the weighting. Domain knowledge holds that RI is closely related to the convective burst within the TC inner core. Therefore, we conclude that the focus of the model from a meteorological perspective is crucial to making correct predictions.

Furthermore, we demonstrate how the model follows the dynamical TC development process with the bottom-right subfigure in Fig. 3. We observe the formation of the TC eye (high IR1 Tb at the center) indicating the potential intensification of the TC in the last three IR1 satellite images. Importance weighting becomes significantly stronger at the TC center, focusing on the evolution of TC inner-core convection.

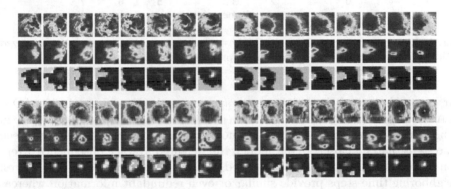

Fig. 3. Cross channel attention. Four randomly sampled instances are visualized, each with three rows; the top, middle, and bottom being the satellite IR1 channel, satellite PMW channel, and model attention map (CCA), respectively. White implies a higher importance for the weighting.

5.4   Sequence Self-attention Analysis

In ADT, an intensity average is taken over a 3-hr window with equal weighting, which is counter-intuitive as the importance of each time step should be different. We inspected the attention weights of the last time step over the time series extracted from the sequence attention component. We assume that since later time steps are closer to the final step, their similarities are higher, analogous to the exponential weighted moving average giving higher weights to more recent information. In Fig. 4 we present the histogram of the time step associated with the highest attention score.

Interestingly, the attention is highest on the first frame and exponentially decays over time. The last four steps have never been given the highest attention score out of all time steps, going against our intuition that later frames are more important. This phenomenon is explained by a key difference between

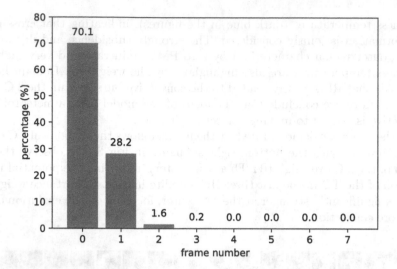

Fig. 4. Histogram of time step with highest attention score. Counts are normalized with respect to total number of instances, presented here as percentages.

our approach toward the weighted average versus the time averaging in ADT. In ADT, as the averaging is applied directly to intensity, due to the changes in atmospheric environment being continuous, intensity remains similar between time steps that are closer. In our model, the weighted average of sequence attention is applied to the hidden maps derived from the input satellite image features. Neighboring time steps provide similar or even redundant information whereas more distant time steps offer more. As the entire 24-h duration is important, the last frame gives more attention to the distant past to retrieve features that may not be sufficiently encoded in the ConvLSTM memory. It is also possible that since the end goal is to predict occurrences of rapid intensification, which is defined as whether the wind speed of the last frame increases by over 35 knots within 24 h, the model looks back in time to determine the difference between features from now and 24 h ago as a point of reference.

6   Conclusion

In this work, we propose a new benchmark for detecting rapid intensification of tropical cyclones, including a benchmark dataset based on TC satellite images with clearly defined, disjoint training, validation, and testing sets. We seek to exploit features from lower-level data such as satellite signals as they offer the most detailed information, in contrast to features utilized in the past derived from trial-and-error heuristics with a mix of measured and simulated data. We point out the insuitability of the commonly-reported Brier skill score given the nature of the high class imbalance of the task, and replace it with the precision-recall area under the curve (PR-AUC) and the Heidke skill score (HSS) to better reflect true model prediction performance. We also explore the potential of applying deep learning methods, which are known to benefit from large quantities of detailed spatiotemporal data which humans are unable to directly process. The proposed model is based on the ConvLSTM network with additional components—cross channel attention (CCA) and sequence self-attention (SSA)—inspired by the Advanced Dvorak Technique, a meteorological technique used to estimate TC intensity. We examine importance weighting from CCA and observe that the model focuses on the TC core where extreme convection occurs, a key factor for rapid intensification. We also evaluate sequence self-attention to determine which time step in a 24-h series is paid the most attention by the model. The overall results suggest that important prediction features are spread across the entire series. In future work, we will better incorporate CCA and SSA to maintain model performance and hopefully incorporate all channels of satellite data.

Acknowledgements. The project was partially supported by the Ministry of Science and Technology in Taiwan via MOST 107 2628 E 002-008-MY3 and 108-2119-M-007-010.

References

1. Bahdanau, D., Cho, K., Bengio, Y.: Neural machine translation by jointly learning to align and translate. arXiv preprint arXiv:1409.0473 (2014)
2. Chen, B.F., Davis, C., Kuo, Y.: An idealized numerical study of shear-relative low-level mean flow on tropical cyclone intensity and size. J. Atmos. Sci. **76**, 2309–2334 (2019)
3. Chen, B., Chen, B.F., Lin, H.T.: Rotation-blended CNNs on a new open dataset for tropical cyclone image-to-intensity regression. In: Proceedings of the 24th ACM SIGKDD International Conference on Knowledge Discovery & Data Mining, pp. 90–99. ACM (2018)
4. Chen, B.F., Chen, B., Lin, H.T., Elsberry, R.L.: Estimating tropical cyclone intensity by satellite imagery utilizing convolutional neural networks. Weather Forecast. **34**(1), 447–465 (2019)
5. Gall, R., Franklin, J., Marks, F., Rappaport, E.N., Toepfer, F.: The hurricane forecast improvement project. Bull. Am. Meteorol. Soc. **94**(3), 329–343 (2013)
6. He, K., Zhang, X., Ren, S., Sun, J.: Deep residual learning for image recognition. In: Proceedings of the IEEE Conference on Computer Vision and Pattern Recognition, pp. 770–778 (2016)
7. Hendricks, E.A., Peng, M.S., Fu, B., Li, T.: Quantifying environmental control on tropical cyclone intensity change. Mon. Weather Rev. **138**(8), 3243–3271 (2010)
8. Kaplan, J., DeMaria, M., Knaff, J.A.: A revised tropical cyclone rapid intensification index for the Atlantic and eastern North Pacific basins. Weather Forecast. **25**(1), 220–241 (2010)
9. Kingma, D.P., Ba, J.: Adam: a method for stochastic optimization. arXiv preprint arXiv:1412.6980 (2014)
10. Luong, M.T., Pham, H., Manning, C.D.: Effective approaches to attention-based neural machine translation. arXiv preprint arXiv:1508.04025 (2015)

11. Miyamoto, Y., Nolan, D.S.: Structural changes preceding rapid intensification in tropical cyclones as shown in a large ensemble of idealized simulations. J. Atmos. Sci. **75**(2), 555–569 (2018)
12. Olander, T.L., Velden, C.S.: The Advanced Dvorak Technique: continued development of an objective scheme to estimate tropical cyclone intensity using geostationary infrared satellite imagery. Weather Forecast. **22**(2), 287–298 (2007)
13. Rappin, E.D., Nolan, D.S.: The effect of vertical shear orientation on tropical cyclogenesis. Q. J. Roy. Meteorol. Soc. **138**(665), 1035–1054 (2012)
14. Rogers, R., Reasor, P., Lorsolo, S.: Airborne Doppler observations of the inner-core structural differences between intensifying and steady-state tropical cyclones. Mon. Weather Rev. **141**(9), 2970–2991 (2013)
15. Rozoff, C.M., Kossin, J.P.: New probabilistic forecast models for the prediction of tropical cyclone rapid intensification. Weather Forecast. **26**(5), 677–689 (2011)
16. Rozoff, C.M., Velden, C.S., Kaplan, J., Kossin, J.P., Wimmers, A.J.: Improvements in the probabilistic prediction of tropical cyclone rapid intensification with passive microwave observations. Weather Forecast. **30**(4), 1016–1038 (2015)
17. Simonyan, K., Vedaldi, A., Zisserman, A.: Deep inside convolutional networks: visualising image classification models and saliency maps. arXiv preprint arXiv:1312.6034 (2013)
18. Xingjian, S., Chen, Z., Wang, H., Yeung, D.Y., Wong, W.K., Woo, W.C.: Convolutional LSTM network: a machine learning approach for precipitation nowcasting. In: Advances in Neural Information Processing Systems, pp. 802–810 (2015)

Model Monitoring and Dynamic Model Selection in Travel Time-Series Forecasting

Rosa Candela[1], Pietro Michiardi[1], Maurizio Filippone[1],
and Maria A. Zuluaga[1,2(✉)]

[1] Data Science Department, EURECOM, Biot, France
{rosa.candela,pietro.michiardi,maurizio.filippone,
maria.zuluaga}@eurecom.fr
[2] Amadeus SAS, Sophia Antipolis, France

Abstract. Accurate travel products price forecasting is a highly desired feature that allows customers to take informed decisions about purchases, and companies to build and offer attractive tour packages. Thanks to machine learning (ML), it is now relatively cheap to develop highly accurate statistical models for price time-series forecasting. However, once models are deployed in production, it is their monitoring, maintenance and improvement which carry most of the costs and difficulties over time. We introduce a data-driven framework to continuously monitor and maintain deployed time-series forecasting models' performance, to guarantee stable performance of travel products price forecasting models. Under a supervised learning approach, we predict the errors of time-series forecasting models over time, and use this predicted performance measure to achieve both model monitoring and maintenance. We validate the proposed method on a dataset of 18K time-series from flight and hotel prices collected over two years and on two public benchmarks.

Keywords: Model monitoring · Model maintenance · Time-series · Forecasting

1 Introduction

Travel industry actors, such as airlines and hotels, nowadays use sophisticated pricing models to maximize their revenue, which results in highly volatile fares [8]. For customers, price fluctuation are a source of worry due to the uncertainty of future price evolution. This situation has opened the possibility to new businesses, such as travel meta-search engines or online travel agencies, providing decision-making tools to customers [32]. In this context, accurate price forecasting over time is a highly desired feature, as it allows customers to take informed decisions about purchases, and companies to build and offer attractive tour packages, while maximizing their revenue margin.

© Springer Nature Switzerland AG 2021
Y. Dong et al. (Eds.): ECML PKDD 2020, LNAI 12460, pp. 513–529, 2021.
https://doi.org/10.1007/978-3-030-67667-4_31

The exponential growth of computer power along with the availability of large datasets has led to a rapid progress in the machine learning (ML) field over the last decades. This has allowed the travel industry to benefit from the powerful ML machinery to develop and deploy accurate models for price time-series forecasting. Development and deployment, however, only represent the first steps of a ML system's life cycle. Currently, it is the monitoring, maintenance and improvement of complex production-deployed ML systems which carry most of the costs and difficulties in time [24, 27]. Model monitoring refers to the task of constantly tracking a model's performance to determine when it degrades, becoming obsolete. Once a degradation in performance is detected, model maintenance and improvement take place to update the deployed model by rebuilding it, recalibrating it or, more generally, by doing model selection.

While it is relatively easy and fast to develop ML-based methods for accurate price forecasting of different travel products, maintaining a good performance over time faces multiple challenges. Firstly, price forecasting of travel products involves the analysis of multiple time-series which are modeled independently, i.e. a model per series rather than a single model for all. According to the 2019 World Air Transport Statistics report, almost 22K city pairs are directly connected by airlines through regular services [15]. As each city pair is linked to a time-series, it is impossible to manually monitor the performance of every associated forecasting model. For scalability purposes, it is necessary to develop methods that can continuously and automatically monitor and maintain every deployed model. Secondly, time-series comprise time-evolving complex patterns, non-stationarities or, more generally, distribution changes over time, making forecasting models more prone to deteriorate over time [1]. Poor estimations of a model's degrading performance can lead to business losses, if detected too late, or to unnecessary model updates incurring system maintenance costs [27], if detected too early. Efficient and timely ways to model monitoring are therefore key to continuously accurate in-production forecasts. Finally, a model's degrading performance also implies that the model becomes obsolete. As a result, a specific model might not always be the right choice for a given series. Since time-series forecasting can be addressed through a large set of different approaches, the task of choosing the most suitable forecasting method requires finding systematic ways to carry out model selection efficiently. One of the most common ways to achieve all of this is cross-validation [3]. However, this approach is only valid at development and cannot be used to monitor and maintain models in-production due to the absence of ground truth data.

In this work we introduce a data-driven framework to continuously monitor and maintain time-series forecasting models' performance in-production, i.e in the absence of ground truth, to guarantee continuous accurate performance of travel products price forecasting models. Under a supervised learning approach, we predict the forecasting error of time-series forecasting models over time. We hypothesize that the estimated forecasting error represents a surrogate measure of the model's future performance. As such, we achieve continuous monitoring by using the predicted forecasting error as a measure to detect degrading performance. Simultaneously, the predicted forecasting error enables model maintenance by allowing to rank multiple models based on their predicted performance, i.e. model comparison, and then select the one with the lowest predicted error measure, i.e. model selection. We refer to it as a model monitoring and model selection framework.

The remaining of this paper is organized as follows. Section 2 discusses related work. Section 3 reviews the fundamentals of time-series forecasting and performance assessment. Section 4 describes the proposed model monitoring and maintenance framework. Section 5 describes our datasets and presents the experimental setup. Experiments and results are discussed in Sect. 6. Finally, in Sect. 7 we summarize our work and discuss key findings.

2 Related Work

Maintainable Industrial ML Systems. Recent works from tech companies [4, 19,24] have discussed their strategies to deal with some of the so-called *technical debts* [27] in which ML systems can incur when in production. These works mainly focus on the hard- and soft-ware infrastructure used to mitigate these *debts*. Less emphasis is given to the specific methods put in place.

Concept Drift. The phenomenon of time-evolving data patterns is known as concept drift. As time-series are not strictly stationary, it is a common problem of time-series forecasting usually addressed through regular model updates. Most works have focused on its detection, what we denote model monitoring, without performing model selection as they are typically limited to a single model [11,20]. The exception to this is the work of [25,26], where a weighted sliding-window is used to combine the forecasts of multiple candidate models into a single value.

Performance Assessment Without Ground Truth. An alternative to cross-validation is represented by information criteria. The rationale consists in quantifying the best trade-off between models' goodness of fit and simplicity. Information criteria are mostly used to compare nested models, whereas the comparison of different models requires to compute likelihoods on the same data. Being fully data-driven, our framework avoids any constraint regarding the candidate models, leading to a more general way to perform model selection. Specifically to time-series forecasting, Wagenmakers et al. [30] achieve performance assessment in the absence of ground truth using a concept similar to ours. They estimate the forecasting error of a new single data point by adding previously estimated forecast errors, obtained from already observed data points. The use of the previous errors makes it sensible to unexpected outlier behaviors of the time-series.

Meta-learning. Meta-learning has been proposed as a way to automatically perform model selection. Its performance has been recently demonstrated in the context of time-series forecasting. Both [2,28] formulate the problem as a supervised learning one, where the meta-learner receives a time-series and outputs the "best" forecasting model. Authors in [7] share our idea that forecasting performance decays in time, thus they train a meta-learner to model the error incurred by the base models at each prediction step as a function of the time-series features. Differently from [28], our approach does not seek to select a different model family for each time-series, and avoids model selection at each time step [7], since these two represent expensive overheads for in-production maintenance. Instead, we maintain a fast forecasting procedure and select the best model for a given time period in the future, which length can be relatively high (6–9 months, for instance).

3   Time-Series Forecasting and Performance Measures

A univariate time-series is a series of data points $\mathcal{T} = \{y_1, \ldots, y_T\}$, each one being an observation of a process measured at a specific time t. Univariate time-series contain a single variable at each time instant, while multivariate time-series record more than one variable at a time. Our application is concerned with univariate time-series, which are recorded at discrete points in time, e.g., monthly, daily, hourly. However, extension to the multivariate setting is straightforward.

Time-series forecasting is the task consisting in the use of these past observations (or a subset thereof) to predict future values $\mathcal{T}_h = \{\hat{y}_{T+1}, \ldots, \hat{y}_{T+h}\}$, with h indicating the forecasting horizon. The number of well-established methods to perform time-series forecasting is quite large. Methods go from classical statistical methods, such as Autoregressive Moving Average (ARMA) and Exponential smoothing, to more recent machine learning models which have shown outstanding performance in different tasks, including time-series forecasting.

The performance assessment of forecasting methods is commonly done using error measures. Despite decades of research on the topic, there is still not an unanimous consensus on the best error measure to use among the multiple available options [14]. Among the most used ones, we find Symmetric Mean Absolute Percentage Error (sMAPE) and Mean Absolute Scaled Error (MASE). These two have been adopted in recent time-series forecasting competitions [21].

4   Monitoring and Model Selection Framework

Let us denote $\mathcal{X} = \{\mathcal{T}^{(i)}, \mathcal{T}_h^{(i)}\}_{i=1}^N$ the input training set. A given input i is formed by the observed time-series $\mathcal{T}^{(i)}$ and h forecasted values, $\mathcal{T}_h^{(i)}$. The values in $\mathcal{T}_h^{(i)}$ are obtained by a given forecasting model which we hereby denote a *monitored model*, g. Let $\mathbf{e}_g = \{e_g^{(i)}\}_{i=1}^N$ be a collection of N performance measures assessing the accuracy of the forecasts $\mathcal{T}_h^{(i)}$ estimated by g. A given performance measure $e_g^{(i)}$ is obtained by comparing the forecasts $\mathcal{T}_h^{(i)}$ from g to the true values.

Lets define a *monitoring model* as a model that is trained to learn a function f mapping the input time-series $\mathcal{X}$ to the target $\mathbf{e}_g$. Given a new set of time-series $\mathcal{X}^* = \{\mathcal{T}^{*(i)}, \mathcal{T}_h^{*(i)}\}_{i=1}^{N^*}$, formed by a time-series of observations $\mathcal{T}^{*(i)}$, $|\mathcal{T}^{*(i)}| = T_i^*$, and h forecasts $\mathcal{T}_h^{*(i)}$ obtained by g, the learned *monitoring model* predicts $\mathbf{e}_g^*$, i.e. the predicted performance measure of g given $\mathcal{X}^*$ (Fig. 1).

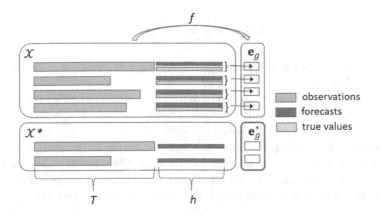

Fig. 1. Illustration of the proposed method. $\mathcal{X}$ and $\mathcal{X}^*$ contain multiple time-series, each of these composed of T_i observations (green) and h forecasts (red) estimated by a *monitored model*, g. $\mathbf{e}_g$ represents the forecasting performance of the *monitored model*. It is computed using the true values (yellow). A *monitoring model* is trained to learn the function f mapping $\mathcal{X}$ to $\mathbf{e}_g$. With the learned f, the *monitoring model* is able to predict $\mathbf{e}_g^*$, the predicted forecasting performance of the *monitored model* given $\mathcal{X}^*$. (Color figure online)

The predicted performance measures $\mathbf{e}_g^*$ represent a surrogate measure of the performance of a given g within the forecasting horizon h. As such it is used for the two tasks: model monitoring and selection. Model monitoring is achieved by using $\mathbf{e}_g^*$ as an alert signal. If the estimated performance measure of the *monitored model* is poor, this means the model has become stale. To achieve model selection, $\mathbf{e}^*$ are used to rank multiple *monitored models* and choose the one with the best performance If the two tasks are executed in a continuous fashion over time, it is possible to guarantee accurate forecasts in an automated way.

In the following, we describe the performance measure $\mathbf{e}$ that we use in our framework, as well as the *monitoring* and *monitored models* that we chose to validate our hypotheses.

4.1 Performance Measure

As previously discussed, performance accuracy of time-series forecasts is measured using error metrics. In this framework, we use the sMAPE. It is defined as:

$$\text{sMAPE} = \frac{1}{h} \sum_{t=1}^{h} 2 \frac{|y_t - \hat{y}_t|}{|y_t| + |\hat{y}_t|}, \tag{1}$$

where h is the number of forecasts (i.e. forecasting horizon), y is the true value and $\hat{y}$ is the forecast.

In the literature, there are multiple definitions of the sMAPE. We choose the one introduced in [9] because it is bounded between 0 and 2; specifically, it has a maximum value of 2 when either y or $\hat{y}$ is zero, and it is zero when the two values are identical. The sMAPE has two important drawbacks: it is undefined when both y, $\hat{y}$ are zero and it can be numerically unstable when the denominator in Eq. 1 is close to zero. In the context of our application, this is not a problem since it is unlikely to have prices with value zero or very close to it.

4.2 Monitoring Models

The formulation of our framework is generic in the sense that any supervised technique that can solve regression problems can be used as a *monitoring model*. In this work, we decided to focus on latest advances in deep learning. We consider four alternative *monitoring models*: Long Short-Term Memory (LSTM) networks, Convolutional Neural Networks (CNNs), Bayesian CNNs and Gaussian processes (GP). The latter two models differ from the former ones in that they also provide uncertainties around the predictions. This can enrich the output provided by the monitoring framework, in that whenever an alert is issued because of poor performance, this is equipped with information about its reliability This section illustrates the basic ideas of each of the selected *monitoring models*.

Long Short-Term Memory Networks. LSTM [13] networks are a type of Recurrent Neural Networks (RNNs) that solve the issue of the vanishing gradient problem [5] present in the original RNN formulation. They achieve this by introducing a cell state into each hidden unit, which memorizes information. As RNNs they are a well-established architecture to model sequential data. By construction, LSTMs can handle sequences of varying length, with no need for extra processing like padding. This is useful in our application, whereby time-series in the datasets have different lengths.

Convolutional Neural Networks. CNNs [17] are particular class of deep neural networks where the weights (filters) are designed to promote local information to propagate from the input to the output at increasing levels of granularity. We use the original LeNet [18] architecture, as it obtains generally good results in image recognition problems, while being considerably faster to train with respect to more modern architectures. CNNs are not originally conceived to work with time-series data. We adapt the architecture to work with time-series by using 1D convolutional filters. Unlike RNNs, this model does not support inputs of variable size, so we to resort to padding: where necessary we append zeros to a time-series to make them uniform in length. We denote this model LeNet.

Bayesian Convolutional Neural Networks. Bayesian CNNs [12] represent the probabilistic version of CNNs, used in applications where quantification of the uncertainty in predictions is needed. Network parameters are assigned a prior distribution and then inferred using Bayesian inference techniques. Due to the intractability of the problem, the use of approximations is required. Here we choose Monte Carlo Dropout [12] as a practical way to carry out approximate Bayesian CNNs. By applying dropout at test time we are able to sample from an approximate posterior distribution over the network weights. We use this technique on the LeNet CNN with 1D filters to produce probabilistic outputs. We denote this model Bayes-LeNet.

Gaussian Processes. GPs [23] form the basis of probabilistic nonparametric models. Given a supervised learning problem, GPs consider an infinite set of functions mapping input to output. These functions are defined as random variables with a joint Gaussian distribution, specified by a mean function and a covariance function, the latter encoding properties of the functions with respect to the input. One of the strengths of GP models is the ability to characterize uncertainty regardless of the size of the data. Similarly to CNNs, in this model input sequences must have the same length, so we resort to padding.

4.3   Monitored Models

Similar to *monitoring models*, given the generic nature of the proposed framework, there is no constraint on the type of *monitored models* that can be used. Any time-series forecasting method can be monitored. For this proof of concept, we consider six different *monitored models*. We select five of them from the ten benchmarks provided in the M4 competition [21], a recent time-series forecasting challenge. These are: Simple Exponential Smoothing (`ses`), Holt's Exponential Smoothing (`holt`), Dampen Exponential Smoothing (`damp`), Theta (`theta`) and a combination of `ses` - `holt` - `damp` (`comb`). Besides these five methods, we included a simple Random Forest (`rf`), in order to enrich the benchmark with a machine learning-based model. We refer the reader to [6,21] for further details on each of these approaches.

5   Experimental Setup

This section presents the data, provides details about the implementation of our methods to ease reproducibility and concludes by describing the evaluation protocol carried during the experiments.

5.1   Data

Flights and Hotels Datasets. We focus on two travel products: direct flights between city pairs and hotels. Our data is an extract of prices for these two travel products obtained from the Amadeus for Developers API[1], an online web-service which enables access to travel-related data through several APIs. It was collected over a two-years and one-month period. Table 1 presents some descriptive features of the datasets.

Using the service's Flight Low-fare Search API, we collected daily data for one-way flight prices of the top 15K most popular city pairs worldwide. The collection was done in two stages. A first batch, corresponding to the top 1.4K pairs (FLIGHTS), was gathered for the whole collection period. The second batch, corresponding to the remaining pairs (FLIGHTS-EXT), was collected only over the second year. For hotels, we used the Hotel API to collect daily hotel prices for a two-night stay at every destination city contained in the top city pairs used for flight search. These represent 3.2K different time-series.

Both APIs provide information about the available offers for flights/hotels, that meet the search criteria (origin-destination and date, for flights; city, date and number of nights, fixed to 2, for hotels) at the time of search. As such, it is possible to have multiple offers (flights or hotel rooms) for a given search criteria. When multiple offers were proposed, we averaged the different prices to have a daily average flight price for a given city pair, in the case of flights, or daily average hotel price for a given city, in the case of hotels. In the same way, it is possible to have no offers for a given search criteria. Days with no available offers were reported as missing data. Lack of offers can be caused by sold outs, specific flight schedules (e.g. no daily flights for a city pair) or seasonal patterns (e.g. flights for a part of the year or seasonal hotel closures). More rarely, they could even be due to a failure in the query sent to the API. As a result, the number of available observations is smaller than the length of the collection period (see Table 1).

Public Benchmarks. In addition to travel products data, we decided to include data coming from publicly available benchmarks. Benchmark data are typically curated and avoid problems present in real data, such as those previously discussed regarding missing data, allowing for an objective assessment and more controlled setup for experimentation. We included two sets from the M4 time-series forecasting challenge competition [21] dataset, YEARLY and WEEKLY. Table 1 presents statistics on the number of time-series and the available number observations per time-series for these two datasets. Here, the number of available observations is equivalent to the time-series length as no time-series contains missing values.

5.2   Implementation

The LSTM network was implemented in Tensorflow. It is composed of one hidden layer with 32 hidden nodes. It is a dynamic LSTM, in that it allows the input sequences to have variable lengths, by dynamically creating the graph during execution. The two CNN-based *monitoring models* use the LeNet architecture.

---

[1] https://developers.amadeus.com/.

Table 1. Information about number of time-series, and minimum (min-obs), maximum (max-obs), mean (mean-obs) and standard deviation (std-obs) of the available number of time-series observations per dataset.

Name	# time-series	min-obs	max-obs	mean-obs	std-obs
FLIGHTS	1,415	431	745	734	23
FLIGHTS-EXT	13,810	50	347	346	13
HOTELS	3,207	1	658	368	128
YEARLY	23,000	13	835	31	25
WEEKLY	359	80	2,597	1022	706

We modified both convolutional and pooling layers with 1D filters, given that the input of the model consists in sequences of one dimension. We added dropout layers to limit overfitting. In the Bayesian CNN, we applied a dropout rate of 0.5, also at testing time, to obtain 100 Monte Carlo samples as approximation of the true posterior distribution. The GP model used the implementation of Sparse GP Regression from the GPy library[2]. The inducing points [29] were initialized with K-means and were then fixed during optimization. We used a variable number of inducing points depending on the size of the input and a RBF kernel with Automatic Relevance Determination (ARD). In all experiments we used 75% data for training and 25% for test and the Adam optimizer with default learning rate [16]. Only in the dataset FLIGHTS-EXT we used mini-batches of size 128 to speed up the training. For the *monitored models*, we used the implementation available from the M4 competition benchmark Github repository[3] and we used the Python sklearn package [22] implementation of R andom Forest. All code has been made publicly available[4].

5.3   Evaluation Protocol

For flight and hotel data we set $h = \{90, 180\}$, which means we are predicting the price for h days ahead. These are two commonly used values in travel, representing 3 and 6 months ahead of the planned trip, so it is important to have accurate predictions over those horizons. For the M4 competition datasets, we use the horizon given by the challenge organizers: $h = 6$ for YEARLY and $h = 13$ for WEEKLY. For each dataset, we reserve the first T_i data points of the i-th time-series, where T_i depends on the time-series's length, as input of the *monitored models* to obtain h forecasts. Where missing values were found, in flights or hotels, these were replaced with the nearest non-missing value in the past. We build $\mathcal{X}$ and $\mathcal{X}^*$, by taking 75% and 25% from the total number of time-series, respectively. We thus compute the forecasting errors e using the sMAPE in Eq. 1

---

[2] http://github.com/SheffieldML/GPy.
[3] https://github.com/M4Competition/M4-methods.
[4] https://github.com/robustml-eurecom/model_monitoring_selection.

for the training set $\mathcal{X}$. Finally, we predict the performance measure $\mathbf{e}^*$ for the time-series in $\mathcal{X}^*$, using the four *monitoring models*.

We compare our model monitoring and selection framework with the standard cross-validation method, which we here denote `baseline`, where a model's estimated performance is obtained "offline" at training time with the available data. Specifically, given T observations, we use the last h observations as validation set to evaluate the model. This implies to reduce the number of observations available to train the forecasting models, which can be problematic when either T is small or h is large.

6   Experiments and Results

We first study the proposed framework's ability to achieve model monitoring (Sect. 6.1). Then, we demonstrate how the predicted forecasting errors can be used to carry out model selection and how it positions w.r.t state-of-the-art methods doing the same task (Sect. 6.2). In Sect. 6.3, we illustrate the performance of the joint model monitoring and selection framework in our target application.

6.1   Model Monitoring Performance

We evaluate if the *monitoring models'* predicted sMAPEs can be used for model monitoring by estimating if the predicted measure represents a good estimate of a *monitored model*'s future forecasting performance. We assess the quality of the predicted forecasting errors by estimating the root mean squared error (RMSE) between the predicted sMAPEs and the true sMAPEs, for every *monitored model*. The true sMAPE is obtained using the *monitored model*'s predictions and the time-series' observations in through Eq. 1. As a reference, we report the `baseline` RMSE, which is obtained by comparing the estimated sMAPE at training with the observed values at testing. Figure 2 left summarizes the obtained results on all datasets.

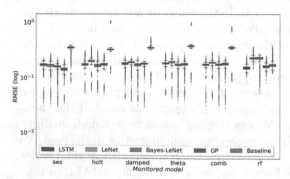

The overall average error incurred by the *monitoring models* is low. This suggests that the forecasting error predictions are accurate, meaning that it is reliable to carry out model monitoring. When compared to it, the *monitoring models* consistently perform better than standard cross-validation when estimating the future performance of the forecasting *monitored models*. There is an exception to this when the *monitored model* is the Random Forests

Fig. 2. RMSE between predicted and measured forecasting error (sMAPE) on all datasets (log scale). The reported baseline RMSE is obtained by comparing the estimated sMAPE at training with the observed values at testing.

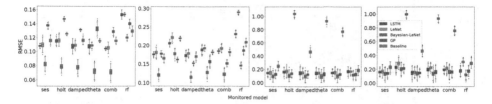

Fig. 3. RMSE between predicted and measured forecasting error (sMAPE). From left to right FLIGHTS and FLIGHTS-EXT (top) with 1) $h = 90$, 2) $h = 180$, hotels 3) $h = 90$, 4) $h = 180$.

Table 2. RMSE between predicted and true sMAPEs for flights and hotel time-series.

Monitoring model	Flights		Hotels	
	$h = 90$	$h = 180$	$h = 90$	$h = 180$
LSTM	0.116 ± 0.017	0.151 ± 0.031	0.193 ± 0.021	0.182 ± 0.039
LeNet	0.117 ± 0.017	0.155 ± 0.031	0.209 ± 0.039	0.224 ± 0.062
Bayes-LeNet	$\mathbf{0.084 \pm 0.017}$	$\mathbf{0.100 \pm 0.035}$	$\mathbf{0.135 \pm 0.022}$	$\mathbf{0.148 \pm 0.044}$
GP	0.136 ± 0.007	0.126 ± 0.028	0.164 ± 0.014	0.165 ± 0.036
baseline	0.119 ± 0.006	0.604 ± 0.328	0.190 ± 0.020	0.609 ± 0.302

(`rf`). In this case, the `baseline` is not the worst performing approach. However, it is still surpassed in performance by both LSTM and GP.

Figure 3 details the results obtained for flights and hotels time-series. Table 2 stratifies the results for travel product time-series in terms of the forecasting horizon. results show that Bayes-LeNet obtains the lowest RMSEs, whereas GPs follows closely and reports lower standard deviation. Overall, our approach outperforms the `baseline` for large forecasting horizons, e.g. $h = 180$, while the methods get closer as the forecasting horizon decreases. This is consistent with our hypothesis that data properties change over time. Using a validation set composed of time points close to the unseen data gives consistent information about the model's performance, because the two sets of data (validation and unseen data) have similar properties. However, increasing h has the effect of pushing away the validation time points from the unseen data. In this case, it is better to rely on the forecast error prediction rather than on an error measure obtained during training.

6.2 Model Selection Performance

In this experiment, we assess the capacity of the proposed method to assist model selection in the absence of ground truth. *Monitored models* are ranked by estimating the average predicted sMAPE over a given time-series and ordering the resulting values in ascending order. In this way, we obtain a list of *monitored models* from the best to the worst one. The best performing *monitored model* is

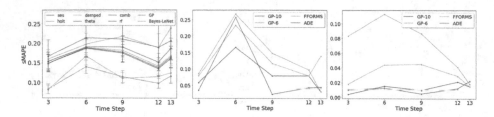

Fig. 4. Measured average forecasting performance(sMAPE) using the proposed method for model selection in the WEEKLY dataset with fixed forecasting models over the whole horizon. Average performance with Bayes-LeNet and GPs as *monitoring models* (left). Error bars denote standard deviation. Using GPs as *monitoring model* with six (GP-6) and ten *monitored models* (GP-10), worst (center) and best (right) model selection performances in comparison with ADE and FFORMS.

selected. We compare the ground truth ranking with the one obtained by each of the *monitoring models* and the baseline. We apply a Wilcoxon test [31] to the ranking results to verify if there are significant differences between each of the ranked *monitored models*. Table 3 presents obtained results in hotels and flights.

Overall, the obtained rankings are consistent with the ground truth, proving the ability of the method to carry out model selection, by identifying the model with the lowest error measure. Moreover, comparing our approach with the baseline, we find that our framework largely outperforms the latter, in that the ranking resulting from the baseline is very different from the true one. Even in predictions with a small forecasting horizon ($h = 90$), the baseline's ranking performance remains sub-optimal . Looking at the four *monitoring models*, we find that they have a different behavior depending on the dataset. Specifically, GPs result to be slightly more reliable than Bayesian-LeNet, as the latter in some cases swapped the first and second model of the ranking. LSTM's performance is close to the two probabilistic models, although the latter two globally have a better performance in terms of RMSE (see Table 2).

Having showed the reliability of the rankings, we evaluate if these can be effectively used to maintain accurate forecasts over time by doing model selection at fixed periods of time. Specifically, given a forecasting horizon, we divide it in smaller periods. At each time point, we use the predicted forecasting error to rank the *monitored models* and thus perform model selection by picking the best ranked model. We use the public benchmark data to guarantee curated data and we limit the experiments to the best two *monitoring models*, Bayesian-LeNet and GPs (Table 2). We compare our model selection with the results obtained using the same *monitored model* along the forecasting horizon. Figure 4 left shows the average forecasting performance, measured through the real sMAPE, on the WEEKLY dataset. The proposed model selection scheme allows to have the lower forecasting errors, i.e. a better performance, along the whole forecasting horizon. Among the two *monitoring models*, GPs result in smoother curves.

Finally, we compare with two state-of-the-art meta-learning methods, arbitrated dynamic ensembler [7], ADE, and Feature-based FORecast-Model Selec-

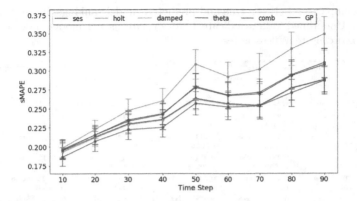

Fig. 5. Average forecasting performance in terms of sMAPE using the proposed model monitoring and selection framework (GPs as *monitoring model*) and using forecasting fixed models over the whole horizon. Error bars denote the standard deviation.

tion [28], FFORMS, with the best performing *monitoring model* in our approach. The characteristics of these two methods allows them to be used to achieve good forecasting model's performance. FFORMS uses 12 different base models, whereas ADE uses up to 40 different models. To remain competitive with these two methods that use a larger number of base models, we add three standard forecasting models, Arima (arima), Random Walk (rwf) and TBATS (tbats) [10], and a feed-forward neural network (nn), to our set of *monitored models*. We present sMAPE results over two time-series from the WEEKLY dataset: one where our method performs worst (Fig. 4 center) and the one where it performs best (Fig. 4 right). We show the results of our approach using the original six *monitored models* and the enlarged set. Using the original six *monitored models*, our performance is worse than the two meta-learning models. However, by enlarging the set of *monitored models*, our method performs better than FFORMS and achieves a performance comparable to ADE with much less monitored/base models.

6.3   Model Monitoring and Selection Performance

Finally, we illustrate the performance of the proposed model monitoring and selection framework by using it to guarantee continuous price forecasting accuracy of our two travel products: flights and hotels. In this context, the predicted sMAPE is used as a surrogate measure of the quality of the forecasts estimated by the *monitored models*. When the predicted sMAPE surpasses a given threshold, model selection is performed. Otherwise, the *monitored model* is kept. We use the best performing *monitoring model*, GPs. Since this is a probabilistic method, in addition to having a high predicted sMAPE, we add the condition of having a low uncertainty in the prediction. In our experiments, we set the sMAPE threshold at 0.02 for flights and 0.01 for hotels. The uncertainty was set at 0.01 for both. For this experiment, we removed rf from the *monitored*

Table 3. Comparison between true and predicted model rankings, in ascending order of sMAPE. Underlined values indicate pairs of forecasting models not significantly different, according to Wilcoxon test.

Ground Truth		Monitoring models								Baseline	
		LSTM		LeNet		Bayes-LeNet		GPs			
model	sMAPE	model	sMAPE	model	sMAPE	model	sMAPE	model	sMAPE	model	sMAPE
HOTELS - h = 180											
model	sMAPE	model	sMAPE	model	sMAPE	model	sMAPE	model	sMAPE	model	sMAPE
1 damp	0.244 (0.153)	ses	208 (0.015)	ses	0.208 (0.032)	ses	0.212 (0.087)	damp	0.230 (0.119)	ses	0.326 (0.202)
2 ses	0.246 (0.164)	damp	0.220 (0.033)	damp	0.211 (0.056)	damp	0.224 (0.130)	ses	0.231 (0.121)	rf	0.413 (0.333)
3 theta	0.269 (0.217)	theta	0.233 (0.059)	theta	0.231 (0.024)	comb	0.249 (0.166)	comb	0.251 (0.149)	damp	0.462 (0.391)
4 comb	0.270 (0.207)	comb	0.234 (0.057)	rf	0.236 (0.047)	theta	0.268 (0.234)	theta	0.252 (0.160)	comb	0.746 (0.569)
5 rf	0.316 (0.300)	holt	0.280 (0.124)	comb	0.278 (0.145)	rf	0.324 (0.329)	rf	0.291 (0.207)	theta	0.938 (0.620)
6 holt	0.325 (0.277)	rf	0.292 (0.210)	holt	0.298 (0.189)	holt	0.325 (0.162)	holt	0.299 (0.190)	holt	1.047 (0.660)
HOTELS - h = 90											
model	sMAPE	model	sMAPE	model	sMAPE	model	sMAPE	model	sMAPE	model	sMAPE
1 damp	0.242 (0.175)	ses	0.203 (0.022)	ses	0.217 (0.065)	ses	0.238 (0.088)	damp	0.221 (0.137)	comb	0.237 (0.166)
2 ses	0.243 (0.174)	damp	0.218 (0.026)	damp	0.223 (0.073)	damp	0.239 (0.122)	comb	0.238 (0.155)	ses	0.239 (0.177)
3 comb	0.253 (0.189)	theta	0.223 (0.022)	theta	0.227 (0.063)	comb	0.259 (0.108)	theta	0.240 (0.151)	damp	0.250 (0.194)
4 theta	0.254 (0.190)	comb	0.224 (0.030)	comb	0.229 (0.047)	theta	0.263 (0.132)	ses	0.244 (0.180)	theta	0.251 (0.201)
5 holt	0.275 (0.217)	holt	0.244 (0.052)	holt	0.252 (0.096)	holt	0.282 (0.190)	holt	0.265 (0.185)	holt	0.277 (0.235)
6 rf	0.293 (0.285)	rf	0.254 (0.103)	rf	0.263 (0.059)	rf	0.298 (0.176)	rf	0.266 (0.191)	rf	0.311 (0.296)
FLIGHTS - h = 180											
model	sMAPE	model	sMAPE	model	sMAPE	model	sMAPE	model	sMAPE	model	sMAPE
1 rf	0.238 (0.163)	rf	0.203 (0.007)	rf	0.199 (0.039)	rf	0.219 (0.075)	rf	0.213 (0.108)	rf	0.259 (0.200)
2 ses	0.247 (0.144)	theta	0.217 (0.026)	ses	0.215 (0.012)	theta	0.220 (0.097)	theta	0.226 (0.098)	damp	0.277 (0.150)
3 theta	0.248 (0.175)	ses	0.218 (0.024)	damp	0.216 (0.074)	ses	0.233 (0.098)	ses	0.227 (0.100)	ses	0.278 (0.151)
4 damp	0.249 (0.144)	damp	0.219 (0.022)	theta	0.217 (0.042)	damp	0.240 (0.090)	damp	0.229 (0.098)	theta	0.281 (0.155)
5 comb	0.250 (0.148)	comb	0.221 (0.027)	comb	0.219 (0.016)	comb	0.241 (0.094)	comb	0.231 (0.054)	comb	0.283 (0.160)
6 holt	0.260 (0.162)	holt	0.223 (0.034)	holt	0.222 (0.036)	holt	0.250 (0.088)	holt	0.238 (0.119)	holt	0.299 (0.199)
FLIGHTS - h = 90											
model	sMAPE	model	sMAPE	model	sMAPE	model	sMAPE	model	sMAPE	model	sMAPE
1 comb	0.174 (0.102)	comb	0.154 (0.081)	theta	0.160 (0.067)	comb	0.151 (0.086)	damp	0.159 (0.073)	ses	0.187 (0.110)
2 damp	0.175 (0.106)	damp	0.155 (0.076)	comb	0.164 (0.085)	damp	0.161 (0.086)	comb	0.160 (0.082)	theta	0.188 (0.109)
3 theta	0.176 (0.105)	theta	0.157 (0.042)	holt	0.166 (0.076)	theta	0.163 (0.087)	theta	0.162 (0.086)	damp	0.189 (0.110)
4 ses	0.177 (0.106)	ses	0.158 (0.028)	rf	0.174 (0.066)	holt	0.188 (0.074)	ses	0.163 (0.094)	comb	0.190 (0.112)
5 holt	0.179 (0.113)	holt	0.159 (0.036)	ses	0.183 (0.087)	ses	0.212 (0.070)	holt	0.171 (0.119)	holt	0.195 (0.118)
6 rf	0.232 (0.150)	rf	0.200 (0.025)	damp	0.212 (0.044)	rf	0.287 (0.083)	rf	0.210 (0.094)	rf	0.207 (0.137)

models pool as it is the method giving the poorest performance. It is important to remark that differently from other approaches removing a method from the *monitored models* pool simply requires to stop generating forecasts with the removed model. No re-training of the *monitoring models* is required.

Figure 5 illustrates the results obtained in terms of the average performance (sMAPE) for HOTELS with forecasting horizon $h = 90$. Our experiment here is quite restrictive, in the sense that no *monitored model* is re-trained along the forecasting period. In this way, we show that even under this restrictive setting the proposed framework is able to improve the performance of simple models. This suggests that through the use of this framework it is possible to extend the moment where *monitored models* need to be re-trained by simply using the ranking information to pick a new model. Delaying model re-training represents important cost savings.

7 Conclusions

In this paper we introduce a data-driven framework to constantly monitor and compare the performance of deployed time-series forecasting models to guarantee accurate forecasts of travel products' prices over time. The proposed approach predicts the forecasting error of a forecasting model and considers it as a surrogate of the model's future performance. The estimated forecasting error is hence used to detect accuracy deterioration over time, but also to compare the performance of different models and carry out dynamic model selection by simply ranking the different forecasting models based on the predicted error measure and selecting the best. In this work, we have chosen to use the sMAPE as forecasting performance measure, since it is appropriate for our application but, it cannot be used in settings where the time-series could present zero-valued observations. However, the framework is general enough that any other measure could be used instead.

The proposed framework has been designed to guarantee accurate price forecasts of different travel products price and it is conceived for travel applications that might be already deployed. As such, it was undesirable to propose a method that performs forecasting and monitoring altogether, as in meta-learning, since this would require deprecating already deployed models to implement a new system. Instead, thanks to the proposed fully data-driven approach, *monitoring models* are completely independent of those doing the forecasts, i.e. the *monitored models*, thus allowing a transparent implementation of the monitoring and selection framework.

Although our main objective is to guarantee stable accurate price forecasts, the problem we address is relevant beyond our concrete application. Sculley *et al.* [27] introduced the term hidden technical debt to formalize and help reason about the long term costs of maintainable ML systems. According to their terminology, the proposed model monitoring and selection framework addresses two problems: 1) the monitoring and testing of dynamic systems, which is the task of continuously assessing that a system is working as intended; and 2) the production management debt, which refers to the costs associated to the maintenance of a large number of models that run simultaneously. Our solution represents a simple, flexible and accurate alternative to these problems.

References

1. Aiolfi, M., Timmermann, A.: Persistence in forecasting performance and conditional combination strategies. J. Econometr. **135**(1), 31–53 (2006)
2. Ali, A.R., Gabrys, B., Budka, M.: Cross-domain meta-learning for time-series forecasting. Proc. Comput. Sci. **126**, 9–18 (2018)
3. Arlot, S., Celisse, A., et al.: A survey of cross-validation procedures for model selection. Stat. Surv. **4**, 40–79 (2010)
4. Baylor, D., et al.: TFX: a tensorflow-based production-scale machine learning platform. In: Proceedings of the 23rd ACM SIGKDD International Conference on Knowledge Discovery and Data Mining, KDD 2017, pp. 1387–1395 (2017)
5. Bengio, Y., Simard, P., Frasconi, P.: Learning long-term dependencies with gradient descent is difficult. IEEE Trans. Neural Netw. **5**(2), 157–166 (1994)
6. Breiman, L.: Random forests. Mach. Learn. **45**(1), 5–32 (2001). https://doi.org/10.1023/A:1010933404324
7. Cerqueira, V., Torgo, L., Pinto, F., Soares, C.: Arbitrated ensemble for time series forecasting. In: Ceci, M., Hollmén, J., Todorovski, L., Vens, C., Džeroski, S. (eds.) ECML PKDD 2017. LNCS (LNAI), vol. 10535, pp. 478–494. Springer, Cham (2017). https://doi.org/10.1007/978-3-319-71246-8_29
8. Chen, Y., Cao, J., Feng, S., Tan, Y.: An ensemble learning based approach for building airfare forecast service. In: 2015 IEEE International Conference on Big Data (Big Data) (2015)
9. Chen, Z., Yang, Y.: Assessing forecast accuracy measures. Preprint Series 2010, 2004-10 (2004)
10. De Livera, A.M., Hyndman, R.J., Snyder, R.D.: Forecasting time series with complex seasonal patterns using exponential smoothing. J. Am. Stat. Assoc. **106**(496), 1513–1527 (2011)
11. Ferreira, J.A., Loschi, R.H., Costa, M.A.: Detecting changes in time series: a product partition model with across-cluster correlation. Sig. Process. **96**, 212–227 (2014)
12. Gal, Y., Ghahramani, Z.: Dropout as a Bayesian approximation: representing model uncertainty in deep learning. In: Proceedings of the 33rd International Conference on Machine Learning, ICML 2016, pp. 1050–1059 (2016)
13. Hochreiter, S., Schmidhuber, J.: Long short-term memory. Neural Comput. **9**(8), 1735–1780 (1997)
14. Hyndman, R.J., Koehler, A.B.: Another look at measures of forecast accuracy. Int. J. Forecast. **22**(4), 679–688 (2006)
15. International Air Transport Association: World air transport statistics. World air transport statistics (2019)
16. Kingma, D.P., Ba, J.: Adam: a method for stochastic optimization. CoRR abs/1412.6980 (2014). http://arxiv.org/abs/1412.6980
17. LeCun, Y., Bottou, L., Bengio, Y., Haffner, P.: Gradient-based learning applied to document recognition. Proc. IEEE **86**(11), 2278–2324 (1998)
18. LeCun, Y., Haffner, P., Bottou, L., Bengio, Y.: Object recognition with gradient-based learning. Shape, Contour and Grouping in Computer Vision. LNCS, vol. 1681, pp. 319–345. Springer, Heidelberg (1999). https://doi.org/10.1007/3-540-46805-6_19

19. Lin, J., Kolcz, A.: Large-scale machine learning at twitter. In: Proceedings of the 2012 ACM SIGMOD International Conference on Management of Data, pp. 793–804. ACM (2012)
20. Liu, S., Yamada, M., Collier, N., Sugiyama, M.: Change-point detection in time-series data by relative density-ratio estimation. In: Gimel'farb, G., et al. (eds.) Structural, Syntactic, and Statistical Pattern Recognition, pp. 363–372 (2012)
21. Makridakis, S., Spiliotis, E., Assimakopoulos, V.: The M4 competition: results, findings, conclusion and way forward. Int. J. Forecast. **34**(4), 802–808 (2018)
22. Pedregosa, F., et al.: Scikit-learn: machine learning in Python. J. Mach. Learn. Res. **12**, 2825–2830 (2011)
23. Rasmussen, C.E., Williams, C.K.I.: Gaussian Processes for Machine Learning. MIT Press, Cambridge (2006)
24. Ré, C., Niu, F., Gudipati, P., Srisuwananukorn, C.: Overton: a data system for monitoring and improving machine-learned products (2019)
25. Saadallah, A., Moreira-Matias, L., Sousa, R., Khiari, J., Jenelius, E., Gama, J.: BRIGHT—drift-aware demand predictions for taxi networks. IEEE Trans. Knowl. Data Eng. **32**(2), 234–245 (2020)
26. Saadallah, A., Priebe, F., Morik, K.: A drift-based dynamic ensemble members selection using clustering for time series forecasting. In: Brefeld, U., Fromont, E., Hotho, A., Knobbe, A., Maathuis, M., Robardet, C. (eds.) ECML PKDD 2019. LNCS (LNAI), vol. 11906, pp. 678–694. Springer, Cham (2020). https://doi.org/10.1007/978-3-030-46150-8_40
27. Sculley, D., et al.: Hidden technical debt in machine learning systems. Adv. Neural Inf. Process. Syst. **28**, 2503–2511 (2015)
28. Talagala, T.S., Hyndman, R.J., Athanasopoulos, G.: Meta-learning how to forecast time series. Monash Econometrics and Business Statistics Working Papers 6/18, Monash University, Department of Econometrics and Business Statistics (2018)
29. Titsias, M.: Variational learning of inducing variables in sparse Gaussian processes. In: Artificial Intelligence and Statistics, pp. 567–574 (2009)
30. Wagenmakers, E.J., Grünwald, P., Steyvers, M.: Accumulative prediction error and the selection of time series models. J. Math. Psychol. **50**(2), 149–166 (2006)
31. Wilcoxon, F.: Individual comparisons by ranking methods. Biometr. Bull. **1**(6), 80–83 (1945)
32. Wohlfarth, T., Clemencon, S., Roueff, F., Casellato, X.: A data-mining approach to travel price forecasting. In: 2011 10th International Conference on Machine Learning and Applications and Workshops (2011)

Learning to Simulate on Sparse Trajectory Data

Hua Wei[1(✉)], Chacha Chen[1], Chang Liu[2], Guanjie Zheng[1], and Zhenhui Li[1]

[1] Pennsylvania State University, University Park, PA 16802, USA
{hzw77,gjz5038,jessieli}@ist.psu.edu,
cjc6647@psu.edu
[2] Shanghai Jiao Tong University, Shanghai, China
only-changer@sjtu.edu.cn

Abstract. Simulation of the real-world traffic can be used to help validate the transportation policies. A good simulator means the simulated traffic is similar to real-world traffic, which often requires dense traffic trajectories (i.e., with high sampling rate) to cover dynamic situations in the real world. However, in most cases, the real-world trajectories are sparse, which makes simulation challenging. In this paper, we present a novel framework *ImIn-GAIL* to address the problem of learning to simulate the driving behavior from sparse real-world data. The proposed architecture incorporates data interpolation with the behavior learning process of imitation learning. To the best of our knowledge, we are the first to tackle the data sparsity issue for behavior learning problems. We investigate our framework on both synthetic and real-world trajectory datasets of driving vehicles, showing that our method outperforms various baselines and state-of-the-art methods.

Keywords: Imitation learning · Data sparsity · Interpolation

1 Introduction

Simulation of the real world is one of the feasible ways to verify driving policies on autonomous vehicles and transportation policies like traffic signal control [22,23,25] or speed limit setting [27] since it is costly to validate them in the real world directly [24]. The driving behavior model, i.e., how the vehicle accelerates/decelerates, is the critical component that affects the similarity of the simulated traffic to the real-world traffic [7,9,14]. Traditional methods to learn the driving behavior model usually first assumes that the behavior of the vehicle is only influenced by a small number of factors with predefined rule-based relations, and then calibrates the model by finding the parameters that best fit the observed data [5,16]. The problem with such methods is that their assumptions oversimplify the driving behavior, resulting in the simulated driving behavior far from the real world.

In contrast, imitation learning (IL) does not assume the underlying form of the driving behavior model and directly learns from the observed data (also

© Springer Nature Switzerland AG 2021
Y. Dong et al. (Eds.): ECML PKDD 2020, LNAI 12460, pp. 530–545, 2021.
https://doi.org/10.1007/978-3-030-67667-4_32

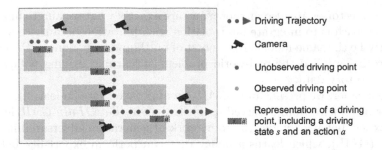

Fig. 1. Illustration of a driving trajectory. In the real-world scenario, only part of the driving points can be observed and form a sparse driving trajectory (in red dots). Each driving point includes a driving state and an action of the vehicle at the observed time step. Best viewed in color.

called demonstrations from expert policy in IL literature). With IL, a more sophisticated driving behavior policy can be represented by a parametcrized model like neural nets and provides a promising way to learn the models that behave similarly to expert policy. Existing IL methods (e.g., behavior cloning [13,21] and generative adversarial imitation learning [3,4,18,30]) for learning driving behavior relies on a large amount of behavior trajectory data that consists of dense vehicle driving states, either from vehicles installed with sensors, or roadside cameras that capture the whole traffic situation (including every vehicle driving behavior at every moment) in the road network.

However, in most real-world cases, the available behavior trajectory data is sparse, i.e., the driving behavior of the vehicles at every moment is difficult to observe. It is infeasible to install sensors for every vehicle in the road network or to install cameras that cover every location in the road network to capture the whole traffic situation. Most real-world cases are that only a minimal number of cars on the road are accessible with dense trajectory, and the driving behavior of vehicles can only be captured when the vehicles drive near the locations where the cameras are installed. For example, in Fig. 1, as the cameras are installed only around certain intersections, consecutive observed points of the same car may have a large time difference, resulting in a sparse driving trajectory. As data sparsity is considered as a critical issue for unsatisfactory accuracy in machine learning, directly using sparse trajectories to learn the driving behavior could make the model fail to learn the behavior policy at the unobserved states.

To deal with sparse trajectories, a typical approach is to interpolate the sparse trajectories first and then learn the model with the dense trajectories [10,28,31]. This two-step approach also has an obvious weakness, especially in the problem of learning behavior models. For example, linear interpolation is often used to interpolate the missing points between two observed trajectory points. But in real-world cases, considering the interactions between vehicles, the vehicle is unlikely to drive at a uniform speed during that unobserved time period, hence the interpolated trajectories may be different from the true trajectories. However, the true trajectories are also unknown and are exactly what we aim to imitate. A better approach is to integrate interpolation with imitation because they should inherently be the same model. To the best of our knowledge, none of the existing literature has studied the real-world problem of learning driving policies from sparse trajectory data.

In this paper, we present *ImIn-GAIL*, an approach that can learn the driving behavior of vehicles from observed sparse trajectory data. *ImIn-GAIL* learns to mimic expert behavior under the framework of generative adversarial imitation learning (GAIL), which learns a policy that can perform expert-like behaviors through rewarding the policy for deceiving a discriminator trained to classify between policy-generated and expert trajectories. Specifically, for the data sparsity issue, we present an interpolator-discriminator network that can perform both the interpolation and discrimination tasks, and a downsampler that draws supervision on the interpolation task from the trajectories generated by the learned policy. We conduct experiments on both synthetic and real-world data, showing that our method can not only have excellent imitation performance on the sparse trajectories but also have better interpolation results compared with state-of-the-art baselines. The main contributions of this paper are summarized as follows:

- We propose a novel framework *ImIn-GAIL*, which can learn driving behaviors from the real-world sparse trajectory data.
- We naturally integrate the interpolation with imitation learning that can interpolate the sparse driving trajectory.
- We conduct experiments on both real and synthetic data, showing that our approach significantly outperforms existing methods. We also have interesting cases to illustrate the effectiveness on the imitation and interpolation of our methods.

2   Preliminaries

Definition 1 (Driving Point). *A driving point $\tau^t = (s^t, a^t, t)$ describes the driving behavior of the vehicle at time t, which consists of a driving state s^t and an action a^t of the vehicle. Typically, the state s^t describes the surrounding traffic conditions of the vehicle (e.g., speed of the vehicle and distance to the preceding vehicle), and the action $a^t \sim \pi(a|s^t)$ the vehicle takes at time t is the magnitude of acceleration/deceleration following its driving policy $\pi(a|s^t)$.*

Definition 2 (Driving Trajectory). *A driving trajectory of a vehicle is a sequence of driving points generated by the vehicle in geographical spaces, usually represented by a series of chronologically ordered points, e.g. $\tau = (\tau^{t_0}, \cdots, \tau^{t_N})$.*

In trajectory data mining [11,12,32], a *dense trajectory* of a vehicle is the driving trajectory with high-sampling rate (e.g., one point per second on average), and a *sparse trajectory* of a vehicle is the driving trajectory with low-sampling rate (e.g., one point every 2 min on average). In this paper, the observed driving trajectory is a sequence of driving points with large and irregular intervals between their observation times.

Problem 3. In our problem, a vehicle observes state s from the environment, take action a following policy π^E at every time interval Δt, and generate a raw driving trajectory τ during certain time period. While the raw driving trajectory is dense (i.e., at a high-sampling rate), in our problem we can only observe a set of sparse trajectories $\mathcal{T}_E$ generated by expert policy π^E as expert trajectory, where $\mathcal{T}_E = \{\tau_i | \tau_i = (\tau_i^{t_0}, \cdots, \tau_i^{t_N})\}$, $t_{i+1} - t_i \gg \Delta t$ and $t_{i+1} - t_i$ may be different for different observation time i. Our goal is to learn a parameterized policy π_θ that imitates the expert policy π^E.

3   Method

In this section, we first introduce the basic imitation framework, upon which we propose our method (*ImIn-GAIL*) that integrates trajectory interpolation into the basic model.

3.1   Basic GAIL Framework

In this paper, we follow the framework similar to GAIL [4] due to its scalability to the multi-agent scenario and previous success in learning human driver models [8]. GAIL formulates imitation learning as the problem of learning policy to perform expert-like behavior by rewarding it for "deceiving" a classifier trained to discriminate between policy-generated and expert state-action pairs. For a neural network classifier $\mathcal{D}_\psi$ parameterized by ψ, the GAIL objective is given by $max_\psi min_\theta \mathcal{L}(\psi, \theta)$ where $\mathcal{L}(\psi, \theta)$ is:

$$\mathcal{L}(\psi, \theta) = \mathbb{E}_{(s,a)\sim\tau\in\mathcal{T}_E} \log \mathcal{D}_\psi(s, a) + \mathbb{E}_{(s,a)\sim\tau\in\mathcal{T}_G} \log(1 - \mathcal{D}_\psi(s, a)) - \beta H(\pi_\theta)$$
$$(1)$$

where $\mathcal{T}_E$ and $\mathcal{T}_G$ are respectively the expert trajectories and the generated trajectories from the interactions of policy π_θ with the simulation environment, $H(\pi_\theta)$ is an entropy regularization term.

- *Learning ψ*: When training $\mathcal{D}_\psi$, Eq. (1) can simply be set as a sigmoid cross entropy where positive samples are from $\mathcal{T}_E$ and negative samples are from $\mathcal{T}_G$. Then optimizing ψ can be easily done with gradient ascent.
- *Learning θ*: The simulator is an integration of physical rules, control policies and randomness and thus its parameterization is assumed to be unknown. Therefore, given $\mathcal{T}_G$ generated by π_θ in the simulator, Eq. (1) is non-differentiable w.r.t θ. In order to learn π_θ, GAIL optimizes through reinforcement learning, with a surrogate reward function formulated from Eq. (1) as:

$$\tilde{r}(s^t, a^t; \psi) = -\log(1 - \mathcal{D}_\psi(s^t, a^t))$$
$$(2)$$

Here, $\tilde{r}(s^t, a^t; \psi)$ can be perceived to be useful in driving π_θ into regions of the state-action space at time t similar to those explored by π^E. Intuitively, when

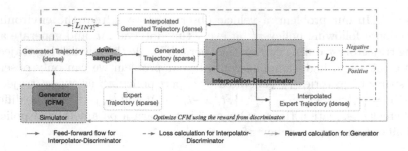

Fig. 2. Proposed *ImIn-GAIL* Approach. The overall framework of *ImIn-GAIL* includes three components: generator, downsampler, and interpolation-discriminator. Best viewed in color.

the observed trajectory is dense, the surrogate reward from the discriminator in Eq. (2) is helpful to learn the state transitions about observed trajectories. However, when the observed data is sparse, the reward from discriminator will only learn to correct the observed states and fail to model the behavior policy at the unobserved states. To relieve this problem, we propose to interpolate the sparse expert trajectory within the based imitation framework.

3.2 Imitation with Interpolation

An overview of our proposed Imitation-Interpolation framework (*ImIn-GAIL*) is shown in Fig. 2, which consists of the following three key components.

Generator in the Simulator. Given an initialized driving policy π_θ, the dense trajectories $\mathcal{T}_G^D$ of vehicles can be generated in the simulator. In this paper, the driving policy π_θ is parameterized by a neural network which will output an action a based on the state s it observes. The simulator can generate driving behavior trajectories by rolling out π_θ for all vehicles simultaneously in the simulator. The optimization of the driving policy is optimized via TRPO [17] as in vanilla GAIL [4].

Downsampling of Generated Trajectories. The goal of the downsampler is to construct the training data for interpolation, i.e., learning the mapping from a sparse trajectory to a dense one. For two consecutive points (i.e., τ^{t_s} and τ^{t_e} in generated sparse trajectory $\mathcal{T}_G$), we can sample a point τ^{t_i} in $\mathcal{T}_G^D$ where $t_s \leq t_i \leq t_e$ and construct training samples for the interpolator. The sampling strategies can be sampling at certain time intervals, sampling at specific locations or random sampling and we investigate the influence of different sampling rates in Sect. 4.5.

Interpolation-Discriminator. The key difference between *ImIn-GAIL* and vanilla GAIL is in the discriminator. While learning to differentiate the expert trajectories from generated trajectories, the discriminator in *ImIn-GAIL* also learns to interpolate a sparse trajectory to a dense trajectory. Specifically, as is shown in Fig. 3, the proposed interpolation-discriminator copes with two sub-tasks in an end-to-end way: *interpolation* on sparse data and *discrimination* on dense data.

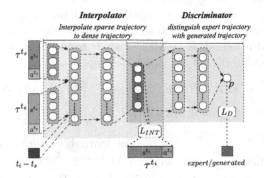

Fig. 3. Proposed interpolation-discriminator network.

Interpolator Module. The goal of the interpolator is to interpolate the sparse expert trajectories $\mathcal{T}_E$ to the dense trajectories $\mathcal{T}_E^D$. We can use the generated dense trajectories $\mathcal{T}_G^D$ and sparse trajectories $\mathcal{T}_G$ from previous downsampling process as training data for the interpolator.

For each point τ^{t_i} to be interpolated, we first concatenate state and action and embed them into an m-dimensional latent space:

$$h_s = \sigma(Concat(s^{t_s}, a^{t_s})W_s + b_s), h_e = \sigma(Concat(s^{t_e}, a^{t_e})W_e + b_e) \quad (3)$$

where K is the feature dimension after the concatenation of s^{t_e} and a^{t_e}, $W_s \in \mathbb{R}^{K \times M}$, $W_e \in \mathbb{R}^{K \times M}$, $b_s \in \mathbb{R}^M$ and $b_e \in \mathbb{R}^M$ are weight matrix to learn, σ is ReLU function (same denotation for the following σ). Here, considering t_s and t_e may have different effects on interpolation, we use two different embedding weights for t_s and t_e.

After point embedding, we concatenate h_s and h_e with the time interval between t_s and t_i, and use a multi-layer perception (MLP) with L layers to learn the interpolation.

$$\begin{aligned}
h_{in} &= \sigma(Concat(h_s, h_e, t_i - t_s)W_0 + b_0) \\
h_1 &= \sigma(h_{in}W_1 + b_1), h_2 = \sigma(h_1 W_2 + b_2), \cdots \\
h_L &= tanh(h_{L-1}W_L + b_L) = \hat{\tau}^{t_i}
\end{aligned} \quad (4)$$

where $W_0 \in \mathbb{R}^{(2M+1) \times N_0}$, $b_0 \in \mathbb{R}^{N_0}$ are the learnable weights; $W_j \in \mathbb{R}^{N_j \times N_{j+1}}$ and $b_j \in \mathbb{R}^{N_{j+1}}$ are the weight matrix for hidden layers ($1 \leq j \leq L-1$) of interpolator; $W_L \in \mathbb{R}^{N_j \times K}$ and $b_L \in \mathbb{R}^K$ are the weight matrix for the last layer of interpolator, which outputs an interpolated point $\hat{\tau}^{t_i}$. In the last layer of interpolator, we use $tanh$ as all the feature value of τ^{t_i} is normalized to $[-1, 1]$.

Discriminator Module. When sparse expert trajectories $\mathcal{T}_E$ are interpolated into dense trajectories $\mathcal{T}_E^D$ by the interpolator, the discriminator module learns to differentiate between expert dense trajectories $\mathcal{T}_E^D$ and generated dense trajectories $\mathcal{T}_D^D$. Specifically, the discriminator learns to output a high score when encountering an interpolated point $\hat{\tau}^{ti}$ originated from $\mathcal{T}_E^D$, and a low score when encountering a point from $\mathcal{T}_G^D$ generated by π_θ. The output of the discriminator $\mathcal{D}_\psi(s, a)$ can then be used as a surrogate reward function whose value grows larger as actions sampled from π_θ look similar to those chosen by experts.

The discriminator module is an MLP with H hidden layers, takes h_L as input and outputs the probability of the point belongs to $\mathcal{T}_E$.

$$h_1^D = \sigma(h_L W_1^D + b_1^D), h_2^D = \sigma(h_1^D W_2^D + b_2^D), \cdots$$
$$p = Sigmoid(h_{H-1}^D W_H^D + b_H^D) \tag{5}$$

where $W_i^D \in \mathbb{R}^{N_{i-1}^D \times N_i^D}$, $b_i^D \in \mathbb{R}^{N_i^D}$ are learnable weights for i-th layer in discriminator module. For $i = 1$, we have $W_1^D \in \mathbb{R}^{K \times N_1^D}$, $b_1^D \in \mathbb{R}^{N_1^D}$, K is the concatenated dimension of state and action; for $i = H$, we have $W_H^D \in \mathbb{R}^{N_{H-1}^D \times 1}$, $b_H^D \in \mathbb{R}$.

Loss Function of Interpolation-Discriminator. The loss function of the Interpolation-Discriminator network is a combination of interpolation loss $\mathcal{L}_{INT}$ and discrimination loss $\mathcal{L}_D$, which interpolates the unobserved points and predicts the probability of the point being generated by expert policy π^E simultaneously,:

$$\mathcal{L} = \lambda \mathcal{L}_{INT} + (1 - \lambda)\mathcal{L}_D = \lambda \mathbb{E}_{\tau^t \sim \tau \in \mathcal{T}_G^D}(\hat{\tau}^t - \tau^t) +$$
$$(1 - \lambda)[\mathbb{E}_{\tau^t \sim \tau \in \mathcal{T}_G} \log p(\tau^t) + \mathbb{E}_{\tau^t \sim \tau \in \mathcal{T}_E} \log(1 - p(\tau^t))] \tag{6}$$

where λ is a hyper-parameter to balance the influence of interpolation and discrimination, $\hat{\tau}^t$ is the output of the interpolator module, and $p(\tau)$ is the output probability from the discriminator module.

3.3   Training and Implementation

Algorithm 1 describes the *ImIn-GAIL* approach. In this paper, the driving policy is parameterized with a two-layer fully connected network with 32 units for all the hidden layers. The policy network takes the driving state s as input and outputs the distribution parameters for a Beta distribution, and the action a will be sampled from this distribution. The optimization of the driving policy is optimized via TRPO [17]. Following [3,8], we use the features in Table 1 to represent the driving state of a vehicle, and the driving policy takes the drivings state as input and outputs an action a (i.e., next step speed). For the interpolation-discriminator network, each driving point is embedded to a 10-dimensional latent space, the interpolator module uses a three-layer fully connected layer to interpolate the trajectory and the discriminator module contains a two-layer fully connected layer. Some of the important hyperparameters are listed in Table 2.

Algorithm 1: Training procedure of *ImIn-GAIL*

Input: Sparse expert trajectories $\mathcal{T}_E$, initial policy and
    interpolation-discriminator parameters θ_0, ψ_0
Output: Policy π_θ, interpolation-discriminator $InDNet_\psi$

1 **for** $i \longleftarrow 0, 1, \ldots$ **do**
2 | Rollout dense trajectories for all agents
      $\mathcal{T}_G^D = \{\tau | \tau = (\tau^{t_0}, \cdots, \tau^{t_N}), \ \tau^{t_j} = (s^{t_j}, a^{t_j}) \sim \pi_{\theta_i}\}$;
3 | (Generator update step)
4 | • Score τ^{t_j} from $\mathcal{T}_G^D$ with discriminator, generating reward using Eq. 2;
5 | • Update θ in generator given $\mathcal{T}_G^D$ by optimizing Eq. 1;
6 | (Interpolator-discriminator update step)
7 | • Interpolate $\mathcal{T}_E$ with the interpolation module in *InDNet*, generating dense
      expert trajectories $\mathcal{T}_E^D$;
8 | • Downsample generated dense trajectories $\mathcal{T}_G^D$ to sparse trajectories $\mathcal{T}_G$;
9 | • Construct training samples for *InDNet*
10 | • Update *InDNet* parameters ψ by optimizing Eq. 6

Table 1. Features for a driving state

Feature type	Detail features
Road network	Lane ID, length of current lane, speed limit
Traffic signal	Current phase of traffic signal
Ego vehicle	Velocity, position in current lane, distance to the next traffic signal
Leading vehicle	Relative distance, velocity and position in the current lane
Indicators	Leading in current lane, exiting from intersection

4 Experiment

4.1 Experimental Settings

We conduct experiments on CityFlow [29], an open-source traffic simulator that supports large-scale vehicle movements. In a traffic dataset, each vehicle is described as (o, t, d, r), where o is the origin location, t is time, d is the destination location and r is its route from o to d. Locations o and d are both locations on the road network, and r is a sequence of road ID. After the traffic data is fed into the simulator, a vehicle moves towards its destination based on its route. The simulator provides the state to the vehicle control method and executes the vehicle acceleration/deceleration actions from the control method.

Dataset. In experiment, we use both synthetic data and real-world data.

Synthetic Data. In the experiment, we use two kinds of synthetic data, i.e., traffic movements under ring road network and intersection road network, as shown in Fig. 4. Based on the traffic data, we use default simulation settings

Table 2. Hyper-parameter settings for *ImIn-GAIL*

Parameter	Value	Parameter	Value
Batch size for generator	64	Batch size for *InDNet*	32
Update epoches for generator	5	Update epoches for *InDNet*	10
Learning rate for generator	0.001	Learning rate for *InDNet*	0.0001
Number of layers in generator	4	Balancing factor λ	0.5

of the simulator to generate dense expert trajectories and sample sparse expert trajectories when vehicles pass through the red dots.

- *Ring*: The ring road network consists of a circular lane with a specified length, similar to [19,26]. This is a very ideal and simplified scenario where the driving behavior can be measured.
- *Intersection*: A single intersection network with bi-directional traffic. The intersection has four directions (West→East, East→West, South→North, and North→South), and 3 lanes (300 m in length and 3 m in width) for each direction. Vehicles come uniformly with 300 vehicles/lane/hour in West↔East direction and 90 vehicles/lane/hour in South↔North direction.

Real-world Data We also use real-world traffic data from two cities: Hangzhou and Los Angeles. Their road networks are imported from OpenStreetMap[1], as shown in Fig. 4. The detailed descriptions of how we preprocess these datasets are as follows (Table 3):

- $LA_{1\times4}$. This is a public traffic dataset collected from Lankershim Boulevard, Los Angeles on June 16, 2005. It covers an 1×4 arterial with four successive intersections. This dataset records the position and speed of every vehicle at every 0.1 s. We treat these records as dense expert trajectories and sample vehicles' states and actions when they pass through intersections as sparse expert trajectories.
- $HZ_{4\times4}$. This dataset covers a 4×4 network of Gudang area in Hangzhou, collected from surveillance cameras near intersections in 2016. This region has relatively dense surveillance cameras and we sampled the sparse expert trajectories in a similar way as in $LA_{1\times4}$.

Data Preprocessing. To mimic the real-world situation where the roadside surveillance cameras capture the driving behavior of vehicles at certain locations, the original dense expert trajectories are processed to sparse trajectories by sampling the driving points near several fixed locations unless specified. We use the sparse trajectories as expert demonstrations for training models. To test the imitation effectiveness, we use the same sampling method as the expert data and then compare the sparse generated data with sparse expert data. To test

---

[1] https://www.openstreetmap.org.

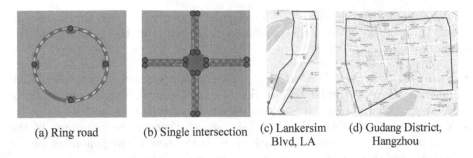

(a) Ring road   (b) Single intersection   (c) Lankersim   (d) Gudang District,
                                                       Blvd, LA   Hangzhou

Fig. 4. Illustration of road networks. (a) and (b) are synthetic road networks, while (c) and (d) are real-world road networks.

Table 3. Statistics of dense and sparse expert trajectory in different datasets

	Ring	Intersection	$LA_{1\times4}$	$HZ_{4\times4}$
Duration (seconds)	300	300	300	300
# of vehicles	22	109	321	425
# of points (dense)	1996	10960	23009	87739
# of points (sparse)	40	283	1014	1481

the interpolation effectiveness, we directly compare the dense generated data with dense expert data.

4.2   Compared Methods

We compare our model with the following two categories of methods: calibration-based methods and imitation learning-based methods.

Calibration-Based Methods. For calibration-based methods, we use Krauss model [7], the default car-following model (CFM) of simulator SUMO [6] and CityFlow [29]. Krauss model has the following forms:

$$v_{safe}(t) = v_l(t) + \frac{g(t) - v_l(t)t_r}{\frac{v_l(t)+v_f(t)}{2b} + t_r} \tag{7}$$

$$v_{des}(t) = \min[v_{safe}(t), v(t) + a\Delta t, v_{max}] \tag{8}$$

where $v_{safe}(t)$ the safe speed at time t, $v_l(t)$ and $v_f(t)$ is the speed of the leading vehicle and following vehicle respectively at time t, $g(t)$ is the gap to the leading vehicle, b is the maximum deceleration of the vehicle and t_r is the driver's reaction time. $v_{des}(t)$ is the desired speed, which is given by the minimum of safe speed, maximum allowed speed, and the speed after accelerating at a for Δt. Here, a is the maximum acceleration and Δt is the simulation time step.

We calibrate three parameters in Krauss model, which are the maximum deceleration of the vehicle, the maximum acceleration of the vehicle, and the maximum allowed speed.

- **Random Search (*CFM-RS*)** [2]: The parameters are chosen when they generate the most similar trajectories to expert demonstrations after a finite number of trial of random selecting parameters for Krauss model.
- **Tabu Search (*CFM-TS*)** [16]: Tabu search chooses the neighbors of the current set of parameters for each trial. If the new CFM generates better trajectories, this set of parameters is kept in the Tabu list.

Imitation Learning-Based Methods. We also compare with several imitation learning-based methods, including both traditional and state-of-the-art methods.

- **Behavioral Cloning (*BC*)** [21] is a traditional imitation learning method. It directly learns the state-action mapping in a supervised manner.
- **Generative Adversarial Imitation Learning (*GAIL*)** is a GAN-like framework [4], with a generator controlling the policy of the agent, and a discriminator containing a classifier for the agent indicating how far the generated state sequences are from that of the demonstrations.

4.3   Evaluation Metrics

Following existing studies [3, 8, 30], to measure the error between learned policy against expert policy, we measure the position and the travel time of vehicles between generated dense trajectories and expert dense trajectories, which are defined as:

$$RMSE_{pos} = \frac{1}{T}\sum_{t=1}^{T}\sqrt{\frac{1}{M}\sum_{i=1}^{m}(l_i^t - \hat{l}_i^t)^2}, \ RMSE_{time} = \sqrt{\frac{1}{M}\sum_{i=1}^{m}(d_i - \hat{d}_i)^2} \quad (9)$$

where T is the total simulation time, M is the total number of vehicles, l_i^t and $\hat{l}_i^t$ are the position of i-th vehicle at time t in the expert trajectories and in the generated trajectories relatively, d_i and $\hat{d}_i$ are the travel time of vehicle i in expert trajectories and generated trajectories respectively.

4.4   Performance Comparison

In this section, we compare the dense trajectories generated by different methods with the expert dense trajectories, to see how similar they are to the expert policy. The closer the generated trajectories are to the expert trajectories, the more similar the learned policy is to the expert policy. From Table 4, we can see that *ImIn-GAIL* achieves consistently outperforms over all other baselines across synthetic and real-world data. *CFM-RS* and *CFM-RS* can hardly achieve

Table 4. Performance w.r.t Relative Mean Squared Error (RMSE) of time (in seconds) and position (in kilometers). All the measurements are conducted on dense trajectories. Lower the better. Our proposed method *ImIn-GAIL* achieves the best performance.

	Ring		Intersection		$LA_{1\times4}$		$HZ_{4\times4}$	
	time (s)	pos (km)	time (s)	pos (km)	time (s)	pos (km)	time (s)	pos (km)
CFM-RS	343.506	0.028	39.750	0.144	34.617	0.593	27.142	0.318
CFM-TS	376.593	0.025	95.330	0.184	33.298	0.510	175.326	0.359
BC	201.273	0.020	58.580	0.342	55.251	0.698	148.629	0.297
GAIL	42.061	0.023	14.405	0.032	30.475	0.445	14.973	0.196
ImIn-GAIL	**16.970**	**0.018**	**4.550**	**0.024**	**19.671**	**0.405**	**5.254**	**0.130**

satisfactory results because the model predefined by CFM could be different from the real world. Specifically, *ImIn-GAIL* outperforms vanilla *GAIL*, since *ImIn-GAIL* interpolates the sparse trajectories and thus has more expert trajectory data, which will help the discriminator make more precise estimations to correct the learning of policy.

4.5   Study of *ImIn-GAIL*

Interpolation Study. To better understand how interpolation helps in simulation, we compare two representative baselines with their two-step variants. Firstly, we use a pre-trained non-linear interpolation model to interpolate the sparse expert trajectories following the idea of [20,28]. Then we train the baselines on the interpolated trajectories.

Table 5 shows the performance of baseline methods in *Ring* and *Intersection*. We find out that baseline methods in a two-step way show inferior performance. One possible reason is that the interpolated trajectories generated by the pre-trained model could be far from the real expert trajectories when interacting in the simulator. Consequently, the learned policy trained on such interpolated trajectories makes further errors.

In contrast, *ImIn-GAIL* learns to interpolate and imitate the sparse expert trajectories in one step, combining the interpolator loss and discriminator loss, which can propagate across the whole framework. If the trajectories generated by π_θ is far from expert observations in current iteration, both the discriminator and the interpolator will learn to correct themselves and provide more precise reward for learning π_θ in the next iteration. Similar results can also be found in $LA_{1\times4}$ and $HZ_{4\times4}$, and we omit these results due to page limits.

Table 5. RMSE on time and position of our proposed method *ImIn-GAIL* against baseline methods and their corresponding two-step variants. Baseline methods and *ImIn-GAIL* learn from sparse trajectories, while the two-step variants interpolate sparse trajectories first and trained on the interpolated data. *ImIn-GAIL* achieves the best performance in most cases.

	Ring		Intersection	
	time (s)	position (km)	time (s)	position (km)
CFM-RS	343.506	0.028	39.750	0.144
CFM-RS (two step)	343.523	0.074	73.791	0.223
GAIL	42.061	0.023	14.405	0.032
GAIL (two step)	98.184	0.025	173.538	0.499
ImIn-GAIL	**16.970**	**0.018**	**4.550**	**0.024**

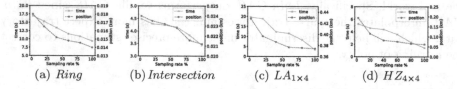

(a) *Ring*   (b) *Intersection*   (c) $LA_{1 \times 4}$ (d) $HZ_{4 \times 4}$

Fig. 5. RMSE on time and position of our proposed method *ImIn-GAIL* under different level of sparsity. As the expert trajectory become denser, a more similar policy to the expert policy is learned.

Sparsity Study. In this section, we investigate how different sampling strategies influence *ImIn-GAIL*. We sample randomly from the dense expert trajectories at different time intervals to get different sampling rates: 2%, 20%, 40%, 60%, 80%, and 100%. We set the sampled data as the expert trajectories and evaluate by measuring the performance of our model in imitating the expert policy. As is shown in Fig. 5, with denser expert trajectory, the error of *ImIn-GAIL* decreases, indicating a better policy imitated by our method.

4.6   Case Study

To study the capability of our proposed method in recovering the dense trajectories of vehicles, we showcase the movement of a vehicle in *Ring* data learned by different methods.

We visualize the trajectories generated by the policies learned with different methods in Fig. 6. We find that imitation learning methods (*BC, GAIL,* and *ImIn-GAIL*) perform better than calibration-based methods (*CFM-RS* and *CFM-TS*). This is because the calibration based methods pre-assumes an existing model, which could be far from the real behavior model. On the contrast, imitation learning methods directly learn the policy without making unrealistic formulations of the CFM model. Specifically, *ImIn-GAIL* can imitate the position of the expert trajectory more accurately than all other baseline methods.

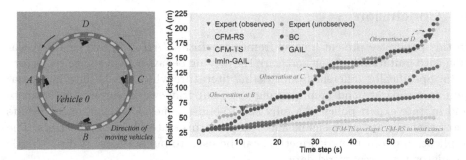

Fig. 6. The generated trajectory of a vehicle in the *Ring* scenario. Left: the initial position of the vehicles. Vehicles can only be observed when they pass four locations *A*, *B*, *C* and *D* where cameras are installed. Right: the visualization for the trajectory of *Vehicle* 0. The x-axis is the timesteps in seconds. The y-axis is the relative road distance in meters. Although vehicle 0 is only observed three times (red triangles), *ImIn-GAIL* (blue points) can imitate the position of the expert trajectory (grey points) more accurately than all other baselines. Better viewed in color.

The reason behind the improvement of *ImIn-GAIL* against other methods is that in *ImIn-GAIL*, policy learning and interpolation can enhance each other and result in significantly better results.

5   Related Work

Parameter Calibration. In parameter calibration-based methods, the driving behavior model is a prerequisite, and parameters in the model are tuned to minimize a pre-defined cost function. Heuristic search algorithms such as random search, Tabu search [16], and genetic algorithm [5] can be used to search the parameters. These methods rely on the pre-defined models (mostly equations) and usually fail to match the dynamic vehicle driving pattern in the real-world.

Imitation Learning. Without assuming an underlying physical model, we can solve this problem via imitation learning. There are two main lines of work: (1) behavior cloning (BC) and Inverse reinforcement learning (IRL). BC learns the mapping from demonstrated observations to actions in a supervised learning way [13,21], but suffers from the errors which are generated from unobserved states during the simulation. On the contrast, IRL not only imitates observed states but also learns the expert's underlying reward function, which is more robust to the errors from unobserved states [1,15,33]. Recently, a more effective IRL approach, GAIL [4], incorporates generative adversarial networks with learning the reward function of the agent. However, all of the current work did not address the challenges of sparse trajectories, mainly because in their application contexts, e.g., game or robotic control, observations can be fully recorded every time step.

6   Conclusion

In this paper, we present a novel framework *ImIn-GAIL* to integrate interpolation with imitation learning and learn the driving behavior from sparse trajectory data. Specifically, different from existing literature which treats data interpolation as a separate and preprocessing step, our framework learns to interpolate and imitate expert policy in a fully end-to-end manner. Our experiment results show that our approach significantly outperforms state-of-the-art methods. The application of our proposed method can be used to build a more realistic traffic simulator using real-world data.

Acknowledgments. The work was supported in part by NSF awards #1652525 and #1618448. The views and conclusions contained in this paper are those of the authors and should not be interpreted as representing any funding agencies.

References

1. Abbeel, P., Ng, A.Y.: Apprenticeship learning via inverse reinforcement learning. In: ICML (2004)
2. Asamer, J., van Zuylen, H.J., Heilmann, B.: Calibrating car-following parameters for snowy road conditions in the microscopic traffic simulator VISSIM. IET Intell. Transp. Syst. **7**(1), 114–121 (2013)
3. Bhattacharyya, R.P., Phillips, D.J., Wulfe, B., Morton, J., Kuefler, A., Kochenderfer, M.J.: Multi-agent imitation learning for driving simulation. In: IEEE/RSJ International Conference on Intelligent Robots and Systems (IROS). IEEE (2018)
4. Ho, J., Ermon, S.: Generative adversarial imitation learning. In: NeurIPS (2016)
5. Kesting, A., Treiber, M.: Calibrating car-following models by using trajectory data: methodological study. Transp. Res. Rec. **2088**(1), 148–156 (2008)
6. Krajzewicz, D., Erdmann, J., Behrisch, M., Bieker, L.: Recent development and applications of SUMO - Simulation of Urban MObility. Int. J. Adv. Syst. Meas. **5**(3&4), 128–138 (2012)
7. Krauss, S.: Microscopic modeling of traffic flow: investigation of collision free vehicle dynamics. Ph.D. thesis (1998)
8. Kuefler, A., Morton, J., Wheeler, T., Kochenderfer, M.: Imitating driver behavior with generative adversarial networks. In: IEEE Intelligent Vehicles Symposium (IV). IEEE (2017)
9. Leutzbach, W., Wiedemann, R.: Development and applications of traffic simulation models at the Karlsruhe Institut fur Verkehrwesen. Traffic Eng. Control **27**(5), 270–278 (1986)
10. Li, S.C.X., Marlin, B.M.: A scalable end-to-end gaussian process adapter for irregularly sampled time series classification. In: NeurIPS (2016)
11. Liu, Y., Zhao, K., Cong, G., Bao, Z.: Online anomalous trajectory detection with deep generative sequence modeling. In: ICDE (2020)
12. Lou, Y., Zhang, C., Zheng, Y., Xie, X., Wang, W., Huang, Y.: Map-matching for low-sampling-rate GPS trajectories. In: SIGSPATIAL. ACM (2009)
13. Michie, D., Bain, M., Hayes-Miches, J.: Cognitive models from subcognitive skills. IEEE Control Eng. Ser. **44** (1990)
14. Nagel, K., Schreckenberg, M.: A cellular automaton model for freeway traffic. J. de Physique I **2**(12), 2221–2229 (1992)

15. Ng, A.Y., Russell, S.J., et al.: Algorithms for inverse reinforcement learning. In: ICML (2000)
16. Osorio, C., Punzo, V.: Efficient calibration of microscopic car-following models for large-scale stochastic network simulators. Transp. Res. Part B: Methodol. **119**, 156–173 (2019)
17. Schulman, J., Levine, S., Moritz, P., Jordan, M.I., Abbeel, P.: Trust region policy optimization. In: ICML (2015)
18. Song, J., Ren, H., Sadigh, D., Ermon, S.: Multi-agent generative adversarial imitation learning. In: NeurIPS (2018)
19. Sugiyama, Y., et al.: Traffic jams without bottlenecks-experimental evidence for the physical mechanism of the formation of a jam. New J. Phys. **10**(3), 1–8 (2008)
20. Tang, X., et al.: Joint modeling of dense and incomplete trajectories for citywide traffic volume inference. In: The World Wide Web Conference. ACM (2019)
21. Torabi, F., Warnell, G., Stone, P.: Behavioral cloning from observation. In: IJCAI (2018)
22. Wei, H., et al.: PressLight: learning max pressure control to coordinate traffic signals in arterial network. In: KDD (2019)
23. Wei, H., et al.: CoLight: learning network-level cooperation for traffic signal control. In: CIKM (2019)
24. Wei, H., Zheng, G., Gayah, V., Li, Z.: A survey on traffic signal control methods. arXiv preprint arXiv:1904.08117 (2019)
25. Wei, H., Zheng, G., Yao, H., Li, Z.: IntelliLight: a reinforcement learning approach for intelligent traffic light control. In: KDD (2018)
26. Wu, C., Kreidieh, A., Vinitsky, E., Bayen, A.M.: Emergent behaviors in mixed-autonomy traffic. In: Conference on Robot Learning (2017)
27. Wu, Y., Tan, H., Ran, B.: Differential variable speed limits control for freeway recurrent bottlenecks via deep reinforcement learning. arXiv preprint arXiv:1810.10952 (2018)
28. Yi, X., Zheng, Y., Zhang, J., Li, T.: ST-MVL: filling missing values in geo-sensory time series data. In: IJCAI. AAAI Press (2016)
29. Zhang, H., et al.: CityFlow: a multi-agent reinforcement learning environment for large scale city traffic scenario. In: International World Wide Web Conference (2019)
30. Zheng, G., Liu, H., Xu, K., Li, Z.: Learning to simulate vehicle trajectories from demonstrations. In: ICDE (2020)
31. Zheng, K., Zheng, Y., Xie, X., Zhou, X.: Reducing uncertainty of low-sampling-rate trajectories. In: ICDE (2012)
32. Zheng, Y.: Trajectory data mining: an overview. ACM Trans. Intell. Syst. Technol. (TIST) **6**(3), 1–41 (2015)
33. Ziebart, B.D., Maas, A.L., Bagnell, J.A., Dey, A.K.: Maximum entropy inverse reinforcement learning. In: AAAI, vol. 8 (2008)

Learning from Crowds via Joint Probabilistic Matrix Factorization and Clustering in Latent Space

Wuguannan Yao[1](✉), Wonjung Lee[1], and Junhui Wang[2]

[1] Department of Mathematics, City University of Hong Kong, Kowloon, Hong Kong
satie.yao@my.cityu.edu.hk, lee.wonjung@cityu.edu.hk
[2] School of Data Science, City University of Hong Kong, Kowloon, Hong Kong
j.h.wang@cityu.edu.hk

Abstract. Learning from noisy labels is getting trendy in the era of big data. However, in crowdsourcing practice, it is still a challenging task to extract ground truth labels from noisy labels obtained from crowds. In this paper, we propose a latent variable model built on probabilistic logistic matrix factorization model and classical Gaussian mixture model for inferring ground truth labels from noisy, crowdsourced ones. The proposed model incorporates item heterogeneity in contrast to previous works and allows for vector space embeddings of both items and worker labels. Moreover, we derive a tractable mean-field variational inference algorithm to approximate the model posterior. Meanwhile, related MAP approximation problem to the model posterior is also investigated to identify links to existing works. Empirically, we demonstrate that the proposed method achieves good inference accuracy while preserving meaningful uncertainty measures in the embeddings, and therefore better reflects the intrinsic structure of data.

Keywords: Crowdsourcing · Label aggregation · Latent variable models · Variational inference

1 Introduction

Data quality is crucial for classification. In practice it is not an easy job to obtain gold standard labels to train accurate classifiers. A large amount of time and a big budget are required to label huge unlabeled datasets. Crowdsourcing arises as a cost-saving paradigm to collect labels as training data from hundreds and thousands of people through large-scale online platforms. However, the practice is not perfect since annotators from these platforms typically have different backgrounds and their abilities to provide accurate labels vary. The collected labels are subject to annotator-specific and item-specific noise. Common concerns in crowdsourcing practice regarding annotator heterogeneity include identification of "spammers", who submit their works without deliberate thinking and expect to be paid effortlessly [17,24], or "adversaries", who intentionally give wrong

© Springer Nature Switzerland AG 2021
Y. Dong et al. (Eds.): ECML PKDD 2020, LNAI 12460, pp. 546–561, 2021.
https://doi.org/10.1007/978-3-030-67667-4_33

answers [10]. As another largely ignored perspective, the properties of tasks can be extremely different, in the sense that some tasks may be extremely difficult or ambiguous in nature whereas some may be trivially simple.

Majority voting is a simple but effective method to combine noisy annotations, and is statistically optimal when all annotators follow the same underlying labeling distribution. However, this is not generally true and improved performance is expected by explicitly modeling annotator behaviors and task difficulties using principled probabilistic methods.

To our best knowledge, the problem is first considered in [8], in which a simple latent variable model (henceforth DS model) is proposed. The labeling accuracy of each annotator is parameterized by a "confusion matrix" and could be estimated by EM scheme. However, DS model does not account for item heterogeneity and the observed noisy labels are assumed to be generated purely based on ground truth labels. Also, DS model consumes large number of parameters ($\mathcal{O}(MK^2)$) and the estimation can be difficult when the number of annotators, M, and classes, K, become larger. This issue is typically handled by restricting the confusion matrices, e.g. the "one-coin" assumption considered in [12].

Several probabilistic models were proposed to extend vanilla DS model to incorporate item specific properties. Specific examples include GLAD [23], a model for binary labels with a single parameter accounting for difficulties, and CUBAM [22], a more generalized multivariate model for binary labels. For more works in this line of research, see also [3,11,18] and references therein.

In deep learning community, [26] proposed a generative framework of variational auto-encoder style and some variants to improve expressing power. Also, as noticed by [19], DS model can be equivalently formulated as a restricted Boltzmann machine with a single hidden node with categorical distribution.

In this paper, we consider to model multinominal noisy labels from crowds and propose an architecture combining exponential family matrix factorization and Gaussian mixture model (GMM) for clustering. In contrast to existing methods, the proposed one is a parsimonious adaptation for multinomial labels and incorporates both annotator specific parameters and item specific latent features. The model draws connections among some existing probabilistic annotation models and generative factor analytic models, e.g. [9,16]. An efficient fully Bayesian inference procedure is devised based on a modified lower bound. Besides, maximum a posteriori (MAP) approximate inference is investigated for better understanding the nature of the algorithm.

2   The Model

2.1   Problem Setting

Given N items that can be divided into K classes, we denote by a one-hot random vector $\mathbf{y}_n \in \{0,1\}^K$ the unknown true label of item n, i.e., $\mathbf{y}_n = \mathbf{e}_k$ represents that "item n is of class k". Additionally, we have access to annotated labels from M annotators, denoted as $\mathbf{r}_{nm} \in \{0,1\}^K$. The full set of predicted labels is collected into a tensor $\mathbf{R} \in \{0,1\}^{N \times M \times K}$. Due to the nature of crowdsourcing, each item is typically labeled by only a subset of annotators, and hence $\mathbf{R}$ might be incomplete. Particularly, we use $\Omega \subset [N] \times [M]$ to represent the observed subset of indices, and $\mathbf{R}_\Omega$ to denote the observed entries in $\mathbf{R}$. In addition, let $\Omega_{\bullet m} = \{n : (n,m) \in \Omega\}$ be subset of items labeled by annotator m, and $\Omega_{n\bullet}$ be similarly defined as the subset of annotators who label item n.

2.2 Proposed Model

To model the generation process of $\mathbf{R}$, we assume that each of $\mathbf{y}_n$ is sampled independently from multinomial distribution with parameter $\boldsymbol{\pi}$, i.e.,

$$\mathbf{y}_n | \boldsymbol{\pi} \sim \mathcal{M}(\mathbf{y}_n | \boldsymbol{\pi}), \tag{1}$$

where $\boldsymbol{\pi} \in \Delta_{K-1}$ is a discrete probability vector over K states, and Δ_{K-1} is the $(K-1)$-dimensional probability simplex. Moreover, let there be a latent feature, $\mathbf{b}_n \in \mathbb{R}^L$, associated with each item and following a Gaussian distribution conditional on ground truth label $\mathbf{y}_n$, i.e.,

$$\mathbf{b}_n | y_{n\ell} = 1 \sim \mathcal{N}(\mathbf{b}_n | \boldsymbol{\mu}_\ell, \boldsymbol{\Lambda}_\ell^{-1}). \tag{2}$$

Equivalently, $\mathbf{b}_n$ is assumed to be sampled marginally from a Gaussian mixture model (GMM) where the ground truth label $\mathbf{y}_n$ serves as the latent class indicator.

During the annotation process, each annotator is assumed to act like a multinomial logistic regression model based on the perceived latent features $\mathbf{b}_n$, rather that ground truth labels, compared with DS model. Specifically, for item n, its annotation probability by annotator m is modeled as

$$\mathsf{p}(r_{nmk} = 1 | \mathbf{b}_n, \mathbf{a}_{mk}, c_{mk}) \propto \exp(\mathbf{a}_{mk}^\mathsf{T} \mathbf{b}_n + c_{mk}), \tag{3}$$

where $\mathbf{a}_{mk} \in \mathbb{R}^L$ and $c_{mk} \in \mathbb{R}$ are coefficient and intercept terms of the kth discriminant function of annotator m. With (restricted) multinomial logistic operator, $\mathcal{S} : \mathbb{R}^{K-1} \to \Delta_{K-1}$, the full likelihood can be written in a compact form as

$$\mathsf{p}(\mathbf{R}_\Omega | \mathbf{A}, \mathbf{B}) = \prod_{(n,m) \in \Omega} \mathcal{M}(\mathbf{r}_{nm} | \mathcal{S}(\mathbf{A}_m \mathbf{b}_n)), \tag{4}$$

where $\mathbf{B} \in \mathbb{R}^{N \times L}$ and $\mathbf{A} \in \mathbb{R}^{M \times (K-1) \times L}$ denote the collection of corresponding individual variables or parameters, namely $\mathbf{b}_n \in \mathbb{R}^L$ and $\mathbf{A}_m \in \mathbb{R}^{(K-1) \times L}$. The intercept terms are henceforth omitted for simplicity and ease of presentation, by assuming that annotators display no particular labeling tendencies.

Then it is clear that the crowdsourcing problem is cast into an exponential family PCA framework [7,14], with a structured GMM prior over item features $\mathbf{B}$. The multinomial likelihood accounts for reconstruction fidelity of raw labels and the prior promotes a clustered structure in latent space and extract information of the unknown labels.

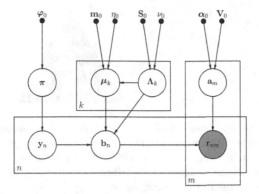

Fig. 1. Graphical representation of the proposed model architecture. Shaded nodes are the observed variables.

To adopt a fully Bayesian formulation, for each annotator $m \in [M]$, we further assign usual independent Gaussian prior distributions over regression coefficients as

$$\tilde{\mathbf{a}}_m = \text{vec}(\mathbf{A}_m^\mathsf{T}) \sim \mathcal{N}(\tilde{\mathbf{a}}_m | \tilde{\boldsymbol{\alpha}}_0, \mathbf{V}_0^{-1}). \tag{5}$$

For GMM parameters $\Theta = \{\boldsymbol{\pi}, \boldsymbol{\mu}, \boldsymbol{\Lambda}\}$, conjugate Dirichlet and Gaussian-Wishart priors are considered, i.e.,

$$\boldsymbol{\pi} \sim \mathcal{D}(\boldsymbol{\pi}|\boldsymbol{\varphi}_0), \quad \boldsymbol{\mu}_\ell, \boldsymbol{\Lambda}_\ell \sim \mathcal{N}(\boldsymbol{\mu}_\ell | \mathbf{m}_0, (\eta_0 \boldsymbol{\Lambda}_\ell)^{-1}) \mathcal{W}(\boldsymbol{\Lambda}_\ell | \mathbf{S}_0, \nu_0); \tag{6}$$

where $\ell \in [K]$. The parameters with subscripts 0 are fixed as hyperparameters. A full graphical representation of the proposed model architecture is shown in Fig. 1.

Intuitively, the main rationale behind the formulation lies in the belief that annotators make their decisions based on perceptions of item-specific properties rather than purely on ground truth labels. The information reflected by different label patterns from crowds could be valuable for discovering relationships across items and classes. The constructed latent item features are designed to capture variability that cannot be explained by different class memberships, such as heterogeneous difficulties of tasks, and allows for correlation among annotations, even conditional on ground truth. The logistic coefficients encode annotators' understandings about each class, and hence are interpreted as crowd label embeddings.

2.3 Relationships with Other Models

Comparing with our model given in Eq. 3, DS model can be summarized as a hierarchical latent variable model with likelihood

$$\mathsf{p}(\mathbf{r}_{nm}|\mathbf{y}_n, \boldsymbol{\Phi}_m) = \mathcal{M}(\mathbf{r}_{nm}|\boldsymbol{\Phi}_m \mathbf{y}_n), \tag{7}$$

where $\boldsymbol{\Phi}_m = [\boldsymbol{\phi}_{m1}, \dots, \boldsymbol{\phi}_{mK}] \in \mathbb{R}^{K \times K}$ is the column-stochastic confusion matrix with $\boldsymbol{\Phi}_{mk\ell} = \mathsf{p}(r_{nmk} = 1|y_{n\ell} = 1)$ indicating the probability that annotator m

Fig. 2. Difference characterizations of annotator confusion probabilities under DS model (left) and our generative model (right). In DS model, a single point $\phi_{mk} \in \Delta_{K-1}$ is used to model the underlying distribution of annotations for a class-k object. In our proposal, the induced logistic-Gaussian distribution reflects variability in annotator accuracy for different items in presence of noise.

assigns label k to a an object belonging to class ℓ. Clearly the formulation resembles a probabilistic version of factor model proposed in [2]. We make clear the relationship between the proposed architecture and DS model as the following proposition.

Proposition 1. *Suppose in the proposed model, it is chosen that $L = K-1$ and $\Lambda_\ell = \lambda_0 I$ is fixed for all $\ell \in [K]$, then in the limit $\lambda_0 \to \infty$, the model reduces to a reparameterization of DS model. That is, for every $\{\Phi_m\}_{m=1}^M$ in DS parameter space, there exists a set of parameters $\{A_m\}_{m=1}^M$ and $\{\mu_\ell\}_{\ell=1}^K$ such that the marginalized conditional matches the likelihood in DS, i.e.*

$$p(\mathbf{r}_{nm}|\mathbf{y}_n, \mathbf{A}_m, \boldsymbol{\mu}) = \int p(\mathbf{r}_{nm}|\mathbf{A}_m, \mathbf{b}_n)p(\mathbf{b}_n|\mathbf{y}_n, \boldsymbol{\mu})\,\mathrm{d}\mathbf{b}_n = \mathbf{r}_{nm}^\mathsf{T}\boldsymbol{\Phi}_m\mathbf{y}_n,$$

for every $(\mathbf{r}_{nm}, \mathbf{y}_n)$.

Collapsing down the Gaussian components effectively connects $\mathbf{y}_n$ with $\{\mathbf{r}_{nm}\}_{m\in[M]}$. It is clear that if we choose $L < K-1$, the model constitutes a reduced rank formulation of confusion matrices. We depict the difference of annotator characterization in DS model and our generative assumption in Fig. 2.

Moreover, we identify that the proposed model is effectively a parsimonious version of mixture of latent trait analyzer (MLTA) [9]. Specifically, MLTA for unordered categorical data can be considered as a hierarchical model with priors over latent variables $\mathbf{y}_n \sim \mathcal{M}(\mathbf{y}_n|\boldsymbol{\pi})$ and $\mathbf{b}_n \sim \mathcal{N}(\mathbf{b}_n|\mathbf{0}, \mathbf{I})$, and likelihood

$$\mathbf{r}_{nm}|\mathbf{b}_n, y_{n\ell} = 1 \sim \mathcal{M}(\mathbf{r}_{nm}|\mathcal{S}(\mathbf{A}_{m\ell}\mathbf{b}_n + \mathbf{c}_{m\ell})), \tag{8}$$

where $\mathbf{A}_{m\ell}$ and $\mathbf{c}_{m\ell}$ are deterministic parameters. Using reparameterization trick of Gaussian distribution and conditioning on $y_{n\ell} = 1$, the natural parameters in likelihood (Eq. 4) can be rewritten as

$$\mathbf{A}_m\mathbf{b}_n + \mathbf{c}_m = \mathbf{A}_m\Lambda_\ell^{-1/2}\Lambda_\ell^{1/2}\boldsymbol{\varepsilon}_n + \mathbf{A}_m\boldsymbol{\mu}_\ell + \mathbf{c}_m, \tag{9}$$

where $\varepsilon_n \sim \mathcal{N}(\varepsilon_n | \mathbf{0}, \mathbf{\Lambda}_\ell^{-1})$ is the conditional noise vector. Redefining $\tilde{\mathbf{A}}_{m\ell} = \mathbf{A}_m \mathbf{\Lambda}_\ell^{-1/2}$, $\tilde{\mathbf{c}}_{m\ell} = \mathbf{A}_m \boldsymbol{\mu}_\ell + \mathbf{c}_m$, and $\tilde{\varepsilon}_n = \mathbf{\Lambda}_\ell^{1/2} \varepsilon_n$ recovers the structure of MLTA with corresponding restrictions. Note that the sphered noise $\tilde{\varepsilon}_n \sim \mathcal{N}(\tilde{\varepsilon}_n | \mathbf{0}, \mathbf{I}_L)$ is now independent of $\mathbf{y}_n$ and serves as the continuous latent features in MLTA.

3   Model Inference

Since the exact model posterior is not available in closed form, mainly due to loss of conjugacy and difficulty of marginalization of nuisance latent variables, we resort to mean-field variational inference based on a modified bound to approximate the posterior. The MAP approximation scheme is then obtained by collapsing down part of the variational distributions to delta measures.

To facilitate the presentation of concrete algorithms, let the restricted label representation be $\mathbf{r}'_{nm} \in \{0,1\}^{K-1}$, which is the first $(K-1)$ coordinates of $\mathbf{r}_{nm}$. The posterior expectation $\mathbb{E}_q(\cdot)$ is abbreviated as $\langle \cdot \rangle$ for notational convenience. Also, we denote as $\varXi = \{\mathbf{A}, \mathbf{B}, \mathbf{Y}, \varTheta\}$ all the latent variables to be inferred.

3.1   The Modified ELBO

To address the intractability of inference arising from non-conjugate model structure, we adopt the treatment with Böhning's quadratic bound [5,6] for log-sum-exp (LSE) function appearing in multinomial likelihood. The bound is previously used for fully probabilistic inference of Bayesian logistic regression [15] and mixture of factor analyzers [13]. Specifically, Böhning's bound is formulated as a quadratic function

$$\tilde{\varPsi}(\boldsymbol{\eta}; \boldsymbol{\xi}) = \frac{1}{2}\boldsymbol{\eta}^\mathsf{T} \mathbf{Q}\boldsymbol{\eta} - \mathbf{h}(\boldsymbol{\xi})^\mathsf{T}\boldsymbol{\eta} + d(\boldsymbol{\xi}) \geq \varPsi(\boldsymbol{\eta}), \tag{10}$$

where $\mathbf{Q}$ is a constant centering matrix satisfying $\mathbf{Q} \preceq \nabla^2 \varPsi(\boldsymbol{\eta})$ for all $\boldsymbol{\eta} \in \mathbb{R}^{K-1}$ and $h(\boldsymbol{\xi})$ and $d(\boldsymbol{\xi})$ are functions solely depending on deterministic variational parameters $\boldsymbol{\xi}$.

Although the Böhning's bound is not tight in general as it utilizes a constant curvature, it bypasses the repeatedly expensive computation of second order statistics and hence dramatically simplifies the computation than other tight approximate inference methods, e.g. [1].

Replacing each $\varPsi$ appearing in the multinomial likelihood by its upper bound leads to a "Gaussianized" likelihood

$$\log \tilde{\mathsf{p}}(\mathbf{r}_{nm} | \mathbf{A}_m, \mathbf{b}_n; \boldsymbol{\xi}_{nm}) = \mathbf{r}'_{nm} \mathbf{A}_m \mathbf{b}_n - \tilde{\varPsi}(\mathbf{A}_m \mathbf{b}_n; \boldsymbol{\xi}_{nm}) \leq \log \mathsf{p}(\mathbf{r}_{nm} | \mathbf{A}_m, \mathbf{b}_n) \tag{11}$$

where $\boldsymbol{\xi}_{nm}$ is a local variational parameter specific to each label. Given a fixed set of variational parameters $\boldsymbol{\xi}$, the modified likelihood can be written in two equivalent forms, which are quadratic in terms of $\tilde{\mathbf{a}}_m$ and $\mathbf{b}_n$, respectively. Specifically, expanding each $\tilde{\varPsi}$ gives the following two equivalent representations

$$\log \tilde{p}(\mathbf{r}_{nm}|\mathbf{A}_m, \mathbf{b}_n; \boldsymbol{\xi}_{nm}) = -\frac{1}{2}\mathbf{b}_n^\mathsf{T}\mathbf{A}_m^\mathsf{T}\mathbf{Q}\mathbf{A}_m\mathbf{b}_n + \mathbf{z}_{nm}^\mathsf{T}\mathbf{Q}\mathbf{A}_m\mathbf{b}_n - d(\boldsymbol{\xi}_{nm}) \quad (12)$$

$$= -\frac{1}{2}\tilde{\mathbf{a}}_m^\mathsf{T}\tilde{\mathbf{B}}_n\mathbf{Q}\tilde{\mathbf{B}}_n^\mathsf{T}\tilde{\mathbf{a}}_m + \mathbf{z}_{nm}^\mathsf{T}\mathbf{Q}\tilde{\mathbf{B}}_n^\mathsf{T}\tilde{\mathbf{a}}_m - d(\boldsymbol{\xi}_{nm}), \quad (13)$$

where $\mathbf{z}_{nm} = \mathbf{Q}^{-1}(\mathbf{r}'_{nm} + \mathbf{h}(\boldsymbol{\xi}_{nm}))$ is a Gaussianized pseudo measurement and $\tilde{\mathbf{B}}_n \in \mathbb{R}^{(K-1)L \times (K-1)}$ is a block diagonal matrix given by $\mathbf{I}_{K-1} \otimes \mathbf{b}_n$.

Then the modified ELBO defined by

$$\tilde{\mathcal{F}}[\mathsf{q}; \boldsymbol{\xi}] = \mathbb{E}_{\mathsf{q}(\Xi)} \log \tilde{p}(\mathbf{R}_\Omega|\Xi; \boldsymbol{\xi}) - \mathcal{J}_{\mathrm{KL}}[\mathsf{q}(\Xi)\|\mathsf{p}(\Xi)] \quad (14)$$

minorizes the original model ELBO, i.e. $\tilde{\mathcal{F}}[\mathsf{q}; \boldsymbol{\xi}] \leq \mathcal{F}[\mathsf{q}]$. The variational inference problem is then transferred to the modified bound, which allows for closed form updates for each block of latent variables.

3.2 Optimal Posterior Distributions

Standard treatment in mean-field variational inference (refer to e.g. [4]) can be applied to $\tilde{\mathcal{F}}$ under the structured mean-field variational family. As the modified bound is quadratic in each $\tilde{\mathbf{a}}_m$ and $\mathbf{b}_n$, we equivalently have a Gaussian observation model on transformed data up to some constants. This recovers a conjugate model structure, which naturally allows for Gaussian posterior distributions, assuming factorization between $\mathsf{q}(\tilde{\mathbf{a}}_m)$ and $\mathsf{q}(\mathbf{b}_n)$.

Making use of the formulation in Eq. 13, the optimal posterior with respect to modified bound over $\tilde{\mathbf{a}}_m$ can be written up to an additive constant as

$$\log \mathsf{q}^*(\tilde{\mathbf{a}}_m; \boldsymbol{\xi}) = \sum_{n \in \Omega_{\bullet m}} \langle \log \mathcal{N}(\mathbf{z}_{nm}|\tilde{\mathbf{B}}_n^\mathsf{T}\tilde{\mathbf{a}}_m, \mathbf{Q}^{-1})\rangle$$

$$+ \log \mathcal{N}(\tilde{\mathbf{a}}_m|\mathbf{m}_0, \mathbf{V}_0^{-1}) + \mathrm{const}, \quad (15)$$

which recovers an Gaussian observation model with pseudo measurement $\mathbf{z}_{nm}$ and design matrix $\langle \tilde{\mathbf{B}}_n \rangle$. Standard operations yield the optimal solution as $\mathsf{q}^*(\tilde{\mathbf{a}}_m; \boldsymbol{\xi}) = \mathcal{N}(\tilde{\mathbf{a}}_m|\boldsymbol{\alpha}_m, \mathbf{V}_m^{-1})$ with posterior parameters

$$\mathbf{V}_m = \sum_{n \in \Omega_{\bullet m}} \langle \tilde{\mathbf{B}}_n \mathbf{Q} \tilde{\mathbf{B}}_n^\mathsf{T}\rangle + \mathbf{V}_0, \quad (16)$$

$$\mathbf{V}_m\boldsymbol{\alpha}_m = \sum_{n \in \Omega_{\bullet m}} \langle \tilde{\mathbf{B}}_n \rangle \mathbf{Q}\mathbf{z}_{nm} + \mathbf{V}_0\boldsymbol{\alpha}_0. \quad (17)$$

The block diagonal nature of $\tilde{\mathbf{B}}_n$ simplifies evaluation of the expectation terms. The quadratic term is evaluated as $\langle \tilde{\mathbf{B}}_n \mathbf{Q} \tilde{\mathbf{B}}_n^\mathsf{T}\rangle = \mathbf{Q} \otimes \langle \mathbf{b}_n \mathbf{b}_n^\mathsf{T}\rangle$ and $\langle \tilde{\mathbf{B}}_n \rangle = (\mathbf{I}_{K-1} \otimes \langle \mathbf{b}_n \rangle)$ due to linearity.

As we shall see, the approximation over $\mathbf{b}_n$ is again Gaussian and hence $\langle \mathbf{b}_n \mathbf{b}_n^\mathsf{T}\rangle$ can be efficiently evaluated. The posterior result for $\tilde{\mathbf{a}}_m$ is analogous to usual Bayesian logistic regression with Böhning's bound, only different by posterior precision terms arising from the posterior uncertainty of $\mathbf{B}$.

For latent item features, the optimal solution can be analogously written as

$$\log q^*(\mathbf{b}_n; \boldsymbol{\xi}) = \sum_{m \in \Omega_{n\bullet}} \langle \log \mathcal{N}(\mathbf{z}_{nm} | \mathbf{A}_m \mathbf{b}_n, \mathbf{Q}^{-1}) \rangle$$

$$+ \sum_\ell \langle y_{n\ell} \rangle \langle \log \mathcal{N}(\mathbf{b}_n | \boldsymbol{\mu}_\ell, \boldsymbol{\Lambda}_\ell^{-1}) \rangle + \text{const.} \quad (18)$$

Similar results then follow, i.e. $q^*(\mathbf{b}_n; \boldsymbol{\xi}) = \mathcal{N}(\mathbf{b}_n | \boldsymbol{\beta}_n, \mathbf{H}_n^{-1})$ with

$$\mathbf{H}_n = \sum_{m \in \Omega_{n\bullet}} \langle \mathbf{A}_m^\mathsf{T} \mathbf{Q} \mathbf{A}_m \rangle + \sum_\ell \langle y_{n\ell} \rangle \langle \boldsymbol{\Lambda}_\ell \rangle \quad (19)$$

$$\mathbf{H}_n \boldsymbol{\beta}_n = \sum_{m \in \Omega_{n\bullet}} \langle \mathbf{A}_m \rangle \mathbf{Q} \mathbf{z}_{nm} + \sum_\ell \langle y_{n\ell} \rangle \langle \boldsymbol{\Lambda}_\ell \boldsymbol{\mu}_\ell \rangle. \quad (20)$$

Unfortunately, full evaluation of term $\langle \mathbf{A}_m^\mathsf{T} \mathbf{Q} \mathbf{A}_m \rangle$ is not as easy as $\langle \tilde{\mathbf{B}}_n \mathbf{Q} \tilde{\mathbf{B}}_n^\mathsf{T} \rangle$. Also note that this result is not true when intercept coefficients are included, which requires explicit adjustment in the pseudo-measurement $\mathbf{z}_{nm}$ and computation of quadratic term $\langle \mathbf{A}_m \mathbf{Q} \mathbf{A}_m \rangle$. Empirically we find a first order approximation works well.

The computation of remaining factors, namely latent class memberships $\mathbf{Y}$ and GMM parameters $\Theta = (\boldsymbol{\pi}, \boldsymbol{\mu}, \boldsymbol{\Lambda})$, are rather similar as the usual inference procedure in a Bayesian GMM (BGMM) [15]. However, appropriate adjustments are required to account for posterior uncertainty arising from full probabilistic inference of $q^*(\mathbf{B}; \boldsymbol{\xi})$.

Specifically, for latent class memberships, it can be derived that $q^*(\mathbf{y}_n) = \mathcal{M}(\mathbf{y}_n | \boldsymbol{\gamma}_n)$ for each $n \in [N]$ with adjusted posterior responsibility $\boldsymbol{\gamma}_n$ given by

$$\gamma_{n\ell} \propto \exp \left\{ -\frac{1}{2} \text{tr}(\nu_\ell \mathbf{S}_\ell \mathbf{H}_n^{-1}) - \frac{L\eta_\ell}{2} + \langle \log \det(\boldsymbol{\Lambda}_\ell) \rangle \right.$$

$$\left. - \frac{\nu_\ell}{2}(\boldsymbol{\beta}_n - \mathbf{m}_\ell)^\mathsf{T} \mathbf{S}_\ell (\boldsymbol{\beta}_n - \mathbf{m}_\ell) + \langle \log \pi_\ell \rangle \right\}, \quad (21)$$

where the first term is from posterior uncertainties of $q^*(\mathbf{b}_n)$ and is not included in BGMM. Intuitively, it tracks whether the posterior covariance is well aligned with a particular component precision. The probability not compatible with a Gaussian component will contribute more to other classes.

For GMM parameters, $q^*(\boldsymbol{\pi})$ depends on $\boldsymbol{\gamma}$ only and the posterior uncertainty from $q^*(\mathbf{B}; \boldsymbol{\xi})$ does not affect factors $q^*(\boldsymbol{\mu}_\ell | \boldsymbol{\Lambda}_\ell)$ due to linear dependence. Therefore, compared with inference of BGMM, the only affected factor is $q^*(\boldsymbol{\Lambda}_\ell) = \mathcal{W}(\boldsymbol{\Lambda}_\ell | \mathbf{S}_\ell, \nu_\ell)$, which depends on second order statistics of $q^*(\mathbf{B})$. The posterior scale matrix is adapted as

$$\mathbf{S}_\ell^{-1} = \sum_n \gamma_{n\ell} \mathbf{H}_n^{-1} + \sum_n \gamma_{n\ell}(\boldsymbol{\beta}_n - \bar{\boldsymbol{\beta}}_\ell)(\boldsymbol{\beta}_n - \bar{\boldsymbol{\beta}}_\ell)^\mathsf{T}$$

$$+ \frac{\eta_0 \sum_n \gamma_{n\ell}}{\eta_\ell}(\bar{\boldsymbol{\beta}}_\ell - \mathbf{m}_0)(\bar{\boldsymbol{\beta}}_\ell - \mathbf{m}_0)^\mathsf{T} + \mathbf{S}_0^{-1}, \quad (22)$$

and $\bar{\beta}_\ell = (\sum_n \gamma_{n\ell})^{-1}(\sum_n \gamma_{n\ell}\beta_n)$ is the weighted average of posterior means of item features.

To tighten the modified bound, variational parameters are also refreshed along with the factors. The optimal solution is obtained by optimizing the modified bound with respect to each ξ_{nm}. The solution is given by $\xi_{nm}^* = \langle \mathbf{A}_m \mathbf{b}_n \rangle = \langle \mathbf{A}_m \rangle \langle \mathbf{b}_n \rangle$, where the second equality is a consequence of mean-field assumption.

3.3 MAP Approximation

We consider MAP approximation over all approximate factors except for the discrete $q(\mathbf{y}_n)$ due to its special role in crowdsourcing practice. Optimal MAP approximations can be obtained by ignoring the infinite entropy contribution of delta approximations and solving the folded-in optimization problem with respect to all parameters $\{\boldsymbol{\alpha}, \boldsymbol{\beta}, \boldsymbol{\gamma}, \hat{\Theta}\}$.

For simplicity, let the component covariances be fixed as $\lambda_0^{-1}\mathbf{I}_L$. Updating the blocks $\{\boldsymbol{\gamma}, \boldsymbol{\pi}, \boldsymbol{\mu}\}$, with fixed $\boldsymbol{\beta}$, is achieved by solving sub-problem

$$\max_{\gamma, \pi, \mu} \sum_{n,\ell} \gamma_{n\ell} \left(\log \pi_\ell - \frac{\lambda_0}{2} \|\beta_n - \mu_\ell\|^2 \right) + \sum_n \mathbb{H}[\gamma_n], \tag{23}$$

where the target is exactly the ELBO for MAP fitting of a deterministic location-mixture of Gaussians with data matrix $\boldsymbol{\beta}$. Therefore standard results implies optimal $\boldsymbol{\gamma}_n$ is given by $\gamma_{nk} \propto \pi_k \mathcal{N}(\beta_n | \mu_k, \lambda_0^{-1}\mathbf{I}_L)$.

Secondly, the folded-in MAP sub-problem for $\boldsymbol{\beta}$ is naturally linked to regularized Bregman projection [7], due to Bregman divergence representation of exponential family distributions. Specifically, with $\boldsymbol{\gamma}$ fixed, updating $\{\boldsymbol{\alpha}, \boldsymbol{\beta}\}$ boils down to solve

$$\min_{\alpha, \beta} \sum_{(n,m) \in \Omega} D_{-\mathbb{H}}[\mathbf{r}_{nm} \| \mathcal{S}(\boldsymbol{\alpha}_m \boldsymbol{\beta}_n)] + \frac{\lambda_0}{2} \sum_{n,\ell} \gamma_{n\ell} \|\beta_n - \mu_\ell\|_2^2, \tag{24}$$

where $D_{-\mathbb{H}}$ is the Bregman divergence defined based on negative entropy function $-\mathbb{H}(\boldsymbol{\pi})$ and $\mathbf{r}_{nm}$ is understood as a degenerated discrete probability. The prior precision λ_0 acts as a regularization coefficient controlling the strength of shrinkage towards cluster centers. Moreover, if we restrict $q(\mathbf{y}_n)$ to be a degenerated discrete measure, the simplified problem constitutes an exponential family extension to the formulation proposed in [25]. Computations of $\boldsymbol{\alpha}$ and $\boldsymbol{\beta}$ can be done with an alternating algorithm and easily parallelized. It worth noting that each sub-problem is convex due to negative semi-definiteness of Hessians.

4 Experiments

In this section we present the experimental results of the proposed model mainly on label aggregation task, compared with two baseline methods, DS and majority voting heuristic. Computationally, it worth mentioning that as finite mixture models, the discrete posterior probability in both our proposed architecture and DS model are only identifiable up to a label permutation. This issue is resolved by initializing using majority voting in our experiments.

4.1   Evaluation Metrics

Since the main goal of labeling models is to infer ground truth labels from noisy ones provided by crowds, we consider aggregation accuracy measured with respect to ground truth labels associated with each item. Secondly, to better assess the fidelity of predictive uncertainty and demonstrate the proposed model is more advantageous in terms of uncertainty calibration, we consider symmetric KL divergence with degenerated labels $\mathbf{y}_n$ replaced by some oracle prediction $\mathsf{q}^*(\mathbf{y}_n)$, when we have full knowledge of the generative parameters. That is,

$$\mathsf{CalibSKL} = \frac{1}{2} \sum_n \mathcal{J}_{\mathrm{KL}}[\mathsf{q}^*(\mathbf{y}_n)\|\mathsf{q}(\mathbf{y}_n)] + \mathcal{J}_{\mathrm{KL}}[\mathsf{q}(\mathbf{y}_n)\|\mathsf{q}^*(\mathbf{y}_n)].$$

Empirically we found the measure is highly correlated with aggregation cross entropy measured against degenerated true labels, which is subject to sampling noise. The estimation error of confusion matrices is also investigated. For our model, the estimates are constructed through a MC estimation of the mean of induced logistic-Gaussian distribution. The error of estimating a single confusion matrix is measured in terms of mean symmetric KL divergence

$$\mathsf{ErrSKL} = \frac{1}{2M} \sum_{m,\ell} \hat{\pi}_\ell \mathcal{J}_{\mathrm{KL}}[\hat{\phi}_{m\ell}\|\phi^*_{m\ell}] + \pi^*_\ell \mathcal{J}_{\mathrm{KL}}[\phi^*_{m\ell}\|\hat{\phi}_{m\ell}].$$

4.2   Synthetic Data

Experimental Setup. We generate synthetic data according to two assumptions, namely the DS assumption and our generative assumption (denoted as LR). Under our generative assumption, we first uniformly generate K points as $\{\boldsymbol{\mu}_k\}$ on an L dimensional hypersphere. A location-GMM model is then formulated using fixed component parameters $\{(\boldsymbol{\mu}_k, \lambda_0^{-1}\mathbf{I})\}_{k=1}^K$ and mixing probabilities $\boldsymbol{\pi} \propto \mathbf{1}_K$. We leverage on linear discriminant analysis as an oracle decision rule to generate coefficients of noisy annotators. Specifically, let $(\mathbf{A}_*, \mathbf{c}_*)$ be the set of coefficients obtained from oracle rule, and for each annotator, we generate annotator coefficients from a Gaussian distribution centered at $(\mathbf{A}_*, \mathbf{c}_*)$. Hierarchical sampling is adopted according to our generative model to obtain full annotations. Under DS assumption, we generate proper confusion matrices directly and confirm that the two assumptions in [27] regarding overall accuracy and class distinguishability are satisfied.

We use $N = 300$ and $M = 50$ as a base configuration under both assumptions for demonstration. Following "missing at random" assumption of observed labels, entries of generated rating tensor are randomly masked with a specified missing probability.

Due to nonidenfiability, for MAP inference, our model is fitted with a fixed component precision matrix $\lambda_0 \mathbf{I}_L$, where λ_0 is taken as a tunable hyperparameter. Other parameters are fitted in mean-field Bayes (MFB) or MAP fashion. For datasets generated under our generative assumption, the full precision matrix is freed as adaptive and also updated with MFB/MAP criterion. For the reduced rank (RR) case, we run the proposed method with a two dimensional latent space ($L = 2$) for datasets with $K = 3, 5$ and $L = 4$ for the datasets with $K = 10$. For the full rank (FR) case, the proposed model is fitted with $L = K-1$ for each scenario.

Annotator Reliability. We evaluate the effects of overall reliability of annotators, by manipulating the generative parameters. For DS assumption, the reliability is measured by the average of diagonal elements in confusion matrices. For our generative assumption, we alter $\|\boldsymbol{\mu}_\ell\|$ to represent change in overall reliability. Note that this characterization of reliability is different from annotator variability, which is reflected by scale of $\mathsf{p}(\tilde{\mathbf{a}})$ in our model. Figure 3, Fig. 4 and Fig. 5 summarize the aggregation accuracy, estimation error and calibration loss, respectively, on different reliability scenarios under both assumptions.

In terms of aggregation accuracy, we can see that the proposed model achieves comparable accuracy with DS model under both assumptions, while gaining advantages under our generative assumption. The margin that the proposed model improves DS model increases as the number of class becomes larger, especially for LR-K10 scenario. This conforms with our prior belief that DS model is less efficient for large K.

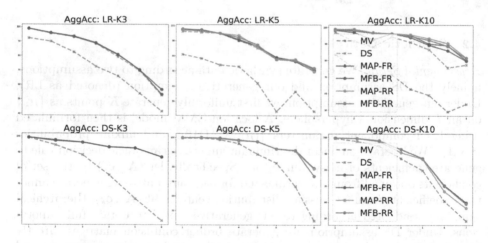

Fig. 3. Aggregation accuracy for synthetic data under our generative assumption and DS assumption. Overall reliability deteriorates along horizontal axis.

Moreover, the comparable aggregation accuracy is achieved with better estimation, in terms of error measured by symmetric KL divergence, as demonstrated in Fig. 4. We can expect better aggregation results when generalizing the fitted model to streaming-in data from the same set of annotators.

Another interesting phenomenon worth mentioning is the impact of item-specific random effect. Under DS generative assumption, one can observe that

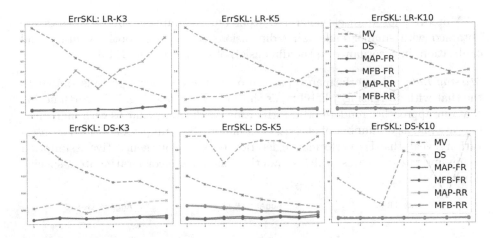

Fig. 4. Estimation accuracy for synthetic data under our generative assumption and DS assumption. Overall reliability deteriorates along horizontal axis.

as overall reliability increases and hence the annotators becomes more "homogeneous", the performance gap between probabilistic methods and voting heuristic tends to vanish. However, it is not the case in presence of item-specific randomness, where the performance gap remains. This empirically verifies that voting is sub-optimal under such scenarios.

Further, smaller calibration loss of our method indicates that it assigns lower probability to misclassified items and demonstrates its ability of preserving item ambiguity in the inferred posterior distribution. In contrast, the higher calibration loss of DS indicates more "confident" errors.

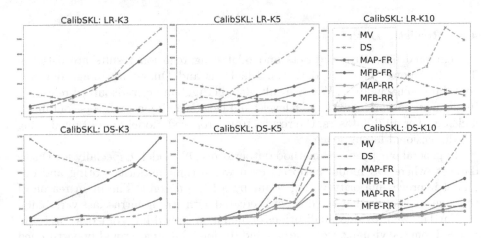

Fig. 5. Calibration accuracy for synthetic data under our generative assumption and DS assumption. Overall reliability deteriorates along horizontal axis.

Moreover, for dataset generated from our assumption, better results are obtained with smaller embedding dimensions in terms of both accuracy and calibration, in accordance to the dimension of ground truth GMM.

Missing Rate. The performance pattern of increasing missing rate is similar to the that with decreasing reliability, as shown in Fig. 6 and Fig. 7. It worth noting that our approaches render better estimation results for scenarios with high missing rate and lager number of classes, even under DS assumptions. Another observation reveals that DS performance deteriorates under our generative assumption as missing rate decreases, which shows the benefits of incorporating item-specific noise.

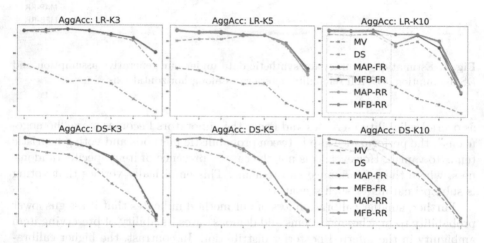

Fig. 6. Aggregation accuracy for synthetic data under our generative assumption and DS assumption. Missing rate increases from 20% to 90% along horizontal axis.

4.3 Realistic Data

We explored 6 real toy datasets and related aggregation results are listed in Table 1. Among these toy datasets, Bluebirds and Ducks are image datasets containing binary labels; WSD, Adult, and TlkAgg5 are crowdsourced relevance judgements containing 3, 4 and 5 levels; and Dogs is an image datasets containing 4 classes. For all 6 datasets, the results from proposed model are obtained using a 2 dimensional latent space.

In general our proposed method outperforms DS model. Especially for Ducks dataset, where DS model performs even worse than majority voting and our model improves majority voting result by a large margin. The main reason of this phenomenon is that the dataset is created with strong intra-class variability.

Further, we provide visualizations of final item embeddings of Dogs dataset in Fig. 8. From the visualization one can find the four classes are roughly partitioned into two groups, reflecting the fact that the pair of classes within a group are more similar than those in different groups. This is also reflected in the block structure of mean confusion matrices obtained from DS model plotted in Fig. 8.

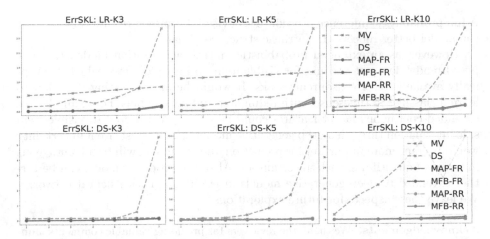

Fig. 7. Estimation accuracy for synthetic data under our generative assumption and DS assumption. Missing rate increases from 20% to 90% along horizontal axis.

Table 1. Aggregation accuracy for real datasets. Tlkagg5 is accessed at https:// research.yandex.com/datasets/toloka.

	Bluebirds [22]	Ducks [22]	Dogs [28]	WSD [21]	Adult [20]	TlkAgg5
DS	**89.81**	61.25	84.14	98.87	75.68	84.34
MAP	**89.81**	80.00	**85.01**	**99.44**	**78.38**	89.06
MFB	**89.81**	**80.83**	**85.01**	98.31	76.58	81.42
MV	75.93	67.92	81.66	**99.44**	75.68	**90.46**

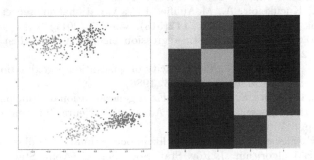

Fig. 8. Final 2D embeddings for Dogs dataset and estimated confusion from DS model.

5   Discussion

In summary, we propose a principled probabilistic model for ground truth inference in crowdsourced multinomial labeling tasks. In fact, the model is general and applies to clustering and visualizing other types of high-dimensional categorical data in a reduced latent space and could be generalized to other applications.

A full probabilistic inference procedure is devised. Empirical results validate that the model better reflects the intrinsic structure of data.

However, as an extended probabilistic matrix factorization model, the proposed model does not account for visible item and user features, which are typically available in crowdsourcing tasks. It would be an interesting direction to learn a classifier in this sense, with redundant and noisy labels. For example, the extended continuous latent features could be directly connected to visible item side information through a deep generative model. The intrinsic item uncertainty would become more important for prediction and the model will be advantageous in terms of quantifying such uncertainties. Also, the proposed model can be further extended to deep generative models in conjunction with neural networks. We leave these aspects for future explorations.

Acknowledgements. We thank the reviewers for providing valuable comments. Junhui Wang's research is supported in part by HKRGC Grants GRF-11303918 and GRF-11300919.

References

1. Ahmed, A., Xing, E.: On tight approximate inference of the logistic-normal topic admixture model. In: Proceedings of the 11th Tenth International Workshop on Artificial Intelligence and Statistics (2007)
2. Bhattacharya, A., Dunson, D.B.: Simplex factor models for multivariate unordered categorical data. J. Am. Stat. Assoc. **107**(497), 362–377 (2012)
3. Blei, D.M., Jordan, M.I.: Modeling annotated data. In: Proceedings of the 26th Annual International ACM SIGIR Conference on Research and Development in Information Retrieval, pp. 127–134. ACM (2003)
4. Blei, D.M., Kucukelbir, A., McAuliffe, J.D.: Variational inference: a review for statisticians. J. Am. Stat. Assoc. **112**(518), 859–877 (2017)
5. Böhning, D.: Multinomial logistic regression algorithm. Ann. Inst. Stat. Math. **44**(1), 197–200 (1992)
6. Böhning, D., Lindsay, B.G.: Monotonicity of quadratic-approximation algorithms. Ann. Inst. Stat. Math. **40**(4), 641–663 (1988)
7. Collins, M., Dasgupta, S., Schapire, R.E.: A generalization of principal components analysis to the exponential family. In: Advances in Neural Information Processing Systems, pp. 617–624 (2002)
8. Dawid, A.P., Skene, A.M.: Maximum likelihood estimation of observer error-rates using the EM algorithm. J. Roy. Stat. Soc.: Ser. C (Appl. Stat.) **28**(1), 20–28 (1979)
9. Gollini, I., Murphy, T.B.: Mixture of latent trait analyzers for model-based clustering of categorical data. Stat. Comput. **24**(4), 569–588 (2014)
10. Jagabathula, S., Subramanian, L., Venkataraman, A.: Identifying unreliable and adversarial workers in crowdsourced labeling tasks. J. Mach. Learn. Res. **18**(1), 3233–3299 (2017)
11. Kajino, H., Tsuboi, Y., Kashima, H.: A convex formulation for learning from crowds. In: 36th AAAI Conference on Artificial Intelligence (2012)
12. Karger, D.R., Oh, S., Shah, D.: Budget-optimal crowdsourcing using low-rank matrix approximations. In: 2011 49th Annual Allerton Conference on Communication, Control, and Computing (Allerton), pp. 284–291. IEEE (2011)

13. Khan, M.E., Bouchard, G., Murphy, K.P., Marlin, B.M.: Variational bounds for mixed-data factor analysis. In: Advances in Neural Information Processing Systems, pp. 1108–1116 (2010)

14. Mohamed, S., Ghahramani, Z., Heller, K.A.: Bayesian exponential family PCA. In: Advances in Neural Information Processing Systems, pp. 1089–1096 (2009)

15. Murphy, K.P.: Machine Learning: A Probabilistic Perspective. MIT Press, Cambridge (2012)

16. Rai, P., Wang, Y., Guo, S., Chen, G., Dunson, D., Carin, L.: Scalable Bayesian low-rank decomposition of incomplete multiway tensors. In: International Conference on Machine Learning, pp. 1800–1808 (2014)

17. Raykar, V.C., Yu, S.: Eliminating spammers and ranking annotators for crowd-sourced labeling tasks. J. Mach. Learn. Res. **13**, 491–518 (2012)

18. Raykar, V.C., et al.: Learning from Crowds. J. Mach. Learn. Res. **11**, 1297–1322 (2010)

19. Shaham, U., et al.: A deep learning approach to unsupervised ensemble learning. In: International Conference on Machine Learning, pp. 30–39 (2016)

20. Sheng, V.S., Provost, F., Ipeirotis, P.G.: Get another label? Improving data quality and data mining using multiple, noisy labelers. In: Proceedings of the 14th ACM SIGKDD International Conference on Knowledge Discovery and Data Mining, pp. 614–622 (2008)

21. Snow, R., O'connor, B., Jurafsky, D., Ng, A.Y.: Cheap and fast-but is it good? Evaluating non-expert annotations for natural language tasks. In: Proceedings of the 2008 Conference on Empirical Methods in Natural Language Processing, pp. 254–263 (2008)

22. Welinder, P., Branson, S., Perona, P., Belongie, S.J.: The multidimensional wisdom of crowds. In: Advances in Neural Information Processing Systems, pp. 2424–2432 (2010)

23. Whitehill, J., Wu, T., Bergsma, J., Movellan, J.R., Ruvolo, P.L.: Whose vote should count more: optimal integration of labels from labelers of unknown expertise. In: Advances in Neural Information Processing Systems, pp. 2035–2043 (2009)

24. Xu, A., Feng, X., Tian, Y.: Revealing, characterizing, and detecting crowdsourcing spammers: a case study in community Q&A. In: 2015 IEEE Conference on Computer Communications, pp. 2533–2541. IEEE (2015)

25. Yang, B., Fu, X., Sidiropoulos, N.D.: Learning from hidden traits: joint factor analysis and latent clustering. IEEE Trans. Sig. Process. **65**(1), 256–269 (2016)

26. Yin, L., Han, J., Zhang, W., Yu, Y.: Aggregating crowd wisdoms with label-aware autoencoders. In: Proceedings of the 26th International Joint Conference on Artificial Intelligence, pp. 1325–1331. AAAI Press (2017)

27. Zhang, Y., Chen, X., Zhou, D., Jordan, M.I.: Spectral methods meet EM: a provably optimal algorithm for crowdsourcing. In: Advances in Neural Information Processing Systems, pp. 1260–1268 (2014)

28. Zhou, D., Basu, S., Mao, Y., Platt, J.C.: Learning from the wisdom of crowds by minimax entropy. In: Advances in Neural Information Processing Systems, pp. 2195–2203 (2012)

Prediction of Global Navigation Satellite System Positioning Errors with Guarantees

Alejandro Kuratomi[✉], Tony Lindgren[✉], and Panagiotis Papapetrou[✉]

Stockholm University, Borgarfjordsgatan 12, 16455 Kista, Sweden
{alejandro.kuratomi,tony,panagiotis}@dsv.su.se

Abstract. Intelligent Transportation Systems employ different local-ization technologies, such as the Global Navigation Satellite System. This system transmits signals between satellite and receiver devices on the ground which can estimate their position on earth's surface. The accuracy of this positioning estimate, or the *positioning error estimation*, is of utmost importance for the efficient and safe operation of autonomous vehicles, which require not only the position estimate, but also an estimation of their operation margin. This paper proposes a workflow for positioning error estimation using a random forest regres-sor along with a post-hoc conformal prediction framework. The latter is calibrated on the random forest out-of-bag samples to transform the obtained positioning error estimates into predicted integrity intervals, which are confidence intervals on the positioning error prediction with at least 99.999% confidence. The performance is measured as the number of ground truth positioning errors inside the predicted integrity intervals. An extensive experimental evaluation is performed on real-world and synthetic data in terms of root mean square error between predicted and ground truth positioning errors. Our solution results in an improvement of 73% compared to earlier research, while providing prediction statistical guarantees.

Keywords: GNSS · Positioning error estimation · Random forest · Linear regression · Feature selection · Conformal prediction

1 Introduction

Intelligent Transport Systems (ITS) refer to technologies aiding transportation safety and mobility, consisting of autonomous driving devices, highway speed and traffic control systems, among others [1]. For an appropriate operation, ITS require accurate vehicle localization, which is available through a Global Nav-igation Satellite System (GNSS) composed of orbital navigation satellites pro-viding geo-spatial positioning [9]. GNSS is conformed by satellite constellations: (1) Global Positioning System (GPS), from the United States; (2) Globalnaya Navigazionnaya Sputnikovaya Sistema (GLONASS), from Russia; (3) Galileo Satellite System, from Europe; and (4) BeiDou Satellite System, from China [5].

© Springer Nature Switzerland AG 2021
Y. Dong et al. (Eds.): ECML PKDD 2020, LNAI 12460, pp. 562–578, 2021.
https://doi.org/10.1007/978-3-030-67667-4_34

The electromagnetic signals transmitted from satellites to ground receivers and the position estimates are affected by different error sources including ionosphere electron content, troposphere temperature, humidity, pressure, receiver and satellite clock offset, dilution of precision (DOP) [12][1] and other unmodelled phenomena, such as *multipath*, also known as *urban canyon* effect [23], which is caused by the reflection of the broadcasted signal onto surfaces surrounding the receiver [9]. In Tang et al. [12] findings indicate that multipath errors have great impact on positioning error, of up to a 100 m according to Zimmermann et al. [13], while Bauer et al. [14] concludes that multipath error is incorrectly modeled and expensive to compensate for.

Real Time Kinematic Differential GNSS (RTK DGNSS) is a technique that uses two satellites and two ground receivers. The latter consist of a master receiver with a known static location, and a rover receiver, whose location is unknown. The master's location is given by a reference positioning system or an already localized static station. By applying double differences between the signals received at both master and rover to calculate the position vector (x, y and z-ground coordinates [6]), ionospheric and troposhperic errors are considerably reduced, but other unmodelled errors remain [2]. The position vector between them and the known master location are added to localize the rover.

The motivation for this paper is the fact that ITS require accurate vehicle localization, and this in turn needs accurate position and positioning error estimates. Obtaining accurate position estimation is crucial due to ITS systems dependency on position knowledge to perform real-time decisions. Not knowing the current position is highly restrictive, e.g., a self-driving vehicle without accurate position estimates imposes risks to pedestrians and other vehicles, and should hence not operate. Moreover, positioning error estimation is as important as the position estimation. A hypothetical autonomous vehicle that estimates its own position but not the position estimate accuracy, may incur in a hazardous operation for itself, its occupants and/or pedestrians, as the system trusts its position estimate because it cannot evaluate its quality. The common safety margins for ITS applications are demanding, and confidence levels from 99.999% up to 99.9999999% are accepted to minimize risk [3,4]. These high levels of confidence are known as *integrity*.

This work was supported by Waysure Sweden AB, which is a high accuracy GNSS service provider in Sweden interested in obtaining real-time high accuracy GNSS positioning error estimation. This has previously been investigated by Kuratomi [15], where decision trees were used to select a set of relevant features from the real-world dataset, which was then used as input to a Support Vector Regressor (SVR) to obtain an estimation of positioning error. Nonetheless, no guarantees were provided to the regression output (e.g., in the form of integrity intervals) and its performance could be improved.

Contributions. In this paper, we propose a framework for GNSS positioning error estimation with statistical guarantees on the predicted error values,

---

[1] It is the error margin in the positioning of the receiver due to the spatial distribution of the satellites.

using conformal prediction on random forest regressors. More concretely, our contributions include:

- **A random forest conformal prediction framework**, called NAVEEG, capable of obtaining positioning error estimates and their corresponding integrity intervals while preserving relevant feature information.
- **High accuracy** positioning error estimation, which is achieved by our proposed framework implemented using RTK DGNSS information, on benchmarks on a real-world dataset involving tests on a truck and a lawnmower.
- **High integrity intervals** of 99.999% confidence level are provided for the predicted positioning error in the real-world dataset, which are obtained through the post-hoc conformal prediction stage to the RF regression algorithm. This is in compliance to typical safety requirements in ITS.
- **Repeatability** is achieved by also performing experimental benchmarks on a synthetic dataset generated based on the real one, which is available along with our code at an anonymous github repository[2].

The remainder of the paper is organized as follows: Section 2 summarizes the related work, followed by the problem formulation. Section 3 presents the methodological approach, the feature selection methods applied, and the machine learning model implemented with the post-hoc conformal prediction framework. Finally, Sect. 4 presents the dataset and our obtained results, while Sect. 5 summarizes our paper and discusses directions for future work.

2   Background

In this section we outline the related work in the area of GNSS positioning error estimation, followed by the problem formulation.

2.1   Related Work

Positioning Error Estimation. In Karlsson et al. [5], data from a GNSS sensor in a Volvo vehicle and a reference position system is gathered. A model based on Autoregression (AR) models and Gaussian Mixture Models (GMM) is proposed to analyze the positioning errors. The model estimates environmental conditions surrounding the test vehicle, based on which, a sub-model is trained that clusters data into subsets, each with similarities in AR model coefficients. The final objective is to simulate positioning error distributions that are compared to the real positioning error distributions. The simulations show a good match, but prediction performance is not shown (due to confidentiality).

Moreover, in El Abbous et al. [7], positioning error distribution is characterized using data obtained from several static positioning tests with a variety of obstructing objects between the receiver and the satellites in orbit. Data is collected and analyzed to calculate error average and standard deviation, which

---

[2] https://github.com/alku7660/gnss_position_error_guarantees.

are then related to the cost function of a Quasi Optimal Satellite Selection algorithm. The higher the cost, the higher average and standard deviation used to estimate the positioning error. The error between the estimated error distributions and the real error distributions lies between 10 and 30 cm. The method implemented relies on the estimation of parameters of assumed normal distributions based on the satellite geometrical cost function in a satellite selection method. As authors explain, variables like multipath phenomena are not easily estimated with parametric models, which may lead to positioning errors.

Kuratomi [15], proposes sky obstruction estimates in the RTK DGNSS feature vector, to try to estimate multipath phenomena. This feature vector is used by decision trees and SVR. The results indicate a performance inside the range described by El Abbous et al. [7] with the maximum RMSE of estimated positioning errors at 29 cm and an added sky obstruction feature which was not statistically significant. However, more complex and accurate methods such as ensemble learning models are suggested. Furthermore, in Yang et al. [16], a real-time model is built using Long Short-Term Memory (LSTM). Mean Absolute Error (MAE) between ground truth error values and predicted error values is experimentally measured, indicating a maximum MAE of 3%. However, the paper does not provide the exact data and the model does not measure confidence intervals for error prediction, which is important for ITS applications [3,4].

Conformal Prediction. Boström et al. [21,22] present a confidence interval generator framework designed as a post-hoc algorithm, adaptable to RF predictor models. The single value output from the machine learning model may then be transformed into a confidence interval, with a chosen significance level. The framework avoids the need to reserve instances for calibration by using Out of Bag (OOB) samples, allowing a more robust model training and testing phases. The interval is also adapted to consider each single instance prediction difficulty in the dataset, i. e., it generates a larger confidence interval for a harder to classify instance, based on the variance of decision trees predictions in the RF.

2.2   Problem Formulation

Let $\mathcal{D} = \{\mathcal{D}_1, \ldots, \mathcal{D}_N\}$ be a dataset of N test-runs. A test-run is a vehicle drive experiment to collect GNSS data, which may have thousands of instances, denoted as $\boldsymbol{x}_{it}$, each with a time stamp t. Hence, each test-run is defined as a set of instances $\mathcal{D}_i = \{\boldsymbol{x}_{it}\}$, $\forall t \in \{1, \ldots, m_i\}$, with m_i being the maximum number of available time stamps in $\mathcal{D}_i$. For each instance $\boldsymbol{x}_{it}$ the respective target positioning error at time t is defined as y_{it}. Note that the positioning errors are obtained from the RTK DGNSS process and a reference positioning system.

Moreover, let $f_i(\boldsymbol{x}_{it})$ be the test-run i predictor function that maps the instance vector $\boldsymbol{x}_{it}$ to a predicted value $\hat{y}_{it}$ so as to minimize $\epsilon_{it} = \hat{y}_{it} - y_{it}$, where ϵ_{it} corresponds to the error between the prediction and target error of $\boldsymbol{x}_{it}$. The total error of a given test-run $\mathcal{D}_i$ is calculated as the RMSE as seen in Eq. 1.

$$RMSE_i = \sqrt{\frac{1}{m_i} \sum_{t=1}^{m_i} \epsilon_{it}^2} \tag{1}$$

The complexity of the predictor function may be reduced by decreasing the number of features that are irrelevant or noisy in x_{it} [8]. This may lead to improved predictive performance in terms of $RMSE_i$.

Hence, in this paper, we apply six different feature selection methods $\mathcal{I} = \{baseline, backward, forward, stepwise, lasso, experts\}$. Let x_{it}^k be the instance vector containing the features returned by the kth feature selection process in $\mathcal{I}$, with $k = \{1, \ldots, |\mathcal{I}|\}$. For each k, a predictor function $f_i(x_{it}^k)$ is identified with a corresponding ϵ_{it}^k and $RMSE_i^k$. The final feature selection process minimizing the RMSE throughout the possible processes in $\mathcal{I}$ is given by Eq. 2:

$$k_i^* = \arg \min_k RMSE_i^k, \forall i \in \{1, 2, \ldots, N\} \tag{2}$$

For a given prediction $f(x_{it}^k)$, a confidence interval (a, b) with significance level $\delta \leq 0.001\%$ is generated around each predicted instance value [21] as follows:

$$P(a \leq f(x_{it}^k) \leq b) \leq 1 - \delta \tag{3}$$

Conformal prediction performance in test-run $\mathcal{D}_i$ is defined as Ω_{int_i}, and is measured as the number of instances inside the predicted confidence interval limits (a, b) over the total number of instances in $\mathcal{D}_i$, as defined in Eq. 4:

$$\Omega_{int_i}^k := \frac{\sum_{t=1}^{m_i} h(f(x_{it}^k))}{m_i}, h(f(x)) = \begin{cases} 1, & a \leq f(x) \leq b \\ 0, & otherwise \end{cases} \tag{4}$$

We finally define a predictor function $F(x_t)$ for dataset $\mathcal{D}$ containing all instances from all test-runs. The goal of this paper is to define such functions $f_i(x_{it})$ and $F(x_t)$ for GNSS positioning error prediction with high accuracy. Moreover, for each prediction $f_i(x_{it})$ in all $\mathcal{D}_i$, we want to define an integrity interval (a_{it}, b_{it}) with significance level $\delta \leq 0.001\%$.

3   NAVEEG: A Positioning Error Estimation Framework with Guarantees

We propose a Positioning Error Estimation framework, which we call NAVEEG, that provides guarantees on the error estimations using Conformal Prediction. The framework comprises three main components: (1) data pre-processing, (2) feature selection, and (3) conformal prediction. In Fig. 1, we outline the main steps of NAVEEG. Next, we provide further details for each step.

3.1   Data Preprocessing

For each test-run in $\mathcal{D}$ we first perform a K-Fold random split (for our experiments, we set K = 5) before applying feature selection. Additionally, every

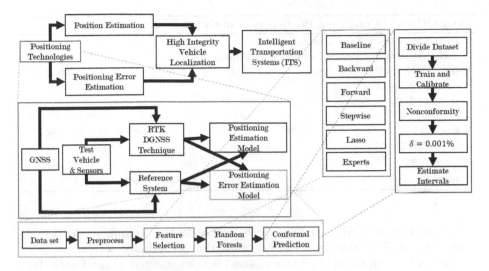

Fig. 1. General research framework implemented.

instance is assumed to be time-independent with other instances, i.e., the time dimension is not considered and predictions are obtained per instance. Moreover, min-max normalization is applied to each continuous data variable, while for categorical variables we apply one-hot encoding.

3.2 Feature Selection

We apply six feature selection techniques, as defined in $\mathcal{I}$ (Sect. 2.2), i.e., *baseline*, *backward*, *forward*, *stepwise*, *lasso regression*, and *experts*. Note that *baseline* includes all features, while *experts* includes the most relevant features according to a set of GNSS experts. The remaining four techniques are linear regression-based feature selection methods.

Assuming the linearity property holds among the independent and dependent variables in the dataset, we compute the linear regression p-value for the independent features, which is an estimate of the feature relevance to the dependent response variable [18]. A p-value lower than the significance threshold indicates that the variable is relevant to the target estimation.

More concretely, the feature selection techniques we employ are as follows:

1. **Baseline:** uses all features;
2. **Backward:** one-by-one eliminates the least significant feature until a set of significant features remains;
3. **Forward:** one-by-one adds the most significant feature, starting from zero, until a set of significant features is built;
4. **Stepwise:** one-by-one adds the most significant feature to a preliminary feature set. The preliminary feature set is then run with another linear regression model to evaluate if all preliminary features are significant. If one is

---

Algorithm 1: Conformal Prediction Framework for Random Forests

---

input : Dataset Z; new instance x_n
output: Integrity interval $\hat{y}_n^\delta$ for new instance x_n

1 $[Z^t, Z^c] = \text{Split}(Z)$ // Split Z into Z^t and Z^c (OOB samples)
2 $model^t = \text{Train}(Z^t)$ // Train $model^t$ on Z^t
3 $\hat{y} = \text{Predict}(model^t, Z^c)$ // Predict on Z^c (OOB samples) with $model^t$
4 $\alpha = |y - \hat{y}|$ // Calculate nonconformity scores α
5 $\text{Sort}(\alpha)$ // sort α in descending order
6 $\alpha^\delta = \text{Find}(\alpha, \delta)$ // find α^δ which separates the top δ fraction
7 $\hat{y}_n = \text{Predict}(x_n)$ // predict $\hat{y}_n$ from new instance x_n
8 $\hat{y}_n^\delta = \hat{y}_n \pm \alpha^\delta$ // estimate the integrity interval $\hat{y}_n^\delta$

---

not (p-value > 0.05) the latest added feature to the preliminary feature set is removed. The algorithm iterates until the preliminary set of significant features has run through all features and is then selected as the feature set;

5. **Lasso:** applies linear model regression with L1 regularization to decrease the coefficient value of irrelevant variables;

6. **Experts:** a team of GNSS experts is consulted to get a feature ranking. The most important features are used, based on a Pareto 80-20 ranking score.

3.3 Random Forest (RF)

The next step is to employ a regression RF using the previously selected features. A regression RF provides predictor functions for regression tasks with categorical or continuous features. It is able to handle high dimensionality data while maintaining a low bias with a relatively low variance compared to single decision trees. As shown in Boström et al. [21,22], RF allows an easy conformal prediction implementation. Although the data records are time evolving, the model we aim to create does not consider time dependency, assuming that the data at a given time t suffices to predict the positioning error at that time.

Hence, the selected features from the previous step are evaluated using the RF regression *feature importance* method, which is based on variance or impurity reduction, averaged throughout the trees in the forest [20]. The features with the highest importance are then used for building the regression RF, which is then forwarded to the third step of NAVEEG.

3.4 Conformal Prediction

Finally, a conformal prediction framework is applied to the RF to define prediction integrity intervals. Conformal prediction is an algorithm that delivers confidence intervals with regards to any model's prediction output by using a sorting score known as the nonconformity measure. Its outcome is the real interval (a, b) for which $P(a \leq X \leq b) \leq 1 - \delta$ for a given significance level δ and instance [21]. This allows the system to obtain an integrity interval of $\delta = 0.001\%$, for the positioning error prediction values, for every instance.

The main steps of the conformal prediction framework that is applied to the regression RF for each test instance are based on Boström et al. [21], and are outlined in Algorithm 1.

4   Empirical Evaluation

In this section, we present the datasets used in our experimental evaluation, we outline the experimental setup and present our results.

4.1   Datasets

We used one real-world dataset, as well as a synthetically generated dataset.

GNSS Dataset. The dataset consists of 23 files, each with the features and the target value for a test-run. The test-runs are distributed into 5 d, and 2 vehicles, a truck and a lawnmower. The features are summarized in Table 1. Due to the RTK DGNSS data processing steps, each instance may not have an appropriate time stamp or a proper ground truth value as the system might be unavailable (due to initialization process). For this reason, the data must be filtered to eliminate instances with incorrect time stamps or without ground truth. The total number of initial instances considering all 23 JSON files is 228894. A total of 33603 instances have an incorrect time stamp, and an additional 13257 lack a correct ground truth position error. The final dataset contains 182034 instances.

The total number of features after applying min-max normalization and one-hot encoding (Sect. 3.1) ranges between 220 and 260 per test-run, while the feature vector length varies according to the number of available satellites in space at any given time t. The preprocessing has been performed using Python 3.6 and the JSON, Numpy, and Pandas libraries.

Synthetic Dataset. We additionally generated a synthetic dataset by multiplying a sampled observation from a normally distributed variable with mean 10 and variance 1, and then adding an observation from another variable with normal distribution with mean 0 and variance 1 to every feature. A total of 220 features have been generated with 10000 instances. The dataset is available at the github repository mentioned in the introduction.

4.2   Results

For the real-world dataset, we carried out the following experimental benchmarks: (1) comparison of the performance of RF with respect to different feature selection techniques; (2) assessment of the impact of integrity intervals introduced by the conformal prediction framework using RF; (3) comparison of all methods with respect to RMSE, and finally (4) a study on the important variables identified by the algorithms. We present four test-run results out of

Table 1. Features available from the real GNSS dataset.

Feature	Description	Data type
Number Satellites	Number of satellite measurements available	Discrete
Constellation	GPS, GLONASS or Galileo	Categorical
Ambiguity lock	Fixed or unknown	Categorical
Cycle slip	Discontinuity in the receiver's phase lock on the satellite signal	Discrete
Multipath	Multipath estimator based on CNO	Discrete
CNO	Carrier to noise ratio	Continuous
RAIM	Receiver autonomous integrity monitoring. Indicates if a satellite induces high error	Discrete
Used	Indicates if a satellite is used	Discrete
Master position	Master reference position on earth's surface	Continuous
LSR	Least Square Residuals per satellite	Continuous
PDOP	Position dilution of precision error	Continuous
Tracking type	pll, costas, or no tracking	Categorical
Elevation	Angle over horizon. Maximum 90°	Continuous
Azimuth	Horizontal angle to north plane	Continuous
Prediction covariance	EKF coordinates position covariance	Continuous
Difference	EKF prediction minus EKF correction	Continuous
Innovation	Innovation value from EKF algorithm	Continuous

the 23 test-runs available from the real-world dataset. The remaining test-run results are in the github repository. Note that RF is implemented using *Scikit-Learn* package *RandomForestRegressor*, with a total of 100 trees, a minimum of 10 samples required for split, and a minimum of 1 instance per node. Next, we provide more details of these benchmarks and discuss our findings.

Feature Selection. Figure 2 depicts the RF performance in terms of RMSE for each feature set for the four selected tests, defined as Truck test 2, Truck test 4, Lawnmower test 4, and Lawnmower test 11. Each method has 5 points in the plot, each corresponding to 1 of the 5 folds in the K-Fold Cross Validation process. We observe that the RMSE performance does not significantly vary for most methods. The results, provided from Figures 1 to 23 in folder *RMSE vs Dim Reduction Figures* in the repository indicate a similar performance with a smaller set of features. *Exp* corresponds to the features selected by the GNSS experts. The results are similar for the rest of the test-runs, indicating highly correlated or noisy features which have been filtered through these selections. The results of each fold are also analyzed in terms of prediction performance and integrity levels.

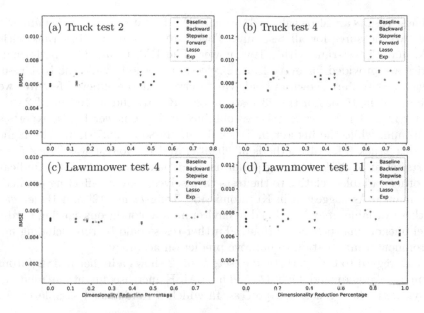

Fig. 2. RF Regressor RMSE vs. Dimensionality reduction percentage (due to feature selection).

Each of the pre-selected features is ranked according to the RF feature relevance. Figure 24, available at the repository, shows the normalized score of features used by the RF regression model. The experts features are, according to RMSE and Ω_{int}, the best at predicting positioning error and did not include *Azimuth* and *Least Squares Residuals (LSR)*. *Elevation* and *CNO* are more relevant. There is a high variance in the weights, and a study should be done to verify significant statistical difference among features' relevance.

RF and Conformal Prediction. Figure 3 shows the Stanford diagram on one fold for Truck tests 1 and 4, and for Lawnmower tests 4 and 11. The diagram indicates the predicted positioning error versus the ground truth positioning error. Each blue line corresponds to each instance's predicted integrity interval with a significance level of 0.001% or confidence (integrity) of 99.999%. The red dashed line is the $y = x$ line, and the closer the center of each blue line or predicted integrity interval is to it, the better the prediction. If the blue line contains the red line, then it is considered a correct prediction. We observe a higher integrity interval variation among test-runs. Figure 3b is considerably different to Fig. 3c, with wider spread of instances and smaller integrity intervals. Lawnmower test 11 (Fig. 3d) fold 1 shows a relatively different performance among methods, where backward presents a compact and accurate set of integrity intervals along the red line, while the rest a broader range, and stepwise the widest integrity intervals. This is expected, as Fig. 2 showed that most folds of backward method had the lowest RMSE for the test-runs, while the stepwise folds a higher RMSE.

These results are also evaluated through RMSE and Ω_{int}. Table 2 summarizes these two measures for all test-runs. There are a total of 8 test-runs with a truck, and 15 test-runs with a lawnmower. The RMSE and Ω_{int} performance metrics are provided for every feature selection method, while the best results for each test-run are shown in bold. We observe that the experts' features were the best set in 16 out of the 23 test-runs for RMSE and in 10 out of 23 test-runs for Ω_{int}. The lowest RMSE was obtained in Lawnmower 1 with experts set at 4.62 mm, while the highest in Truck 6 with lasso at 28.75 mm. The highest Ω_{int} was Truck 6 for all methods at 99.99%, and the lowest was Truck 5 with experts set at 57.96%. A strong performance dependency on each test indicates that other variables relative to the tests themselves, may be affecting prediction performance. As suggested in Kuratomi [15], Tang *et al.* [12] and Bauer *et al.* [14] these variables may be related to environmental conditions causing hard-to-model phenomena such as multipath. Further tests could be run including more environmental information to improve prediction accuracy.

With regard to the synthetic dataset, Table 2 shows a higher RMSE (around 41 mm in average) and high Ω_{int}. The RMSE increase might be related to the synthetic data generation process, in which a randomized sample from all

Table 2. RMSE and interval accuracy results for every test. "T.", "L." and "Synth" stand for Truck, Lawnmower and Synthetic dataset respectively.

RMSE (mm) & Integrity Interval inclusion Percentage [Ω_{int}] (%)

Test	Baseline		Backward		Stepwise		Forward		Lasso		Experts	
	RMSE	Ω_{int}	RMSE	Ω_{int}	RMSE	Ω_{int}	RMSE	Ω_{int}	RMSE	Ω_{int}	RMSE	Ω_{int}
T. 1	4.94	98.89	5.02	98.54	5.18	98.37	**4.84**	98.81	4.94	98.70	4.84	**98.95**
T. 2	6.28	79.17	6.43	77.95	6.69	75.11	6.37	77.32	6.27	77.88	**6.26**	**80.86**
T. 3	8.04	71.19	8.01	69.22	8.34	66.78	8.24	**71.65**	8.03	70.37	**7.88**	69.34
T. 4	8.53	**69.24**	**8.27**	68.83	8.66	68.19	8.39	68.23	8.48	66.13	8.31	69.05
T. 5	5.28	65.57	5.41	65.02	5.66	**67.56**	5.46	65.12	5.47	63.93	**5.03**	57.96
T. 6	27.73	**99.99**	28.44	**99.99**	25.30	**99.99**	27.27	**99.99**	28.75	**99.99**	26.16	**99.99**
T. 7	13.23	**99.41**	12.88	99.33	12.74	98.90	13.15	99.35	**12.24**	99.34	12.90	99.16
T. 8	8.22	90.85	8.30	**90.87**	8.73	89.51	8.33	90.42	8.21	90.60	**8.17**	90.75
L. 1	4.89	93.00	4.90	92.82	4.84	92.60	4.79	94.70	4.88	93.71	**4.62**	**94.86**
L. 2	8.31	85.36	8.16	83.59	8.13	85.98	8.39	85.50	8.81	84.83	**7.59**	**89.19**
L. 3	6.68	85.08	6.88	87.46	7.95	93.79	**6.58**	84.06	6.81	86.18	7.40	**99.32**
L. 4	5.34	64.61	5.16	66.56	5.61	**67.26**	5.29	64.24	5.30	64.13	**5.03**	64.01
L. 5	7.23	98.99	7.06	**99.22**	7.08	89.67	7.49	95.88	7.21	98.94	**6.78**	97.97
L. 6	11.65	98.04	11.79	97.77	15.37	95.91	11.61	97.11	11.70	98.07	**6.69**	**98.29**
L. 7	6.89	88.34	7.12	91.44	7.73	**93.33**	7.11	89.82	7.01	86.27	**6.76**	92.55
L. 8	7.03	62.17	7.02	62.68	7.14	60.29	7.04	61.86	7.10	62.13	**6.60**	**65.96**
L. 9	9.47	**92.42**	11.46	87.23	10.45	89.99	9.67	90.94	9.54	92.38	**9.20**	91.87
L. 10	8.14	70.61	8.96	72.23	9.56	70.50	8.21	70.74	8.27	**72.93**	**7.66**	72.92
L. 11	7.41	73.76	**6.34**	79.03	8.53	**80.33**	7.54	77.48	7.44	77.24	7.34	74.07
L. 12	7.66	64.01	7.59	64.92	7.99	61.50	7.85	64.44	7.70	60.84	**7.03**	**65.18**
L. 13	7.00	67.70	6.95	66.53	7.61	70.22	7.06	68.36	7.03	67.27	**6.65**	**69.06**
L. 14	15.85	99.53	15.19	99.61	21.85	**99.75**	**14.82**	99.15	16.03	99.26	15.01	99.40
L. 15	11.34	97.32	10.44	97.37	13.17	98.01	10.83	93.59	11.34	**98.22**	**10.43**	92.89
Synth	**40.16**	**99.98**	40.24	99.97	41.44	**99.98**	40.92	99.97	40.74	**99.98**	40.91	99.97

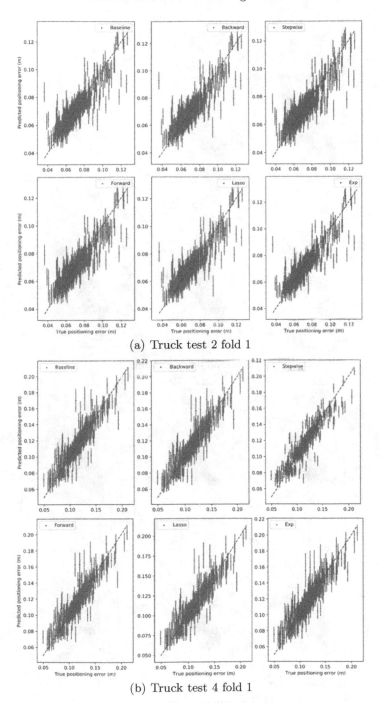

(a) Truck test 2 fold 1

(b) Truck test 4 fold 1

Fig. 3. Stanford diagram with varying conformal prediction integrity intervals (blue lines) for each test instance in the fold, for every feature selection method. (Color figure online)

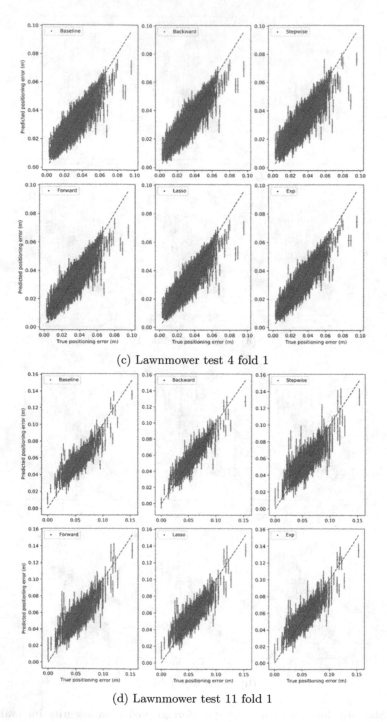

(c) Lawnmower test 4 fold 1

(d) Lawnmower test 11 fold 1

Fig. 3. (*continued*)

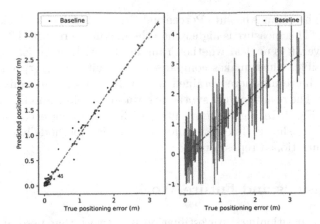

Fig. 4. Stanford diagram for baseline method, which achieved best result on fold 1 in the synthetic dataset.

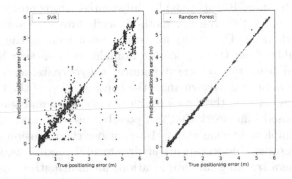

Fig. 5. SVR and RF Stanford diagrams for all instances and no integrity intervals.

test-runs (with different multipath influence levels) is modified and used to train and test the model.

Comparison with Previous Results. A Support Vector Regression (SVR) was implemented with the hyperparameters used in Kuratomi [15]. Using a 5-Fold Cross Validation scheme to measure RMSE, for both SVR and RF Regression with all the test-runs combined.

The RF algorithm performed better than SVR using the same set of features. Kuratomi [15] obtained a 35.4 mm RMSE, which differs to these results due possibly to the dataset and preprocessing. A total of 12 features are used: *CNO, number of satellites, elevation, difference East, difference North, difference Up, LSR, azimuth, innovation East, innovation North, innovation Up* and *pdop* which matches the feature selection obtained from the combined RF feature relevance and linear based methods. RF had an average RMSE of 9.44 mm which is 91.4% lower than SVR, which is 109.17 mm, 73.3% lower compared to Kuratomi [15] best RMSE. The Mean Absolute Percentual Error (MAPE) is calculated for this case at 18.8%. This measure is higher than the maximum reported 3% by Yang *et al.* [16], however, it is unclear whether Yang used MAPE or MSE, and their data was not accesible, which makes comparisons difficult. However, it is important to note that the results here obtained were achieved without considering time dependencies. The results on the synthetic dataset for fold 1 and baseline features are shown in Fig. 4. In Fig. 5 the results for all positioning error predictions of SVR and RF are shown. The results for all selection methods and folds can be found in the mentioned repository.

5   Conclusions and Future Work

We proposed a positioning error estimation framework that used an RF regressor in conjunction with conformal prediction that fits the requirements for an implementation in the field of GNSS localization services and ITS. The obtained performance in terms of RMSE provides substantial improvements compared to a recent state-of-the-art competitor method, while providing statistical guarantees on the predictions. Directions for future work include the exploration of additional variables from the application domain, such as velocities and acceleration obtained from the vehicle dynamics, a hyperparameter and program optimization for improved performance, the consideration of the time dimension and dependencies between the data variables, as well as a real-time implementation of the proposed framework that can possibly result in lower error intervals. This real-time implementation could also consider the inclusion of information obtained from sensors able to capture environmental conditions, such as having an infrared camera tracking satellite locations for multipath estimation.

Acknowledgements. We thank the company Waysure Sweden AB for supporting this research project, providing the real-world dataset, and their GNSS experts input for the feature selection process.

References

1. Belhajem, I., Maissa, Y.B., Tamtaoui, A.: Improving vehicle localization in a smart city with low cost sensor networks and support vector machines. Mob. Netw. Appl. **23**(4), 854–863 (2017). https://doi.org/10.1007/s11036-017-0879-9
2. Wang, L., Li, Z., Zao, J., Zhou, K., Wang, Z., Yuan, H.: Smart device-supported BDS/GNSS real-time kinematic positioning for sub-meter-level accuracy in urban location-based services. Sensors J. **16**, 1–15 (2016)
3. Wörner, M., Schuster, F., Dölitzscher, F., Keller, C.G., Haueis, M., Dietmayer, K.: Integrity for autonomous driving: a survey. In: IEEE/ION Position, Location and Navigation Symposium (PLANS), pp. 666–671 (2016)
4. Blomenhofer, H., Ehret, W., Su, H., Blomenhofer, E.: Sensitivity analysis of the GALILEO integrity performance dependent on ground sensor station network. In: ION GNSS 18th International Technical Meeting of Satellite Division, pp. 1361–1373 (2005)

5. Karlsson, E., Mohammadiha, N.: A statistical GPS error model for autonomous driving. In: Proceedings of the IEEE Intelligent Vehicles Symposium, pp. 754–759 (2018)
6. Parakkal, P.G., Variyar., S.V.V.: GPS based navigation system for autonomous car. In: 2017 International Conference on Advances in Computing, Communications and Informatics (ICACCI), pp. 1888–1893 (2017)
7. El Abbous, A., Samanta., N.: A modeling of GPS error distributions. In: European Navigation Conference (ENC), pp. 119–127 (2017)
8. Lee, G., Rodriguez, C., Madabhushi, A.: An empirical comparison of dimensionality reduction methods for classifying gene and protein expression datasets. In: Măndoiu, I., Zelikovsky, A. (eds.) ISBRA 2007. LNCS, vol. 4463, pp. 170–181. Springer, Heidelberg (2007). https://doi.org/10.1007/978-3-540-72031-7_16
9. Joardar, S., Siddique., T.A., Alam, S., Hossam-E-Haider, M.: Analyses of different types of errors for better precision in GNSS. In: 3rd International Conference on Electrical Engineering and Information and Communication Technology, pp. 1–6 (2016)
10. Molnar, C.: Interpretable Machine Learning. https://christophm.github.io/interpretable-ml-book/limo.html. Accessed 2 Mar 2020
11. Radi, A., Nassar., S., Khedr, M., El-Sheimy, N., Molinari, R., Guerrier, S.: Improved stochastic modelling of low-cost GNSS receivers positioning errors. In: IEEE/ION Position, Location and Navigation Symposium (PLANS), pp. 108–117 (2018)
12. Tang, D., Lu, D., Baigen, C., Wang, J.: GNSS localization propagation error estimation considering environmental conditions. In: 16th International Conference on Intelligent Transportation Systems Telecommunications (ITST), pp. 1–7 (2018)
13. Zimmermann, F., Schmitz, B., Klingbeil, L., Kuhlmann, H.: GPS multipath analysis using Fresnel zones. Sensors J. **19**, 25 (2018)
14. Bauer, S., Obst, M., Wanielik, G.: 3D environment modeling for GPS multipath detection in urban areas. In: International Multiconference on Systems, Signals and Devices (SSD), pp. 1–5 (2012)
15. Kuratomi, A.: GNSS position error estimated by machine learning techniques with environmental information input. M.Sc. Mechatronics, KTH, Sweden (2019)
16. Yang, S., Tabatowski-Bush, B., Xiang, W.: Build up a real-time LSTM positioning error prediction model for GPS sensors. In: 90th IEEE Vehicular Technology Conference (VTC), pp. 1–5 (2019)
17. Fan, C., Zhang, Y., Pan, Y., Li, X., Zhang, C., Yuan, R., et. al.: Multi-horizon time series forecasting with temporal attention learning. In: 25th SIGKDD Conference on Knowledge Discovery and Data Mining (ADST), pp. 2527–2535 (2019)
18. Amrhein, V., Korner-Nievergelt, F., Roth, T.: The earth is flat (p > 0.05): significance thresholds and the crisis of unreplicable research. PeerJ. **5**, e3544 (2015)
19. Suzuki, T., Kitamura, M., Yoshiharu, A., Hashizume, T.: High accuracy GPS and GLONASS positioning by multipath mitigation using omnidirectional infrared camera. In: IEEE International Conference on Robotics and Automation, pp. 311–316 (2011)
20. Pereira, S., et al.: Enhancing interpretability of automatically extracted machine learning features: application to a RBM-random forest system on brain lesion segmentation. Med. Image Anal. **44**, 228–244 (2018)
21. Boström, H., Linusson, H., Löfström, T., Johansson, U.: Accelerating difficulty estimation for conformal regression forests. Ann. Math. Artif. Intell. **81**(1), 125–144 (2017). https://doi.org/10.1007/s10472-017-9539-9

22. Bostrom, H., Asker, L., Gurung, R., Karlsson, I., Lindgren, T., Papapetrou, P.: Conformal prediction using random survival forests. In: 16th IEEE International Conference on Machine Learning and Applications (ICMLA), Cancun, pp. 812–817 (2017). https://doi.org/10.1109/ICMLA.2017.00-57
23. Gong, H., Chen, C., Bialostozky, E., Lawson, C.: A GPS/GIS method for travel mode detection in New York City. Comput. Environ. Urban Syst. **36**, 131–139 (2012)

Author Index

Printed in the United States
By Bookmasters